Coal Geology of China

'Coal' and 'China' to some extent have become synonymous. China is by far the largest user of coal in the world. In 2016, coal production in China amounted to 3.21 billion tons, about half of the total global coal production. Coal consumption accounts for more than 65% of primary energy consumption in China. The Chinese coal industry greatly contributes to the economic development in China, the second largest economy in the world. However, periodically, ubiquitous images of smog blanketing major Chinese cities are viewed all over the world. Coal combustion is one of the important contributors to smog, which is considered to be a major environmental and human health problem for China and other countries. News stories also highlight the periodic coal mine disasters that kill hundreds of Chinese coal miners annually. The need to address these and other human health, environmental, and mine safety issues and to maximize resource recovery and use justifies a vigorous coal research effort. This book brings together experts on almost every aspect of coal geology, coal production, composition and use of the coal and its by-products, and coal's environmental and human health impacts.

The chapters in this book were originally published in a special issue of the *International Geology Review*.

Shifeng Dai is a Professor at the China University of Mining and Technology, China. His research fields include coal geology and coal geochemistry. He is the Editor-in-Chief of the *International Journal of Coal Geology*, the Chief Scientist of the National Key Basic Research Program of China, Changjiang Scholar Professor of Ministry of Education (China) and the National Science Fund for Distinguished Young Scholars of China.

Robert B. Finkelman retired in 2005 after 32 years with the U.S. Geological Survey. He is currently a Research Professor in the Geosciences Department at the University of Texas at Dallas, USA, and an Adjunct Professor at the China University of Geosciences (Beijing). He is an internationally recognized scientist widely known for his work on coal chemistry and as a leader of the emerging field of Medical Geology.

Coal Geology of China

Edited by
Shifeng Dai and Robert B. Finkelman

Routledge
Taylor & Francis Group

LONDON AND NEW YORK

First published 2019
by Routledge
2 Park Square, Milton Park, Abingdon, Oxon, OX14 4RN, UK

and by Routledge
52 Vanderbilt Avenue, New York, NY 10017, USA

First issued in paperback 2020

Routledge is an imprint of the Taylor & Francis Group, an informa business

British Library Cataloguing-in-Publication Data
A catalogue record for this book is available from the British Library

ISBN 13: 978-0-367-57118-4 (pbk)
ISBN 13: 978-1-138-32722-1 (hbk)

Typeset in Myriad Pro
by codeMantra

Publisher's Note
The publisher accepts responsibility for any inconsistencies that may have arisen during the conversion of this book from journal articles to book chapters, namely the possible inclusion of journal terminology.

Disclaimer
Every effort has been made to contact copyright holders for their permission to reprint material in this book. The publishers would be grateful to hear from any copyright holder who is not here acknowledged and will undertake to rectify any errors or omissions in future editions of this book.

Contents

Citation Information

The chapters in this book were originally published in the journal *International Geology Review*, volume 60, issues 5–6 (April 2018). When citing this material, please use the original page numbering for each article, as follows:

Chapter 1
Coal geology in China: an overview
Shifeng Dai and Robert B. Finkelman
International Geology Review, volume 60, issues 5–6 (April 2018) pp. 531–534

Chapter 2
Coal production in China: past, present, and future projections
Xiangfei Bai, Hua Ding, Jinjing Lian, Dong Ma, Xiaoyu Yang, Nanxiang Sun, Wenlin Xue and Yijun Chang
International Geology Review, volume 60, issues 5–6 (April 2018) pp. 535–547

Chapter 3
The geologic settings of Chinese coal deposits
Zengxue Li, Dongdong Wang, Dawei Lv, Ying Li, Haiyan Liu, Pingli Wang, Ying Liu, Jianqiang Liu and Dandan Li
International Geology Review, volume 60, issues 5–6 (April 2018) pp. 548–578

Chapter 4
The health impacts of coal use in China
Robert B. Finkelman and Linwei Tian
International Geology Review, volume 60, issues 5–6 (April 2018) pp. 579–589

Chapter 5
Valuable elements in Chinese coals: a review
Shifeng Dai, Xiaoyun Yan, Colin R. Ward, James C. Hower, Lei Zhao, Xibo Wang, Lixin Zhao, Deyi Ren and Robert B. Finkelman
International Geology Review, volume 60, issues 5–6 (April 2018) pp. 590–620

Chapter 6
Geochemistry of uranium in Chinese coals and the emission inventory of coal-fired power plants in China
Jian Chen, Ping Chen, Duoxi Yao, Wenhui Huang, Shuheng Tang, Kejian Wang, Wenzhong Liu, Youbiao Hu, Qingguang Li and Ruwei Wang
International Geology Review, volume 60, issues 5–6 (April 2018) pp. 621–637

Chapter 7

Emission controls of mercury and other trace elements during coal combustion in China: a review

Yongchun Zhao, Jianping Yang, Siming Ma, Shibo Zhang, Huan Liu, Bengen Gong, Junying Zhang and Chuguang Zheng

International Geology Review, volume 60, issues 5–6 (April 2018) pp. 638–670

Chapter 8

A review on the applications of coal combustion products in China

Jing Li, Xinguo Zhuang, Xavier Querol, Oriol Font and Natalia Moreno

International Geology Review, volume 60, issues 5–6 (April 2018) pp. 671–716

Chapter 9

Mineralogy and geochemistry of ash and slag from coal gasification in China: a review

Shuqin Liu, Chuan Qi, Zhe Jiang, Yanjun Zhang, Maofei Niu, Yuanyuan Li, Shifeng Dai and Robert B. Finkelman

International Geology Review, volume 60, issues 5–6 (April 2018) pp. 717–735

Chapter 10

Stone coal in China: a review

Shifeng Dai, Xue Zheng, Xibo Wang, Robert B. Finkelman, Yaofa Jiang, Deyi Ren, Xiaoyun Yan and Yiping Zhou

International Geology Review, volume 60, issues 5–6 (April 2018) pp. 736–753

Chapter 11

CO_2 storage in coal to enhance coalbed methane recovery: a review of field experiments in China

Zhejun Pan, Jianping Ye, Fubao Zhou, Yuling Tan, Luke D. Connell and Jingjing Fan

International Geology Review, volume 60, issues 5–6 (April 2018) pp. 754–776

Chapter 12

Resources and geology of coalbed methane in China: a review

Yong Qin, Tim A. Moore, Jian Shen, Zhaobiao Yang, Yulin Shen and Geoff Wang

International Geology Review, volume 60, issues 5–6 (April 2018) pp. 777–812

For any permission-related enquiries please visit:
http://www.tandfonline.com/page/help/permissions

Notes on Contributors

Xiangfei Bai Coal Chemistry Branch, Coal Science Technology Research Institute Corporation Limited, China; College of Environmental Engineering, North China Institute of Science and Technology, Langfang, China; and Beijing Research Institute of Coal Chemistry, China Coal Research Institute

Yijun Chang Shanxi Fenwei Energy Corporation Limited, Taiyuan, China

Jian Chen School of Earth and Environment, Anhui University of Science and Technology, Huainan, China

Ping Chen School of Earth and Environment, Anhui University of Science and Technology, Huainan, China

Luke D. Connell CSIRO Energy Business Unit, Canberra, Australia

Shifeng Dai College of Geoscience and Surveying Engineering; State Key Laboratory of Coal Resources and Safe Mining; and School of Resources and Geoscience, China University of Mining and Technology, China

Hua Ding Coal Chemistry Branch, Coal Science Technology Research Institute Corporation Limited, China; Beijing Research Institute of Coal Chemistry, China Coal Research Institute; and State Key Laboratory of High Efficient Mining and Clean Utilization of Coal Resource, China

Jingjing Fan College of Geoscience and Surveying Engineering, China University of Mining and Technology, Xuzhou, China

Robert B. Finkelman Geosciences Department, University of Texas at Dallas, USA and the China University of Geosciences (Beijing)

Oriol Font Institute of Environmental Assessment and Water Research, CSIC, Madrid, Spain

Bengen Gong State Key Laboratory of Coal Combustion, School of Energy and Power Engineering, Huazhong University of Science & Technology, Wuhan, China

James C. Hower Center for Applied Energy Research, University of Kentucky, Lexington, USA

Youbiao Hu School of Earth and Environment, Anhui University of Science and Technology, Huainan, China

Wenhui Huang School of Energy Resources, China University of Geosciences (Beijing)

Yaofa Jiang Jiangsu Institute of Architectural Technology, Xuzhou, China

Zhe Jiang School of Chemistry and Environmental Engineering, China University of Mining and Technology (Beijing)

Dandan Li College of Earth Science and Engineering, Shandong University of Science and Technology, Qingdao, China

Jing Li Key Laboratory of Tectonics and Petroleum Resources (China University of Geosciences), Ministry of Education

Qingguang Li School of Earth and Environment, Anhui University of Science and Technology, Huainan, China

NOTES ON CONTRIBUTORS

Ying Li Coalfield Geological Planning Institute, Shandong Bureau of Coal Geology, China

Yuanyuan Li School of Chemistry and Environmental Engineering, China University of Mining and Technology (Beijing)

Zengxue Li College of Earth Science and Engineering, Shandong University of Science and Technology, Qingdao, China

Jinjing Lian Coal Chemistry Branch, Coal Science Technology Research Institute Corporation Limited, China; Beijing Research Institute of Coal Chemistry, China Coal Research Institute; and State Key Laboratory of High Efficient Mining and Clean Utilization of Coal Resource, China

Haiyan Liu College of Earth Science and Engineering, Shandong University of Science and Technology, Qingdao, China

Huan Liu State Key Laboratory of Coal Combustion, School of Energy and Power Engineering, Huazhong University of Science & Technology, Wuhan, China

Jianqiang Liu College of Earth Science and Engineering, Shandong University of Science and Technology, Qingdao, China

Shuqin Liu School of Chemistry and Environmental Engineering, China University of Mining and Technology (Beijing)

Wenzhong Liu School of Earth and Environment, Anhui University of Science and Technology, Huainan, China

Ying Liu College of Earth Science and Engineering, Shandong University of Science and Technology, Qingdao, China

Dawei Lv College of Earth Science and Engineering, Shandong University of Science and Technology, Qingdao, China

Dong Ma Coal Chemistry Branch, Coal Science Technology Research Institute Corporation Limited, China; Beijing Research Institute of Coal Chemistry, China Coal Research Institute; and State Key Laboratory of High Efficient Mining and Clean Utilization of Coal Resource, China

Siming Ma State Key Laboratory of Coal Combustion, School of Energy and Power Engineering, Huazhong University of Science & Technology, Wuhan, China

Tim A. Moore School of Earth, Environmental and Biological Sciences, Queensland University of Technology, Brisbane, Australia and Cipher Consulting, Australia

Natalia Moreno Institute of Environmental Assessment and Water Research, CSIC, Madrid, Spain

Maofei Niu School of Chemistry and Environmental Engineering, China University of Mining and Technology (Beijing)

Zhejun Pan CSIRO Energy Business Unit, Canberra, Australia

Chuan Qi School of Chemistry and Environmental Engineering, China University of Mining and Technology (Beijing)

Yong Qin Key Laboratory of Coalbed Methane Resources and Reservoir Formation (Ministry of Education of China), China University of Mining and Technology; and School of Mineral Resources and Geoscience, China University of Mining and Technology, Xuzhou, China

Xavier Querol Institute of Environmental Assessment and Water Research, CSIC, Madrid, Spain

Deyi Ren College of Geoscience and Surveying Engineering, and State Key Laboratory of Coal Resources and Safe Mining, China University of Mining and Technology, Beijing, China

Jian Shen Key Laboratory of Coalbed Methane Resources and Reservoir Formation (Ministry of Education of China), China University of Mining and Technology; and School of Mineral Resources and Geoscience, China University of Mining and Technology, Xuzhou, China

Yulin Shen Key Laboratory of Coalbed Methane Resources and Reservoir Formation (Ministry of Education of China), China University of Mining and Technology; and School of Mineral Resources and Geoscience, China University of Mining and Technology, Xuzhou, China

Nanxiang Sun Coal Chemistry Branch, Coal Science Technology Research Institute Corporation Limited, China; Beijing Research Institute of Coal Chemistry, China Coal Research Institute; and State Key Laboratory of High Efficient Mining and Clean Utilization of Coal Resource, China

Yuling Tan CSIRO Energy Business Unit, Canberra, Australia; State Key Laboratory of Geomechanics and Geotechnical Engineering, Institute of Rock and Soil Mechanics, Chinese Academy of Sciences, Beijing, China; and University of Chinese Academy of Sciences, Beijing, China

Shuheng Tang School of Energy Resources, China University of Geosciences (Beijing)

Linwei Tian School of Public Health, The University of Hong Kong, China

Dongdong Wang College of Earth Science and Engineering, Shandong University of Science and Technology, Qingdao, China

Geoff Wang School of Chemical Engineering, the University of Queensland, Brisbane, Australia

Kejian Wang School of Earth and Environment, Anhui University of Science and Technology, Huainan, China

Pingli Wang College of Earth Science and Engineering, Shandong University of Science and Technology, Qingdao, China

Ruwei Wang CAS Key Laboratory of Crust-Mantle Materials and Environment, School of Earth and Space Sciences, University of Science and Technology of China, Hefei, China

Xibo Wang College of Geoscience and Surveying Engineering, and State Key Laboratory of Coal Resources and Safe Mining, China University of Mining and Technology, Beijing, China

Colin R. Ward School of Biological, Earth and Environmental Sciences, University of New South Wales, Sydney, Australia

Wenlin Xue Shanxi Fenwei Energy Corporation Limited, China

Xiaoyun Yan State Key Laboratory of Coal Resources and Safe Mining, and College of Geoscience and Survey Engineering, China University of Mining and Technology, Beijing, China

Jianping Yang State Key Laboratory of Coal Combustion, School of Energy and Power Engineering, Huazhong University of Science & Technology, Wuhan, China; and School of Energy Science and Engineering, Central South University, Changsha, China

Xiaoyu Yang Coal Chemistry Branch, Coal Science Technology Research Institute Corporation Limited, China; Beijing Research Institute of Coal Chemistry, China Coal Research Institute; and State Key Laboratory of High Efficient Mining and Clean Utilization of Coal Resource, China

Zhaobiao Yang Key Laboratory of Coalbed Methane Resources and Reservoir Formation (Ministry of Education of China), China University of Mining and Technology; and School of Mineral Resources and Geoscience, China University of Mining and Technology, Xuzhou, China

Duoxi Yao School of Earth and Environment, Anhui University of Science and Technology, Huainan, China

Jianping Ye China National Offshore Oil Corporation, Beijing, China

Junying Zhang State Key Laboratory of Coal Combustion, School of Energy and Power Engineering, Huazhong University of Science & Technology, Wuhan, China

Shibo Zhang State Key Laboratory of Coal Combustion, School of Energy and Power Engineering, Huazhong University of Science & Technology, Wuhan, China

Yanjun Zhang School of Chemistry and Environmental Engineering, China University of Mining and Technology (Beijing)

Lei Zhao College of Geoscience and Surveying Engineering, and State Key Laboratory of Coal Resources and Safe Mining, China University of Mining and Technology, Beijing, China

Lixin Zhao College of Geoscience and Surveying Engineering, China University of Mining and Technology, Beijing, China

Yongchun Zhao State Key Laboratory of Coal Combustion, School of Energy and Power Engineering, Huazhong University of Science & Technology, Wuhan, China

Chuguang Zheng State Key Laboratory of Coal Combustion, School of Energy and Power Engineering, Huazhong University of Science & Technology, Wuhan, China

Xue Zheng State Key Laboratory of Coal Resources and Safe Mining, and College of Geoscience and Survey Engineering, China University of Mining and Technology, Beijing, China

Fubao Zhou Key Laboratory of Gas and Fire Control for Coal Mines, China University of Mining and Technology, Xuzhou, China

Yiping Zhou Yunnan Institute of Coal Geology Prospection (Emeritus), China

Xinguo Zhuang Key Laboratory of Tectonics and Petroleum Resources (China University of Geosciences), Ministry of Education

Coal geology in China: an overview

Shifeng Dai ⓘ and Robert B. Finkelman ⓘ

China has abundant coal resources, and currently is and will continue to be the largest coal producer for the foreseeable future. According to the World Energy Council (2017), China holds an estimated 114.5 Bt of coal reserve as of 2016, the third largest in the world behind the United States and Russia and about 13% of the world's total reserves. From 1949 to 2016, coal production in China increased rapidly (Figure 1), although it has decreased a bit since 2013. In 2016, coal production in China amounted to 3.21 Bt (Enerdata 2017), about half of the total global coal production. China is also the largest user of coal in the world, consuming about as much coal as the rest of the world combined. Coal consumption accounts to more than 65% of the primary energy consumption in China. Also, the coal-bearing areas in China occur in almost every region of the country (Figure 2) and coal formation occurred in many geologic periods, including Middle Devonian, Carboniferous and Early Permian (C–P$_1$), Late Permian (P$_2$), Late Triassic (T$_3$), Early and Middle Jurassic (J$_{1-2}$), Late Jurassic and Early Cretaceous (J$_3$–K$_1$), and Eogene and Neogene (E–N) (Han and Yang 1980; Han et al. 1996; Dai et al. 2012; Figure 2).

'Coal' and 'China' to some extent have become synonymous. The Chinese coal industry greatly contributes to the economic development in China. However, periodically ubiquitous images of smog blanketing major Chinese cities are viewed all over the world. Coal combustion is one of the important contributors to smog, which is considered to be a major environmental and human health problem for China and other countries. News stories also highlight the periodic coal mine disasters that kill hundreds of Chinese coal miners annually. The need to address these and other human health, environmental, and mine safety issues and to maximize resource recovery and use justifies a vigorous coal research effort.

Although it is widely known that China is largest coal producer and consumer in the world, what is less well known is the fact that China has quietly become the world's leading country for research and development on coal. Although more and more papers related to coal geology have been published in international journals, many of the Chinese coal publications are found in the Chinese literature limiting access to and awareness of this literature to all but Chinese scientists. This Special Issue of *International Geology Review* is intended to bring this wealth of information to English-speaking audiences. This Special Issue includes 11 papers, which provide a cross-section of current research on different aspects of the coal geology in China. They represent contributions from 70 individual authors, ranging from PhD students and early-career researchers to scientists with extensive experience who are leaders in their field. Although it is a special issue for coal geology in China, some papers have been prepared by teams drawn from several different countries, reflecting the increasingly global nature of coal and coal-related research.

The Special Issue encompasses a series of review papers, drawing on the experience of the group of researchers to provide overviews of particular topics. The first of these (Li et al. 2017b) discusses the geological setting of coal deposits in China by reviewing their distribution, coal-forming ages, coal-forming environments, and characteristics of major coal deposit formed in different ages. This is followed by an overview of China's coal production, processing, and standardization of coal quality management over the past 30 years (Bai et al. 2017); they also analyse the present

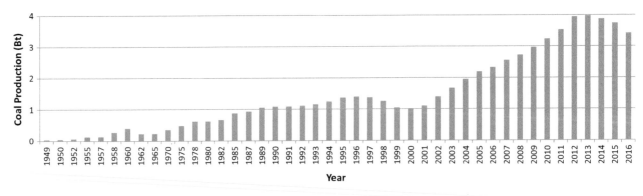

Figure 1. China's coal productions (Bt) from 1949 to 2016.

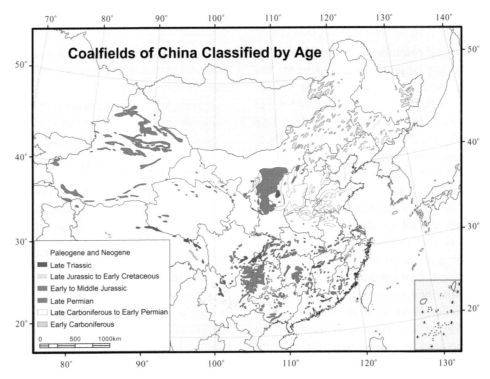

Figure 2. Coal deposits in China classified by geological age (modified from Dai *et al.* 2012).

situation and future projections of China's coal production. Coal can be considered as a source for critical elements and in some cases, base metal Al (Dai and Finkelman 2017). A comprehensive review paper by Dai *et al.* (2016) describes the progress and future prospects of some Chinese coals as a promising source of critical elements, including Ge, rare earth elements and Y, Ga, Al, V, Se, Zr, Nb, Au, Ag, Re, and U. Coal use can result in environmental degradation and widespread and severe health problems. Some of the elements of particular concern in some coals are As, Hg, F, Se, Be, and U, largely because of their potential for adverse environmental and human health impact. Finkelman and Tian (2017) review the health problems cause by coal combustion and mining in China, including fluorosis, arsenosis, lung cancer, selenosis, and premature deaths,

as well as the possible positive impact on iodine deficiency disease. In view of its dual roles in resource recovery and environmental impact, a critical review of geochemistry of uranium in Chinese coals and its atmospheric emission by coal-fired plants in China is included in this compilation (Chen *et al.* 2017), covering abundance and enrichment origin of U, as well as the amount released into atmosphere from Chinese coal-fired power plants.

The emission and control of toxic trace elements, especially mercury, during coal combustion have been extensively investigated in China in recent years. Zhao *et al.* (2017) review the emission characteristics and control strategies of mercury and other trace elements (As, F, Se, and Cr) during coal combustion. These authors also discuss the removal performance of various sorbents (e.g., activated carbon, fly ash, calcium-based sorbents, metal oxides, and mineral sorbents) for trace element removal. Coal-fired power plants in China produce large volumes of coal combustion products (CCPs), of which disposal usually occupies large areas of useful cultivated land and may cause serious environmental issues. On the other hand, these CCPs could also be utilized as recoverable resources, and this is discussed in a review paper (Li et al. 2017a), along with the potential for economic recovery of an increasingly significant group of elements from coal products for use in modern industrial applications. Following on these reviews is a paper (Liu *et al.* 2017) that describes mineralogy and geochemistry of ash and slag from coal gasification in China. It covers the transformation of minerals typically found in coal and the effects of operating conditions on mineral transformation and melting/slagging behaviour of minerals during gasification process, as well as trace element migration behaviour in commercialized gasifiers.

Commercial exploitation of coalbed methane (CBM) as a natural gas resource is an important milestone in the history of the global energy industry. CBM resources less than 2000 m in depth in China are estimated to be 36.81 trillion cubic meters. Qin *et al.* (2017) comprehensively review the distribution of CBM resources, geological factors affecting CBM enrichment, and technical challenges of economically extracting gas from low-rank coal reservoirs and stimulating economic gas flow from deep low-permeability reservoirs. Coal seams, especially deep unminable coal seams, are one of the viable geological target for CO_2 storage to mitigate greenhouse gas emissions. Although facing some technical challenges, this storage can not only reduce the cost of pumping and injection but also can enhance recovery of CBM. The paper by Pan *et al.* (2017) summarize the main properties of the target coal seams, the well technologies used, the injection programmes, monitoring techniques, and key findings in China.

Stone coal in China is defined as a high-ash (usually >50%), low heat value (usually 3.5–10.5 MJ/kg), anthracite-rank combustible rock. It has been widely used as a fuel in rural areas of southern China and is the source of vanadium extraction. The final paper by Dai *et al.* (2017) is concerned with this special rock in China, including its occurrence, uses, geochemical and mineralogical properties, and Byproduct potential, as well as its adverse effects on environment and human health during its utilization.

The editors would like to thank all of the authors who contributed papers to the Special Issue, not only for making their efforts in organizing high-level review papers available for publication but also for demonstrating the wide variety of research objectives and coal-geology applications that have been and will be incorporated into past and future studies in a number of disciplines.

Sincere thanks are expressed to colleagues who served as reviewers, some on several occasions, for the papers that were submitted. These reviewers provided numerous constructive comments that helped many of the authors to improve the quality of their papers and generally reinforced the high standard of the work submitted.

Finally, we want to say that although this volume focuses on coal issues in China, these coal-related issues are of importance to any country that mines or burns coal. Exchange of information and collaboration of scientists from different countries and different disciplines on coal-related issues are essential for the most efficient, environmentally sensitive use of this important natural resource.

Contributions to constructive comments by Professor Deyi Ren and Yiping Zhou are also gratefully acknowledged. We would also like to thank the National Key Basic Research Program of China (No. 2014CB238902), the '111' Project (No. B17042), and the National Natural Science Foundation of China (No. 41420104001), which financially supported guest editors' international travel for discussion on various aspects of this issue and supported some papers included in this Special Issue.

Disclosure statement

No potential conflict of interest was reported by the authors.

ORCID

Shifeng Dai ⓘ http://orcid.org/0000-0002-9770-1369
Robert B. Finkelman ⓘ http://orcid.org/0000-0002-7295-5952

References

Bai, X., Ding, H., Lian, J., Ma, D., Yang, X., Sun, N., Xue, W., and Chang, Y., 2017, Coal production in China: Past, present, and future projections: International Geology Review. doi:10.1080/00206814.2017.1301226

Chen, J., Chen, P., Yao, D., Huang, W., Tang, S., Wang, K., Liu, W., Hu, Y., Li, Q., and Wang, R., 2017, Geochemistry of uranium in Chinese coals and the emission inventory of coal-fired power plants in China: International Geology Review. doi:10.1080/00206814.2017.1295284

Dai, S., and Finkelman, R.B., 2017, Coal as a promising source of critical elements: Progress and future prospects: International Journal of Coal Geology. doi:10.1016/j.coal.2017.06.005

Dai, S., Ren, D., Chou, C.-L., Finkelman, R.B., Seredin, V.V., and Zhou, Y., 2012, Geochemistry of trace elements in Chinese coals: A review of abundances, genetic types, impacts on human health, and industrial utilization: International Journal of Coal Geology, v. 94, p. 3–21. doi:10.1016/j.coal.2011.02.003

Dai, S., Yan, X., Ward, C.R., Hower, J.C., Zhao, L., Wang, X., Zhao, L., Ren, D., and Finkelman, R.B., 2016, Valuable elements in Chinese coals: A review: International Geology Review. doi:10.1080/00206814.2016.1197802

Dai, S., Zheng, X., Wang, X., Finkelman, R.B., Jiang, Y., Ren, D., Yan, X., and Zhou, Y., 2017, Stone coal in China: A review: International Geology Review. doi:10.1080/00206814.2017.1378131

Enerdata, 2017. https://yearbook.enerdata.net/coal-lignite/coal-production-data.html. Accessed 4th, August 2017.

Finkelman, R.B., and Tian, L., 2017, The health impacts of coal use in China: International Geology Review. doi:10.1080/00206814.2017.1335624

Han, D., Ren, D., Wang, Y., Jin, K., Mao, H., and Qin, Y., 1996, Coal petrology of China: Xuzhou, China: China University of Mining & Technology Press, 599. (in Chinese with English abstract).

Han, D., and Yang, Q., Eds., 1980, Coal geology of China: Beijing: Publishing House of China Coal Industry. Vol. 2. (in Chinese).

Li, J., Zhuang, X., Querol, X., Font, O., and Moreno, N., 2017a, A review on the applications of coal combustion products in China: International Geology Review. doi:10.1080/00206814.2017.130999

Li, Z., Wang, D., Lv, D., Li, Y., Liu, H., Wang, P., Liu, Y., Liu, J., and Li, D., 2017b, The geologic settings of Chinese coal deposits: International Geology Review. doi:10.1080/00206814.2017.1324327

Liu, S., Qi, C., Jiang, Z., Zhang, Y., Niu, M., Li, Y., Dai, S., and Finkelman, R.B., 2017, Mineralogy and geochemistry of ash and slag from coal gasification in China: A review: International Geology Review. doi:10.1080/00206814.2017.1287013

Pan, Z., Ye, J., Zhou, F., Tan, Y., Connell, L.D., and Fan, J., 2017, CO_2 storage in coal to enhance coalbed methane recovery: A review of field experiments in China: International Geology Review. doi:10.1080/00206814.2017.1373607

Qin, Y., Moore, T.A., Shen, J., Yang, Z., and Wang, G., 2017, Resources and geology of coalbed methane in China: A review: International Geology Review, under review.

World Energy Council, 2017. https://www.worldenergy.org/wp-content/uploads/2017/03/WEResources_Coal_2016.pdf. Access on 5th August 2017

Zhao, Y., Yang, J., Ma, S., Zhang, S., Liu, H., Gong, B., Zhang, J., and Zheng, C., 2017, Emission controls of mercury and other trace elements during coal combustion in China: A review: International Geology Review. doi:10.1080/00206814.2017.1362671

Coal production in China: past, present, and future projections

Xiangfei Bai, Hua Ding, Jinjing Lian, Dong Ma, Xiaoyu Yang, Nanxiang Sun, Wenlin Xue and Yijun Chang

ABSTRACT

General information of China's coal production, processing, and standardization of coal quality management over the past 30 years is introduced in this article. Analysis of the present situation and future projections of China's coal production are also presented. Coal production in China increased year by year from 1980 and reached the maximum of 3.974 Gt in 2013. Provinces including Shanxi, Inner Mongolia, and Shaanxi are the main coal producing regions in China at present and Xinjiang will become another important coal producer in the future. The production of low-rank bituminous coal and lignite increased rapidly in the past 10 years. Coal consumption for electricity generation accounts for about 50% of product coal in China, much lower than that in developed countries. On the contrary, the proportion of coal consumption for coke-making is higher than other main coal producing countries, up to about 15% of total coal production. China became a net importer of coal since 2009 and the imports had reached 3.27 Mt in 2013. Coal washing rate in China has remained at a lower level for many years, much lower than that in developed countries (such as America and Europe). As for coal quality management, China has the world's most systematic management standards for coal quality evaluation and specifications for feed coal in various utilizations. In recent years, with the strengthening of coal management, China has set a series of commercial coal quality standards in which civil coal standards were also included. China is facing the problem of decline in coal demand in recent years. Traditional coal production has suffered from excess capacity and higher inventory in China since 2012. The development of coal industry in China will change to rely more on quality and efficiency, developing scientific capacity, and achieving clean utilization.

1. Introduction

China is the largest coal producing and consuming country in the world. Coal consumption accounts for more than 65% of primary energy consumption in China (Wang et al. 2015). From 2006 to 2010, coal production and the main coal consumption industries in China contributed about 15% to the GDP and 18% to the GDP value added (Xie et al. 2012a). On the other hand, effects of various pollutants mobilized during coal utilization on the environment and human health (Dai et al. 2012b) and the impact of carbon emissions on global climate (Yan and Yang 2010; Guo 2011; Whitmarsh et al. 2011; Ang and Su 2016), as well as endemic diseases caused by in-door coal-combustion in southwestern China (such as fluorosis, arsenosis, lung cancer; Finkelman et al. 2002; Dai et al. 2007; Tian et al. 2008), are of wide concern, and have become important obstacles to the further development of the coal industry. Since the beginning of the twenty-first century, China has been facing severe environmental challenges in the sustainable development of coal industry (Xu et al. 2012; Luo et al. 2016). In this article, we review the general situation of coal production and coal quality management in China for the past more than 30 years. Prospects for coal production in China and the changing trend of quality of coal for coking and steam coal (defined in GB/T 3715 2007) are also presented, which may help to understand China's future coal production.

2. Coal production and quality control of china in the past 36 years

2.1. Changes in coal production in china over the years

Table 1 lists China raw coal production from 1980 to 2015.

Table 1. Raw coal production from 1980 to 2015 (Mt).

Year	Production	Year	Production
1980	620.13	1998	1232.51
1981	621.63	1999	1043.63
1982	666.32	2000	999.17
1983	714.53	2001	1105.59
1984	789.23	2002	1415.31
1985	872.28	2003	1727.94
1986	894.04	2004	2122.61
1987	928.09	2005	2365.15
1988	979.88	2006	2528.55
1989	1054.15	2007	2691.64
1990	1079.30	2008	2802.00
1991	1084.28	2009	2973.00
1992	1114.55	2010	3428.45
1993	1151.38	2011	3516.00
1994	1229.53	2012	3945.13
1995	1292.18	2013	3974.32
1996	1374.08	2014	3874.92
1997	1325.25	2015	3744.97

Data resource: National Bureau of Statistics of the People's Republic of China.

As shown in Table 1, since 1980, China's coal production had increased year by year on the whole. Since the 1990s China had become the largest coal producers in the world. In the late 1990s, China coal industry experienced the problem of overcapacity, which was strongly related to the weak economic situation at that time.

Some coal mines were reorganized and some small-scale coal mines were shut down in Shanxi and other provinces, which lead to a brief downturn in coal production (Zhang *et al.* 2011). After 2001, China's coal production increased rapidly, and reached a peak of 3.974 Gt in 2013. Since 2014, with the slowdown of China's economic growth, another coal production downward trend appeared.

2.1.1. Raw coal production in china's main coal producing provinces

Figure 1 shows coalmines in the main coal producing provinces in China. Table 2 lists raw coal production in China's main coal producing provinces since 2002, including Shanxi, Inner Mongolia, Shaanxi, Anhui, Heilongjiang, Xinjiang, Shandong, Henan, and Guizhou provinces.

Shanxi, Inner Mongolia, and Shaanxi Province are the main coal producing provinces in China, and the total raw coal production in these three provinces accounts for more than 60% of the country's total raw coal production.

Shanxi has played an important role in China's coal industry. On one hand, coal production in Shanxi province was the highest over a long period of time. On

Figure 1. Coal mines in main coal producing provinces in China.

Table 2. Raw coal production in China's main coal producing provinces (Mt).

Year	Shanxi	Inner Mongolia	Shaanxi	Anhui	Heilongjiang	Xinjiang	Shandong	Henan	Guizhou
2002	362.61	111.96	101.97	64.73	67.17	30.98	128.37	100.52	50.02
2003	449.53	150.43	116.14	70.99	81.04	34.83	144.72	126.33	78.03
2004	496.13	203.15	132.83	76.01	93.68	41.89	140.78	154.21	97.57
2005	573.58	264.32	154.91	78.35	99.30	37.39	130.86	149.57	107.95
2006	594.55	299.17	164.35	82.73	108.68	42.99	137.88	183.13	118.17
2007	643.73	415.03	172.46	93.70	111.16	41.53	136.84	188.56	108.65
2008	670.69	500.70	215.15	117.82	103.94	63.58	134.92	192.29	117.99
2009	632.27	599.69	310.09	125.68	104.43	95.71	140.49	196.03	136.91
2010	756.77	786.79	378.28	131.50	104.91	101.30	148.92	179.09	159.55
2011	898.50	985.31	418.12	136.76	102.55	121.74	153.46	182.45	156.01
2012	913.93	1109.36	435.57	147.14	95.42	140.15	145.01	148.07	181.07
2013	969.42	1045.61	494.49	139.60	80.17	155.76	151.45	147.21	191.18
2014	976.70	980.00	511.00	127.99	68.11	143.02	148.00	136.00	185.00
2015	975.00	909.57	522.24	135.30	66.78	146.44	144.00	133.11	170.01

Data resource: Statistical Department of the State Administration of Production Safety Supervision and Administration, Statistical Summary of Coal Industry, Internal Publication; Shanxi Fenwei Energy Consulting Co. Ltd., Internal Research Report.

the other hand, Shanxi is the most important source of coal for coking and anthracite products. In addition, this province has a regional advantage which is located in the middle of China, close to many important population and economic centres.

Inner Mongolia is another coal-rich province in China. There are some large mining areas with wide coal distribution with shallow depth and flat-lying beds which are suitable for large-scale surface mining. There are large strip coal mines such as Yimin, Huolinhe, Yuanbaoshan, and Jungar (Figure 2). Since 2011, coal production in Inner Mongolia exceeded that in Shanxi and Inner Mongolia became China's largest coal-producing region.

In addition to Shanxi, Inner Mongolia, and Shaanxi province, Xinjiang autonomous region is rich in coal with 1890 Gt of forecast resource and 314.5 Gt of total proven reserves (Anon 2010). There are some coal mines in which the single coal seam is more than 50-m thick, such as in the Wucaiwan

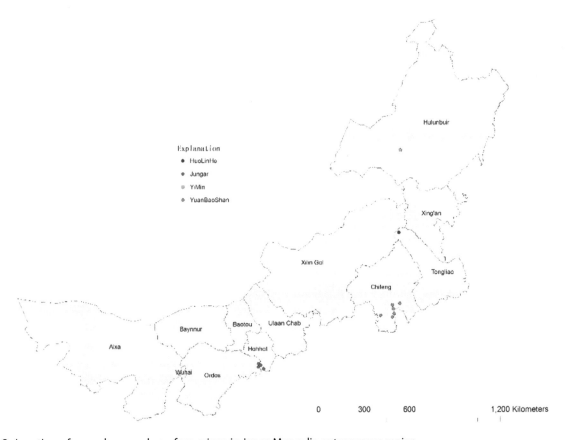

Figure 2. Location of some large scale surface mines in Inner Mongolia autonomous region.

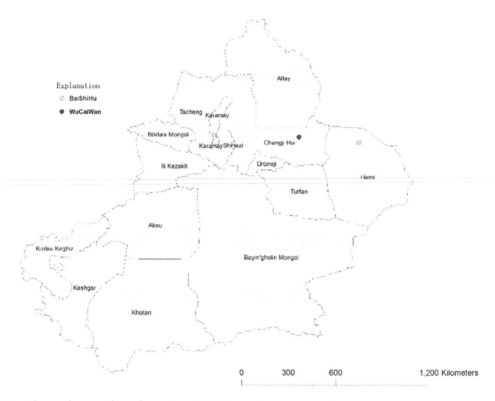

Figure 3. Location of some large scale surface mines in Xinjiang autonomous region.

surface mine in the Zhundong coal field and the Baishihu surface mine in the Hami region (Figure 3), which are suitable for large-scale mining. Most of Xinjiang coals have low ash yield and sulphur content. For example, the average ash yield of product coal from the South Surface Mine of the Zhundong coal field is less than 6.0%, and the average sulphur content is less than 0.5%. Xinjiang autonomous region will become one of the main coal producing areas in the future.

2.1.2. Production of different categories of coal in china

According to the Chinese national standard GB/T 5751 'Chinese Classification of Coal', coals are divided into anthracite, bituminous coal, and lignite in China. Bituminous coal is further divided into meagre coal, meagre lean coal, lean coal, coking coal, fat coal, gas-fat coal, 1/3 coking coal, gas coal, 1/2 medium caking coal, weakly caking coal, non-caking coal, and long flame coal by volatile yield (dry, ash-free basis), caking index, and plastometric indexes. Anthracite is classified into three subcategories by volatile and hydrogen content (dry, ash-free basis), named anthracite 1, anthracite 2, and anthracite 3, and coded as WY1, WY2, and WY3, respectively. Lignite includes two subcategories named lignite 1 and lignite 2, which are differentiated by light transmittance and gross calorific value (moist, ash-free basis) and are the lower-rank lignite and higher-rank

lignite respectively (GB/T 5751). This classification, which is a technical classification and used widely in coal production and research by government, enterprises, and institutes, is different from either ISO 11,760 'Classification of Coal', which is not intended to be used for commercial purposes, and ASTM D388-15 'Standard Classification of Coals by Rank' (ISO 11760 2005; ASTM D388-15 2012). Table 3 shows the rough comparison between coal classification of China and that of USA.

The classification of ASTM D388-15 does not apply to certain coals that are rich in inertinite or liptinite (ASTM D388-15 2012). In China, most of the weakly caking coal and non-caking coal that occur abundantly in the Early–Middle Jurassic coal-bearing basin in Northwest China and North China and are used widely in electricity generation and coal chemistry industry are rich in inertinite and have a low rank (Dai *et al*. 2012c, 2015b; Jiang *et al*. 2015). The volatile matter of these coals are lower than the vitrinite-rich coal with similar coal rank, the former is generally classified as long flame coal based on GB/T 5751. On the other hand, at least some gas fat coals, which are distributed mainly in East China, are rich in liptinite and have low or medium rank. These coals almost could not be classified in any class or group properly by ASTM D388-15. And by the comparison shown in Table 3, we can find that the Chinese classification of coal is probably better suited for coal

Table 3. Rough comparison of coal classification between China and USA.

GB/T 5751 Chinese classification of coal			ASTM D388-15 standard classification of coals by rank		
Key indexes				Key indexes	
V_{daf} (%)	$G_{R.I}$	Class	Class/group	V_{dmmf} (%)	$Q_{gr,m,mmf}$ (MJ/kg)
≤3.5	–	WY1	Meta-anthracite	≤2	–
>3.5–6.5	–	WY2	Anthracite	>2–8	–
>6.5–10.0	–	WY3	Semianthracite	>8–14	–
>10.0–20.0	≤5	PM			
>10.0–20.0	>5–20	PS	Low volatile bituminous coal	>14–22	–
>10.0–20.0	>20–50	SM			
>10.0–20.0	>50–65	SM			
>10.0–20.0	>65	JM			
>20.0–28.0	>50–65	JM	Medium volatile bituminous coal	>22–<31	–
>20.0–28.0	>65	JM			
>20.0–37.0	>85	FM			
>28.0–37.0	>65	1/3JM	High volatile A bituminous coal	>31	≥32.557
>37.0	>85	QF	High volatile B bituminous coal		
>37.0	>65	QM	High volatile C bituminous coal		≥30.23~<32.56
>37.0	>50–65	QM			
>37.0	>35–50	QM			≥26.74~<30.23
>28.0–37.0	>50–65	QM			≥24.42~<26.74
>20.0–37.0	>30–50	1/2ZN	–	–	–
>20.0–37.0	>5–30	RN	–	–	–
>20.0–37.0	≤5	BN	–	–	–
>37.0	>5–35	CY	Subbituminous A coal	–	≥24.42~<26.74
>37.0	≤5	CY	Subbituminous B coal		≥22.09~<24.41
>37.0	–	HM2	Subbituminous C coal		≥19.30–22.09
>37.0	–	HM1	Lignite A	–	≥14.65~<19.30
>37.0	–		Lignite B	–	<14.650

V_{daf}: volatile matter on the dry, ash free basis; $G_{R.I}$: caking index; V_{dmmf}: volatile matter on the dry, mineral matter free basis; $Q_{gr,\,m,\,mmf}$: gross calorific value on the moist, mineral matter free basis; WY1: anthracite with V_{daf} ≤3.5% and H_{daf} ≤2.0%, one of anthracite with the highest metamorphism degree; WY2: anthracite with 3.5%< V_{daf} ≤6.5% and 2.0%< H_{daf} ≤3.0%, one of anthracite with medium metamorphism degree; WY3: anthracite with 6.5%< V_{daf} ≤10.0% and H_{daf} >3.0%, one of anthracite with relatively lower metamorphism degree; PM: meagre coal; PS: meagre lean coal; SM: lean coal; JM: coking coal; FM: fat coal; 1/3JM: 1/3 coking coal; QF: gas-fat coal; QM: gas coal; 1/2ZN: 1/2 medium caking coal; RN: weakly caking coal; BN: non-caking coal; CY: long flame coal; HM2: lignite with relatively higher rank; HM1: lignite with the lowest rank.

for coking. Among 12 categories of bituminous coals, a total of eight categories of coal for coking are classified individually, which are meagre lean coal, lean coal, coking coal, fat coal, gas-fat coal, 1/3 coking coal, gas coal, and 1/2 medium caking coal, respectively, although they account for only about 25% of coal resources in China.

Table 4 Lists the production of each category of raw coal in China since 2002.

It is shown in Table 4 that the production of low-rank coals, including long flame coal, non-caking coal, and weakly caking coal, has been higher than that of coal for coking, and even showed a rapidly increasing trend since 2012. On the contrary, production of coal for coking, especially lean coal and fat coal, has increased slowly since 2012. This is partially related to the low proportion of coal for coking reserves, and that, as a whole, the mining conditions of coal for coking mines are more complicated than that of low-rank bituminous coal. Production of anthracite is relatively stable. With the rapid development of China's economy since 2000, production of lignite increased significantly. Since 2010, production of lignite in China has reached more than 300 Mt/a, this is mainly due to the high intensity exploitation of lignite in eastern Inner Mongolia in recent years.

2.1.3. Consumption of coal for different uses since 1995

Table 5 lists consumption of coal for different uses in China since 1995. Figure 4 shows coal production and main consumptions in these years.

Coal for industry accounts for more than 90% of the national coal consumption (Table 5). About 50% of the coal is used to generate electricity, which is significantly lower than that of most developed countries. According to the International Energy Agency (IEA) statistics, the world's consumption of coal for electricity generation accounted for 62.7% of coal production in 2011, with an average annual increase of 0.4% during 1990–2011. In general, consumption of coal for electricity generation in developed countries is mostly above 80%. For example, in 2011, the proportion of coal used for electricity generation in America was 91% and 76.2% in EU (Anon 2014). According to 'Notice of action plan about upgrading coal and electricity energy saving and emission reduction' issued by the National Development and Reform Commission of the People's

Table 4. Production of each categories of coal (Mt).

Year	Anthracite	Meagre lean coal	Lean coal	Coking coal	Fat coal	1/3 Coking coal	Gas-fat coal	Gas coal	Meagre coal	1/2 Medium caking coal	Weakly caking coal	Non-caking coal	Long flame coal	Lignite
2002	249.6	37.5	48.92	124.18	64.89	118.38	67.99	182.76	61.59	0.46	82.83	74.25	140.42	52.65
2003	304.1	59.04	50.58	163.56	86.93	137.71	78.18	204.24	79.44	0.45	85.63	89.09	162.26	62.91
2004	463.83	58.24	64.03	181.95	91.81	140.16	87.88	222.01	96.82	0.94	90.05	99.19	199.35	79.72
2005	486.95	62.05	60.57	215.6	101.76	147.43	96.79	210.86	109.33	3.45	105.22	138.84	225.64	97.64
2006	442.02	66.36	67.17	231.27	95.37	146.03	119.37	220.37	134.71	3.59	101.28	156.47	244.7	105.11
2007	439.23	73.21	73.54	221.56	94.22	135.78	75.82	282.09	141.45	0.41	97.41	223.19	348.71	146.41
2008	446.69	74.36	71.12	222.99	103.06	158.52	78.8	293.99	154.9	1.24	98.37	237.63	475.65	196.34
2009	426	73.66	72.61	190.58	90.96	160.16	105.77	316.94	214.19	0.21	83.37	289.28	362.41	265.65
2010	495.98	87.99	70.17	261.37	110.18	180.32	75.98	404.12	174.61	2.37	86.89	441.79	579.75	319.09
2011	516.29	99.95	81.27	259.45	111.16	168.53	86.81	433.31	188.47	2.01	109.35	815.85	512.53	339.81
2012	469.74	119.09	70.74	294.69	115.6	165.01	66.43	470.33	244.97	1.86	96.62	593.83	776.75	371.4
2013	450.67	101.97	76.07	279.91	119.12	189.1	92.78	459.71	256.96	37.54	110.66	710.57	769.58	299.09
2014	421.68	101.05	71.97	273.72	112.53	176.36	87.53	450.28	249.2	35.39	113.48	675.94	746.61	271.74
2015	396.87	104.16	82.98	262.22	129.01	171.75	88.57	406.19	234.25	33.06	138.73	635.88	725.27	253.93

Data resource: Shanxi Fenwei Energy Consulting Co. Ltd., Internal Research Report.

Table 5. Consumption of coal for different uses in China (Mt).

		1995	2000	2005	2010	2014
Total consumption		1376.77	1320.00	2167.23	3122.37	4116.14
Industry	Electricity generation	444.40	558.11	1032.64	1537.43	1845.25
	Heating	58.87	87.94	135.42	175.53	224.45
	Coke-making	183.96	164.96	316.67	499.50	519.97
	Coal to gas	7.64	9.60	12.77	10.40	9.48
	Others	480.84	372.40	528.59	737.46	1196.85
	Total	1175.71	1193.01	2026.09	2960.32	3904.97
Agriculture, forestry, animal husbandry, fishery, water conservancy industry		18.57	16.48	23.15	17.11	25.79
Construction		4.40	5.37	6.04	7.19	9.14
Transportation, warehousing and postal services		13.15	11.32	8.15	6.39	5.58
Wholesale, retail and accommodation, food and beverage industry		9.77	8.15	8.74	19.70	37.67
Residential consumption		135.30	79.07	87.39	91.59	92.53
Others		19.87	6.61	7.66	20.07	40.46

Data resource: National Bureau of Statistics of the People's Republic of China.

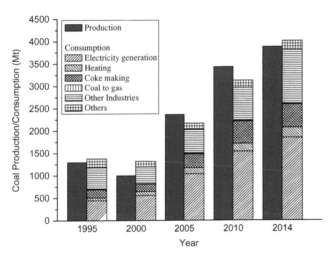

Figure 4. Coal production and main consumption in China.

Republic of China, by 2020, the consumption of coal for electricity generation in China will be more than 60% (The National Development and Reform Commission of the People's Republic of China 2014). Electricity generation, with the advantage in controlling pollution, is the predominant use of coal in developed countries.

Compared with other main coal producing countries worldwide, the proportion of coal for coke-making has been relatively higher in China for many years. Consumption of cleaned coal for coke-making, most of which are obtained by jigging, heavy medium, or flotation coal preparation and used as the feed of metallurgical coke production, accounted for about 12–15% of total coal consumption for the past two decades. From 2002, production of 'cleaned coal for coking' (defined in GB/T 3715 2007) had increased annually with a fluctuating growth rate because of the increased demand of the iron and steel industry. Since 2010, with the rapid growth of Chinese coking industry, production of coking coal exceeded 450 Mt/a, as shown in Table 6. In recent years, some coals have been known as potential raw source of valuable metals including Ge, Ga, rare earth elements, Al, Nb, and Zr (Seredin and Finkelman 2008; Seredin 2012; Hower *et al.* 2016). Although some rare metal

Table 6. Production of cleaned coal for coking in China (Mt).

Year	Production	Year	Production
2002	192.41	2009	404.99
2003	233.37	2010	446.88
2004	286.68	2011	468.71
2005	343.12	2012	476.64
2006	374.58	2013	508.43
2007	413.85	2014	502.76
2008	410.26	2015	496.27

Data resource: Shanxi Fenwei Energy Consulting Co. Ltd., Internal Research Report.

(including U, Ge, Al, and Ga) have been or is being currently industrially extracted from coal ash in China (Dai *et al.* 2016), for example, Ge in the Lincang and Wulantuga (Zhuang *et al.* 2006; Qi *et al.* 2007; Dai *et al.* 2012c), Al and Ga in the Jungar (Seredin 2012; Dai *et al.* 2012a), and REE in coal deposits in south-western China (Seredin and Dai 2012), the proportion of metalliferous coals is small.

2.2. China coal imports and exports since 1995

China is the world's largest coal producing and con-suming country. With the rapid development of the national economy, China turned into a net importer of coal from a net exporter of coal. Table 7 lists China coal import and export volume since 1995.

As shown in Table 7 from the mid-1990s to the early-2000s, China's coal export volume showed a rising trend in general. After 2003, with the rapid development of the Chinese economy, demand for coal increased, and coal production was in short supply. Since then, coal export volume decreased year by year and import increased shar-ply. China became a net importer of coal in 2009 and the

Table 7. China coal import and export volume (Mt).

Year	Import	Export	Net import
1995	1.20	28.62	0
1996	3.20	29.03	0
1997	2.00	30.72	0
1998	1.58	32.29	0
1999	1.67	37.41	0
2000	2.18	55.07	0
2001	2.66	90.13	0
2002	11.26	83.90	0
2003	11.10	94.03	0
2004	18.61	86.66	0
2005	26.22	71.73	0
2006	38.11	63.27	0
2007	51.02	53.19	0
2008	40.34	45.43	0
2009	125.84	22.40	103.44
2010	183.07	19.11	163.96
2011	182.10	14.66	167.44
2012	288.41	9.28	279.13
2013	327.02	7.51	319.51
2014	291.22	5.74	285.48
2015	204.10	5.33	198.77

Data resource: National Bureau of Statistics of the People's Republic of China.

import volume reached a peak of 327 Mt in 2013. Changes in China's coal imports and exports are closely related to the economic development (Wang and Wang 2016).

2.3. Coal quality management in china

2.3.1. Raw coal washing rate

Most coal mines in China are underground, some with complicated geological conditions, particularly those in southwestern China (Peng *et al.* 2011; Zhang *et al.* 2012; Yuan 2016). The raw coals of some mines in China have high ash yields and sulphur contents. For example, the ash yield of coal seam from the Chengzihe Formation is up to about 40% in some mines in Heilongjiang province and the total sulphur content of coal seam from the Taiyuan Formation is often more than 2% in most mines in North China. Particularly, some late Permian coals preserved within carbonate successions from southern China, contain super-high organic sulphur (>4%) and total sulphur con-tents (up to >10%) (Chou 2012); these include Guiding coals from Guizhou province (Lei *et al.* 1994; Dai *et al.* 2015a), Heshan and Fusui coals from Guangxi (Shao *et al.* 2003; Zeng *et al.* 2005; Dai *et al.* 2013), Yanshan coals from Yunnan province (Dai *et al.* 2008). However, due to lack of attention to coal quality, the raw coal washing rate has been low for a long time. Table 8 lists the volume of wash-ing coal and its ratio in China over the past 30 years.

The coal washing rate is more than 90% in Europe and 75% for bituminous in America (Anon 2016). From Table 8, we find that coal washing rate in China is much lower than that in developed countries. It was less than 50% until 2010. In accordance with requirements in 'China air pollution prevention action plan (2013)', the coal washing rate in China will be above 70% by 2017 (Ministry of Environmental Protection of the People's Republic of China 2013).

On the other hand, the coal washing rate in large-scale state-owned enterprises is higher. Table 9 lists volume and ratio of washing coal in preparation plants of key state-owned enterprises in China.

Table 8. The volume of washing coal and its ratio in China since 1985.

Year	Volume (Mt)	(%)
1985	142.94	16.4
1990	190.88	17.7
1995	201.72	15.6
2000	336.65	33.7
2005	703.00	32.0
2010	1650	50.9
2012	2050	56.0
2013	2214	55.8
2014	2420	62.5

Data resource: National Bureau of Statistics of the People's Republic of China.

Table 9. Volume and ratio of washing coal in preparation plants of key state-owned enterprises.

		1995	2000	2005	2007	2008
Total	Volume (Mt)	201.72	239.89	503.81	589.88	690.20
	Ratio (%)	65.17	72.33	73.94	74.59	73.88
Coal for coking	Input (Mt)	131.04	133.05	235.05	247.75	288.17
	Output (Mt)	76.44	81.22	151.05	155.15	166.08
	Yield (%)	58.33	61.04	64.26	62.62	57.63
Steam coal	Input (Mt)	70.68	106.84	268.76	342.13	422.03
	Output (Mt)	55.02	92.30	221.45	284.85	343.86
	Yield (%)	77.84	86.39	82.40	83.26	81.48

Data resource: Statistical Department of the State Administration of Production Safety Supervision and Administration, Statistical Summary of Coal Industry, Internal Publication.

2.3.2. Standards for coal quality management

China is a major coal producing country with a number of different coal-forming geological environments requiring different mining techniques. The situation is further complicated by substantial variations in coal quality. There are often some controversies between coal production and consumption enterprises about how to properly understand the technical properties of coal products in coal utilization process, for example, how to define 'high calorific coal' or how to make an acceptable specification of feed coal for different boilers. Therefore, China has issued a series of coal quality management standards, including 29 national standards and 18 coal industrial standards, as shown in Table 10. Compared with other coal producing countries, China has a relatively complete coal quality management standard system.

3. Status of production and quality management of coal in china

Since 2014, China's national economic growth has been adjusted, and coal production started to decline as a

Table 10. Standards for coal quality management.

Standard NO.	Title	Publish year
GB/T 5751-2009	Chinese classification of coals	2009
GB/T 17607-1998	Chinese classification of seam coals	1998
GB/T 16772-1997	Codification systems for Chinese coals	1997
GB/T 17608-2006	Division of variety and grading for coal products	2006
GB/T 25209-2010	Commercial coal marking	2010
GB 25960-2010	Rules for steam coal blending	2010
GB/T 26128-2010	Classification and utilization of scarce and special coal resources	2010
GB/T 15224.1-2010	Classification for quality of coal – Part 1: ash	2010
GB/T 15224.2-2010	Classification for quality of coal – Part 2: sulphur content	2010
GB/T 15224.3-2010	Classification for quality of coal – Part 3: calorific value	2010
GB/T 20475.1-2006	Classification for content of harmful elements in coal – Part 1□phosphorus	2006
GB/T 20475.2-2006	Classification for content of harmful elements in coal – Part 2□chlorine	2006
GB/T 20475.3-2012	Classification for content of harmful elements in coal – Part 3□arsenic	2012
GB/T 20475.4-2012	Classification for content of harmful elements in coal – Part 4□Mercury	2012
GB/T 23251-2009	Guideline for coal used in coal chemical conversion	2009
GB/T 397-2009	Specification of coal for coke making	2009
GB/T 7562-2010	Specification of coal used for pulverized coal-fired boiler for power generation	2010
GB/T 7563-2000	Technical condition of coal for cement rotary kiln	2000
GB/T 9143-2008	Specifications of coal for atmospheric fixed-bed gasifier	2008
GB/T 18342-2009	Specifications of coal for chain and travelling grate stoker boiler	2009
GB/T 18512-2008	Specifications of coal used in pulverized coal injection (PCI)	2008
GB/T 23810-2009	Specification of coal for direct liquefaction	2009
GB/T 26126-2010	Specifications of pulverized coal for small and medium-sized industrial boilers	2010
GB/T 29721-2013	Specifications of coal for fluidized bed gasifier	2013
GB/T 29722-2013	Specifications of coal for entrained-flow gasifier	2013
GB/T 31087-2014	General technical requirements for impurities control of commercial coal	2014
GB/T 31091-2014	General technical requirements for management of coal yard	2014
GB/T 18855-2014	Coal water slurry for fuel	2014
GB/T 31356-2014	Quality evaluation and control guide for commercial coal	2014
MT/T 560-2007	Classification for thermal stability value of coal	2007
MT/T 561-2007	Classification for fixed carbon of coal	2007
MT/T 596-2008	Classification for caking index of bituminous coal	2008
MT/T 849-2000	Classification for volatile matter of coal	2000
MT/T 850-2000	Classification for total moisture in coal	2000
MT/T 852-2000	Classification for Hardgrove grindability index of coal	2000
MT/T 853.1-2000	Classification for coal ash fusibility	2000
MT/T 853.2-2000	Classification for coal ash fusibility	2000
MT/T 964-2005	Classification for lead in coal	2005
MT/T 965-2005	Classification for chromium in coal	2005
MT/T 966-2005	Classification for fluorine in coal	2005
MT/T 967-2005	Classification for germanium in coal	2005
MT/T 1028-2006	Classification for selenium in coal	2006
MT/T 1029-2006	Classification for cadmium in coal	2006
MT/T 1074-2008	Classification for alkali metal (potassium, sodium) content in coal	2008
MT/T 1010-2006	Specifications of briqutte coals for fixed-bed gasification	2006
MT/T 1030-2006	Specifications of coals used for sinter	2006
MT/T 1072-2007	Technical guidance for collection of coal data	2007

result. In 2015, the national coal production was 3.74 Gt, falling 3.4% from 2014. In 2016, it was 3.36 Gt, falling 10.2% from 2015. Coal export was reduced by 7.1% year-to-year to 5.33 Mt, while import volume was 204.06 Mt, 29.9% lower from past year.

3.1. Coal production and variety in different regions in 2015

Table 11 shows annual raw coal production of some coal producing provinces in China in 2015. Coal production in Shanxi was 975 Mt, nearly equal to that of the year before, in which there was more than 50% coal for coking. Coal production in Inner Mongolia plunged 7.2% from a year earlier to 909.57 Mt, mainly due to the reduced demand for steam coal in China. Coal production in Guizhou decreased 8.1% over last year to 170.01 Mt, which might due to the production decline of high-sulphur coal. Except for Anhui, Shaanxi, and Xinjiang, raw coal production of other main coal producing provinces decreased in varying degrees compared with 2014.

It is also shown in Table 10 that the coal industry in Shanxi occupies a special position in China. In 2015, Shanxi got back to being China's leading coal producer with a production of 975 Mt, 53% coal for coking, accounting for about 42% of the total coal for coking production in the country. As for other major coal producing provinces, Inner Mongolia and Shaanxi produced mainly steam coal, with only about 5% used for metallurgical coke-making. That is the major reason that China's coking industry depends essentially on

Shanxi coal products. Coal production in China's major producing areas of steel and coke, such as Hebei and Shandong provinces, is lower, but coal for coking accounts of their 60% of raw coal production.

3.2. Production of different categories of coal in 2015

Table 12 shows national production of different categories of coal in 2015.

As can be seen from Table 11, in 2015, the production of primary steam coals, including long flame coal, non-caking coal, and weakly caking coal, account for 40% of raw coal production, near to the 2014 production. The proportion of lignite production is down from the peak of 9.23% in 2010 to 6.78% in 2015. The proportion of anthracite is about 11%, consistent with the previous year.

In 2015, the production of raw coal for coking reached 1.277 Gt, accounting for 33.2% of raw coal production, nearly same as 2014. But the proportion of coal with favourable coking performance, including lean coal, fat coal, and coking coal, was lower, with only about 37% among total coal for coking production. While weakly coking coals had a higher proportion, such as gas coal and gas fat coal. According to the literature, consumption of cleaned coals for coke-making should be 496 to 513 Mt in 2015 (Weng et al. 2016). Based on this estimation, average cleaned coal recovery of washed raw coals for coke-making in 2015 was approximately 40%.

3.3. Status of coal quality management and standardization

Since 2012, hazy weather had occurred frequently in most areas of China and utilization of raw, non-cleaned coal was considered to be one of the main causes (Miao and Zhou 2014; Wang 2015a). Pollutant emission control

Table 11. Coal production of some provinces in China in 2015 (Mt).

Province	Raw coal	Coal for coking
Shanxi	975.00	520.03
Inner Mongolia	909.57	54.91
Shaanxi	522.24	20.98
Xinjiang	146.44	26.08
Guizhou	170.01	76.35
Shandong	144.00	129.98
Anhui	135.30	131.30
Henan	133.11	55.62
Hebei	83.84	54.49
Liaoning	51.76	11.34
Heilongjiang	66.78	53.84
Sichuan	63.56	20.24
Ningxia	74.44	9.61
Yunnan	48.84	22.23
Gansu	43.99	1.12
Hunan	33.95	0.85
Chongqing	34.78	13.83
Jilin	26.23	8.33
Jiangxi	21.20	9.62
Jiangsu	19.14	19.14
Hubei	8.75	0.99

Data resource: Shanxi Fenwei Energy Consulting Co. Ltd., Internal Research Report.

Table 12. Production of each categories of coal (by GB/T 5751) in 2015 (Mt).

Category	Raw coal production
Anthracite	396.87
Meagre coal	234.25
Meagre lean coal	104.16
Lean coal	82.98
Coking coal	262.22
Fat coal	129.01
1/3 Coking coal	171.75
Gas coal	406.19
Gas fat coal	88.57
1/2 Medium caking coal	33.06
Weakly caking coal	138.73
Non-caking coal	635.88
Long flame coal	725.27
Lignite	253.93

Data resource: Shanxi Fenwei Energy Consulting Co. Ltd., Internal Research Report.

Table 13. Standards for coal quality management set or released in 2015.

Standard no.	Title	Published date	Implementation date
GB/T 31426-2015	Coal water slurry for gasification	15 May 2015	1 July 2015
GB/T 31861-2015	Clean fuel for industrial furnace briquette	11 September 2015	1 December 2015
GB/T 31862-2015	Commercial coal quality lignite	11 September 2015	1 December 2015
	Commercial coal quality civil bulk coal	Draft completed and submitted to SAC	
	Commercial coal quality civil briquette	Draft completed and submitted to SAC	

for residential coal combustion in winter in northern China had been increasingly emphasized. Governance of bulk coal quality, especially for rural residents, had been strengthened through encouraging the use of low-sulphur clean anthracite briquettes and this accelerated standardization process of civil coal-quality management (Hou 2013). Table 13 shows standards for coal quality management drafted or released in 2015.

4. Projections of coal production and quality management in china

4.1. Main factors that affect coal production and consumption

China's coal production will be influenced by economic development, industry restructuring, environment pollution concerns, water resource availability, mining technology and other factors such as transportation. Due to the multiple factors, including the global financial crisis, lower energy prices, and the domestic economic slowdown, China is facing the problem of declining coal demand in recent years. Traditional coal production has suffered from excess capacity and high inventory in China since 2012. The development of the coal industry in China will change to rely more on quality and efficiency, developing scientific capacity, and achieving clean coal utilization (Xie et al. 2012b; Lu 2016).

The research results of domestic institutions shows that the proportion of coal in China's primary energy consumption would gradually decline from the current 66% to 62% by 2020, 55% by 2030, and 50% by 2050. However, some studies proposed that the peak of China's coal consumption would occur in 2020 with the amount of 4.06 Gt/a (The National Development and Reform Commission of the People's Republic of China 2015). There are even some reports showing that China's coal consumption had already reached the peak in 2013 (Wu 2016).

However, China's coal resources are relatively abundant and inexpensive compared with oil and natural gas. With the progress of clean coal utilization, the status of Chinese coal as the primary energy would not be changed for the foreseeable future.

4.2. Projections of coal quality and management in china

In China, coals from different regions, with various coal rank, are characterized by significantly different technical properties. In general, in the Carboniferous-Permian coal regions of North China, the sulphur content is lower in the upper-group coal seams (mainly in the Shanxi Formation) and higher in the under-group coal seams (mainly in the Taiyuan Formation) (Liu et al. 2001; Gao et al. 2005; Luo et al. 2005). North China is the main coal for coking and anthracite-producing area. Therefore, the sulphur content of China's raw coal for coking and anthracite coal tends to increase with mining depth. So, it will be necessary to improve the raw coal washing rate and promote desulphurization and denitrification technology in the coking and gasification industry for reasonable exploitation and utilization of coal resources in the North China coal region. Meanwhile, the caking and coking performance of coal for coking will change with the gradual exhaustion of resource in the existing mines, the construction of new mines of coal for coking, and the mining depth. The overall trend of caking property of China's coal for coking according to relevant mine development planning and coal seam geological characteristics is shown in Table 14.

The Northwest Jurassic coal region, which includes five provinces and autonomous regions, i.e. Shaanxi, Ningxia, Gansu, Qinghai, and Xinjiang, is the most important steam-coal resource area in China, especially for low-rank coals

Table 14. Prediction of the proportion of strong caking coal among coal for coking in 2015–2030 in different regions.

Province	2015–2020 (%)	2020–2025 (%)	2025–2030 (%)
Anhui	31	35	35
Guizhou	63	60	65
Hebei	81	81	82
Henan	94	83	21
Heilongjiang	36	36	46
Shandong	23	26	28
Shanxi	70	72	61
Total	60%	62%	56%

Data resource: Shanxi Fenwei Energy Consulting Co. Ltd., Internal Research Report.

which include lignite, long flame coal, and non-caking coal by GB/T 5751 'Chinese Classification of Coal'. Coals in this region possess low-ash and low-sulphur contents on the whole. For example, the average ash yield of product coal from Shenfu coal field in Shaanxi is lower than 10.0%, and the average sulphur content is not more than 0.50%. The Northwest region will become the main source of steam coal with China's western development plan. Taking into account the Chinese government plan to improve coal washing rate and clean-coal technology, it can be predicted that the ash and sulphur content of Chinese steam coal will remain low in general, which would be conducive to achieve the target of ultra-low emission for Chinese coal-fired generating units.

As to coal-quality management, a series of national standards for commercial coal will be drafted in order to adapt to the new environmental requirements with the increasingly stringent environmental standards. Some local governments and enterprises using coal have been working on the development of local coal-product standards suitable for local conditions and are stricter than the national standards. Clean manufacturing will be the main goal for coal production and consumption in China.

5. Conclusions

(1) Over the past 30 years, coal production in China developed rapidly. Raw coal production had reached a peak of 3.974 Gt/a in 2013.

(2) Shanxi, Inner Mongolia, and Shaanxi are the main coal-producing provinces in China, but there are significant differences in the coal product structure.

(3) China has the world's most complete standard system for coal quality management. In the future, in order to meet the increasingly stringent environmental requirements, coal-quality management standardization will still be the primary technological means for China's coal production.

(4) China is facing the problem of decline in coal demand in recent years. Traditional coal production has suffered from excess capacity and higher inventory in China since 2012. The development of coal production in China will change to rely more on quality and efficiency, developing scientific capacity and achieving clean coal utilization.

Acknowledgements

This work was supported by the National Key Research and Development Program of China (Grant 2016YFB0600301).

Disclosure statement

No potential conflict of interest was reported by the authors.

Funding

This work was supported by the National Key Research and Development Program of China [grant number 2016YFB0600301].

References

Ang, B.W., and Su, B., 2016, Carbon emission intensity in electricity production: A global analysis: Energy Policy, v. 94, p. 56–63. 10.1016/j.enpol.2016.03.038

Anon, 2010, Evaluation on potential of Coal Resource in Xinjiang Uyger Autonomous Region, Internal Report: Xinjiang Bureau of Coal Geological Exploration (in Chinese).

Anon, 2014, Coal information trading network. http://www.meitanwang.com/meitan/c3/684812.html (in Chinese).

Anon, 2016, Annual Research and Consultation Report of Market Research and Investment Strategy on China Industry: Sansheng Consulting (in Chinese).

ASTM D388-15, 2012, Standard Classification of Coals by Rank, West Conshohocken, PA: ASTM International.

Chou, C.-L., 2012, Sulfur in coals: A review of geochemistry and origins: International Journal of Coal Geology, v. 100, p. 1–13. 10.1016/j.coal.2012.05.009

Dai, S., Jiang, Y., Ward, C.R., Gu, L., Seredin, V.V., Liu, H., Zhou, D., Wang, X., Sun, Y., Zou, J., and Ren, D., 2012a, Mineralogical and geochemical compositions of the coal in the Guanbanwusu Mine, Inner Mongolia, China: further evidence for the existence of an Al (Ga and REE) ore deposit in the Jungar Coalfield: International Journal of Coal Geology, v. 98, p. 10–40. 10.1016/j.coal.2012.03.003

Dai, S., Li, W., Tang, Y., Zhang, Y., and Feng, P., 2007, The sources, pathway, and preventive measures for fluorosis in Zhijin County, Guizhou, China: Applied Geochemistry, v. 22, p. 1017–1024. 10.1016/j.apgeochem.2007.02.011

Dai, S., Ren, D., Chou, C.-L., Finkelman, R.B., Seredin, V.V., and Zhou, Y., 2012b, Geochemistry of trace elements in Chinese coals: a review of abundances, genetic types, impacts on human health, and industrial utilization: International Journal of Coal Geology, v. 94, p. 3–21. 10.1016/j.coal.2011.02.003

Dai, S., Ren, D., Zhou, Y., Chou, C.-L., Wang, X., Zhao, L., and Zhu, X., 2008, Mineralogy and geochemistry of a superhigh-organic-sulfur coal, Yanshan Coalfield, Yunnan, China: evidence for a volcanic ash component and influence by submarine exhalation: Chemical Geology, v. 255, p. 182–194. 10.1016/j.chemgeo.2008.06.030

Dai, S., Seredin, V.V., Ward, C.R., Hower, J.C., Xing, Y., Zhang, W., Song, W., and Wang, P., 2015a, Enrichment of U–Se–Mo–Re–V in coals preserved within marine carbonate successions: geochemical and mineralogical data from the Late Permian Guiding Coalfield, Guizhou, China: Mineralium Deposita, v. 50, p. 159–186. 10.1007/s00126-014-0528-1

Dai, S., Wang, X., Seredin, V.V., Hower, J.C., Ward, C.R., O'Keefe, J.M.K., Huang, W., Li, T., Li, X., Liu, H., Xue, W., and Zhao, L., 2012c, Petrology, mineralogy, and geochemistry of the Ge-rich coal from the Wulantuga Ge ore deposit, Inner Mongolia, China: new data and genetic implications: International Journal of Coal Geology, v. 90–91, p. 72–99. 10.1016/j.coal.2011.10.012

Dai, S., Yan, X., Ward, C.R., Hower, J.C., Zhao, L., Wang, X., Zhao, L., Ren, D., and Finkelman, R.B., 2016, Valuable elements in Chinese coals: A review: International Geology Review, p. 1–31, 10.1080/00206814.2016.1197802

Dai, S., Yang, J., Ward, C.R., Hower, J.C., Liu, H., Garrison, T.M., French, D., and O'Keefe, J.M.K., 2015b, Geochemical and mineralogical evidence for a coal-hosted uranium deposit in the Yili Basin, Xinjiang, northwestern China: Ore Geology Reviews, v. 70, p. 1–30. 10.1016/j.oregeorev.2015.03.010

Dai, S., Zhang, W., Ward, C.R., Seredin, V.V., Hower, J.C., Li, X., Song, W., Wang, X., Kang, H., Zheng, L., Wang, P., and Zhou, D., 2013, Mineralogical and geochemical anomalies of late Permian coals from the Fusui Coalfield, Guangxi Province, southern China: influences of terrigenous materials and hydrothermal fluids: International Journal of Coal Geology, v. 105, p. 60–84. 10.1016/j.coal.2012.12.003

Finkelman, R.B., Orem, W., Castranova, V., Tatu, C.A., Belkin, H.E., Zheng, B., Lerch, H.E., Maharaj, S.V., and Bates, A.L., 2002, Health impacts of coal and coal use: possible solutions: International Journal of Coal Geology, v. 50, p. 425–443. 10.1016/S0166-5162(02)00125-8

Gao, L., Liu, G., Chou, C.-L., Zheng, L., and Zheng, W., 2005, The study of sulfur geochemistry in Chinese coals: Bulletin of Mineralogy, Petrology and Geochemistry, v. 24, p. 79–87. (in Chinese with English abstract)

GB/T 3715, 2007, Terms relating to properties and analysis of coal: Beijing, Standards Press of China. (in Chinese with English abstract)

GB/T 5751, 2009, Chinese Classification of Coal: Beijing, Standards Press of China. (in Chinese with English abstract)

Guo, J., 2011, On China's Energy Saving and Emission Reduction and International Law Analysis about Global Climate Change: Energy Procedia, v. 5, p. 2568–2575. 10.1016/j.egypro.2011.03.441

Hou, T., 2013, 52 suburban villages reserve on high-quality low sulfur coal: Shijiazhuang Daily 2013.09.14 (001), (in Chinese).

Hower, J.C., Eble, C.F., Dai, S., and Belkin, H.E., 2016, Distribution of rare earth elements in eastern Kentucky coals: indicators of multiple modes of enrichment?: International Journal of Coal Geology, v. 160–161, p. 73–81. 10.1016/j.coal.2016.04.009

ISO 11760: 2005 (E), Classification of Coal.

Jiang, Y., Zhao, L., Zhou, G., Wang, X., Zhao, L., Wei, P., and Song, H., 2015, Petrological, mineralogical, and geochemical compositions of Early Jurassic coals in the Yining Coalfield, Xinjiang, China: International Journal of Coal Geology, v. 152, Part A, p. 47–67. 10.1016/j.coal.2015.07.011

Lei, J., Ren, D., Tang, Y., Chu, X., and Zhao, R., 1994, Sulfur accumulating model of superhigh organosulfur coal from Guiding, China: Chinese Science Bulletin, v. 39, p. 1817–1821.

Liu, D., Yang, Q., Tang, D., Kang, X., and Huang, W., 2001, Geochemistry of sulfur and elements in coals from the Antaibao surface mine, Pingshuo, Shanxi Province, China: International Journal of Coal Geology, v. 46, p. 51–64. 10.1016/S0166-5162(00)00040-9

Lu, X., 2016, Present situation and suggestion for clean coal development and utilization in China: Coal Engineering, v. 48, p. 8–10. (in Chinese with English abstract) 10.11799/ce201603003.

Luo, H., Zhang, B., Lu, L., and Pei, Y., 2016, Preliminary study of China's total coal consumption control scheme based on air pollution control: Climate Change Research, v. 12, p. 172–178. (in Chinese with English abstract) 10.12006/j.issn.1673-1719.2015.180.

Luo, Y., Li, W., Jiang, Y., and Bai, X., 2005, Distribution of sulfur in coals of China: Coal Conversion, v. 28, p. 14–18(in Chinese with English abstract)

Miao, R., and Zhou, F., 2014, Green revolution of China coal consumption: Coal Economic Research, v. 34, p. 63–65. (in Chinese with English abstract) 10.13202/j.cnki.cer.2014.11.013.

Ministry of Environmental Protection of the People's Republic of China, 2013, Air pollution prevention and control action plan (in Chinese).

Peng, H., Cai, Q., Zhou, W., Shu, J., and Li, G., 2011, Study on stability of surface mine slope influenced by underground mining below the endwall slope: Procedia Earth and Planetary Science, v. 2, p. 7–13. 10.1016/j.proeps.2011.09.002

Qi, H., Hu, R., and Zhang, Q., 2007, Concentration and distribution of trace elements in lignite from the Shengli coalfield, inner Mongolia, China: implications on origin of the associated Wulantuga germanium deposit: International Journal of Coal Geology, v. 71, 129–152. 10.1016/j.coal.2006.08.005

Seredin, V., and Finkelman, R.B., 2008, Metalliferous coals: a review of the main genetic and geochemical types: International Journal of Coal Geology, v. 76, p. 253–289. 10.1016/j.coal.2008.07.016

Seredin, V.V., 2012, From coal science to metal production and environmental protection: a new story of success: International Journal of Coal Geology, v. 90–91, p. 1–3. 10.1016/j.coal.2011.11.006

Seredin, V.V., and Dai, S., 2012, Coal deposits as potential alternative sources for lanthanides and yttrium: International Journal of Coal Geology, v. 94, p. 67–93. 10.1016/j.coal.2011.11.001

Shao, L., Jones, T., Gayer, R., Dai, S., Li, S., Jiang, Y., and Zhang, P., 2003, Petrology and geochemistry of the high-sulphur coals from the Upper Permian carbonate coal measures in the Heshan Coalfield, southern China: International Journal of Coal Geology, v. 55, p. 1–26. 10.1016/S0166-5162(03)00031-4

The National Development and Reform Commission of the People's Republic of China, 2014, Notice of action plan about upgrading coal and electricity energy saving and emission reduction (2014-2020) (in Chinese).

The National Development and Reform Commission of the People's Republic of China, 2015, Coal consumption peak may be brought forward to 2020 (in Chinese).

Tian, L., Dai, S., Wang, J., Huang, Y., Ho, S.C., Zhou, Y., Lucas, D., and Koshland, C.P., 2008, Nanoquartz in Late Permian C1 coal and the high incidence of female lung cancer in the Pearl River Origin area: a retrospective cohort study: BMC Public Health, v. 8, p. 398. 10.1186/1471-2458-8-398

Wang, C., and Wang, W., 2016, Development of import-export coal trade by port on mainland China: spatial pattern, evolution and dynamics: Resources Science, v. 38, p. 631–644. (in Chinese with English abstract) 10.18402/resci.2016.04.06.

Wang, J., 2015a, Problem and Countermeasure Research of Domestic Clean Coal Electricity Generation: Coal Technology, v. 34, p. 331–332. (in Chinese with English abstract) 10.13301/j.cnki.ct.2015.05.128.

Wang, W., Li, S., and Han, J., 2015, Analysis of the main global coal resource countries' supply-demand structural trend and coal industry outlook: China mining magazine, v. 24, p. 5–9(in Chinese with English abstract)

Weng, X., Gao, X., and Pan, D., 2016, Coal consumption in iron and steel industry: fact and prediction in 2013-2015, projection in 2016: Metallurgical economics and management, v. 1, p. 12–13(in Chinese)

Whitmarsh, L., Seyfang, G., and O'Neill, S., 2011, Public engagement with carbon and climate change: To what extent is the public 'carbon capable'?: Global Environmental Change, v. 21, p. 56–65. 10.1016/j.gloenvcha.2010.07.011

Wu, X., 2016, Different opinions on coal consumption peak: China Energy News, 2016.01.25(001) (in Chinese).

Xie, H., Liu, H., and Wu, G., 2012a, Quantitative analysis on the contribution of coal to the development of national economy: Energy of China, v. 34, p. 5–9. (in Chinese with English abstract) 10.3969/j.issn.1003-2355.2012.04.001.

Xie, H., Wang, J., Shen, B., Liu, J., Jiang, P., Zhou, H., Liu, H., and Wu, G., 2012b, New idea of coal mining: scientific mining and sustainable mining capacity: Journal of China Coal Society, v. 37, p. 1069–1079. (in Chinese with English abstract) 10.13225/j.cnki.jccs.2012.07.010.

Xu, Z., Hou, H., Zhang, S., Ding, Z., Ma, C., Gong, Y., and Liu, Y., 2012, Effects of mining activity and climatic change on ecological losses in coal mining areas: Transactions of the Chinese Society of Agricultural Engineering, v. 28, p. 232–240. (in Chinese with English abstract) 10.3969/j.issn.1002-6819.2012.05.039.

Yan, Y., and Yang, L., 2010, China's foreign trade and climate change: A case study of CO_2 emissions: Energy Policy, v. 38, p. 350–356. 10.1016/j.enpol.2009.09.025

Yuan, L., 2016, Control of coal and gas outbursts in Huainan mines in China: A review: Journal of Rock Mechanics and Geotechnical Engineering, v. 8, p. 559–567. doi:10.1016/j.jrmge.2016.01.005

Zeng, R., Zhuang, X., Koukouzas, N., and Xu, W., 2005, Characterization of trace elements in sulphur-rich Late Permian coals in the Heshan coal field, Guangxi, South China: International Journal of Coal Geology, v. 61, p. 87–95. 10.1016/j.coal.2004.06.005

Zhang, J., Fu, M., Geng, Y., and Tao, J., 2011, Energy saving and emission reduction: A project of coal-resource integration in Shanxi Province: China: Energy Policy, v. 39, p. 3029–3032.

Zhang, Y., Gao, L., Yang, W., Zhang, Z., Zhang, P., Liu, C., and Fang, F., 2012, Hydro-geological conditions and mine inflow water forecast of the western long beach in Junggar coalfield: Energy Procedia, v. 16, p. 915–920. 10.1016/j.egypro.2012.01.146

Zhuang, X., Querol, X., Alastuey, A., Juan, R., Plana, F., Lopez-Soler, A., Du, G., and Martynov, V.V., 2006, Geochemistry and mineralogy of the Cretaceous Wulantuga high germanium coal deposit in Shengli coal field, Inner Mongolia, Northeastern China: International Journal of Coal Geology, v. 66, p. 119–136. 10.1016/j.coal.2005.06.005

The geologic settings of Chinese coal deposits

Zengxue Li, Dongdong Wang, Dawei Lv, Ying Li, Haiyan Liu, Pingli Wang, Ying Liu, Jianqiang Liu and Dandan Li

ABSTRACT

China has abundant coal resources, which are extensively developed in marine-influenced, continental, and transitional environments. The coal-bearing strata in China have complicated geneses and distributions. There were eight major coal-forming periods in China, namely the Terreneuvian, Mississippian, Pennsylvanian–Cisuralian–Guadalupian, Lopingian, Late Triassic, Early–Middle Jurassic, and Early Cretaceous Epochs, as well as the Palaeogene–Neogene Periods. The distributions of the coal-bearing strata formed during these different coal-forming periods displayed obvious regional and regularity characteristics. In accordance with their plate-tectonic settings and coal-bearing characteristics, the coal-bearing basins in China can be divided into six types. The coal-bearing basins that were formed during the Palaeozoic Era were mainly large epicontinental sea basins. During the early Palaeozoic Era, the shallow sea was the most important coal-forming sedimentary environment. In addition, coastal delta and delta-detrital coast systems were the most important coal-forming sedimentary environments in the late Palaeozoic Era. The coal-bearing basins that were formed during the Late Triassic Epoch were mainly offshore basins, and coastal, coast-delta, coastal alluvial, and coastal inter-mountainous plain environments dominated their coal-forming sedimentary environments. Coast–bay and lagoon–estuary systems comprised additional main coal-forming sedimentary environments. The coal-bearing basins that formed during the Early–Middle Jurassic Epochs were large- and medium-sized inland lake basins, in which alluvial-lake delta systems recorded the best coal formation processes, followed by lakeshore sedimentary environments. The coal-bearing basins that formed during the Early Cretaceous Epoch and Palaeogene–Neogene Periods were mainly small-sized continental basin groups. Coal-forming processes mainly occurred in the lake-delta swamp environments during lake siltation stages when the filling evolution of the lake basins occurred. In addition, this study proposed a new coal-forming process that occurs during transgression events. Finally, this study generally summarized the characteristics and evolution theories of coal-forming processes in China and especially focused on the characteristics of the evolution of coal-forming sedimentary environments.

1. Introduction

Coal geology in China is characterized by its rich and varied coal deposits, eight coal-forming periods, complex tectonic framework, and various types of sedimentary environments. In its geologic history, sediments in China have mainly formed within continental environments, followed by transitional environments, and then marine-influenced environments. The coal resources formed in continental environments are the most abundant, followed by those formed in marine-continental transitional and marine-influenced environments.

Among the global coal geological systems, the characteristics of coal geology in China are both globally comparable and unique. The formation periods of China's coal-bearing processes were basically consistent with the main global coal-forming periods, although they differed in the sizes and intensity levels of their coal deposits. The coal-bearing basin types, basin backgrounds, peat accumulations, and coal-forming intensities differed significantly between various periods of coal formation, which led to differences in the spatial distribution and characteristics of coals formed in different periods. In this study, the sedimentation types of coal-bearing processes in China are divided into interactive marine and continental deposits within plates, marine-continental transitional deposits on plate edges, inland depressed-basin deposits, and inland rifted-basin deposits.

The differences in the characteristics and types of coal-forming basins in China depended on the background of the basins' tectonic events during the different stages of crustal evolution, their tectonic-palaeogeographic background, the tectonic events that occurred during the basin-forming period, and the nature of the basins' basements. Based on the degree of tectonic stability of the basins during the coal-forming period, these coal basins can be divided into stable coal-forming basins, transitional coal-forming basins, and active coal-forming basins.

Coal-forming intensity varied greatly during different periods in China due to differences in the types of coal-forming basins, the tectonic activity that occurred during the coal-forming periods, the continuity of peat swamp development, and palaeogeographic features during the peat accumulation periods. The strength of the coal-forming processes in different areas of China, as well as the locations of the basins, also varied.

This article describes the different aspects of China's coal-forming periods and plant types, as well as its coal basin types, distribution and evolution, the spatial and temporal distribution of coal-bearing strata, its coal-forming geological background, sedimentary sources, coal-forming processes, and some aspects of recent research progress, thus allowing readers to gain a better understanding of China's coal geological characteristics.

2. Coal-forming periods in China

2.1. Coal-forming plants and coal-forming periods in China

Coal formation has occurred since multicellular organisms first appeared on the Earth. The coal that formed during the Proterozoic Era in China was mainly composed of fungi and, in particular, algae. During the Cambrian Period, a large number of fungi and algal-type plants bred rapidly in water, resulting in the coal in the Early Palaeozoic Era containing a large number of laminated carbonized algae. The coal seam roof contains large amounts of the remains of fungi and algae in South China (Han and Yang 1980). This type of coal (sapropelic) was also formed during the Cambrian, Ordovician, and Silurian Periods. However, it later became sapanthracite, which is the highest metamorphic stage of sapropelic coal (Han and Yang 1980; O'Keefe et al. 2013). This type of coal is widely distributed in South China, such as in southern Shaanxi, western Zhejiang, and throughout other provinces.

Humic coal was formed after Devonian plants colonized the land and formed peat swamps. These terrestrial plants had experienced continuous development since the Devonian. In the Devonian, the earliest known land plants grew along the coast of a shallow epeiric sea. In the Middle Devonian Epoch in China, coal-bearing deposits were of littoral or neritic origin, and Devonian coals thus contain a great deal of cuticles and microspores; hence, they are called cutinitic liptobiolith or 'Luquanite' (Han 1989). These coals are classified as cutinitic liptobiolith, sporinite-rich durain, cutinite-rich durain, and sporinitic liptobiolith. Their vitrinite content is very low, and they are dominated by collodetrinite, collotelinite, and corpogelinite (Dai et al. 2006).

The coal-forming plants differed within each of the coal-forming periods following the Devonian Period due to plant evolution. In particular, when a historically large crustal movement occurred, the terrain, climate, and atmospheric CO_2 and O_2 contents, among others (Beerling and Woodward 2001), continuously changed, which not only promoted the evolution of plants in each stage, but also led to different degrees of accumulation occurring during each period. Moreover, due to the influence of tectonic movements on land–sea distribution, the depositional effects and intensities of the coal-forming processes during each period were distinctly different (Figure 1) (Han and Yang 1980).

The coal-forming processes during each of the geologic periods of China experienced significantly different development histories (Figure 1). There were eight periods with strong coal-forming processes in China. These included the Terreneuvian, Mississippian, Pennsylvanian–Cisuralian–Guadalupian, Lopingian, Late Triassic, Early–Middle Jurassic, and Early Cretaceous Epochs, as well as the Palaeogene–Neogene Periods. Among these periods, the four coal-forming periods of the Pennsylvanian–Cisuralian–Guadalupian Epochs, Lopingian Epoch, Late Triassic Epoch, and Early–Middle Jurassic Epoch displayed the strongest coal-forming processes. Coals formed during the eight geological periods were mainly humic. The coals formed during the Late Palaeozoic Era were mainly formed by the plant residue of lycopodiatae, sphenopsida, filicopsida, pteridospermae, codaitopsida, and others. The coal-forming plants in the Mesozoic Era were mainly the gymnosperms pinopsida, ginkgopsida, cycadopsida, and bennettiopsida, as well as the filicopsida of pteridophyte. The coal formed during the Cenozoic Era was mainly formed by angiosperms. These changes in coal-forming materials were closely related to the development of plants (Han and Yang 1980).

The eight coal-forming periods in China are comparable and consistent with the main global coal-forming periods. Almost every coal-forming period dating back to the Cambrian (Figure 2) recorded differences in its coal-forming intensity, rank, and composition. This

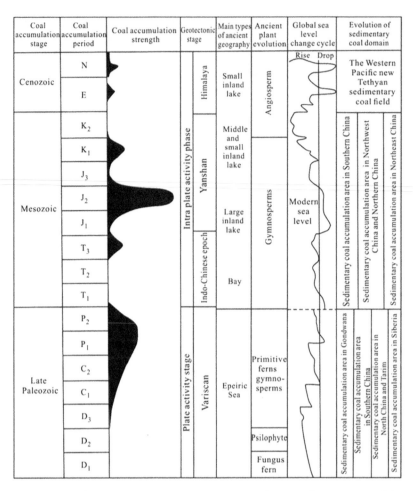

Figure 1. China's geological period coal accumulating strength evolution (Cheng and Lin 2001). (Global sea level change curve according to Vail *et al*. 1977.)

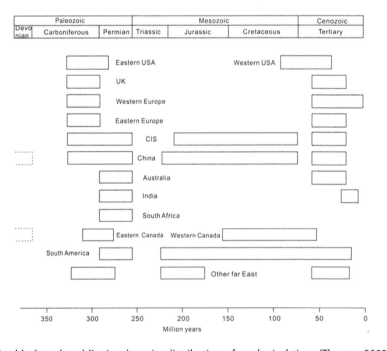

Figure 2. The world's major black coal and lignite deposits distribution of geological time (Thomas 2002).

provided the basic materials for systematic and comprehensive research on the coal-forming processes in each historical geologic period, thus making China the ideal place to research the geology of coal.

2.2. Geographical distribution of coal-bearing strata for each coal-forming period in China

The coal-bearing strata formed during the Devonian and Mississippian Periods were mainly distributed within South China (Figure 3). The coal-bearing strata formed during the Pennsylvanian–Cisuralian–Guadalupian Epochs were widely distributed in both North and South China. Initially, they were mainly distributed in North China and then followed the migration of the coal depositional environment southwards over time. The coal-bearing strata formed during the Mesozoic Period were mainly distributed in South China, west of the Lvliang Mountains in North China, and then followed the migration of the coal depositional environment northwards over time. The coal-bearing strata formed during the Cenozoic Palaeogene Period were mainly distributed in the area east of the Lvliang Mountains in North China, whereas the coal-bearing strata formed during the Neogene Period were mainly distributed in the eastern coastal and southern areas of South China (Zhang 1995b).

3. Distribution, type, and characteristics of the main coal-bearing basins in China

Global coal resources are distributed in two large belts. The first stretches across the Eurasian continent, from the UK to East Germany, Poland, Russia, and the northern region of China; the other stretches across the middle of North America and includes the coalfields in Canada and USA. The coal resources in the southern hemisphere are mainly distributed in temperate zones, namely the resource-rich countries of Australia, South Africa, and Botswana, among others (Fettweis 1979; Jones 1980).

3.1. Distribution of the coal-bearing basins in China during each coal-forming period

In China, coal-bearing basins are widely distributed due to diverse tectonic activities. The majority of the coal basins have become denudation residue basins, which are a kind of transformed sedimentary basin that have been subjected to intense epigenetic tectonic movements by complex geological evolutionary processes. A great many prototype basins have become stretching mountain ranges or plains due to complex geological processes. The coal-bearing series have become separated, and the coal-bearing strata in many areas have nearly been denuded. Some of the isolated basins or

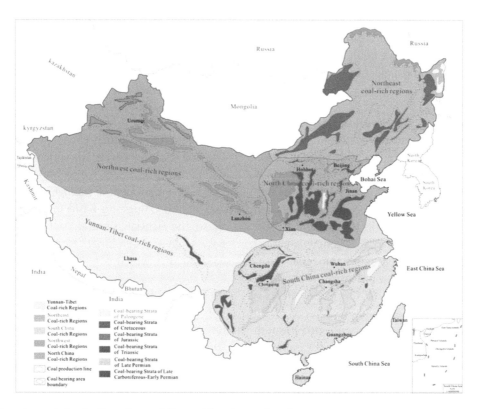

Figure 3. China's main coal distribution phase of the coal-bearing rock series.

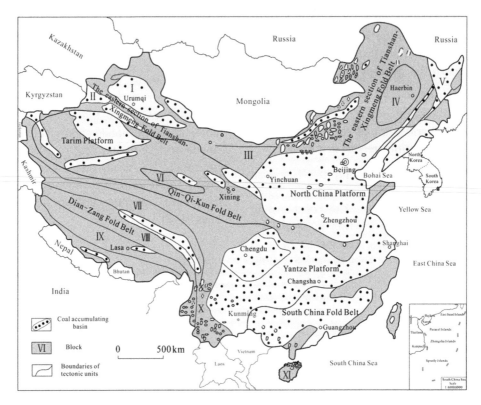

Figure 4. The Chinese Mainland tectonic framework and coal accumulating basin distribution (modified by Mang 1994). I, Junggar Block; II, Yili Block; III, Alxa Block; IV, Songliao Block; V, Jiamusi block; VI, Qaidam Block; VII, Northern Qiangtang–Changdu Block; VIII, Southern Qiangtang–Baoshan Block; IX, Lhasa–Tengchong Block; X, Lanping–Simao Block; XI, Middle Hainan Block(Qiongzhong Block).

basin groups may originally have been uniform deposit basins. At the present time, it is still difficult to determine the exact full extent of the basins, and the boundaries of most of the basins have been inferred based on their palaeogeography and palaeotectonic characteristics (Figure 4).

(1) Terreneuvian Epoch coal-bearing basins
The coal-bearing basins that were formed during this period were mainly distributed throughout South China. In these regions, including Jixi, Ningguo, Hexian, and Taiping in southeastern Anhui, western Zhejiang, Yushan, and Xiushui in Jiangxi, southern Shaanxi, Xichuan, and Neixiang in Henan, and Hushan in Hubei Province (Figure 5), typical Terreneuvian Epoch coal-bearing strata, which contained thin coal seams produced by weak coal-forming processes, were deposited. Overall, research on the coal formation effects of the Terreneuvian Epoch is relatively rudimentary, with detailed research having only been performed in southern Shaanxi and western Zhejiang (Han and Yang 1980).

(2) Permian–Carboniferous Period coal-bearing basins
The coal-bearing basins that formed during this period were mainly distributed in eastern China. For example,

the coal basins in North China were primarily formed during the Pennsylvanian–Cisuralian–Guadalupian Epochs, and the coal basins in South China were primarily formed during the Mississippian and Lopingian Epochs, yielding a total area measuring 2,400,000 km^2. These are the most important large coal basins and contain abundant coal resources.

(3) Late Triassic Epoch coal-bearing basins
The coal-bearing basins that formed during this period mainly occur in South China and include the Sichuan, Central Yunnan, Central Jiangxi, and Changdu Basins, as well as the Qiangtang Basin in northern Tibet. Among these, the Sichuan, Central Yunnan, and Central Jiangxi Basins have a combined total area of approximately 233,000 km^2, which contain abundant coal resources.

(4) Early–Middle Jurassic Epoch coal-bearing basins
The coal-bearing basins that formed during this period were mainly distributed in central, northern, and southwestern China. Among these, the Ordos, Junggar, Turpan–Hami, and North Tarim Basins span the largest areas, yielding a total area of 620,000 km^2, and contain abundant coal resources. South China only contains sporadic small basins of

22

Figure 5. The location map of China's provinces and the places mentioned in the text.

Jurassic Period coals. The Ordos Basin has associated energy and mineral deposits of coal, oil, gas, and uranium. Therefore, it is considered to be a large energy basin, and many studies have been conducted there. In addition, there are also some basin groups distributed in the northern and northeastern sections of North China. These mainly include Datong, western Beijing, and western Liaoning, as well as the Daqing Mountains and the Great Khingan, Dayangshu, Tianshifu–Shansonggang, and Hexi Corridor Basin groups. They comprise a total area of 185,000 km^2, thus making them important coal-bearing basins.

(5) Early Cretaceous Epoch coal-bearing basins
The coal-bearing basins that were formed during this period mainly occur in northern China in the form of basin groups or scattered small basins; they occur in both the northwestern and northeastern regions, but mainly occur in the northeastern area. The northwestern area mainly includes Gansu and the northwestern section of Inner Mongolia, such as the Tulu–Tuomatan, Sandaomingshui, Nanquan, Jiuquan, and Huicheng Basins. The northwestern area mainly includes the eastern section of Inner Mongolia, the northern section of Hebei, and the three provinces of northeastern China (Liaoning, Jilin, and Heilongjiang). The basins in these areas mainly include the Erlian–Hailar, Songliao, and

Sanjiang–Muling Basins. The coal-bearing basins that formed during this period were mainly small basin groups, with only the Jixi–Hegang Basin spanning a large area. The main basin groups include the Erlian–Hailar Basin group, which consists of more than 70 large and small basins that comprise a total distribution area of 250,000 km^2 and feature abundant coal resources associated with oil and gas. Therefore, they are also a basin group that has been closely studied. The basins and basin groups that were formed during the Cretaceous Period have a total distribution area of 450,000 km^2 and are important coal-bearing basins in China.

(6) Palaeogene–Neogene Period coal-bearing basins
The coal-bearing basins that were formed during this period were widely distributed in the offshore and coastal areas near the Pacific Ocean in eastern China, as well as in the area south of the Hengduan Mountains to the area north of the Beibu Gulf. North China is dominated by the coal-bearing basins that were formed during the Palaeogene Period, whereas South China is dominated by the coal-bearing basins that were formed during the Neogene Period. The East Yunnan–West Yunnan Basin group, which is located in the area south of the Hengduan Mountains, is the largest basin group and contains abundant coal resources. These coal-bearing strata were mainly formed during the

Miocene Epoch. The basins formed during the Palaeogene and Neogene Periods have a combined total distribution area of 550,000 km^2 and contain abundant coal resources.

Overall, the coal-bearing basins that were formed during the Terreneuvian Epoch were mainly distributed in South China. The coal-bearing basins that were formed during the Pennsylvanian–Cisuralian–Guadalupian Epochs were mainly distributed in North China. The coal-bearing basins that were formed during the Mississippian and Lopingian Epochs were mainly distributed in South China, and the coal-bearing basins that were formed during the Late Triassic Epoch were mainly distributed in South China. The coal-bearing basins that were formed during the Early–Middle Jurassic Epochs were mainly distributed in central and northwestern China. The coal-bearing basins that were formed during the Palaeogene Period were mainly distributed in northeastern China. The coal-bearing basins that were formed during the Neogene Period were mainly distributed in southwestern China, primarily in Yunnan Province. From the perspective of coal resources, the coal-bearing basins that were formed during the Early–Middle Jurassic Epochs and the Permo-Carboniferous Periods contain the most abundant coal resources. These are followed by the coal-bearing basins that were formed during the Early Cretaceous Epoch and the Palaeogene–Neogene Periods, whereas the coal-bearing basins that were formed during the Late Triassic Epoch contain the smallest coal resources.

3.2. Types and characteristics of coal-bearing basins in China

The formation, evolution, and coal-forming processes of China's coal-bearing basins were controlled by the evolutionary background of China's palaeo-continental plates. The locations of the coal-bearing basins with respect to the tectonic plates varied with changes in geological periods, the nature of plate margins, the nature of the crust (i.e. continental or oceanic crust), and the effects of geodynamics on the formation of the basins, all of which were closely related to the coal-bearing basins and their tectonic histories. On the Chinese mainland, with the evolution of palaeo-plates, different types of coal-bearing basins regularly developed in different tectonic locations. Each type of coal-bearing basin records its own coal-forming histories (Wang and Li 1998). Therefore, studies of these histories can be used to enrich current coal geology theories and to guide future investigations of coal resources.

In this paper, in accordance with the locations of tectonic plates and the characteristics of coal-bearing basins, the coal-bearing basins in China were divided into six types: 1) inner craton depression coal-bearing basin; 2) Earth suture zone coal-bearing basin; 3) foreland coal-bearing basin; 4) post-collision orogeny coal-bearing basin; 5) intra-continental rift coal-bearing basin; and 6) active marginal belt coal-bearing basin (Wang and Li 1998). Table 1 records the characteristics and distributions of the six types of coal-forming basins.

4. Characteristics of China's coal sedimentology compared with that of typical regions of the world

There were eight major coal-forming periods in China; each period recorded evidence of distinct depositional environments and coal-forming histories and thus has distinctive characteristics. This article summarizes the coal-forming environment and sedimentary characteristics of the eight coal-forming periods, namely the Terreneuvian, Mississippian, Pennsylvanian–Cisuralian–Guadalupian, Lopingian, Late Triassic, Early–Middle Jurassic, and Early Cretaceous Epochs, as well as the Palaeogene–Neogene Periods. Then, according to the coal-forming periods mentioned above, the coal sedimentary characteristics of the typical coal-forming regions of the world are analysed and compared to those in China.

4.1. Coal-forming deposit characteristics during the Terreneuvian Epoch

The Terreneuvian Epoch is one of the eight coal-forming periods in China. The coal formed during this period is known as stone coal, which was mainly formed by the remains of bacteria and algae within a shallow sea. During the depositional period of the Terreneuvian Epoch (in which stone coal was mainly deposited in the Hetang Formation), the southeastern part of South China was Cathaysia, the western part was Sichuan–Yunnan ancient land, the northern part was Huaiyang ancient land, and the central region was the submarine Jiangnan Uplift. Among these ancient lands was a wide shallow sea. Within the depression area between the Jiangnan Uplift and Cathaysia, as the ancient land rose and underwent erosion, large quantities of sediment were transported to the depression between the Jiangnan Uplift and Cathaysia; these sediments formed giant flysch deposits, which are mainly composed of sandstone and mudstone. This kind of environment is not conducive to forming biological blooms of algae, and thus stone coal is rarely seen. However, in northern

Table 1. Type, coal-forming characteristics, and distribution of the coal-bearing basins in China (modified by Wang and Li 1998).

Basin type		Basin characteristics	Coal-forming characteristics	Basin distribution
Inner craton depression coal-bearing basin		Gentle subsidence and flat terrain, mainly developed epicontinental and inland shelf seas. Almost no faulting and volcanism. Stable and slow terrigenous-endogenous deposits. Sediments very thin, stable and wide distribution. Basically the same depocentre and subsidence centre, with little change.	Coal seams thin and stable, concentrated distribution. Good coal-bearing properties, and high coal-bearing coefficient. Major development in marine- continental transitional environment. Coal-forming environment migration along the coastline.	Basins in northern China, Yangtze- Huaxia, Tarim regions, etc.
Continental accretion zone coal-bearing basin	Earth suture zone coal-bearing basin	The surrounding folds rose under the squeezing effects of the plates, and the plates sank to form a basin. Tectonic activity, frequent sea land changes, strong magmatic activity, rapid shallow water settlement.	Active tectonics, frequent water invasion, and retreat of large numbers of coal seams with small thickness and poor stability. Intense magmatic activity, high degree of coal seam metamorphism. Very poor coal-forming conditions.	Junggar Basin and basins in northern and southern margin of the northern China region, etc.
	Foreland coal-bearing basin	Squeezing-type groove sedimentary basin, parallel to the fold belt, formed by cratonic lithosphere kink. Provenance mainly includes orogenic belts. High sediment thickness.	Coal seams mainly formed in marine-continental transitional environments. Gentle slope belt with wide coal formation, high thickness, and good stability. Good coal-forming conditions.	Northern Tianshan Mountain basins, Sichuan basin, etc.
Post-collision orogeny coal-bearing basin		Include piedmont and intermountain basins, with stable tectonics. Mainly developed rivers, alluvial fans, fan deltas, deltas, and lake sedimentary systems. Mountains surrounding the basin provide the source.	Piedmont alluvial fan-(fan) delta and lake shore belt are the main coal-forming environments. Usually, coal seams with poor stability, and high thickness. Upper and lower lacustrine mudstones usually develop coal measures.	Tianshan Mountain–Yinshan Mountain tectonic belt (e.g. Turpan–Hami Basin)
Intra-continental rift coal-bearing basin		Basin size varies greatly. Mainly half-graben basins, with en echelon distribution. Subsidence and sedimentation centres development and migration along the fracture. Mainly developed river-lake sedimentary systems. Active tectonics and poor stability strata.	Fan delta and lakeshore belt are the main coal-forming environments. Gentle slopes underwent good coal-forming performance. Coal seams exhibit poor stability, with only local areas possessing great thickness.	Basins widely distributed from the north to south in eastern China, etc.
Active marginal belt coal-bearing basin		Continental stretching basin, controlled by basement normal faults. Violent volcanic activity. Includes back-arc basins and inter-arc basins. Large deposit thicknesses and complex lithology, containing marine and volcaniclastic deposits. Mainly developed alluvial fans, rivers, limnetic facies, and shore epeiric sea facies.	Depressions among alluvial fans, river flood plains, limnetic facies, and shore epeiric sea plains are the main coal-forming environments. High thickness of coal-bearing strata with many coal seams, small thickness of each coal seam, and poor stability. Very poor coal- forming conditions.	The east China Sea basins, northern South China Sea basins, etc.

Guangdong and southern Jiangxi, a region containing thick siliceous sedimentary rock, stone coal did form. The Jiangnan Uplift zone and the Yangtze sag record evidence of a shallow sea, with a wide-ranging, quiet sedimentary environment. This environment underwent relatively slow subsidence, which was conducive to the accumulation of algae blooms and thus the broad development of stone coal. This surrounding stone coal mainly developed thin horizontal bedding; in the stone coal, organic matter and minerals record homogeneous distribution and a lack of benthic biota, from which it can be speculated that the stone coal was mainly formed under the wave base in a shallow sea and that the coal-rich belt was formed on the near-shore side. Some stone coal also formed in the shoreface zone between the shallow water and the wave base (Han and Yang 1980).

During the Cambrian Period in northern China, stone coal was only formed and spread widely throughout the area during the Terreneuvian Epoch. This stone coal was the result of the synsedimentation of fungous algal organic remains and siliceous gel deposits, followed by complicated diagenesis and metamorphism. The formation of stone coal was clearly controlled by the quiet anoxic viscous flow basin environment, and mainly developed on the basin slope, since the low-lying area of the basin slope was most conducive to the development of stone coal (Zhou 1990; Jiang et al. 1994). Zhou (1990) argued that the conditions of stone coal formation were dominated by three decisive factors: 1) the abundance of fungous-algae; 2) palaeogeographic location, i.e. the slope section of basins with anoxic viscous flow; and 3) sedimentation, i.e. the sedimentary conditions or burial history. Jiang et al. (1994) believed that the genetic mechanism of stone coal was as follows: during a period of sea level rise, the current brings in nutrient-rich water; in addition, in slowly accumulating anoxic sedimentary environments, bacteria and algae remain accumulated; after undergoing diagenesis and low-grade metamorphism, these remains form stone coal.

The stone coal is enriched in many elements, including V, Mo, Ni, Co, U, P, Ba, Au, and Ag (Han and Yang 1980; Jiang et al. 1994; Jr et al. 1994; Belkin and Luo 2008), as well as some potentially harmful elements, such as As and F (Luo 2011).

4.2. Coal-forming deposit characteristics during the Permo-Carboniferous

In China, Permo-Carboniferous coal deposits are among the most important and most widespread coal-bearing strata. During this time, these two regions were separate small-scale continents located on opposite sides of the equator. The seam geometry and quality of the coal were directly controlled by geography and changes in their depositional environments. However, the coal formation and the migration of thicker coal zones were indirectly controlled by palaeoclimate and sea-level fluctuations that occurred in response to palaeo-tectonic influences (Liu 1990).

The coal-bearing strata formed during the Permo-Carboniferous Period of China were mainly developed in northern and South China, which comprised large epicontinental sea depression basins during this period. The palaeogeographical pattern during this coal-forming period, moving successively from the continental to marine environment, was as follows: alluvial fan-braided river, meandering stream-lake, debris coastal zone (including deltas, barrier coast, and barrier-free coast), shore and shallow sea deposits, and carbonate deposits in shallow seas. Among these, the debris coastal zone comprised the most favourable coal-forming belt. In this area, the tectonic activity in the material source area and the regional progressive and regressive action of the seawater controlled the formation and migration of the debris coastal zone. Therefore, this activity controlled the formation and migration of the coal-rich belt, together with the activity of the sedimentary basement tectonics, the progressive and regressive action of seawater, and the migration of the lithofacies zone. The coastal delta system and the delta-debris coastal system are the most important coal-forming environments, which formed coal-rich centres in regions such as Shanxi, Kailuan, Fengfeng, and Central Henan–Eastern Henan, Huainan, and Huaibei Cities in North China, as well as Liupanshui, Zhijin–Nayong, and Huayingshan, among others, in South China (Liu 1990; Wang 1996; Shang 1997).

4.2.1. Coal-forming deposit characteristics during the Mississippian Epoch in China

During the Mississippian Epoch, coal-forming processes and the resultant coal-bearing rocks were widely distributed throughout South China. The main coal-forming rock series during this period was mainly formed in the Datangian Stage of the Carboniferous Period. The early period of the Datangian Stage was a transgression period, which produced a limited distribution of coal-bearing strata. The middle period of the Datangian Stage was a regressive period featuring good coal-bearing properties, which produced widely developed minable coal seams, or local minable coal seams. The eastern section of South China was the main coal-forming area, and regions in the middle of Hunan Province, including Xinhua, Lengshuijiang, Lianyuan, and Shuangfeng, comprised the coal-rich belt during this period. The coal thickness generally ranges from 2 to 3 m, with a maximum thickness of 5.26 m. The western section of South China featured poor coal-bearing properties, and regions such as the Lower Yangtze area, Qinghai-Tibet, western Yunnan, and western Sichuan areas exhibit the poorest coal properties. The coal seam in South China formed during this period was mainly developed in delta depositional environments, followed by meandering stream-lake, alluvial fan-braided river, and tidal flat-beach environments (Yi 1980; Zhang 1995b; Zheng 2008).

4.2.2. Coal-forming deposit characteristics during the Pennsylvanian–Cisuralian–Guadalupian Epochs in China

The coal-forming processes of the Pennsylvanian–Cisuralian–Guadalupian Epochs occurred mainly in North China. The tectonics in North China during this period were stable for a long period of time and thus developed the most important coal-bearing strata in North China. The total thickness of coal seams in most areas is greater than 5–10 m. Additionally, coalbeds with thicknesses greater than 15–25 m were formed in Datong and Taiyuan of Shanxi, Junggar Banner of Inner Mongolia, Daqingshan, Hebei, Beijing, and Tianjin, as well as in the areas of western Shandong, Anhui Huainan, and Huaibei, which record excellent coal quality and all kinds of coal types.

The coal-forming depositional environment during the Late Palaeozoic Era in North China exhibited different characteristics during different geologic periods, and many experts and researchers have performed in-depth studies of these coals (Han and Yang 1980; Li et al. 1989, 1995, 2003, 2006; Cheng and Jiang 1990; Liu 1990; Lin et al. 1991; Chen 2000; Dai et al. 2002; Shao et al. 2006a, 2007, 2014a; Wu et al. 2008; Zhang 1995b). During the deposition of the Benxi Formation during the Pennsylvanian Epoch, the basement of the North China Basin became inclined towards the east and seawater invaded from the east to the west. This continuous tectonic sinking produced a deep covering of water, and the overall depositional environment was that of an epicontinental sea with bays and tidal flats. Additionally, strong transgressive action occurred, thus causing weaker coal-forming processes. During the deposition of the Taiyuan Formation during the

Figure 6. The palaeogeographic map of Early Permian in China.

Pennsylvanian–Cisuralian Epochs, the basins became basically inclined towards the south and seawater in the South China Sea invaded towards the north. The debris materials on the northern margin highland migrated and were deposited towards the south. North China can be divided into a northern margin, northern zone, middle zone, and southern zone. Alluvial fan and braided river deposits were developed in the near-source zone of the northern margin. Meandering stream-lake and coastal delta deposits were widely developed within the northern zone. The middle zone was mainly dominated by debris shoreline deposits. The southern zone recorded universally distributed shallow-water carbonate deposits. The northern margin and northern zone of the basin recorded a wide alluvial-delta deposition facies belt oriented in a nearly E–W direction, as well as widely developed peat bogs. Therefore, this was the most favourable sedimentary facies zone for coal formation. The largest coal-rich centre was formed in the Daqingshan and Junggar Banner of Inner Mongolia, as well as in the northern and central areas of Shanxi. During the Guadalupian Epoch, North China experienced an overall regression towards the south and formed a widely developed alluvial-delta sedimentary facies zone. The subsided depression zone was usually in a state of shallow water cover, which was favourable for coal-forming processes and thus formed many coal-rich belts. As regression proceeded, the alluvial-delta sedimentary facies zone migrated towards the middle and even the southern regions of the basin, and the coal-rich belt migrated along with it. The most favourable coal-forming environment during the Permo-Carboniferous Period in North China was the delta depositional system, which was followed by river and tidal flat-lagoon depositional systems (Figure 6).

Following the later period of the Pennsylvanian Epoch, the area of North China maintained the same tectonic palaeogeographic pattern in the South China Sea for a long period of time. It displayed an increasingly obvious differential uplift-subsidence process along the S–N direction, controlled the sedimentary facies and coal-rich belts that were spread along the E–W direction, and continuously migrated along the S–N direction. During the late Guadalupian Epoch and the early Lopingian Epoch, the northern section of the middle of the basin was an inland lake basin, and the southern section of the middle of the basin was a bay-delta environment. The delta advanced to the Henan and Anhui areas in the southern section of the basin, causing coal-forming processes, which were influenced by the seawater and damp maritime climate, to occur in the southern areas. The coal-rich belt also migrated towards the south at the same time. During the Lopingian Epoch, the overall area of North China was raised to form land, and the climate thus became dry. At that point, coal-forming processes ceased.

4.2.3. Coal-forming deposit characteristics during the Lopingian Epoch in China

Coal-forming processes during the Lopingian Epoch were widely distributed throughout South China (Liu 1990; Xiao 1990; Shao *et al.* 1994, 1998, 2003a, 2013b; Wu *et al.* 2008; Zhang 1995b). During the early period (Longtanian) of the Lopingian Epoch, coal-forming processes recorded a wide distribution and large intensity, and the coal reserves are ranked first among all the various coal-forming periods during this time. During the Changxingian Stage of the Lopingian Epoch, coal-forming processes were limited to regions such as eastern Yunnan, western Guizhou, and southern Sichuan in the eastern part of the Kangtien ancient land in the western part of South China. In addition, coal-forming processes were relatively weak during this time.

The upper Yangtze Region was located to the west of the Yunkai ancient land–Xuefeng submarine uplift in the central and eastern parts of South China. This region developed good coal-bearing strata and formed the most important coal-rich belt in South China, namely the Sichuan–Yunnan–Guizhou coal-rich belt. The Chongqing and Liupanshui–Zhijin–Nayong areas, which were located in the western part of Guizhou, were the most representative zones of this coal-rich belt. The cumulative thickness of the coal seam in the coal-rich centre was greater than 40 m (Zhang 1995b). The lower Yangtze Region, which was located to the east of the Yunkai ancient land–Xuefeng submarine uplift, developed a series of uplifts and depressions. Coal-bearing strata were distributed within the depressions, and their coal-bearing properties were inferior to those of the upper Yangtze Region. The coal-rich belt in the Yangtze Region gradually migrated to land from the early Longtanian to the Changxingian; this belt spanned the largest area in the Longtanian and the smallest area in the Changxingian. The scope of the coal-forming processes gradually decreased over time, and the coal-rich belt became more concentrated.

The formation and distribution of coal-rich belts during each stage of the Lopingian Epoch in South China were directly controlled by the palaeogeography, and a series of transgressions and regressions led to the continuous migration of these coal-rich belts. The terrigeneous clastic coastal belt was the main zone in which strong coal-forming processes occurred. The bay depositional assemblages had the best coal-bearing properties. However, the depositional assemblages of the delta also had relatively good coal-bearing properties, and the coal-bearing properties of the coastline depositional assemblages without deltas were inferior to the delta depositional assemblages. The depositional assemblages of the meandering stream-lakes had poor coal-bearing properties, and the depositional assemblages of the coastal shallow seas had the poorest coal-bearing properties (Liu 1990; Xiao 1990; Shao *et al.* 1994, 1998; Zhang 1995b).

4.2.4. New progress in coal geological research in South China

In South China, tectonic activity during the Lopingian Epoch was very complex, and the mechanisms of the formation of coal-bearing basins, sediment sources, and volcanism are not yet fully understood. However, in recent years, some scholars have performed thorough and systematic research and have proposed several new ideas.

Dai *et al.* (2016) argued that the evolution of the mantle plume resulted in an environment of peat accumulation that represented the sedimentary source area in southwestern China. The control of the Emeishan Large Igneous Province (ELIP) on the Lopingian Epoch coal-bearing strata in South China can be described in terms of four aspects. (1) Its topography (high in the west and low in the east), which resulted from the ELIP providing favourable sites for peat accumulation (China Coal Geology Bureau 1996), led to the majority of coal-bearing basins being distributed in the middle and outer zones of the ELIP. (2) The Kangtien Upland was the dominant sediment source region for the Lopingian Epoch coal-bearing strata (China Coal Geology Bureau 1996; Zhou *et al.* 2000; Zhuang *et al.* 2012; Dai *et al.* 2014a). (3) In some cases, volcanic ash resulting from the waning activity of the plume either served as a base for peat accumulation (e.g. as the floor of a coal seam) or terminated peat accumulation (e.g. formed the roof of a coal seam) (Dai *et al.* 2014b). (4) The accumulation of organic matter in both the Lopingian Epoch coal-bearing basins of South China and the Cenozoic coal-bearing basins of the Primorye was accompanied by frequent active volcanism (China Coal Geology Bureau 1996; Zhou *et al.* 2000; Dai *et al.* 2011) and related hydrothermal processes (Zhou and Ren 1992; Ding *et al.* 2001; Zhang *et al.* 2004; Yang 2006; Dai *et al.* 2008, 2013a, 2013b).

The authors of previous papers have argued that the southwestern sedimentary source region of the Lopingian Epoch coal is the Kangtien ancient land, but the authors of some recent studies have claimed that different coal basin sedimentary source areas record substantial differences.

Dai (2014a) argued that the dominant sediment source regions during peat accumulation in the Lopingian Epoch Huayingshan coalfield in Sichuan, southwestern China, were three uplands/uplifts, namely the Hannan Upland, the Dabashan Uplift, and the

Leshan–Longnvsi Uplift, which were the sources of not only coal benches but also roof materials (Dai *et al.* 2014a). In addition, the sediment source region of the Lopingian Epoch Yanshan Coalfield in Yunnan was the northern Vietnam Upland (Dai *et al.* 2008).

In the Guangxi Zhuangzu Autonomous Region, larger proportions of detrital quartz, albite, and clay minerals of terrigeneous origin have been found in the Lopingian Epoch Heshan coals (Dai *et al.* 2013b), thus indicating that the Yunkai Upland experienced a greater input of detrital materials of terrigenous origin than the Heshan coals (Dai *et al.* 2013a). Volcanism has also had a great impact on the growth of peat bogs in this region. For example, the Lopingian Epoch No. 12 coal seam in the Songzao Coalfield of southwestern China was derived from a peat bog, which formed on the residual plain of mafic tuffs (Dai *et al.* 2010). Furthermore, in the Xinde Mine in eastern Yunnan, volcanic ash may have also led to the termination of peat accumulation in the Lopingian Epoch (Dai *et al.* 2014b).

In addition, some scholars found that peat bogs may develop inside ancient land, resulting in the formation of coal seams. The Pennsylvanian Epoch Shuanmazhuang Formation and the Cisuralian Epoch Zahuaigou Formation, which were deposited in a continental environment, are the coal-bearing sequences in the Daqingshan Coalfield, which is located in the interior of the Yinshan ancient land (Qimu 1980; Jia and Wu 1995; Zhong *et al.* 1995; Zhang *et al.* 2000; Zhou and Jia 2000; Wang and Ge 2007; Dai *et al.* 2015a).

4.2.5. Coal-forming deposit characteristics during the Permo-Carboniferous in other typical regions of the world

The Permo-Carboniferous is the most important coal-forming period in the world. Many other areas throughout the world have developed rich Permo-Carboniferous coal-bearing strata.

In addition to China, other parts of the world also contain Lower Carboniferous coal-bearing strata. A series of Mississippian Epoch coalbeds formed in the Kayak Formation and the Mattson Formation of the northern Yukon Territory and northwestern Canada, respectively. The coal-bearing Kayak strata accumulated in a coastal plain setting transgressively overlain by younger marine beds. The Mattson coalbeds appear to have formed in prograding delta and lacustrine environments within a rift basin (Cameron *et al.* 1996). The Lower Carboniferous Walbrzych Formation in the Walbrzych Basin of southwestern Poland was deposited in a delta-fluvial plain environment (Nemec 1985).

In Nova Scotia, Canada, the Pennsylvanian coal seams in the Sydney, Stellarton, and Cumberland Basins were formed in fluvial-dominated depositional environments, which then evolved upwards through distal braided river plains to meandering fluvial plains. The mires developed around the margin of a central basin lake. Thick coals were formed in broad, humid swamps of large flood basins between the unconfined channels of large meandering rivers or basin centres during periods of lower subsidence rates (Rust *et al.* 1987; Kalkreuth *et al.* 1991; Marchioni *et al.* 1996; Kalkreuth 2004). In the Minto Coalfield in New Brunswick, the Pennsylvanian Epoch coal measures were deposited in an upper delta plain (Kalkreuth *et al.* 2000).

In eastern Ukraine and adjacent portions of Russia, thick coal measures were deposited in the Donets Basin from the Serpukhovian to the Moscovian, during which time there were approximately 130 seams, each with a thickness of over 0.45 m. The Serpukhovian coal seams were inter-fingered with lagoonal sediments (Shulga 1981). The Serpukhovian coal seams were accumulated during a regressive–transgressive cycle in a wide shore-zone dissected by rivers discharging into a nearby shallow sea in the central Donets Basin. The floor and roof of the Moscovian coal seams were formed by lacustrine and lagoonal clay stones, respectively (Ritenberg 1972). In the eastern part of the basin, the seam is directly overlain by marine limestone (Uziyuk *et al.* 1972).

In USA, east of the Mississippi River in Kentucky, Virginia, Indiana, Tennessee, Alabama, and Pennsylvania, the Pennsylvanian coal measures were formed in a fluvial environment. In some places, such as Indiana, western Kentucky, Illinois, and Virginia, these coal measures were marine-influenced (Andrews *et al.* 1996; Churnet 1996; Hower and Eble 2004).

In southern South Africa, the Pennsylvanian Epoch coal measures in the Karoo Basin were formed in the Dwyka Glacial Period. During the Dwyka glaciation, valleys with predominantly N–S orientations were scoured into the pre-Karoo floor rocks by glaciers; after the northwards retreat of the ice sheets, these valleys controlled subsequent sedimentation and, to some extent, peat deposition (Cairncross 1989; Cairncross and Cadle 1991). The Cisuralian–Guadalupian Epoch coal measures in the Natal coalfield were deposited in deltaic, fluvial, and shoreline settings (Cadle *et al.* 1991). In the Springbok Flats of the Waterberg and Soutpansberg coalfields in northern South Africa, they were deposited in floodplain settings (Beukes *et al.* 1991).

In southern Africa (e.g. Botswana, Malawi, Mozambique, Namibia, South Africa, Swaziland,

Tanzania, Zambia, and Zimbabwe), coal deposits were mainly formed during the Cisuralian–Guadalupian Epochs. The coal-bearing sediments were deposited in varying tectono-sedimentary basins; their depositional environments were primarily deltaic and fluvial, with some minor shoreline and lacustrine settings (Cairncross 2001).

In Poland, the Carboniferous/Pennsylvanian coals of the Upper Silesia coal basin were formed in a paralic environment (Gmura and Kwiecin´Skab 2002). In Germany, the Pennsylvanian coal measures in the Ruhr Basin were mainly formed in paralic clastic sedimentary environments (Suess *et al.* 2007).

In India, the Cisuralian Epoch Barakar coal measures in the Son-Mahanadi Valley and the Pence-Kanhan Valley were formed in braided fluvial environments. The Late Permian Raniganj coals in the Damodar Valley Basin were formed in a fluviolacustrine environment (Mishra 1996), and the Permian Sattupalli coals of the Godavari Valley were formed in a fluvial environment (Singh *et al.* 2012).

In Antarctica, the Permian Weller Coal Measures are exposed along the edge of the Polar Plateau from the Mawson Glacier to the Mulock Glacier in southern Victoria Land. These rocks record a rapid change from glacial to postglacial conditions, with the establishment of polar peat-forming conditions during the Late Palaeozoic. The Weller Coals were formed in braided streams, meandering streams, and lacustrine environments. Thick coals were usually deposited in meandering stream environments (Isbell and Cúneo 1996).

In eastern Australia, the main Permian coal-bearing basins are the foreland Sydney and Bowen Basins in the east and the interconnected cratonic Cooper and Galilee Basins in the west. The Cisuralian Series in the foreland basins comprise marine sediments, with coal measures restricted to the orogenic and cratonic margins. Owing to the expansion of the eastern orogen, marine deposition was replaced in the Middle Permian by deltaic deposition and in the Late Permian by fluvial sediments and extensive coal measures (Hunt 1988; Michaelsen *et al.* 2000). Cratonic basins were sites of coal measure deposition throughout most of the Permian. These coals were deposited in high-latitude fluvial-lacustrine environments. The sediment accumulation rates and distribution of coal in the Permian basins were mainly controlled by tectonic subsidence (Hunt 1988; Curry *et al.* 1994).

In Western Australia, the Cisuralian Epoch Vasse, Irwin River, and Sue coal measures in the Perth Basin were formed in fluviolacustrine to lacustrine environments. The Permian coal measures in the Collie Basin were formed in braided fluvial and fluviolacustrine to lacustrine environments. The Late Permian Liveringa coal measures in the Canning Basin were formed in fluvial environments with over-bank or marsh facies (representing a marine transgression sequence) (Mishra 1996).

4.3. Coal-forming deposit characteristics during the Late Triassic Epoch

4.3.1. Coal-forming deposit characteristics during the Late Triassic Epoch in China

The coal-forming processes during the Late Triassic Epoch mainly occurred in the moist and rainy regions of South China, especially in the eastern areas of Sichuan, Yunnan, and Tibet (Cheng and Lin 2001; Shao *et al.* 2014b; Zhang 1995b). The northern areas of China experienced mainly arid to semi-arid climates, and the only continental coal-bearing strata, which are represented by the Wayaobu coal-bearing strata on the roof of the Yanchang Formation, were mainly formed in the Ordos Basin during the later period of the Late Triassic Epoch (Tian *et al.* 2011).

During the Late Triassic Epoch, the coal-forming zone in the western section of South China was mainly located in the western parts of Yunnan and Guizhou Provinces. The coal-forming environments in these areas were mainly a coastal plain and a delta plain, in which coal-rich belts were formed (Lu *et al.* 2008; Shao *et al.* 2008, 2014b; Li *et al.* 2011; Yang and Li 2011; Xie *et al.* 2014; Zhou *et al.* 2016). The nappe tectonic belt on the leading edge of the Longmen Mountains was the leading factor that controlled the distribution and migration of the basin facies belt, which formed coal-rich belts in the coastal, lakeshore delta plain, and coastal delta plain areas. The Panzhihua–Xichang region and the central Yunnan Basin were part of the coastal intermountain plain, and the most favourable coal-forming regions were the transtensional rift basins in which the Baoding, Yongren, and Hongni coalfields, which recorded the best coal-bearing properties during the Late Triassic Epoch, were formed (Wang *et al.* 2007a; Lu *et al.* 2008; Shao *et al.* 2008). The southeastern Yunnan and Guizhou Province Zhenfeng Basins were coastal tidal flat environments, which recorded coal-forming characteristics similar to those in the area west of the central Yunnan Basin. The minable coal seam records a sporadic distribution (Zhang 1995b). The Sichuan Basin formed three coal-rich belts during the Late Triassic Epoch, namely the Dayi-Ya'an coal-rich belt in the western section of the basin, the Dazhou–Zigong–Emei coal-rich belt in the middle of the basin, and the Guangyuan coal-rich belt in the northern margin of the basin. Palaeotectonic activity and palaeogeography jointly controlled the development and migration of the coal-rich belts.

The coal-bearing strata deposited during the Late Triassic Epoch in the eastern area of South China were mainly distributed in Hunan, Jiangxi, Fujian, Guangdong, and Zhejiang Provinces, as well as in southern Jiangsu and Anhui and southeastern Hubei Provinces (Zhang 1995b; Zhang et al. 2009; Shao et al. 2014b). During the Late Triassic Epoch, the coal-bearing strata in this area were deposited on the basement, which had been strongly folded, but not fully flattened, by Indosinian movements. Their tectonic features were dominated by NE-trending, narrow and long depressions. The coal-bearing strata were mainly developed within an interactive depositional environment between bay-lagoon and continental facies. Additionally, the coal-bearing strata formed during the Late Triassic Epoch in the eastern area of South China were mainly developed in three depressions in the Lower Yangtze, Hunan–Guangdong–Jiangxi, and Zhejiang–Fujian–Hubei regions. The formation of these coal-bearing strata was closely related to transgressions and regressions; the series was mainly formed during a transgression sequence in the early to middle depositional period, whereas the regression sequence was formed during the middle to late depositional period. The coastal-bay coal-forming environments had optimal coal-bearing properties, and the minable coal seams were distributed continuously within a large area. The zone extending from SE Hunan to Pingxiang in the western Jiangxi area represented the region with the strongest coal-forming processes during this period. The coal-forming processes in the lagoon-estuary environment were weaker. Minable coal seams were discontinuously distributed throughout a larger area. The inter-montane lake basins and inter-montane valleys recorded the poorest coal-bearing properties, and minable coal seams were visible in local areas.

The coal-bearing strata formed during the Late Triassic Epoch in the Qiangtang Basin were mainly developed in the shallow sea detrital shelf, coastal and delta depositional systems. The delta plain was the major coal-forming zone in this area (Wang et al. 2009; Xian et al. 2012).

4.3.2. Coal-forming deposit characteristics during the Triassic in other regions of the world

In South Africa, the Middle Triassic coal measures in the Karoo Basin were formed in fluvial environments (Cadle et al. 1991). The climatic conditions were not suitable for peat formation, and the resulting coal seams tended to be lenticular. In Australia, the Late Triassic (Carnian–Rhaetian) Callide Coal Measures in eastern-central Queensland were deposited in high-gradient alluvial

fans and later in low-sinuosity river environments (Glikson and Fielding 1991). In southern Sweden, the Late Triassic (Rhaetian) coal measures were deposited in coastal plain environments (Petersen et al. 2013).

4.4. Coal-forming deposit characteristics during the Early–Middle Jurassic Epochs

4.4.1. Coal-forming deposit characteristics during the Early–Middle Jurassic Epochs in China

The Early–Middle Jurassic Epochs were the most important periods of the many coal-forming periods in China's history. The coal resources formed during this period account for approximately two-thirds of the total coal resources in China. The coal-bearing strata formed during the Early–Middle Jurassic Epochs were widely distributed throughout the northern region of China, beginning from western Xinjiang in the west and ending at Beipiao in the western part of Liaoning Province in the east. The northwestern region of China developed the most coal-bearing strata. The coal-bearing strata that formed in the inland depressed and graben basins comprise different scales and types, but include the Ordos Basin and Junggar Basin, among others, which are world famous, super-large inland coal-bearing basins (Zhang 1995b; Zhang 1998).

The coal-forming processes that occurred during the Early Jurassic Epoch were strong in the Xinjiang region (Zhang 1982; Shao et al. 2006b, 2009; Shi et al. 2011; Tian and Yang 2011), and the coal-forming processes that occurred during the Middle Jurassic Epoch were strong throughout the entire northern region of China (Cao 1991; Zhang 1995b; Wang 1996; Zhang 1998; Shao et al. 2003b; Wu et al. 2008; Qin et al. 2009). The coal-forming processes that occurred during the Early–Middle Jurassic Epochs were continuous in most areas (and basins), in which the coal-forming processes in the large and super-large coal-bearing basins were of absolute superiority. The regional regularity of coal formation was macroscopically controlled by the tectonic palaeogeography, palaeoclimate, and the properties of the basin basement. The siltation effects of the lakes, or the abandonment of rivers in the basins, led to the formation of peat bogs in large areas, resulting in the creation of large coal formations.

Within the large and super-large depressions of the coal-bearing basins, the wide development of lake-river-delta systems was the most important characteristic of the depositional environment. The zonal differential distribution of sedimentary facies in the basin, along with the development of facies from the basin margins to the deposition centres, was successively alluvial–proluvial facies, fluvial facies, coastal delta facies, and lake facies sedimentary belts. The coal-rich belts were distributed along the margins of the basins, and their scales of development and degrees of stability were controlled by the meandering stream and lake delta rock facies belts. Their coal-rich centres were basically aligned with the developed positions of large lake deltas or meandering streams (Fei *et al.* 1991; Wang 1996, 2012; Wang and Zhang 1999; Zhong *et al.* 2010; Wang *et al.* 2013). The medium- and small-sized inter-montane or hollow coal-bearing basins were formed during the river-filling stages in the early depositional period and during the lake-filling stages in the later depositional period. Strong coal-forming effects usually occurred in the lake basin-filling stages, as well as in the transitional filling stages from the inter-montane valleys to the inter-montane lake basins. The Datong River, Qaidam, Turpan–Hami, Minhe, and Chaoshui Basins in Gansu and Qinghai Provinces are representative of this type of inter-montane basin (Wang and Wu 1994; Zhang 1995a, 1995b; Shao *et al.* 2003b, 2006b; Wang *et al.* 2007b; Deng *et al.* 2009; Chen *et al.* 2010). Their coal-rich belts were typically discontinuously distributed in the centres of the basins, and their spreading directions were consistent with the extensional directions of the basins. The Yining and Yanqi Basins in southern Xinjiang and the northern margin of the Qaidam Basin were also typical inter-montane basins. However, their coal-rich belts were continuously distributed at the margins of

the basins (Wang and Chen 2004; Zhong *et al.* 2010; Shi *et al.* 2011; Li 2012). Within the series of medium- and small-sized inter-montane basins in the eastern sections of the northern region of China, the representative basins were the Beipiao, Jilin Wanhong, and Beijing Basins, as well as the Daqing Mountain Basin in Inner Mongolia. The coal-rich belts in these basins were mainly distributed at the margins of the basins, and their coal seams generally recorded small thicknesses and high stabilities (Zhang 1995b; Zeng and Wang 2013). The filling evolution of this type of basin was significantly influenced by the tectonic movement of the Pacific Plate. The palaeotectonic types of these basins' basements were mainly wavy depressions, and their palaeogeographies were inland inter-montane basins. The presence of lake delta or river environments during the coal formation period may have developed peat bogs, which eventually evolved into coal (Figure 7).

The coals formed during this period are often inertinite-rich. The presence of inertinite-rich coals in the Muli Coalfield on the Tibetan Plateau indicates that they formed within a relatively oxidizing and dry environment, in which a depressed water table existed during peat accumulation (Dai *et al.* 2015b). The presence of inertinite-rich coals of the Huanglong Coalfield in the southern Ordos Basin indicates that intense corruption and biochemical oxidation occurred during the early paludification stage (Zhuang and Wu 1996).

Figure 7. The palaeogeographic map of Early Jurassic and Middle Jurassic in China.

4.4.2. Coal-forming deposit characteristics during the Jurassic in other typical regions of the world

In Denmark, the Lower–Middle Jurassic coal measures from the Island of Bomholm were deposited in a paralic environment within a fault-bounded, subsiding pull-apart basin. The peats were accumulated in low-lying inter-channel environments situated in a lower coastal plain during a transgressive period. In addition, the peat swamps were predominantly freshwater swamps, but were occasionally marine-influenced (Petersen and Nielsen 1995).

In northeastern Greenland, the Middle Jurassic coal measures at Kulhøj were deposited in a floodplain environment related to meandering river channels. Their compositions are dominated by inertinite and vitrinite, and they represent deposits formed in a fresh-water mire. No evidence of marine transgression has been found (Bojesen-Koefoed et al. 2012). In eastern Greenland, Upper Jurassic coal measures were deposited in coastal swamps (Clemmensen and Surlyk 1976).

In Hungary, Early Jurassic coal measures from the Mecsek Basin were deposited in the coastal plain of a delta environment. As with transgression, the central delta plain environment migrated and advanced towards the peripheral delta plain zone (Barbackaab et al. 2015).

In Austria, the Jurassic coal measures in the basement of the Alpine–Carpathian frontal zone were deposited in a flood basin that transitioned to a delta-plain environment. The coals originated in frequently flooded mires and evolved within an oxygenated and acidic environment (Sachsenhofer et al. 2006).

4.5. Coal-forming deposit characteristics during the Early Cretaceous Epoch

4.5.1. Coal-forming deposit characteristics during the Early Cretaceous Epoch in China

The Early Cretaceous Epoch was the third most important coal-forming period in China. These coal-bearing strata were mainly developed in inland fault basins or inter-montane and offshore depression basins, and usually contained thick or extremely thick coal seams. The areas of the coal-forming basins were generally very small; however, they often occurred in groups. These basins have abundant coal resources and are ranked third among the basins that were formed during the Permo-Carboniferous and Jurassic.

The depositional environments of the coal-forming basins during the Early Cretaceous Epoch in the northern region of China were unique. The filling sequences, depositional patterns, and facies belt distributions of the fault basins were mainly distributed in the northeastern region of China and were significantly controlled by the tectonic framework of the basins, especially the faults at the edges of the basins (Zhang 1995b).

The Erlian–Hailar inter-montane fault basin group in the eastern section of Inner Mongolia underwent a very strong coal formation process. The upper and lower areas of the thick lake facies mudstone sections, which represented the period of the largest lake basin development, were usually the two major coal formation units in the basin. The spreading of the coal-rich belt was typically consistent with the lakeshore delta, braided river delta, and alluvial fan sedimentary facies zone. Extremely thick coal seams developed in the coal-rich centre, in which the maximum thickness of a single coal seam exceeds 200 m in the Shengli coalfield (Zhu and Zhang 2000; Jiang et al. 2005; Wang 2012, Dai et al. 2012, 2015c; Hower et al. 2013; Wang et al. 2013; Guo et al. 2014). The Songliao zone, which is located to the east of the Greater Hinggan Mountains and to the west of the Yilan–Yitong Fault, developed a series of fault coal-bearing basin groups. The coal-bearing strata were mainly deposited in alluvial fan, fan delta, lakeshore delta, and lakeshore plain environments. The coal-bearing strata had one to four coal-bearing horizons, and their main coal-bearing horizons were located in the upper and lower areas of the thick lake facies mudstone sections. The eastern belt of the basin group exhibited the best coal-bearing properties, followed by the middle belt and the western belt, which had the poorest coal-bearing properties. The coal-bearing properties in the eastern, middle, and western belts in the southern basins were clearly superior to those of the northern basins. The coal-rich centres in these basins were adjacent to the major fault at the margin of the basin, and their spreading directions were consistent with the trends of the basin (Yang 1987; Zhang et al. 2000; Wu et al. 2008; Cai et al. 2011; Shao et al. 2013a). The Sanjia–Muling River Basin (also known as the Jixi–Hegang Basin), located to the east of Heilongjiang, was an offshore depression basin that developed on the basement of the continental marginal block. A large area of the abandoned delta plain was developed by strong coal formation processes against the background of a large-scale regression during the early period of the Early Cretaceous Epoch (Gao et al. 2012).

The volcanic rock-type fault coal-bearing basins that formed during the Early Cretaceous Epoch in northern Heilongjiang, e.g. the Hola, Heibaoshan–Muerqi, and Dayangshu Basins, formed in a fan delta-lake environment during the intermittent stages of volcanic activity and produced a minimal volume of coal (Zhao et al. 1991; Hou et al. 2008; Liu 2012; Cui et al. 2013). The inter-montane depression basin in the northern and southern areas of Gansu Province recorded relatively weak coal-forming processes. The coal-forming environment was mainly an inland lakeshore delta, and coal seams with poor stability were developed only during the water transgression sequence, prior to the lake-filling stage that occurred in the early period of the basin (Zhang 1995b; Gao and Zhang 2014). The coal-bearing strata that formed in Tibet during the Early Cretaceous Epoch were mainly deposited in alluvial fan, fan delta, and shallow sea environments and thus formed part of the marine-terrigenous facies coal-bearing clastic deposits (Zhang et al. 2013).

4.5.2. Coal-forming deposit characteristics during the Cretaceous in other typical regions of the world

In Canada, Early Cretaceous coal measures formed the Front Ranges and Inner Foothills of the Rocky Mountains, which accumulated within the coastal plains of the Fernie and Moosebar–Clearwater Seas, respectively. The Late Cretaceous and Palaeocene coal measures from the Outer Foothills of the Rocky Mountains originated within the prograding coastal plains during the withdrawal of the Pakowki Sea. Later, the sedimentary environment changed to the alluvial plain environment of the foreland basin during the Laramide orogeny (Bustin and Smith 1993). The Lower Cretaceous coal measures from the interior plains were deposited in a prograding barrier coastline, and their coal seams were recognized to have transgressive or regressive origins (Holz et al. 2002). In the Western Canada Basin, Early Cretaceous coal measures were formed in depositional settings ranging from coastal plains to upper delta plains. The coastal plain coal is characterized by its great lateral continuity and substantial thickness, whereas the coal of the upper delta plains is thin and discontinuous (Kalkreuth et al. 1991; Bustin and Palsgrove 1997; Kalkreuth 2004).

In Central Mongolia, the Lower Cretaceous coal measures of the inter-montane Baganuur Basin were mainly developed in swamps of fluvial, deltaic, and lacustrine environments and do not record any marine transgressions (Dill et al. 2004).

In the Alaska Peninsula, and specifically within the area extending from southwestern Wide Bay to Pavlof Bay, Late Cretaceous (Campanian to Maestrichtian) coal measures were deposited in a marine transgression sequence. The lower coal-bearing strata were deposited in alluvial fan to flood plain environments, the middle strata were deposited in inner-neritic (upper and lower shoreface) continental shelf environments, and the upper strata were deposited in outer-neritic continental shelf to bathyal continental slope environments (Merritt

1986). In New Mexico, the Late Cretaceous (Campanian) coal measures from the San Juan Basin were deposited in a littoral wedge composed of offshore, shoreface, and onshore facies. The continental facies consist of coastal-plain deposits (coals, shales, and sandstones), represented by three main domains, which are progressively distant from the palaeoshoreline, namely the deltaic-plain, intermediate-plain, and alluvial-plain domains (Buillit et al. 2002). In Wyoming, the Late Cretaceous coal-bearing strata from the western part of the Wind River Basin record transitions in depositional environments from coastal deltaic (neritic and paralic) to non-marine inter-montane (limnic) environments. The higher iron and sulphur contents in the coalbeds are attributed to the influences of marine and brackish water on syn- and post-depositional sedimentation. The elevated ash values and silica content in the inter-montane coalbeds resulted from the increased influx of volcaniclastic sediments (Windolph et al. 1996).

4.6. Coal-forming deposit characteristics during the Palaeogene–Neogene periods

4.6.1. Coal-forming deposit characteristics during the Palaeogene–Neogene in China

The coal-bearing basins that formed in China during the Palaeogene were mainly distributed in the areas east of the Greater Hinggan and Taihang Mountains, north of the Qinling Mountains, and in the southwestern area of the Guangxi Autonomous Region. The coal-bearing basins that formed during the Neogene were mainly distributed in Yunnan Province and Taiwan. The coal-forming intensity was most concentrated during the Eocene Epoch of the Palaeogene Period and the Miocene and Pliocene Epochs of the Neogene Period. Additionally, these basins represented a component of the global Pacific Rim coal-forming belt. With the exception of Taiwan, the depositional environments of the coal-bearing basins during the Palaeogene and Neogene Periods were continental facies environments. The majority of these basins were catchment basins. However, the intensities of the sediment source supplies surrounding the basins differed, and the planar configurations of the sedimentary facies recorded asymmetric circularity (Hu 1980; Du 1982; Li et al. 1982, 1998, 1999; He 1983; Zhao 1983; Chen and Sun 1988; Huang 1994; Du et al. 2001; Shu 2006; Luo and Zhang 2013; Xia and Wang 2013; Dai et al. 2014c; Wang et al. 2015; Xu et al. 2015; Lv et al. 2016, 2017). The lake and peat swamp facies were developed during the filling evolution of the basins. The main coal-forming method was swampy lake siltation. Additionally, the coal-bearing basins formed during this period typically only developed a set of coal seams. These basins were influenced by syndepositional faults or depressions, and locations with higher coal-forming intensities were typically the frontal zones of alluvial fan and fan delta environments. Coal-forming conditions were poor among alluvial fans or fan deltas, as well as in sections that were covered with deep water. The deposition, subsidence, and coal-rich centres of most of the coal-bearing basins were similar. In the centres of the basins, the coal seams had simple structures and large thicknesses, whereas at the edges of the basins, the coal seams had complex structures and small thicknesses.

In northeastern China, the Palaeogene Period coal-bearing strata were mainly distributed in the Fushun and Meihe Basins in the Fushun–Mishan fault zone and the Shulan Basin in the Yitong–Jiamusi Fault zone. The Fushun Basin is an inland graben basin; the main coal-bearing strata of the Eocene Series Guchengzi Formation, which record the largest measured single coal thickness of 70 m, were mainly formed in lacustrine bog environments during the lake sedimentation shallowing stage (Zhao and Wen 2016). The Meihe Basin is also an inland graben basin, and the coal-bearing strata of the Palaeogene Period Meihe Group were mainly formed in depressions among alluvial fan, fan delta plain, and lacustrine bog environments during the lake sedimentation shallowing stage (Wu et al. 2008). The Shulan Basin is an inland half-graben basin, which contains the coal-bearing strata of the Palaeocene Epoch Xin'ancun Formation and the Eocene Epoch Shulan Formation. The Xin'ancun Formation coal measures were mainly formed in depressions between alluvial fan and lacustrine bog environments, where the coal-forming effects were very weak. The Shulan Formation coal measures were mainly formed in fluvial environments during the initial expansion stage of the basin and in lakeshore delta environments during the contraction stage of the basin (Zhao and Wen 2016).

The vast majority of the Neogene Period coal-rich basins in Yunnan Province are located in the eastern area. Among these basins, the Xianfeng, Xiaolongtan, and Zhaotong Basins contain most of the coal resources, and record maximum single coal seam thicknesses of 237 m, 223 m, and 140 m, respectively. There are only a small number of coal-rich basins in western Yunnan Province, most of which are poor coal-bearing basins (Liu 2008). Coal measures were mainly formed in shore to shallow lacustrine depressions between alluvial fans and fan delta plain environments (Wu et al. 2003, 2006).

In China, the Cenozoic Era graben coal-bearing basins often developed super-thick coal seams (with single coalbed thicknesses of more than 40 m).

Synsedimentary slumped layers and argillaceous gravity flow dirt bands can often be seen in these super-thick coal seams. Accordingly, scholars in China have established four new coal heterotopic formation sub-models, namely the 'Fushun sub-model' (Wu 1994), the 'Fuxin sub-model' (Wu et al. 1996), the 'Xianfeng sub-model' (Wu et al. 2003, 2003), and the 'Xiaolongtan sub-model' (Wu et al. 2003). On this basis, Chinese scholars further summarized the formation model of the Palaeogene–Neogene Period super-thick coal seams in inland graben basins (Wu et al. 2003).

Cenozoic Era coal-bearing strata were also developed in China's sea area, but relatively little research has been performed there. Li et al. (2012) studied the sedimentary characteristics of the Palaeogene Period coal-bearing strata in the Huanxian coal-bearing basin in Bohai Bay, the Qiongdongnan coal-bearing basin in the northern South China Sea, and the Xihu coal-bearing basin in the East China Sea. The coal depositional systems of the Huangxian Basin include fan delta, braided river, braided river delta, and alluvial fan sedimentary systems, in which the braided river delta system exhibits the best coal-forming effects. It is worth noting that coal seams and oil shales usually exhibit symbiotic associations. The coal-forming environment of the Qiongdongnan Basin consists of fan delta plain, braided river delta plain, lagoon, and tidal flat environments, in which the braided river delta plain is the most conducive to the development of coal seams. According to the results of the analysis of coal petrology and coal properties, these coal seams were formed in an offshore environment and record the characteristics of allochthonous and hypautochthonous coal. The coal-forming environments of the Xihu Basin include coastal swamp, tidal flat, lagoon, and delta environments. In the sea area, the Palaeogene Period coal-bearing basins record good continuity within the basin group and very thick coal-bearing strata, but single coal seams exhibit low thickness and poor stability. In the deep-sea area, due to the burial depth of the coal-bearing strata and the high degree of coal metamorphism, the coal seams contain good hydrocarbon source rocks.

4.6.2. Coal-forming deposit characteristics during the Palaeogene–Neogene in other typical regions of the world

In southwestern Canada, Middle Eocene coal measures from British Columbia were formed in lacustrine or fluviatile-lacustrine environments (Kalkreuth 2004). In Alberta, southwestern Canada, Palaeocene coal measures were formed in alluvial-lacustrine environments (Wolfgang 2004).

In northwestern Canada, Late Eocene coal measures in the inter-montane Rock River Basin of the Yukon Territory were deposited in areas at a great distance from the main river channel in a series of elongate forested swamps, which were periodically inundated by flood water from meandering rivers (Long and Sweet 1994).

In western Washington, USA, some Late Eocene coal-bearing strata accumulated within large flood basins, in cut-off meanders, and along channel margins in point bars and crevasse-splay settings (Burnham 1990).

In Assam, India, the Late Oligocene coal measures from the Tirap coal mine were formed within a tropical delta. The lower two-thirds of the Late Oligocene section represent lower delta plain environments that record only a small degree of brackish water (marine) influence. The upper third of the section represents upper delta plain environments with high sediment flux (Kumar *et al.* 2012).

In western Anatolia, Turkey, the Miocene lignite layers from the Soma coalfield were deposited in alluvial and fluvial-lacustrine environments. The Miocene coal successions were most likely deposited in the slowly subsiding, fault-controlled, karst-based palaeo-valley and the lowlands of the inter-montane palaeo-morphology, which resulted from the Early Tertiary collision of the Eurasian and Anatolian plates (İnciUğur 2002).

In the Karlovo graben, Bulgaria, the thin, coaly Neogene layers represent the transition from a fluvio-deltaic environment to a lacustrine environment. The peat accumulation was terminated by a major flooding event and the establishment of a lake (Zdravkov *et al.* 2006).

5. Basic theories of coal formation in China

5.1. Evolution of the Palaeogeographic characteristics of coal formations

The filling of coal-forming basins during the different coal-forming periods in China formed a unique series of palaeogeographical patterns of coal formations by producing specific sedimentary facies associations or system tracts. In addition, the different coal-forming periods also recorded their own unique palaeogeographical coal formation characteristics.

(1) During the Terreneuvian Epoch, the main coal-forming materials were bacteria and algae, and the main coal-forming environment was the shallow sea. The presence of quiet and shallow seawater and slow tectonic subsidence was conducive to the formation and accumulation of algae blooms and thus the formation of stone coal. The shallow sea under the wave base was the most suitable environment for stone coal formation, and the coal-rich belt typically developed on the near-shore side.

(2) During the Permo-Carboniferous Periods, coastal plains were the main locations where peatification occurred. The main coal-forming depositional environments included coastal plains, coastal deltas, tidal flats, and lagoon barrier islands, and carbonate tidal flats, among others. These depositional systems formed a specific depositional system configuration during certain filling stage-depositional system tracts. The coastal delta or delta-detrital coast systems were the most important coal-forming palaeogeographical environments and typically coincided with coal-forming centres.

(3) During the Late Triassic Epoch, the coal-bearing strata in the large Sichuan–Yunnan offshore basins, which were located in the western region of South China, along with the bay-type offshore basins, which were located in the eastern region of South China, were mainly formed during a regression filling sequence. The main coal-forming depositional environments during this time included coastal, coast-delta, coastal alluvial, and coastal inter-montane plain environments, as well as coast-bay and lagoon-estuary systems. Overall, the coal formation processes were weak, and the thicknesses of the filling rock series of the basins changed significantly. The deposited lithofacies were complicated and usually featured a lack of large-area and stably distributed thick coal seams.

(4) During the Early and Middle Jurassic Epochs, the coal-forming basins mainly comprised large-sized inland depression basins. The coal-bearing strata were formed during the different filling evolution stages of the inland lake basins, and the main coal seam was formed during the filling stages of the lakeshore deltas. The large inland depression basins that were formed during the Early and Middle Jurassic Epochs had fixed the lake systems during the long-term filling evolution of the basins. The depositional system tract consisted of alluvial-lakeshore delta systems. The coast-lakeshore belt and submerged delta systems could be divided from the margins of the basins to the centres of the lakes. The alluvial-lakeshore delta systems had the best coal formation effects, followed by the lakeshore belt systems.

(5) During the Early Cretaceous Epoch and the Palaeogene–Neogene Periods, the coal-forming basins basically consisted of mutually isolated medium- and small-sized basin groups. However, in the offshore depression basins that were formed during the Early Cretaceous Epoch, such as those of the Sanjiang–Muling River and the Erlian–Hailar fault basin group in the eastern section of Inner Mongolia, as well as in many of the Palaeogene and Neogene Periods small fault-depression lake basins distributed throughout the Pacific Rim, the coal-forming intensity was high. In these areas, very thick and extremely thick coal seams were developed and were mainly formed during the lake siltation stages of the filling evolution of the lake basins.

5.2. Transgression event coal-forming theory and model

In China, many large-sized epicontinental sea basins formed during the Palaeozoic Era. In particular, a very large plate-shaped epicontinental sea basin formed in North China, and many coal seams of industrial value were developed during the evolution of the epicontinental sea. The coal-forming process of the epicontinental sea was significantly different from that of the marginal sea, and the geological background of its formation and filling mechanisms had particular characteristics. First, the coal-forming basin formed within the large-sized epicontinental sea and recorded the characteristics of high tectonic stability and uniform tectonic integrity. Although the rising amplitude of the sea surface was not large, large parts of the basin became flooded. This caused the transgression 'process' to become very short relative to the marginal sea basin, without recording a long-term 'gradual and slow' transgression process towards the land. This type of transgression process could be deemed an event, such as the 'transgressive event' described by He *et al.* (1991), with extremely strong isochronism. Second, the overlap of the epicontinental sea basin with land was not obvious and only occurred at the margins of the basin. Relatively consistent superimposition deposition (or aggradation) occurred throughout a large region of the basin.

Through an in-depth study of special coal formation processes, Li *et al.* (2001) defined this as the coal-forming process that occurred during this transgression event. The coal-forming process in those events mainly emphasized its transgressive properties. This type of event also played a role in controlling the coal-forming process. However, the accumulation of peat was not considered to be an event. The coal-forming process of the transgression event occurred after another transgression event, whereas peat accumulation occurred during the sea-level oscillation prior to a large-scale transgression and represented the beginning of a depositional sequence. Therefore, the coal seam and its directly overlying marine limestone represented a type of event deposition combination. The coalification process of the peat occurred entirely within a deep-water environment with reductive conditions, in which its gelatinization was relatively complete. As this process caused the peat to quickly become flooded and remain in a deep-water environment, it was a sudden event, and the transgression surface between the coal seam and the marine limestone was isochronous. The coal seam and its roof marine limestone facies were also isochronous deposition layers. It is believed that the marine limestone and coal seam were each deposited in single depositional environments with broad distribution areas and lateral stability. At the same time, they were also isochronous facies, and could thus be used as stratum comparison markers (Figure 8).

The sedimentary horizon of the coal-forming process in transgression events was isochronous, and complete deposition sequences corresponding to the transgression process between the marine limestone and coal seams did not exist (i.e. the facies sequences were not complete). The coal seam floor was root clay, and there were interruptions between the depositions of the coal seam and root clay, which represented important interfaces that were used to divide the high-resolution sequences and their internal units. If the largest marine flooding surface was located in the coal-forming combination in the transgression event, then there should be a 'condensed section' deposit near this interface. In the present study, it was believed that this section should be a component of the coal seams or marine beds, or the lower section of the marine beds. As a type of new coal-forming model, the theory of the coal-forming process in transgression events can potentially contribute to the enrichment of the basic theory of coal geology (Li *et al.* 2002, 2003; Lv 2009; Lv and Chen 2014; Lv *et al.* 2015).

5.3. Basic regularities of the coal-forming processes of China

China has undergone many coal-forming periods, producing widely distributed coal-bearing strata, within a series of diverse coal-forming depositional environments. Moreover, tectonic activity in China has been complicated. Therefore, vast differences in these coal-

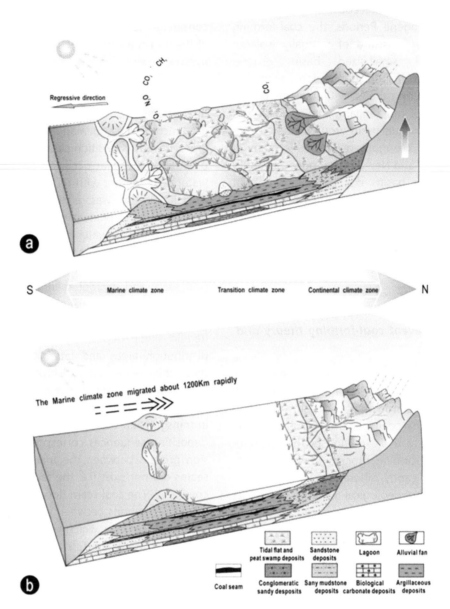

Figure 8. The Early Permian transgression, regression coal-forming, and climate zone migration of epicontinental sea basin in North China.

forming processes have been observed during different periods and within different districts of China. However, there are also certain similarities that can be observed.

In terms of the prime material used for coal formation, there are two different stages: before and after the Devonian. Before the Devonian, coal-forming material mainly comprised bacteria and algae, along with other lower organisms. After the Devonian Period, the prime coal-forming material was mainly terrestrial plants. The prime coal-forming materials during the Late Palaeozoic Era were mainly the plants that bred along coasts and offshore areas. The prime coal-forming materials during the Meso-Cenozoic Era were mainly the phytocoenosium in the open sea or inland, highland, or mountain areas. With the transition of coal-forming plants from low to high altitudes, the coal-forming process migrated from offshore lowlands to inland basins. The coal-forming process thus expanded from the tropical and subtropical zones to temperate zones. The majority of the coal-forming basins formed during the Palaeozoic Era were distributed in the tropical and subtropical moist climate zones, whereas the coal-forming basins formed during the Meso-Cenozoic Era were mainly distributed in the temperate moist climate zones.

From the tectonic-sedimentologic perspective, the main tectonic regions that developed in China had long evolutionary histories. These development processes established the palaeotectonic framework for each historical geological period and controlled the formation, evolution, and distribution of the coal-forming basins. Therefore, these processes controlled the development of coal-bearing strata, as well as the distribution of coal-rich belts and centres (Zhang 1995b).

(1) The coal that formed during the Hercynian and Indosinian periods was mainly concentrated in large epicontinental sea depression basins and used the steady platforms as basements; these include the coal-forming depression basins formed during the Pennsylvanian–Cisuralian–Guadalupian Epochs in North China and the Lopingian Epoch in the Yangtze region of South China. Tectonic activities in the material source areas, along with regional transgressions and regressions, were the main factors controlling the formation and migration of the coal-rich belts in the epicontinental sea and offshore basins. The coastal deltas and coastal plains in the clastic coastal belt sedimentary systems were the most important coal-forming environments; coal-rich belts (centres) were usually located there.

(2) The most important coal-forming basins formed during the early Yanshanian Period were large inland lake basins that used steady old platforms or blocks as basements, such as the Ordos and Junggar Basins. The shore-shallow lake-lakeshore delta systems and the alluvial fan-fan delta systems in the basin margin belts were the most important coal-forming environments, both before and after the large-scale expansions of lake basins, as their coal-rich belts coincide with these environments.

(3) The coal produced from the middle Yanshanian Period to the Himalayan Period mainly accumulated in the medium- and small-sized inland fault basins and depression basins related to basement faults. These basins were often characterized by their very thick to extremely thick coal seams; although their basin areas were small, their coal-bearing properties were good. The Sanjia–Muling River offshore depression basin, which formed on the continental margin block basement during the middle Yanshanian Period, is also well known for its outcrops of several billions of tons of high-quality coking coal resources.

In addition, the tectonic framework of the basements of the basins and their tectonic activity exerted an important influence on coal-forming processes. The large-sized, important coal-forming depressions or coal-bearing basins usually used stable platforms or blocks as supporting bases. During the Palaeozoic Era, the successively superimposed depression basins that developed on the folded basements during coal-forming processes were commonly weak, and thus typically developed smaller scales and poorer coal-bearing properties, due to the stronger tectonic activity embodied by the basements during coal-forming periods.

Under the comprehensive controls of palaeotectonic activity and paleaoclimate, among other factors, the palaeogeographic development and evolution of coal-forming processes in China also displayed certain regularities. The overall trend of the coal-forming palaeogeography evolved from marine facies to continental facies, and the scale of the coal-forming basins evolved from large to small basin groups.

(1) Within the epicontinental sea basins that developed during the Palaeozoic Era, their coal-forming palaeogeographic types evolved from shore and shallow sea-types during the Early Palaeozoic Era to coastal and coastal alluvial plain-types during the Late Palaeozoic Era.

(2) Within the inland depressions and fault basins that mainly developed during the Meso-Cenozoic Era, their palaeogeography evolved from large inland coal-forming basins to small depressions and fault basin groups over time.

(3) The overall trend of palaeo-topography ranged from simple to complex. For example, the topography during the Palaeozoic Era was flat and stable, whereas the topography during the Cenozoic Era was complex and variable. In each of the coal-bearing zones, the coal-rich areas decreased over time; however, the coal seam thicknesses increased over time.

6. Basic conclusions and recognitions

This study systematically introduced the characteristics of coal geology in China, in particular, the coal deposition characteristics and typical theories of different coal-forming periods. The following main conclusions were drawn.

(1) There were eight main important coal-forming periods in China's geological history. Among these, the Pennsylvanian, Cisuralian–Guadalupian, Late Triassic, Early–Middle Jurassic, and Early Cretaceous Epochs experienced

relatively strong coal-forming processes, especially the Early–Middle Jurassic Epochs, which experienced the strongest coal-forming processes.

(2) The coal-bearing strata formed during the Terreneuvian, Mississippian, Lopingian, and Late Triassic Epochs were mainly distributed throughout South China. The coal-bearing strata produced during the Pennsylvanian–Cisuralian–Guadalupian Epochs were mainly distributed throughout North China. The coal-bearing strata formed during the Early–Middle Jurassic Epochs were widely distributed in the northern, central, and southwestern areas of China. The coal-bearing strata formed during the Early Cretaceous Epoch were mainly distributed throughout northeastern China. The coal-bearing strata formed during the Palaeogene Period were mainly distributed throughout northeastern China, and the coal-bearing strata produced during the Neogene Period were mainly distributed throughout the east coast and southwestern areas of South China.

(3) Based on the locations of the tectonic plates and the coal-bearing characteristics of coal-bearing basins, the coal-bearing basins in China can be divided into six types: 1) inner craton depression coal-bearing basins; 2) Earth suture zone coal-bearing basins; 3) foreland coal-bearing basins; 4) post-collision orogeny coal-bearing basins; 5) intra-continental rift coal-bearing basins; and 6) active marginal belt coal-bearing basins.

(4) In China, Early Palaeozoic Era coal-bearing strata were mainly deposited in broad shallow sea environments. The coal-bearing basins formed during the Late Palaeozoic Era were mainly large epicontinental sea basins, in which coastal delta or coastal plain systems were the most important coal-forming sedimentary environments. The coal-bearing basins formed during the Late Triassic Epoch were mainly offshore basins, and coastal plains, delta plains, coastal alluvial plains, and coastal inter-montane plains, as well as coast-bay and lagoon-estuary systems, were the main coal-forming sedimentation environments. The coal-bearing basins formed during the Early–Middle Jurassic Epochs were large- and medium-sized inland lake basins, in which alluvial-lakeshore delta systems experienced the best coal-forming processes, followed by the lakeshore belts. The coal-bearing basins formed during the Early Cretaceous Epoch and Palaeogene–Neogene Periods were mainly small-sized continental basin groups; coal-forming processes mainly occurred in lake-delta swamp environments during the lake siltation stages of the filling evolution of these basins.

(5) This study proposed a new type of coal-forming process, i.e. a coal-forming process in transgression events, which is a process unique to epicontinental sea basins. The combination of coal-forming deposition occurring during a transgression event, along with the identification and evaluation of related event interfaces, can provide a theoretical and practical basis for the division and comparison of high-resolution sequence stratigraphy, as well as the establishment of a high-resolution sequence and isochronous stratigraphic framework for epicontinental sea basins.

(6) This study generally summarized similarities in the characteristics and evolution of coal-forming processes in China from the perspective of coal-forming prime material, tectonic-sedimentologic activity, palaeotectonics, palaeoclimate, and palaeogeography, among others, with special attention paid to the characteristics and evolution of different coal-forming sedimentation environments.

Acknowledgements

We would like to thank all the reviewers, Editor-in-Chief Prof. Robert J. Stern and Guest Editors Prof. Shifeng Dai and Prof. Bob Finkelman for their careful reviews and constructive comments, which greatly improved the paper quality.

Disclosure statement

No potential conflict of interest was reported by the authors.

Funding

This work was supported by the Fund of the National Science Foundation of China (Microstructure and its evolution mechanism of fine grained deposit of coal measures [No.41502151]); Special-thick coal seam genetic mechanism in Middle Jurassic Yan'an Formation in southwest margin of Ordos Basin [No.41402086]; Comparative study on metallogenic mechanism and model of coal and oil shale coexit [No. 41272172]; Provincial College Excellent Young Talents Joint Fund of Natural Science Foundation of Shandong Province (Study on genetic mechanism of Late Paleozoic coal bearing strata volcanic event deposits in Northwestern Shandong [No. ZR2015JL016]); College Scientific Research Project of Shandong Province (The characteristics of inner discontinuity surface of autochthonous accumulate special-thick coal seam and its depositional model [No. J14LH06]).

References

Andrews, W.M., Hower, J.C., Ferm, J.C., Evans, S.D., Sirek, N.S., Warrell, M., and Eble, C.F., 1996, A depositional model for the Taylor coal bed, Martin and Johnson counties, eastern Kentucky: International Journal of Coal Geology, v. 31, no. 1, p. 151–167. doi:10.1016/S0166-5162(96)00015-8

Barbackaab, M., Püspökic, Z., Bodorcd, E., Forgácse, Z., Hámor-Vidóc, M., Pacynaf, G., and McIntoshg, R.W., 2015, Palaeotopography related plant succession stages in a coal forming deltaic succession in early Jurassic in Hungary: Palaeogeography, Palaeoclimatology, Palaeoecology, v. 440, p. 579–593. doi:10.1016/j.palaeo.2015.09.027

Beerling, D.J., and Woodward, F.I., 2001, Vegetation and the terrestrial carbon cycle: Modelling the First 400 million years: Cambridge, Cambridge University Press, p. 1–405.

Belkin, H.E., and Luo, K.L., 2008, Late-stage sulfides and sulfarsenides in Lower Cambrian black shale (stone coal) from the Huangjiawan mine, Guizhou Province, People's Republic of China: Mineralogy and Petrology, v. 92, p. 321–340. doi:10.1007/s00710-007-0201-9

Beukes, N.J., Siepker, E.H., and Naude, F., 1991, Genetic stratigraphy of the Waterberg coalfield. Conference on South Africa's Coal Resources: Geological Society of South Africa, Witbank, v. 11, p. 6–9. [Abstracts.]

Bojesen-Koefoed, J.A., Kalkreuth, W., Petersen, H.I., and Piasecki, S., 2012, A remote coal deposit revisited: Middle Jurassic coals at Kulhøj, western Germania Land, northeast Greenland: International Journal of Coal Geology, v. 98, p. 50–61. doi:10.1016/j.coal.2012.04.006

Buillit, N., Lallier-Vergès, E., Pradier, B., and Nicolas, G., 2002, Coal petrographic genetic units in deltaic-plain deposits of the Campanian Mesa Verde Group (New Mexico, USA): International Journal of Coal Geology, v. 51, no. 2, p. 93–110. doi:10.1016/S0166-5162(02)00076-9

Burnham, R.J., 1990, Some late eocene depositional environments of the coal-bearing puget Group of western washington State: U.S.A. International Journal of Coal Geology, v. 15, no. 1, p. 27–51. doi:10.1016/0166-5162(90)90062-4

Bustin, R.M., and Palsgrove, R., 1997, Lithofacies and depositional environments of the Telkwa coal measures central British Columbia, Canada: International Journal of Coal Geology, v. 34, no. 1, p. 21–51. doi:10.1016/S0166-5162(97)00005-0

Bustin, R.M., and Smith, G.G., 1993, Coal deposits in the front ranges and foothills of the Canadian Rocky Mountains, southern Canadian Cordillera: International Journal of Coal Geology, v. 23, no. 1–4, p. 1–27. doi:10.1016/0166-5162(93)90041-8

Cadle, A.B., Cairncross, B., Christie, A.D.M., and Roberts, D.L., 1991, The permo-triassic coal-bearing deposits of the Karoo Basin, South Africa. Conference on South Africa's Coal Resources: Geological Society of South Africa, Witbank, v. 11, p. 6–9. [Abstracts.]

Cai, H.A., Li, B.F., Shao, L.Y., Xu, D.B., Shao, K., and Zhou, Y.Y., 2011, Sedimentary environments and coal accumulation patterns of the lower cretaceous shahai formation in Fuxin Basin, Liaoning Province: Journal of Palaeogeography, v. 13, no. 5, p. 481–491. [in Chinese with English abstract.]

Cairncross, B., 1989, Palaeodepositional environments and tectonisedimentary controls of the post-glacial Permian coals, Karoo Basin, South Africa: International Journal of Coal Geology, v. 12, p. 365–380. doi:10.1016/0166-5162(89)90058-X

Cairncross, B., 2001, An overview of the Permian (Karoo) coal deposits of southern Africa: African Earth Sciences, v. 33, p. 529–562. doi:10.1016/S0899-5362(01)00088-4

Cairncross, B., and Cadle, A.B., 1991, Fluvial depositional systems in the Vryheid Formation, northern Karoo basin. Conference on South Africa's Coal Resources: Geological Society of South Africa, Witbank, v. 11, p. 6–9. [Abstracts.]

Cameron, A.R., Goodarzi, F., and Potter, J., 1996, Coal and oil shale of Early Carboniferous age in northern Canada: Significance for paleoenvironmental and paleoclimatic interpretations: Palaeogeography, Palaeoclimatology, Palaeoecology, v. 106, no. 1–4, p. 135–155. doi:10.1016/0031-0182(94)90007-8

Cao, Y.L., 1991, A preliminary study on sedimentary characteristics of coal measures in Weixian County coalfield, Hebei Province: Coal Geology of China, v. 3, no. 4, p. 27–29. [in Chinese.]

Chen, H.H., and Sun, S., 1988, Lithofacies and depositional environments of the Tertiary coal-bearing series in Huang Xian basin, Shan Dong Province: Acta Petrologica Sinica, v. 02, p. 50–58. [in Chinese with English abstract.]

Chen, J.Z., Xu, W.C., Xie, H.L., Ye, J.L., Ma, Z.L., Dai, M.F., Wang, J., and Ren, W.Y., 2010, Accumulation rule of a coal seam in Muli coalfield: Journal of Qinghai University, v. 28, no. 5, p. 42–48. [in Chinese with English abstract.]

Chen, S.Y., 2000, Control of sea-level change to coal accumulation in Carboniferous-Permian, North China: Coal Geology & Exploration, v. 28, no. 5, p. 8–11. [in Chinese with English abstract.]

Cheng, A.G., and Jiang, H.Q., 1990, The types and evoluation of Carboniferous-Permian coal-forming environment in the South and North Huaihe: Coal Geology & Exploration, v. 03, p. 1–7. [in Chinese with English abstract.]

Cheng, A.G., and Lin, D.Y., 2001, China coal formation effect of system analysis: Xuzhou, China University of Mining and Technology Press, [in Chinese.]

China Coal Geology Bureau, 1996, Sedimentary environments and coal accumulation of late permian coal formation in western Guizhou, Southern Sichuan, and Eastern Yunnan, China: Chongqing, Chongqing University Press, p. 156–216. [in Chinese with English abstract.]

Churnet, H.G., 1996, Depositional environments of Lower Pennsylvanian coal-bearing siliciclastics of southeastern Tennessee, northwestern Georgia, and northeastern Alabama, USA: International Journal of Coal Geology, v. 31, p. 21–54. doi:10.1016/S0166-5162(96)00010-9

Clemmensen, L.B., and Surlyk, F., 1976, Upper Jurassic coal-bearing shoreline deposits, Hochstetter Forland, east Greenland: Sedimentary Geology, v. 15, no. 3, p. 193–211. doi:10.1016/0037-0738(76)90016-6

Cui, X.Q., Sun, B.L., Fan, J.Y., Wang, Z.M., Jia, H.W., Liu, Y.J., Hao, S.F., and Zeng, F.G., 2013, Sedimentary environment and coal- accumulation pattern for the Upper Jiufengshan Formation in the central of Dayangshu Basin: Journal of China Coal Society, v. 38, no. S, p. 416–423. [in Chinese with English abstract.]

Curry, D.J., Emmett, J.K., and Hunt, J.W., 1994, Geochemistry of aliphatic-rich coals in the Cooper Basin, Australia and Taranaki Basin, New Zealand. implications for the occurrence of potentially oil-generative coals: Geological Society, London, Special Publications, v. 77, no. 1, p. 149–181. doi:10.1144/GSL.SP.1994.077.01.09

Dai, S.F., Chekryzhov, I.Y., Seredin, V.V., Nechaev, V.P., Graham, I.P., Hower, J.C., Ward, C.R., Ren, D.Y., and Wang, X.B., 2016, Metalliferous coal deposits in East Asia (Primorye of Russia and South China): A review of geodynamic controls and styles of mineralization: Gondwana Research, v. 29, no. 1, p. 60–82. doi:10.1016/j.gr.2015.07.001

Dai, S.F., Han, D., and Chou, C.L., 2006, Petrography and geochemistry of the Middle Devonian coal from Luquan, Yunnan Province, China: Fuel, v. 85, no. 4, p. 456–464. doi:10.1016/j.fuel.2005.08.017

Dai, S.F., Hower, J.C., Ward, C.R., Guo, W., Song, H.J., O'Keefe, J.M.K., Xie, P.P., Hood, M.M., and Yan, X.Y., 2015b, Elements and phosphorus minerals in the middle Jurassic inertinite-rich coals of the Muli Coalfield on the Tibetan Plateau: International Journal of Coal Geology, v. 144–145, p. 23–47. doi:10.1016/j.coal.2015.04.002

Dai, S.F., Li, T., Seredin, V.V., Ward, C.R., Hower, J.C., Zhou, Y.P., Zhang, M.Q., Song, X.L., Song, W.J., and Zhao, C.L., 2014b, Origin of minerals and elements in the late Permian coals, tonsteins, and host rocks of the Xinde Mine, Xuanwei, eastern Yunnan, China: International Journal of Coal Geology, v. 121, p. 53–78. doi:10.1016/j.coal.2013.11.001

Dai, S.F., Li, T.J., Jiang, Y.F., Ward, C.R., Hower, J.C., Sun, J.H., Liu, J.J., Song, H.J., Wei, J.P., Li, Q.Q., Xie, P.P., and Huang, Q., 2015a, Mineralogical and geochemical compositions of the Pennsylvanian coal in the Hailiushu Mine, Daqingshan Coalfield, Inner Mongolia, China: Implications of sediment-source region and acid hydrothermal solutions: International Journal of Coal Geology, v. 137, p. 92–110. doi:10.1016/j.coal.2014.11.010

Dai, S.F., Liu, J.J., Ward, C.R., Hower, J.C., Xie, P.P., Jiang, Y.F., Hood, M.M., O'Keefe, J.M.K., and Song, H.J., 2015c, Petrological, geochemical, and mineralogical compositions of the low-Ge coals from the Shengli Coalfield, China: A comparative study with Ge-rich coals and a formation model for coal-hosted Ge ore deposit: Ore Geology Reviews, v. 71, p. 318–349. doi:10.1016/j.oregeorev.2015.06.013

Dai, S.F., Luo, Y.B., Seredin, V.V., Ward, C.R., Hower, J.C., Zhao, L., Liu, S.D., Zhao, C.L., Tian, H.M., and Zou, J.H., 2014a, Revisiting the late Permian coal from the Huayingshan, Sichuan, southwestern China: Enrichment and occurrence modes of minerals and trace elements: International Journal of Coal Geology, v. 122, p. 110–128. doi:10.1016/j.coal.2013.12.016

Dai, S.F., Ren, D.Y., Tang, Y.G., Shao, L.Y., and Li, S.S., 2002, Distribution, isotopic variation and origin of sulfur in coals in the Wuda coalfield, Inner Mongolia, China: International Journal of Coal Geology, v. 51, p. 237–250. doi:10.1016/S0166-5162(02)00098-8

Dai, S.F., Ren, D.Y., Zhou, Y.P., Chou, C.L., Wang, X.B., Zhao, L., and Zhu, X.W., 2008, Mineralogy and geochemistry of a superhigh-organic-sulfur coal, Yanshan Coalfield, Yunnan, China: Evidence for a volcanic ash component and influence by submarine exhalation: Chemical Geology, v. 255, p. 182–194. doi:10.1016/j.chemgeo.2008.06.030

Dai, S.F., Wang, P.P., Ward, C.R., Tang, Y.G., Song, X.L., Jiang, J.H., Hower, J.C., Li, T., Seredin, V.V., Wagner, N.J., Jiang, Y.F., Wang, X.B., and Liu, J.J., 2014c, Elemental and mineralogical anomalies in the coal-hosted Ge ore deposit of Lincang, Yunnan, southwestern China: Key role of N$_2$–CO$_2$ –mixed hydrothermal solutions: International Journal of Coal Geology, v. 152, p. 19–46. doi:10.1016/j.coal.2014.11.006

Dai, S.F., Wang, X., Zhou, Y., Hower, J.C., Li, D., Chen, W., and Zhu, X., 2011, Chemical and mineralogical compositions of silicic, mafic, and alkali tonsteins in the late Permian coals from the Songzao Coalfield, Chongqing, southwest China: Chemical Geology, v. 282, p. 29–44. doi:10.1016/j.chemgeo.2011.01.006

Dai, S.F., Wang, X.B., Chen, W.M., Li, D.H., Chou, C.L., Zhou, Y.P., Zhu, C.S., Li, H., Zhu, X.W., Xing, Y.W., Zhang, W.G., and Zou, J.H., 2010, A high-pyrite semianthracite of Late Permian age in the Songzao Coalfield, southwestern China: Mineralogical and geochemical relations with underlying mafic tuffs: International Journal of Coal Geology, v. 83, p. 430–445. doi:10.1016/j.coal.2010.06.004

Dai, S.F., Wang, X.B., Seredin, V.V., Hower, J.C., Ward, C.R., O'Keefe, J.M.K., Huang, W.H., Li, T., Li, X., Liu, H.D., Xue, W.F., and Zhao, L.X., 2012, Petrology, mineralogy, and geochemistry of the Ge-rich coal from the Wulantuga Ge ore deposit, Inner Mongolia, China: New data and genetic implications: International Journal of Coal Geology, v. 90–91, no. p, p. 72–79. doi:10.1016/j.coal.2011.10.012

Dai, S.F., Zhang, W.G., Seredin, V.V., Ward, C.R., Hower, J.C., Song, W.J., Wang, X.B., Li, X., Zhao, L.X., Kang, H., Zheng, L.C., Wang, P.P., and Zhou, D., 2013a, Factors controlling geochemical and mineralogical compositions of coals preserved within marine carbonate successions: A case study from the Heshan Coalfield, Southern China: International Journal of Coal Geology, v. 109–110, p. 77–100. doi:10.1016/j.coal.2013.02.003

Dai, S.F., Zhang, W.G., Ward, C.R., Seredin, V.V., Hower, J.C., Li, X., Song, W.J., Wang, X.B., Kang, H., Zheng, L.C., Wang, P.P., and Zhou, D., 2013b, Mineralogical and geochemical anomalies of late Permian coals from the Fusui Coalfield, Guangxi Province, Southern China: Influences of terrigenous materials and hydrothermal fluids: International Journal of Coal Geology, v. 105, p. 60–84. doi:10.1016/j.coal.2012.12.003

Deng, W.S., Zhang, L.X., and Cheng, Y.S., 2009, Coal-bearing Strata sedimentary features and coal accumulation process in lower middle series, Jurassic System, Qinghai Province: Coal Geology of China, v. 21, no. S2, p. 14–18. 56. [in Chinese with English abstract]

Dill, H.G., Altangerel, S., Bulgamaa, J., Hongor, O., Khishigsuren, S., Majigsuren, Y., Myagmarsuren, S., and Heunisch, C., 2004, The Baganuur coal deposit, Mongolia: Depositional environments and paleoecology of a Lower Cretaceous coal-bearing intermontane basin in Eastern Asia: International Journal of Coal Geology, v. 60, no. 2–4, p. 197–236. doi:10.1016/j.coal.2003.09.008

Ding, Z., Zheng, B., Zhang, J., Long, J., Belkin, H.E., Finkelman, R.B., Zhao, F., Chen, C., Zhou, D., and Zhou, Y., 2001, Geological and geochemical characteristics of high arsenic coals from endemic arsenosis areas in southwestern Guizhou Province, China: Applied Geochemistry, v. 16, p. 1353–1360. doi:10.1016/S0883-2927(01)00049-X

Du, W.R., 1982, The geological characteristics and coal formation rule of Tertiary lignite field, Yunnan: Yunnan Geology, v. 1, no. 3, p. 234–244. [in Chinese.]

Du, Z.C., Jin, K.K., Yan, C.Y., Liu, L.F., and Cai, X.Y., 2001, High-resolution sequence stratigraphic characteristics and coal-accumulation in Paleogene period of Baise Basin: Journal of China Coal Society, v. 26, no. 5, p. 463–467. [in Chinese with English abstract.]

Fei, S.Y., Lei, Z.M., and Liu, X.W., 1991, A study on the depositional environment and law of coal-forming of the Mid-Jurassic Yan'an formation in the Ordos Basin: Bulletin of the 562 comprehensive geological brigade: Chinese Academy of Geological Sciences, v. 09, p. 77–88. [in Chinese with English abstract.]

Fettweis, G.B., 1979, World coal resources: Methods of assessment and results: Developments in Economic Geology: Amsterdam, Elsevier, Vol. 10, p. 415.

Gao, D., Shao, L.Y., and Li, Z., 2012, Sequence- paleogeography and coal accumulation of Early Cretaceous in Sanjiang basin, northeastern China: Journal of China University of Mining &Technology, v. 41, no. 5, p. 746–752. [in Chinese with English abstract.]

Gao, Z.B., and Zhang, W., 2014, Tuli-tuomatan mining area tectonic characteristics and coal resources of Northern Gansu province: Gansu Science and Technology, v. 30, no. 20, p. 32–37. [in Chinese.]

Glikson, M., and Fielding, C., 1991, The late triassic callide coal measures, Queensland, Australia: Coal petrology and depositional environment: International Journal of Coal Geology, v. 17, no. 3–4, p. 313–332. doi:10.1016/0166-5162(91)90037-J

Gmura, D., and Kwiecin'Skab, B.K., 2002, Facies analysis of coal seams from the cracow sandstone series of the Upper Silesia Coal Basin, Poland: International Journal of Coal Geology, v. 52, p. 29–44. doi:10.1016/S0166-5162(02)00101-5

Guo, B., Shao, L.Y., Zhang, Q., Ma, S.M., Wang, D.D., and Zhou, Q.Y., 2014, Sequence stratigraphy and coal accumulation pattern of the Early Cretaceous coal measures in Hailar Basin, Inner Mongolia: Journal of Palaeogeography, v. 16, no. 5, p. 631–640. [in Chinese with English abstract.]

Han, D.X., 1989, The features of the Devonian coal-bearing deposits in South China, the People's Republic of China: International Journal of Coal Geology, v. 12, no. 1–4, p. 209–223. doi:10.1016/0166-5162(89)90052-9

Han, D.X., and Yang, Q., 1980, China coal geology (part ii): Beijing, China Coal Industry Publishing House, p. 1–415, [in Chinese.]

He, Q.X., Ye, Y.Z., Zhang, M.S., and Li, H., 1991, Transgression model of restricted epiconental sea: Acta Sedimentologica Sinica, v. 9, no. 1, p. 1–10. [in Chinese with English abstract.]

He, X.G., 1983, The distribution law of Tertiary coal field in South China: Yunnan Geology, v. 2, no. 2, p. 122–135. [in Chinese.]

Holz, M., Kalkreuth, W., and Banerjee, I., 2002, Sequence stratigraphy of paralic coal-bearing strata: An overview: International Journal of Coal Geology, v. 48, p. 147–179. doi:10.1016/S0166-5162(01)00056-8

Hou, Y., Lu, J.Y., and Lu, J.H., 2008, Huolapen coalfield coal bearing conditions of Moue Conty Heilongjiang Province: Heilongjiang Science and Technology Information, v. 14, p. 61. [in Chinese.]

Hower, J.C., and Eble, C.F., 2004, Coal facies studies in the eastern United States: International Journal of Coal Geology, v. 58, p. 3–22. doi:10.1016/j.coal.2003.03.001

Hower, J.C., O'Keefe, J.M.K., Wagner, N.J., Dai, S.F., Wang, X.B., and Xue, W.F., 2013, An investigation of Wulantuga coal (Cretaceous, Inner Mongolia) macerals: Paleopathology of faunal and fungal invasions into wood and the recognizable clues for their activity: International Journal of Coal Geologyv. 144, p. 45–53.

Hu, Y.H., 1980, The sedimentary characteristics and genetic types of classification of Tertiary coal-bearing basin, Yunnan: Coal Geology & Exploration, v. 0, p. 1–11. [in Chinese.]

Huang, Z.Y., 1994, The coal formation law research of Paleogene in Northeast China Region: Mineral Rock Geochemistry, v. 01, p. 25–26. [in Chinese.]

Hunt, J.W., 1988, Sedimentation rates and coal formation in the Permian basins of eastern Australia: Australian Journal of Earth Sciences, v. 35, no. 2, p. 259–274. doi:10.1080/14400958808527945

Isbell, J.L., and Cúneo, N.R., 1996, Depositional framework of Permian coal-bearing strata, southern Victoria Land, Antarctica: Palaeogeography, Palaeoclimatology, Palaeoecology, v. 125, p. 217–238. doi:10.1016/S0031-0182(96)00032-6

Jia, B.W., and Wu, Y.Q., 1995, The provenance and stratigraphic significance of volcanic event layers in the Late Paleozoic coal measures from Daqingshan, Inner Mongolia: Journal of Geology and Mineral Resources North China, v. 10, p. 203–213. [in Chinese with English abstract.]

Jiang, J.H., Yin, D.S., and Yan, W., 2005, The Tertiary sedimentary and coal formation characteristics of Jixi basin, Heilongjiang Province: Coal Technology, v. 24, no. 2, p. 99–101. [in Chinese.]

Jiang, Y.H., Yue, W.Z., and Ye, Z.Z., 1994, Characteristics, sedimentary environment and origin of the Lower Cambrian stone-like coal in southern China: Coal Geology of China, v. 6, no. 4, p. 26–31.

Jones, M.P., 1980, World coal resources. Methods of assessment and results: Earth-Science Reviews, v. 16, p. 57–59. doi:10.1016/0012-8252(80)90005-7

Jr, R.M.C., Grauch, R.I., and Murowchick, J.B., 1994, Metals, phosphate and stone coal in the proterozoic and Cambrian of China: The geologic setting of precious metal-bearing Ni-Mo ore beds: Socity of Economic Geology, v. 18, p. 6–11.

Kalkreuth, W., Marchioni, D., and Utting, J., 2000, Petrology, palynology, coal facies, and depositional environments of an upper carboniferous coal seam, Minto Coalfield, New Brunswick, Canada: Canadian Journal of Earth Sciences, v. 37, p. 1209–1228. doi:10.1139/e00-039

Kalkreuth, W.D., 2004, Coal facies studies in Canada: International Journal of Coal Geology, v. 58, p. 23–30. doi:10.1016/j.coal.2003.06.001

Kalkreuth, W.D., Marchioni, D.L., Calder, J.H., Lamberson, M.N., Naylor, R.D., and Paul, J., 1991, The relationship between coal petrography and depositional environments from selected coal basins in Canada: International Journal of Coal Geology, v. 19, no. 1–4, p. 21–76. doi:10.1016/0166-5162(91)90014-A

Kumar, M., Srivastava, G., Spicer, R.A., Spicer, T.E.V., Mehrotra, R.C., and Mehrotra, N.C., 2012, Sedimentology, palynostratigraphy and palynofacies of the late Oligocene Makum Coalfield, Assam, India: A window on lowland tropical vegetation during the most recent episode of significant global warmth: Palaeogeography, Palaeoclimatology, Palaeoecology, v. 342–343, p. 143–162. doi:10.1016/j.palaeo.2012.05.001

Li, B.F., Li, Z., Fu, Z.M., Liu, G.H., Xu, S.L., Liu, Q.G., Lin, M.Q., Xu, J.M., He, B.Z., Du, S.Q., and Song, Z.J., 1989, The depositional filling, coal accumulation characteristics and tectonic evolution of the Late Paleozoic Epeiric Sea in the south of North China: Earth Science, v. 14, no. 4, p. 367–378. [in Chinese with English abstract.]

Li, J.Q., 2012, Coal bearing environment and sedimentary characteristics of Yining Coalfield: Science & Technology Information, v. 04, p. 383. [in Chinese.]

Li, S.T., Huang, J.F., Yang, S.G., Zhang, X.M., Cheng, S.T., Zhao, G.R., Li, D.A., Li, G.L., and Ding, J.L., 1982, Depositional and structural history of the late Mesozoic Huolinhe and its characteristics of coal accumulation: Acta Geologica Sinica, v. 03, p. 244–254. [in Chinese with English abstract.]

Li, Y.J., Liang, W.L., Shao, L.Y., Xu, X.H., and Tong, X., 2011, Late triassic coal-bearing strata sequence stratigraphy and coal accumulation characteristics in Sichuan Province: Coal Geology of China, v. 23, no. 8, p. 32–37. [in Chinese with English abstract.]

Li, Z.X., Wang, M.Z., Yu, J.F., Han, M.L., Li, J.T., and Lv, D.W., 2006, Sequence stratigraphy of late paleozoic coal-bearing measures and the transgressive coal-formed features in Ordos Basin: Acta Sedimentologica Sinica, v. 24, no. 6, p. 834–840. [in Chinese with English abstract.]

Li, Z.X., Wei, J.C., and Han, M.L., 2001, Coal formation in transgressive events - A new pattern of coal accumulation: Advance in Earth Sciences, v. 16, no. 1, p. 120–124. [in Chinese with English abstract.]

Li, Z.X., Wei, J.C., Lan, H.X., and Li, S.C., 1999, Coal accumulation in the lowstand and highstand systems tracts in the Huangxian paleogene faulted Basin: Acta Sedimentologica Sinica, v. 17, no. 2, p. 82–86. [in Chinese with English abstract.]

Li, Z.X., Wei, J.C., and Li, S.C., 1995, the depositional system of fluvial-controlled shallow-water delta and coal-forming analysis in Western Shandong: Coal Geology & Exploration, v. 23, no. 2, p. 7–13. [in Chinese with English abstract.]

Li, Z.X., Wei, J.C., Li, S.C., Han, M.L., and Lan, H.X., 1998, The basin-filling features and sequence division in the Paleogence Huangxian Fault basin, Shandong: Lithofacies Paleogeography, v. 18, no. 04, p. 1–8. [in Chinese with English abstract.]

Li, Z.X., Yu, J.F., and Guo, J.B., 2002, Study on coal accumulation under marine ransgression events in the epicontinental basin of North China: Coal Geology & Exploration, v. 30, no. 5, p. 1–4. [in Chinese with English abstract.]

Li, Z.X., Yu, J.F., Guo, J.B., and Han, M.L., 2003, Analysis on coal formation under transgression events and its mechanism in epicontinental Sea Basin: Acta Sedimentologica Sinica, v. 21, no. 2, p. 288–296. [in Chinese with English abstract.]

Li, Z.X., Zhang, G.C., Li, Y., Liu, H.Y., Lv, D.W., Shen, H.L., and Shang, L.N., 2012, The Paleogene coal-bearing basin and coal0measures distribution of China Sea area: Earth Science Frontiers, v. 19, no. 4, p. 314–326. [in Chinese with English abstract.]

Lin, C.S., Yang, Q., Li, S.T., and Li, Z., 1991, Deltaic sedimentary architecture and accumulation of Taiyuan and Shanxi Formations in Helan-Zhuozi Range, Northwest China: Journal of China Coal Society, v. 16, no. 3, p. 61–76. [in Chinese with English abstract.]

Liu, G., 1990, Permo-Carboniferous paleography and coal accumulation and their tectonic control in North China and South China continental plates: International Journal of Coal Geology, v. 16, p. 73–117. doi:10.1016/0166-5162 (90)90014-P

Liu, Y.J., 2012, The study on sedimentary evolution of Jiufengshan Formation in Dayangshu basin and the accumulation rules of coal: Taiyuan, Taiyuan University of Technology, [in Chinese with English abstract.]

Liu, Y.L., 2008, Coal accumulating law of Neogene basin in Yunnan: Yunnan Coal, v. 1, p. 31–32. [in Chinese with English abstract.]

Long, D.G.F., and Sweet, A.R., 1994, Age and depositional environment of the rock River coal basin, Yukon Territory, Canada: Canadian Journal of Earth Sciences, v. 31, no. 5, p. 865–880. doi:10.1139/e94-079

Lu, J., Shao, L.Y., Ran, L.M., Su, S.C., Wei, K.M., Sun, Y.Z., Chen, J.F., and Yu, X.H., 2008, A sequence stratigraphic analysis and coal accumulation of Late Triassic coal measures in the Baoding Basin: Coal Geology & Exploration, v. 36, no. 6, p. 1–6. [in Chinese with English abstract.]

Luo, K.L., 2011, Arsenic and fluorine contents and distribution patterns of early paleozoic stonelike coal in the daba fold zone and Yangtze Plate, China: Energy Fuels, v. 25, p. 4479–4487. doi:10.1021/ef200737s

Luo, X.Y., and Zhang, Y.Y., 2013, Neogene coal-forming basin characteristics and genetic types in Yunnan Province: Coal Geology of China, v. 25, no. 9, p. 10–17. [in Chinese with English abstract.]

Lv, D.W., Chen, J.T., 2014, Depositional environments and sequence stratigraphy of the Late Carboniferous_Early Permian coal-bearing successions (Shandong Province, China): Sequence development in an epicontinental basin: Journal of Asian Earth Sciences, v. 79, p. 16–30.

Lv, D.W., Zong, R.F., Li, Z.X., Wang, D.D., Liu, H.Y., Wu, X.Y., Wang, X.B., Yu, D.M., Feng, T.T., Zhao, L.Y., Yang, Q., Yong, P.L., 2016, Oil shale paleo-productivity disturbed by sea water in a coal and oil shale bearing succession: A case study from the Paleogene Huangxian basin of Eastern China: Journal of Petroleum Science and Engineering, v. 139, p. 62–70.

Lv, D.W., Wang, D.D., Li, Z.X., Liu, H.Y., Li, Y., 2017, Depositional environment, sequence stratigraphy and sedimentary mineralization mechanism in the coal bed- and oil shale-bearing succession: A case from the Paleogene Huangxian Basin of China: Journal of Petroleum Science and Engineering, v. 148, p. 32–51.

Lv, D.W., 2009, Study of the transgressive event depositon and paleogeography cgaeacteristics of noepaleozoic in north china area: Qingdao, Shandong University of Science and Technology, [in Chinese with English abstract.]

Lv, D.W., Li, Z.X., Wang, D.D., Liu, H.Y., Jia, Q., Wang, P.L., Yu, D.M., and Wu, X.Y., 2015, Discussion on micro-characteristics of transgressive event deposition and its coal-forming mechanism in the late paleozoic epicontinental Sea Basin of North China: Acta Sedimentologica Sinica, v. 33, no. 4, p. 633–672. [in Chinese with English abstract.]

Mang, D.H., 1994, Tectonics of coal basin in China: Beijing, Geological Publishing House, p. 1–181, [in Chinese.]

Marchioni, D., Gibling, M., and Kalkreuth, W., 1996, Petrography and depositional environment of coal seams in the carboniferous Morien Group, Sydney Coalfield, Nova Scotia: Canadian Journal of Earth Sciences, v. 33, no. 6, p. 863–874. doi:10.1139/e96-065

Merritt, R.D., 1986, Depositional environments and resource potential of cretaceous coal-bearing strata at Chignik and Herendeen bay, Alaska Peninsula: Sedimentary Geology, v. 49, no. 1–2, p. 21–49. doi:10.1016/0037-0738(86)90014-X

Michaelsen, P., Henderson, R.A., Crosdale, P.J., and Mikkelsen, S.O., 2000, Facies architecture and depositional dynamics of the upper permian rangal coal measures, Bowen Basin, Australia: Journal of Sedimentary Research, v. 70, no. 4, p. 879–895. doi:10.1306/2DC4093F-0E47-11D7-8643000102C1865D

Mishra, H.K., 1996, Comparative petrological analysis between the Permian coals of India and Western Australia: Paleoenvironments and thermal history: Palaeogeography, Palaeoclimatology, Palaeoecology, v. 125, p. 199–216. doi:10.1016/S0031-0182(96)00031-4

Nemec, W., 1985, Warstwy walbrzyskie (dolny namur) w Zagebiu Walbrzyskim: Analiza aluwialnej sedymentacji w basenie weglowymv: Geologia Sudetica, v. 2, p. 7–73.

O'Keefe, J.M.K., Bechtel, A., Christanis, K., Dai, S.F., DiMichele, W.A., Eble, C.F., Esterle, J.S., Mastalerz, M., Raymond, A.L., Valentim, B.V., Wagner, N.J., Ward, C.R., and Hower, J.C., 2013, On the fundamental difference between coal rank and coal type: International Journal of Coal Geology, v. 118, no. 3, p. 58–87. doi:10.1016/j.coal.2013.08.007

Petersen, H.I., Lindström, S., Therkelsen, J., and Pedersen, G.K., 2013, Deposition, floral composition and sequence stratigraphy of uppermost Triassic (Rhaetian) coastal coals, southern Sweden: International Journal of Coal Geology, v. 116–117, p. 117–134. doi:10.1016/j.coal.2013.07.004

Petersen, H.I., and Nielsen, L.H., 1995, Controls on peat accumulation and depositional environments of a coal-bearing coastal plain succession of a pull-apart basin; a petrographic, geochemical and sedimentological study, Lower Jurassic, Denmark: International Journal of Coal Geology, v. 27, no. S2–4, p. 99–129. doi:10.1016/0166-5162(94)00020-Z

Qimu, D., 1980, The relation between coal-bearing struction and tectonic system in Inner Mongolia: Coal Geology & Exploration, v. 1, p. 26–30. [in Chinese.]

Qin, J.B., Wu, S.F., Wang, X.H., Yang, D.X., and Yu, H.Y., 2009, Coal formation law and resources prediction of Jurassic coalfield in northern ZhangJiakou: Acta Geologica Sinica, v. 83, no. 5, p. 738–747. [in Chinese with English abstract.]

Ritenberg, M.I., 1972, Method of the investigation of facies and cycles (in Russian), in Makedonov, A.V., ed., Correlation of coalbearing sediments and coal seams in the Donets Basin: Nauka, Leningrad, p. 71–95.

Rust, B.R., Gibling, M.R., Best, M.A., Dilles, S.J., and Masson, A.G., 1987, A sedimentological overview of the coal-bearing Morien Group (Pennsylvanian), Sydney Basin, Nova Scotia, Canada: Canada Journal of Earth Sciences, v. 24, p. 1869–1885. doi:10.1139/e87-177

Sachsenhofer, R.F., Bechtel, A., Kuffner, T., Rainer, T., Gratzer, R., Sauer, R., and Sperl, H., 2006, Depositional environment and source potential of Jurassic coal-bearing sediments (Gresten Formation, H?flein gas/condensate field, Austria): Petroleum Geoscience, v. 12, no. 2, p. 99–114. doi:10.1144/1354-079305-684

Shang, G.X., 1997, Late paleozoic coal geology of North China platform: Taiyuan, Shanxi Science and Technology Publishing House, p. 1–405, [in Chinese with English abstract.]

Shao, K., Shao, L.Y., Qu, Y.L., Zhang, Q., Wang, J., Gao, D., Wang, D.D., and Li, Z., 2013a, Study of sequence stratigraphy of the Early Cretaceous coal measures in Northeastern China: Journal of China Coal Society, v. 38, no. 2, p. 423–433. [in Chinese with English abstract.]

Shao, L.Y., Gao, C.X., Zhang, C., Wang, H., Guo, L.J., and Gao, C.H., 2013b, Sequence- palaeogeography and coal accumulation of late permian in Southwestern China: Acta Sedimentologica Sinica, v. 31, no. 5, p. 856–866. [in Chinese with English abstract.]

Shao, L.Y., Dong, D.X., Li, M.P., Wang, H.S., Wang, D.D., Lu, J., Zheng, M.Q., and Cheng, A.G., 2014a, Sequence-paleogeography and coal accumulation of the carboniferous-permian in the North China Basin: Journal of China Coal Society, v. 39, no. 8, p. 1725–1734. [in Chinese with English abstract.]

Shao, L.Y., Gao, D., Luo, Z., and Zhang, P.F., 2009, Sequence stratigraphy and palaeogeography of the Lower and Mfiddle Jurassic coal measures in Turpan-Hami Basin: Journal of Palaeogeography, v. 11, no. 2, p. 215–224. [in Chinese with English abstract.]

Shao, L.Y., Li, Y.J., Jin, F.X., Gao, C.X., Zhang, C., Liang, W.L., Li, G.M., Chen, Z.S., Peng, Z.Q., and Cheng, A.G., 2014b, Sequence stratigraphy and lithofacies palaeogeography of the late triassic coal measures in South China: Journal of Palaeogeography, v. 16, no. 5, p. 613–630. [in Chinese with English abstract.]

Shao, L.Y., Liu, H.M., Tian, B.L., and Zhang, P.F., 1998, Sedimentary evolution and its controls on coal accumulation for the late permian in the Upper Yangtze Area: Acta Sedimentologica Sinica, v. 16, no. 2, p. 55–60. [in Chinese with English abstract.]

Shao, L.Y., Liu, T.J., Ran, M.Y., Luo, Z., and Lu, J., 2006b, Palaeo environments and coal accumulation of Early Jurassic Badaowan Formation in Turpan-Hami oil basin northwestern China: Journal of Liaoning Technical University, v. 25, no. 2, p. 189–191. [in Chinese with English abstract.]

Shao, L.Y., Lu, J., Ran, L.M., Su, S.C., Wei, K.M., Sun, Y.Z., Chen, J.F., and Yu, X.H., 2008, Late triassic sequence stratigraphy and coal accumulation in Baoding Basin of Sichuan Province: Journal of Palaeogeography, v. 10, no. 4, p. 355–361. [in Chinese with English abstract.]

Shao, L.Y., Xiao, Z.H., He, Z.P., Liu, Y.F., Shang, L.J., and Zhang, P.F., 2006a, Palaeogeography and accumulation for coal measures of the Carboniferous-Permian in Qinshui Basin, southern Shanxi Province: Journal of Palaeogeography, v. 8, no. 1, p. 43–52. [in Chinese with English abstract.]

Shao, L.Y., Xiao, Z.H., Lu, J., He, Z.P., Wang, H., and Zhang, P.F., 2007, Permo-Carboniferous coal measures in the Qinshui basin: Lithofacies paleogeography and its control on coal accumulation: Frontiers of Earth Science, v. 1, no. 1, p. 106–115. doi:10.1007/s11707-007-0014-5

Shao, L.Y., Zhang, P.F., Chen, D.Z., and Luo, Z., 1994, Braided delta depositional system and coal accumulation during early late Permian period in eastern Yunnan and western

Guizhou, southwest China: Acta Sedimentologica Sinica, v. 12, no. 4, p. 132–139. [in Chinese with English abstract.]

Shao, L.Y., Zhang, P.F., Gayer, R.A., Chen, J.L., and Dai, S.F., 2003a, Coal in a carbonate sequence stratigraphic frame-work: The Upper Permian Heshan Formation in central Guangxi, Southern China: Journal of the Geological Society, London, v. 160, p. 285–298. doi:10.1144/0016-764901-108

Shao, L.Y., Zhang, P.F., Hilton, J., Gayer, R., Wang, Y.B., Zhao, C. Y., and Luo, Z., 2003b, Paleoenvironments and paleogeo-graphy of the lower and lower middle Jurassic coal mea-sures in the Turpan-Hami oil-prone coal basin, northwestern China: AAPG Bulletin, v. 87, no. 2, p. 335–355. doi:10.1306/09160200936

Shi, H.N., Meng, F.Y., and Tian, J.J., 2011, Coal forming envir-onment and sedimentary characteristics of Badaowan for-mation in Yili basin: Journal of Xi'an University of Science and Technology, v. 31, no. 1, p. 33–38. [in Chinese with English abstract.]

Shu, Z.W., 2006, Coal-forming environmental analysis of Dalian Formation, Paleogene in Yilan Ming area: Coal Geology of China, v. 18, no. 3, p. 20–22. [in Chinese with English abstract.]

Shulga, V.F., 1981, Lower carboniferous coal formations of the donets Basin: Moscow, Nauka, p. 176, [in Russian.]

Singh, P.K., Singh, M.P., Prachiti, P.K., Kalpana, M.S., Manikyamba, C., Lakshminarayana, G., Singh, A.K., and Naik, A.S., 2012, Petrographic characteristics and carbon isotopic composition of Permian coal: Implications on depositional environment of Sattupalli coalfield, Godavari Valley, India: International Journal of Coal Geology, v. 90–91, p. 34–42. doi:10.1016/j.coal.2011.10.002

Suess, M.P., Drozdzewski, G., and Schaefer, A., 2007, Sedimentary environment dynamics and the formation of coal in the Pennsylvanian Parisian foreland in the Ruhr Basin (Germany, Western Europe): International Journal of Coal Geology, v. 69, no. 4, p. 267–287. doi:10.1016/j.coal.2006.05.003

Thomas, L. 2002. Coal Geology: West Sussex, John Wiley & Sons, Ltd., p. 1–256.

Tian, J.J., and Yang, S.G., 2011, Sequence strata and coal accumulation of lower and middle Jurassic formation from southern margin of Junggar Basin, sinking, China: Journal of China Coal Society, v. 36, no. 1, p. 58–64. [in Chinese with English abstract.]

Tian, Y., Li, Z.X., Shao, L.Y., Wang, D.D., and Li, P.M., 2011, Upper triassic series wayaobu formation sequence strati-graphic and coal accumulation characteristic studies in ordos Basin: Coal Geology of China, v. 23, no. 8, p. 13–17. 27. [in Chinese with English abstract.]

Uğur, İ., 2002, Depositional evolution of Miocene coal succes-sions in the Soma coalfield, western Turkey: International Journal of Coal Geology, v. 51, no. 1, p. 1–29. doi:10.1016/S0166-5162(01)00066-0

Uziyuk, V.I., Ginzburg, A.P., Inosova, K.I., Lapteva, A.M., and Fokina, E.I., 1972, Coal petrography method for correlation of sections (in Russian), in Makedonov, A.V., ed., Correlation of coalbearing sediments and coal seams in the donets Basin: Nauka, Leningrad, p. 24–48.

Vail, P.R., Mitchum, R.M., and Thompson, S., 1977, Seismic stratigraphy and global changes of sea level, in Payton, C. E., ed., Seismic stratigraphy applications to hydrocarbon exploration: AAPG Memoir 26, Tulsa, Oklahoma, p. 49–212.

Wang, D.D., 2012, Sequence-palaeogeography and coal-accmnulation of the Middle Jurassic Yan' an Formation in Ordos Basin: Beijing, China University of Mining and Technology (Beijing), p. 1–180, [in Chinese with English abstract.]

Wang, D.D., Shao, L.Y., Zhang, Q., Ding, F., Li, Z.B., and Zhang, W.L., 2013, Analysis of coal-forming characteristics in the lower cretaceous coal-containing strata of the Erlian basin group: Journal of China University of Mining &Technology, v. 42, no. 2, p. 257–265. [in Chinese with English abstract.]

Wang, H., Zhang, F., Wang, B.J., Yin, W.J., and Liu, C.Y., 2009, The structure characteristics and coal-forming features under sequence framework in the late triassic of Qiangtang Basin: Northwestern Geology, v. 42, no. 4, p. 92–101. [in Chinese with English abstract.]

Wang, J.M., Zhang, Y.H., and Zhu, S.B., 2007a, Central Yunnan (Chuxiong) late triassic basin genetic mechanism, coal-forming paleogeographic type and coal looking orienta-tion: Coal Geology of China, v. 19, no. 4, p. 1–4. [in Chinese with English abstract.]

Wang, K.F., Li, W.P., and Zhao, H., 2015, Coal occurrence characteristics of Paleogene coal measure strata and coal accumulation discipline in Shandong Province: Coal Science and Technology, v. 43, no. 3, p. 104–109. [in Chinese with English abstract.]

Wang, Q., and Wu, P., 1994, Sedimentary environment of Jurassic coal bearing strata in the Minhe basin: Coal Geology and Exploration, v. 22, no. 6, p. 5–10. [in Chinese.]

Wang, R.N., and Li, G.C., 1998, Evolution of coal basins and coal-forming laws in china: Beijing, China Coal Industry Publishing House, [in Chinese with English abstract.]

Wang, S.L., and Ge, L.M., 2007, Geochemical characteristics of rare earth elements in kaolin rock from Daqingshan coal-field: Coal Geology & Exploration, v. 35, no. 5, p. 1–5. [in Chinese with English abstract.]

Wang, S.M., 1996, Coal accumulation and coal resource eva-luation of Ordos basin: Beijing, China Coal Industry Publishing House, p. 1–437, [in Chinese with English abstract.]

Wang, S.M., and Zhang, Y.P., 1999, Study on the formation, evolutiong and coal-formation regularity of the Jurassic Ordos Basin: Geoscience Frontiers, v. 6, no. S1, p. 147–155. [in Chinese with English abstract.]

Wang, Y.F., and Chen, Z.B., 2004, Jurassic coal-bearing strata and coal-forming patterns in the Yining Basin: Coal Geology of China, v. 16, no. 2, p. 10–13. [in Chinese with English abstract.]

Wang, Z., Deng, Y.T., Ren, Y.M., and Zhang, Z.X., 2007b, Sedimentary features of the Jurassic system in tidewater basin and the coal-prospecting potentials: Shaanxi Geology, v. 25, no. 1, p. 28–37. [in Chinese with English abstract.]

Windolph, J.F., Warlow, R.C., and Hickling, N.L., 1996, Deposition of deltaic and intermontane Cretaceous and Tertiary coal-bearing strata in the Wind River Basin, Wyoming: Geological Society of America Special Papers. v. 210, p. 123–140.

Wu, C.L., 1994, The genesis model of the coal and extra-thick coal seam in the Fushun Basin: Chinese Science Bulletin, v. 39, no. 23, p. 2175–2177. [in Chinese with English abstract.]

Wu, C.L., Li, S.H., Huang, F.M., and Zhang, R.S., 1996, Analysis on the sedimentary conditions of extra-thick coal seam from Fushun coal field: Coal Geology & Exploration, v. 25, no. 2, p. 1–6. [in Chinese with English abstract.]

Wu, C.L., Li, S.H., Wang, G.F., Liu, G., and Kong, C.F., 2006, Genetic model about the extra-thick and high quality coalbed in Xianfeng Basin, Yunnan Province, China: Acta Sedimentologica Sinica, v. 24, no. 1, p. 1–9. [in Chinese with English abstract.]

Wu, C.L., Li, S.H., Wang, G.F., and Luo, Y.J., 2003, New evidence and new model about allochthonous accumulation of extra-thick coalbeds in continental fault basin, China: Earth Science– Journal of China University of Geosciences, v. 28, no. 3, p. 289–296. [in Chinese with English abstract.]

Wu, K.P., Liu, Y.X., Ji, B.Z., and Lu, Y.P., 2008, Sedimentary coal-accumulating environment and prospecting coal prospect of Meihe Basin: Jilin Geology, v. 27, no. 3, p. 24–29. [in Chinese with English abstract.]

Xia, G.X., and Wang, B.J., 2013, Coal bearing strata sedimentary environment type of Paleogene Hulin Formation in Jidong coal basin: Journal of Jixi University, v. 13, no. 2, p. 150–151. [in Chinese with English abstract.]

Xian, M.L., Wang, H., Ma, J.L., Gou, J.Y., and Wang, S.F., 2012, Late triassic coal-bearing strata depositional system and sequence stratigraphic features in Qiangtang Basin: Coal Geology of China, v. 24, no. 4, p. 1–5. [in Chinese with English abstract.]

Xiao, J.X., 1990, Sedimentary facies and depositional systems of late Late Permian coal measures in western Guizhou: Lithofacies Paleogeography, v. 04, p. 1–8. [in Chinese with English abstract.]

Xie, W., Yang, C.Y., and Peng, S.F., 2014, Contrast of coal accumulation regulation of longtan and xujiahe formation in Sichuan Basin: Coal Technology, v. 33, no. 12, p. 109–110. [in Chinese with English abstract.]

Xu, K.J., Shao, L.Y., Ma, L.J., Ye, Z.R., and Qu, Y.L., 2015, Paleogene-neogene sedimentary environment and coal accumulation pattern analysis in Lake Xingkai Basin, Heilongjiang: Coal Geology of China, v. 27, no. 8, p. 1–7. [in Chinese with English abstract.]

Yang, C.Z., 1987, A study on structural features of coal-forming basin and coal- Formation regularity in the eastern margin of the Song Liao basin, Jilin province: Jilin Geology, v. 01, p. 1–7. [in Chinese with English abstract.]

Yang, J., 2006, Concentrations and modes of occurrence of trace elements in the Late Permian coals from the Puan Coalfield, southwestern Guizhou, China: Environmental Geochemistry and Health, v. 28, p. 567–576. doi:10.1007/s10653-006-9055-z

Yang, M.X., and Li, D.H., 2011, Sedimentary environment and coal Accumulation pattern of Xujiahe Formation in Chongqing Area: Coal Geology of China, v. 23, no. 9, p. 9–13. [in Chinese with English abstract.]

Yi, P.F., 1980, Late of early carboniferous Period paleogeographic and coal forming rule of Eastern China: Coal Geology & Exploration, v. 05, p. 1–8. [in Chinese.]

Zdravkov, A., Kostova, I., Sachsenhofer, R.F., and Kortenski, J., 2006, Reconstruction of paleoenvironment during coal deposition in the Neogene Karlovo graben, Bulgaria: International Journal of Coal Geology, v. 67, p. 79–94. doi:10.1016/j.coal.2005.09.005

Zeng, Y.Y., and Wang, Z.Q., 2013, Sedimentary environment and coal accumulation of Mesozoic Erathem of

Heshanggou Coal Mine in Beipiao Basin: Coal, v. 165, p. 5–7. [in Chinese with English abstract.]

Zhang, C., Wu, Y.D., Ma, W.H., Jin, C.W., Shao, L.Y., and Ma, S.M., 2009, Upper triassic sedimentary environment and coal accumulation features in Zixing-Rucheng Coalfield, Hunan Province: Coal Geology of China, v. 21, no. 12, p. 12–14. 18. [in Chinese with English abstract.]

Zhang, G.S., 1995a, Minhe Basin in Jurassic coal deposits and coal accumulation characteristics analysis: Coal Geology of China, v. 7, no. 2, p. 26–30. [in Chinese with English abstract.]

Zhang, H., 1998, Jurassic coal-bearing strata and coal formation law of Northwest China: Beijing, Geological Publishing House, p. 1–317, [in Chinese.]

Zhang, H., Jia, B., Zhou, A., and Guo, M., 2000, Microscopic characteristics of kaolinite from the partings in the thick coal seam from the Daqingshan Coalfield and their genetic significance: Acta Mineral, v. 20, p. 117–120. [in Chinese with English abstract.]

Zhang, J.Y., Ren, D.Y., Zhu, Y.M., Chou, C.-L., Zeng, R.S., and Zheng, B.S., 2004, Mineral matter and potentially hazardous trace elements in coals from Qianxi Fault Depression Area in southwestern Guizhou, China: International Journal of Coal Geology, v. 57, p. 49–61. doi:10.1016/j.coal.2003.07.001

Zhang, L., Mao, X.D., Xu, B., and Tu, S.Y., 2013, The coal-forming rule and sedimentary facies of lower Cretaceous Duoni coal series, Xizang: Yunnan Geology, v. 32, no. 3, p. 359–361. [in Chinese with English abstract.]

Zhang, T., 1995b, China's main coal formation period depositional environment and coal formation law: Beijing, Geological Publishing House, p. 1–260, [in Chinese with English abstract.]

Zhang, Z.Y., 1982, The paleogeographic environment of early and middle Jurassic coal bearing formation and sedimentary characteristics of Gansu, Qinghai, Xinjiang: Coal Geology and Exploration, v. 02, p. 8–13. [in Chinese.]

Zhao, F.M., Lu, M.Q., Yan, Y.J., Liu, X.M., Wang, C.X., and Wu, D., 1991, The sedimentary structure and the characters of formation in the later Jurassic in the Heibaoshan-Muerqi basin: Journal of Changchun University of Science and Technology, v. 21, no. 4, p. 411–416. [in Chinese with English abstract.]

Zhao, L.Y., 1983, Tertiary coal basin and their coal-bearing characters: Acta Geologica Sinica, v. 03, p. 270–282. [in Chinese with English abstract.]

Zhao, S.P., and Wen, H., 2016, Cenozoic coal accumulation in northeast of China: Heilongjiang Science, v. 7, no. 12, p. 68–69. [in Chinese with English abstract.]

Zheng, Z.H., 2008, Early carboniferous sedimentary environments and coal forming characteristics of Guangxi Province: Southern Land Resources, v. 01, p. 39–40. [in Chinese.]

Zhong, D.K., Wang, Z.M., Zhang, L.J., Li, C., Wu, D., Peng, W.M., and Wang, X.M., 2010, Sedimentary characteristics and evolution of the Jurassic in northern Tarim Basin: Journal of Palaeogeography, v. 12, no. 1, p. 42–48. [in Chinese with English abstract.]

Zhong, R., Sun, S., Chen, F., and Fu, Z., 1995, The discovery of Rhyo-tuffite in the Taiyuan Formation and stratigraphic correlation of the Daqingshan and Datong Coalfield: Acta Geoscientica Sinica, v. 17, no. 3, p. 291–301. [in Chinese with English abstract.]

Zhou, A., and Jia, B., 2000, Analysis of Late Paleozoic conglomerates from Daqing Mountain in Inner Mongolia: Journal of Taiyuan University of Technology, v. 31, p. 498–504. [in Chinese with English abstract.]

Zhou, H.D., 1990, On the mechanism and the characteristic of Early Cambrian "stone coal" in Lower Yangtze Region and their relation to petroliferous potentials: Experimental Petroleum Geology, v. 12, no. 1, p. 36–43. [in Chinese with English abstract.]

Zhou, Q.Y., Shao, L.Y., Chen, Z.X., Li, M.S., and Li, H.R., 2016, Sequence-paleogeography and coal –accumulation of Xujiahe Formation in Jingang mining area of eastern Sichuan basin: Coal Geology & Exploration, v. 44, no. 2, p. 8–13. [in Chinese with English abstract.]

Zhou, Y., Bohor, B.F., and Ren, Y., 2000, Trace element geochemistry of altered volcanic ash layers (tonsteins) in Late Permian coal-bearing formations of eastern Yunnan and western Guizhou Province, China: International Journal of Coal Geology, v. 44, p. 305–324. doi:10.1016/S0166-5162(00)00017-3

Zhou, Y., and Ren, Y., 1992, Distribution of arsenic in coals of Yunnan Province, China, and its controlling factors: International Journal of Coal Geology, v. 20, p. 85–98. doi:10.1016/0166-5162(92)90005-H

Zhu, Y.H., and Zhang, W.Z., 2000, Lower cretaceous sedimentary facies and oil content in Erlian basin: Beijing, Science Press, p. 1–234, [in Chinese with English abstract.]

Zhuang, J., and Wu, J.J., 1996, Middle Jurassic coal accumulation characteristics and coal comprehensive utilization of southern Ordos Basin: Beijing, Geological Publishing House, p. 1–256, [in Chinese with English abstract.]

Zhuang, X., Su, S., Xiao, M., Li, J., Alastuey, A., and Querol, X., 2012, Mineralogy and geochemistry of the Late Permian coals in the Huayingshan coal-bearing area, Sichuan Province, China: International Journal of Coal Geology, v. 94, p. 271–282. doi:10.1016/j.coal.2012.01.002

REVIEW ARTICLE

The health impacts of coal use in China

Robert B. Finkelman ⓘ and Linwei Tian

ABSTRACT

Each year China burns as much coal as the rest of the world. The coal use results in many benefits such as electricity, industrial heat, coke for steel manufacturing, and many valuable chemical products. Unfortunately, coal use also results in environmental degradation and widespread and severe health problems. China is especially vulnerable to these problems because of the abundant coal deposits occurring in every Chinese province, the predominantly rural population, and the widespread use of domestic coal burning. The health problems are attributed to exposure to fine particulate air pollution ($PM_{2.5}$), mobilization of trace elements such as arsenic, fluorine, selenium, and possibly mercury and the release of organic compounds from incomplete combustion of domestic coal fires. Inhalation of minerals such as quartz and pyrite has deleterious health impacts contributing to lung cancer in women and black lung disease in coal miners. The extensive uncontrolled coal fires in China may also be a source of health problems. The overall situation has improved in recent years as uncontrolled fires have been extinguished, improved coal-burning stoves are being introduced, and people are switching to alternative fuel sources. Nevertheless, more needs to be done to minimize the health consequences of coal use in China.

1. Introduction

Modern society would not exist without coal. Coal-fired steam engines and railroad locomotives fuelled the Industrial Revolution. Steel, the skeletons of all of the skyscrapers that populate every city in the world, could not be mass produced without coal-derived coke. Coal is also the primary feedstock for hundreds of chemicals, pharmaceuticals, and other products that we use every day. And for the past century or so, coal has been the primary fuel for the generation of electricity that powers our computers, telephones, TVs, and every other electrical device that civilization has come to rely on. The use of coal has grown steadily during this past century reaching nearly 8 billion tonnes in recent years (https://www.iea.org/publications/freepublica tions/publication/KeyCoalTrends.pdf) about half of this amount used by one country – China.

Unfortunately, the widespread use of coal has resulted in a range of serious health problems that has impacted millions of people. Coal is an extraordinarily complex rock. In addition to the complex carbon matrix, coal contains almost every element in the periodic table, scores of different minerals, water, gases, and oil. Moreover, the conditions under which coal forms is conducive to the concentration of potentially toxic siderophile and chalcophile elements such as arsenic, mercury, selenium, lead, etc. Coal is commonly enriched, with respect to crustal averages, in these potentially harmful elements (Ketris and Yudovich 2009). Coal mining and, especially, coal combustion releases many of these constituents to the environment exposing people to these potentially harmful effluents. Nowhere is this problem more evident than in China.

2. Why China?

China is the most populous country on Earth with some 1.3 billion people. However, the health problems that we will discuss are not communicable; they are not passed from person to person but rather are caused by exposure to components released from coal. Therefore, population density is not a factor. Nor is the general composition of Chinese coals at fault. Dai et al. (2012) has demonstrated that the average trace element composition of Chinese coals is similar to that of U.S. and world coal averages reported by Finkelman (1993) and Ketris and Yudovich (2009).

The reasons why these coal-related health problems are so prevalent in China include the abundant coal deposits occurring in most Chinese provinces, the

Table 1. Examples of health issues in China related to coal burning.

Health issue	Cause	Estimated number of people who have been afflicted
Arseniasis	Exposure to arsenic	Tens of thousands
Fluorosis	Exposure to fluorine	More than 10 million
Selenosis	Exposure to selenium	Several hundred
Mercury poisoning	Exposure to mercury	Unknown
Coal workers' pneumoconiosis	Inhalation of pyrite	Several hundred thousand coal miners
Lung cancer	Inhalation of quartz	Several thousand
Respiratory problems	Exposure to particulates	Hundreds of millions
Various issues	Exposure to organics	Probably millions
Iodine deficiency disorder prevention	Exposure to iodine mobilized by coal combustion	Thousands

highly elevated concentrations of some toxic components in some local coal deposits, the predominantly rural population, deforestation, and the widespread use of domestic coal burning.

3. Health impacts of coal use

A major health concern resulting from coal use is the exposure to respirable particulates (fly ash) and flue gas released from coal-fired power plants. According to Lockwood et al. (2009), this exposure results in extensive respiratory effects such as asthma, cardiovascular effects such as heart attacks, and nervous system effects such as strokes. Berkeley Earth (http://berkeleyearth.org/deaths-per-gigawatt-year/) estimates that air pollution in China, mainly from coal-burning power plants, is responsible for somewhere between 700,000 and 2.2 million premature deaths annually. Chen et al. (2013) concludes that exposure to air pollution from coal burning in China reduces life expectancy by 5.5 years in northern China and, more generally, in a 3-year reduction in life expectancy. The health problems attributed to inhalation of fly ash are a consequence of the combustion process and can best be addressed through engineering solutions that efficiently remove particulates from the escaping flue gas. However, many other health problems are a direct consequence of the chemical and mineralogical composition of the coal in the ground and the being mined and burned; this is the focus of this review.

The earliest evidence of coal mining comes from China where commercial mining existed several thousand years ago (Schobert 1987). Prior to this time, coal in China and elsewhere was undoubtedly extracted from surface exposures and used to cook food and heat homes. It is equally probable that this early coal use in unvented residential settings resulted in a range of health issues from respiratory problems to trace element exposure. There is ample evidence that these health problems in China persist to modern times (see Table 1). In the following is a discussion of some of these coal-related health problems that have been recognized in recent years (Finkelman et al. 1999; Dai et al. 2007, 2012; Zheng et al. 2010).

3.1. Fluorosis from coal combustion

Exposure to excess fluorine can result in a range of health problems. Typical symptoms of fluorosis include mottling of tooth enamel (dental fluorosis: Figure 1) and various forms of skeletal fluorosis, including limited movement of the joints and spinal curvature (Figure 2). Li et al. (1982) conducted an investigation of fluorosis in Guizhou Province. They reported that a large amount of gas and dust containing fluoride was released by domestic coal combustion and quickly combined with vapour in the air to form an aerosol or a hydrofluoric acid mist, which adhered to the surface of corn and chilli. They also pointed out that fluorine concentrations in the soils in the fluorosis area were higher than those in other areas. The authors demonstrated that fluorine concentrations in fresh corn from Zhijin County, Bijie County, and Guiyang City were 3.60, 2.75, and 5.35 mg/kg, respectively, higher than the world average value. Ando et al. (1998) reported a prevalence of 42.10% for skeletal fluorosis in the endemic area of Pengsui County, Hubei Province, in comparison with a prevalence of 0% in the non-endemic area of Zhaoxian County, Jiangxi Province.

Zheng (1992) pointed out that fluorine emissions from fuel made by coal mixed with clay was larger than from the coal alone and the amount of fluorine released from clay was two times more than that from the coal. Research in collaboration with the U.S. Geological Survey on fluorine and other elements in coal and coal ash found that there was a positive

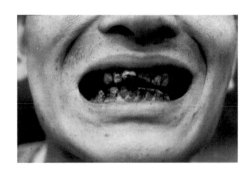

Figure 1. Dental fluorosis.

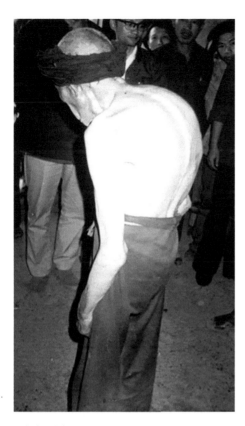

Figure 2. Skeletal fluorosis.

correlation between the concentrations of fluorine and potassium. Because potassium is mainly distributed in clay minerals in coal and coal combustion products, it can be inferred that fluorine is also distributed in the clay minerals in the coal and coal ash and the contents of fluorine increase with increasing proportions of clay (Finkelman *et al.* 2002).

Dai *et al.* (2004) showed that the average fluorine content in the Lopingian coals from western Guizhou Province is from 16.6 to 500 mg/kg, averaging 83.1 mg/kg, which is close to the averages for world coals (80 mg/kg; Ketris and Yudovich 2009) and is a bit lower than that of Chinese coals (130 mg/kg; Dai and Ren 2006). The fluorine content of drinking water and fresh corn is too low to lead to fluorosis in western Guizhou Province (Dai *et al.* 2004). However, the clay used as an additive for coal-burning and as a binder in briquette-making by local residents has a very high content of fluorine, from 101 to 2456 mg/kg (with an average of 1028 mg/kg). The endemic fluorosis is likely caused by fluorine in the clay. The F content of corn and peppers stored in the house is about 1000 and 110 times higher, respectively, than the permitted level of F in foodstuffs according to the Chinese Standard (Dai *et al.* 2007). Further investigations by Dai *et al.* (2007) suggested that in addition to developing F-sequestration technologies, changing the living habits of the residents in the endemic area, for example, washing corn and chillies before cooking

and peeling corn before being pulverized into corn flour, will remove a large proportion of the F, and thus it can play an important role in endemic fluorosis prevention.

Wu *et al.* (2004) demonstrated that coal mixed with clay to form briquettes used as fuel was very common and that the clay was the main source of fluorine causing fluorosis. Based on a survey on the environment and living conditions of fluorosis areas, it was found that local residents had to use the mixture of coal powder and clay as their fuel because the coal powder was cheaper than lump coal.

In recent years, with the improvement in economic conditions, the conditions of fluorosis have greatly changed and the prevalence has gradually decreased (Luo *et al.* 2011). Most of the villages are able to use electric energy, and the usage of bio-energy sources (straw, biogas) has increased. Local residents have used government-funded stoves with chimneys for heating and cooking, and the use of stoves without chimneys has decreased. The staple food of the residents is rice, and corn is primarily used for livestock feed, which greatly reduces the fluoride intake of the residents. However, fluorine-contaminated peppers are now the main source of fluorine intake (Sun *et al.* 2005; Luo *et al.* 2011).

3.2. Arsenosis from coal combustion

Exposure to arsenic can cause serious health problems including hyperpigmentation (flushed appearance, freckles), hyperkeratosis (scaly lesions on the skin, generally concentrated on the hands and feet), Bowen's disease (dark, horny, precancerous lesions of the skin: Figure 3), and squamous cell carcinoma. Endemic arsenosis caused by indoor combustion of high-arsenic coal was first confirmed in Xingren County, Guizhou Province, China, in 1977 and subsequently in Xingyi City, Anlong County, Zhijin Country, and Kaiyang County, all in Guizhou Province and in south Shaanxi Province where local farmers burned 'stone coal' (carbonaceous shale) with high arsenic contents in stoves without chimneys. For references, see Zheng *et al.* (2010). From a survey completed in 2003, there were 142 villages in Guizhou Province and Shaanxi Province using coal containing arsenic higher than 100 ppm. More than 2,400 patients (mostly in Guizhou Province) were found out of nearly 31,000 people tested (Jin *et al.* 2003).

Coal has been produced and used in Guizhou Province for the past 100 years. After the production and utilization of high-As coal, arsenosis became an epidemic disease. The earliest-known arsenosis case was found in 1953 and was called 'Laizi disease', the high-As coal was called Laizi coal, but the local residents did not know that the so-called Laizi disease was caused by arsenic exposure (Zheng *et al.* 1996, 1999).

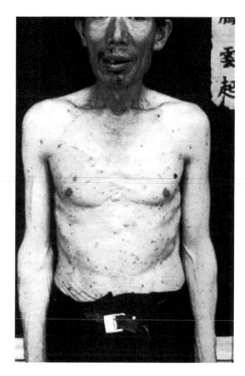

Figure 3. Extensive scaly lesions (hyperkeratosis) are evident on the torso. The dark spot over the left breast was diagnosed as Bowen's disease a precancerous lesion attributed to arsenic exposure.

The practice of using high-As coals indoor without a chimney and good ventilation leads to As pollution of indoor air as well as pollution of crops and/or other foodstuff such as chilli peppers which are preserved and dried indoors over the coal fires. The resident's ingestion of excessive As from polluted indoor air, food, and water was the major cause of arsenosis (Finkelman *et al.* 1999).

The coals have As concentrations typically >100 ppm, commonly >1000 ppm, and in a few selected cases >30,000 ppm (3 wt%) on a whole coal, as-determined basis. The coals that have 3 wt% As are exceptional, as As contents that high are usually restricted to metalliferous samples from ore deposits. The highest concentration of As in coal was 35,000 ppm with much of the arsenic in organic association (Belkin *et al.* 1997). The coals are also highly enriched in Sb (up to 370 ppm), Hg (up to 45 ppm), and Au (up to 570 ppb); U is also high in some samples (up to 65 ppm). A coal-seam channel anthracite from the epidemic disease area, Xingren, Guizhou, contains 2226-ppm As, 3860-ppm Sb, 12.1-ppm Hg, and 7.5-ppm Tl (Dai *et al.* 2006).

Zheng *et al.* (2005) have shown that the higher the As content in the coal and the longer the coal is used, the higher the As in urine and hair and the higher the incidence of the arsenosis. There is a good dose–response relationship between the As exposure and the ratio of arsenosis and but it is much more complicated than that of drinking water arsenosis. The recognition that coal containing high arsenic levels was the primary cause of the arsenosis resulted in a concerted effort to close the mines from which the high-arsenic coals were obtained.

3.3. Selenosis

Intake of excessive selenium can cause selenosis in human and animals, typically manifested in hair and nail loss (Figure 4). Human selenosis in Western Hubei Province could be traced to 1923. A total of 477 cases of human selenosis were reported from 1923 to 1987 and 70% of the cases occurred between 1959 and 1963. Ninety percent of the patients were located in the towns of Shadi, Xintang, and Shuanghe in the Enshi District and the other patients were distributed between Laiwu Town in Enshi, Nantan, and Houmen Towns in Badong County and Shatuo Town in Xuanen County (Zheng *et al.* 2010). Most cases of selenosis were attributed to the residential use of carbonaceous shales (known locally as stone coal) with an average Se content of 143.9 mg/kg (Mou *et al.* 2007) and the maximum content is up to 84,123 mg/kg (Zhu *et al.* 2008).

In the endemic selenosis areas, human activities, such as mining of stone coal for use as a fuel or fertilizer, and discharging lime into cropland to improve soil, caused variable addition of Se to the soil and further accumulation of Se in the food chain.

Hair loss and nail loss were the prime symptoms of endemic selenosis, but disorders of the nervous system, skin, poor dental heath, garlic breath, and paralysis were also reported. Although no health investigations were carried out in the peak incidence years of 1961–1964 in Enshi District, subsequent studies in these areas carried out in the 1970s revealed very high dietary intakes of 3.2–6.8 mg/day with a range of selenium in the blood of 1.3–7.5 mg/L and hair selenium levels of 4.1–100 mg/kg (Yang *et al.* 1983; Tan

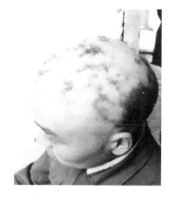

Figure 4. Hair loss due to selenium exposure.

and Hou 1989). Due to decreasing dependence on locally grown foodstuffs in the diet, no human cases of selenium toxicity have been reported since 1987 in these areas but the local animals still frequently suffer hoof and hair loss as a result of the high environmental selenium (Fordyce 2005).

Detailed discussion of coal-related fluorosis, arsenosis, selenosis, and selenosis-related issues can be found in Chen *et al.* (2014).

3.4. Mercury

There is also considerable concern about the health effects of mercury and the proportion of anthropogenic mercury in the environment. So far, there is no direct evidence of health problems caused by mercury released from coal burning in China but there are circumstances where poisoning from mercury released from coal combustion may be occurring. Finkelman *et al.* (1999) report on a village in Guizhou Province, China, where many elderly people suffer from the loss of vision, a symptom that could be attributed to mercury exposure. Mineralogical analysis of the coal being used in the homes of people having visual impairment revealed abundant mercury minerals. Chemical analysis of a coal sample being used in the home of a blind senior citizen indicated a mercury concentration of 55 ppm, about 200 times the average mercury concentration of U.S. and Chinese coals.

Another possible source of exposure to coal-derived uranium (as well as As, Tl, etc.) is the fly ash derived from the Wulantuga coal in Inner Mongolia from which germanium is being extracted (Dai *et al.* 2014).

3.5. Uncontrolled coal fires

Finkelman (2004) noted that uncontrolled release of pollutants from burning coal beds (Figure 5) and waste banks presents potential environmental and human health hazards. The emissions from burning coal beds and waste banks of acidic gases, particulates, organic compounds, and trace elements can contribute to a range of respiratory and other human health problems. Although there are few published reports of health problems caused by these emissions, the potential for problems can be significant. China has more uncontrolled fires that any other country (Stracher *et al.* 2013). Volatile elements such as arsenic, fluorine, mercury, and selenium are commonly enriched in coal deposits. Burning coal beds can volatilize these elements, which then can be inhaled, or adsorbed on crops and foods, taken up by livestock or bioaccumulated in birds and fish. Some of these elements can condense on dust particles that can be inhaled or ingested. In addition, selenium, arsenic, lead, tin, bis-

Figure 5. Uncontrolled coal fire in China (Finkelman 2004). Credit line: Institute Burning coal in Xinjiang Autonomous Region, P. R. China. Photo taken by Zhao Jingyu, Xi'an University of Science and Technology, Xi'an.

muth, fluorine, and other elements condense where the hot gaseous emissions come in contact with ambient air, forming mats of concentrated efflorescent minerals on the surface of the ground. These mats can be leached by rainwater and washed into local water bodies providing other potential routes of exposure.

Liang *et al.* (2014) studied the Wuda coalfield in Inner Mongolia, China, one of the typical coal seam fire areas widespread in northern of China. They found high levels of mercury being released by the coal fire in the near-surface air at the fire zone. They concluded that the continuous and accumulative mercury emission from the coal seam fires may lead to elevated mercury concentrations in the ambient air in adjacent populated urban areas.

3.6. Health impacts of coal mining

Coal mining can mobilize potentially toxic trace elements and organic compounds. These elements and chemicals can degrade the environment and contribute to a wide range of health problems. There have been many studies on these issues including Liang *et al.* (2017), Liao *et al.* (2010), Liu *et al.* (2012), Song *et al.* (2016), Tang *et al.* (2008), Wang *et al.* (2010), and Zheng *et al.* (2011).

In occupational settings, some 440,000 coal miners in China are suffering from Coal Workers Pneumoconiosis (Black Lung Disease) and an estimated 140,000 minors have died of this problem since the 1950s (www.people.com.cn; 18 March 2005) and there are approximately 25,000 new cases each year (Figure 6). Huang *et al.* (2004) have demonstrated a strong correlation between the incidence of black lung disease in the U. S. and the amount of pyrite in the coal. This finding indicates that reduced exposure to respirable pyrite should reduce the incidence of black lung disease in China and elsewhere. Coal miners in China and elsewhere are also at risk of silicosis from inhaling particles of quartz and other silicate minerals liberated in the dust generated by mining operations.

3.7. Lung cancer

3.7.1. Coal smoke and lung cancer

Indoor emissions from household combustion of coal were classified as 'carcinogenic to humans (Group 1)' in 2006, based on sufficient evidence in both humans and experimental animals (Straif et al. 2006). Most of the studies supporting this classification were from China. The carcinogenicity of coal smoke was first noted in the county of Xuan Wei which had the highest rate of female lung cancer mortality of China (Mumford et al. 1987). Within Xuan Wei, there was a large geographic variability in lung cancer rates, which were higher in villages using locally produced bituminous coals, producing especially smoky emissions, than in villages using anthracites and wood. Mouse skin tumorigenicity was stronger for emission from the bituminous coal than anthracite and wood. Inhalation of the bituminous coal smoke also increased the incidence of malignant lung tumours in male and female Kunming mice and Wistar rats (Liang et al. 1988).

3.7.2. Organic carcinogens in coal smoke

Extensive research has been done on the specific chemical components which might be responsible for the carcinogenicity of coal smoke, which is a complex mixture of particulate and gaseous chemical species, including carbon monoxide, nitrogen dioxide, and particulate matter (PM).

The organic extracts of coal smoke have been subjected to extensive chemical analysis and toxicological studies (Mumford et al. 1987, 1990; Chuang et al. 1992). High concentrations of carcinogenic polycyclic aromatic hydrocarbons (PAHs), methylated PAHs, and nitrogen-containing heterocyclic aromatic compounds were found in the emission particles from bituminous coal combustion in open fire pits. The sub-fractions containing alkylated three- and four-ring PAHs was found to contribute to most of the mutagenicity in the PAH fraction in the organic extracts (Chuang et al. 1992). The aromatic fractions from coal combustion contained higher concentrations (Chuang et al. 1992) and more species (Chen 1994) of methylated PAHs than the sample from wood combustion.

3.7.3. Inorganic carcinogens in coal smoke

Some fine particles of refractory inorganic species locked into the organic matrix can be carried into the bulk gas stream with the volatiles during coal devolatilization (Baxter et al. 1997). Some carcinogenic substances in coal, such as Ni, Cr, and quartz, are likely to be released into air, especially during the combustion of lignite and bituminous coals, which have higher contents of volatiles than anthracites. Two examples to emphasize the importance of coal quality in the studies of indoor coal smoke and cancer are the lignite used in Shenyang City of Northern China and the bituminous coal used in Xuan Wei County of Southeast China. Indoor air pollution from coal smoke has been linked to the high female lung cancer rates in these two places (Mumford et al. 1987; Xu et al. 1991). Lignites from Shenbei (Northern Shenyang) Coal Field have high concentrations of Ni (81 ppm) and Cr (86 ppm) and Ni and Cr are found to occur as sulphides (Ren et al. 1999), which has the highest carcinogenic potency relative to other Ni compounds (Oller et al. 1997). Fine quartz grains have also been found in some bituminous coals and the resulting coal smoke in Xuan Wei, China (Tian 2005).

3.7.4. Nanometer-scale quartz in coal smoke

Epidemiological studies in Xuan Wei, China, revealed a closer association of lung cancer with the indoor burning of bituminous coal (as opposed to anthracite or wood; Figure 7) than with tobacco use or occupation (Mumford et al. 1987). Even within the large category of bituminous coal, there was a marked heterogeneity in lung cancer risk associated with specific subtypes of bituminous coal used in Xuan Wei (Lan et al. 2008; Downward et al. 2014). The large variabilities in both coal type and lung cancer risk constitute a unique natural experiment into the chemical carcinogenesis due to coal smoke exposure. Fifteen different coals

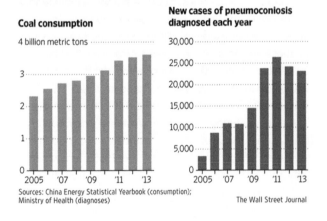

Occupational Hazard

China has been struggling with coal workers' pneumoconiosis, or black lung disease, as the country's coal usage rises.

Coal consumption

New cases of pneumoconiosis diagnosed each year

Sources: China Energy Statistical Yearbook (consumption); Ministry of Health (diagnoses)

The Wall Street Journal

Figure 6. Incidence of black lung disease (coal workers' pneumoconiosis) in China.

Figure 7. Indoor combustion of bituminous coal in Xuan Wei County, Yunnan province, P.R. China.

everyday conversation among average Chinese people. In 2013, the average concentration of $PM_{2.5}$ for China was 54 $\mu g/m^3$, far above the World Health Organization Air Quality Guideline of 10 $\mu g/m^3$ (Brauer *et al.* 2016). Coal burning accounts for 9%–21% of the $PM_{2.5}$ mass in northern and western China probably because of the extensive use of coal in residential heating (Huang *et al.* 2014). Around 37% of the sulphate in $PM_{2.5}$ forms primarily through atmospheric oxidation of SO_2 emitted from coal burning. Besides contributing to the $PM_{2.5}$ mass, coal burning emissions also dominate the hazardous chemical species including PAHs and heavy metals such as lead and arsenic. Exposure to $PM_{2.5}$ contributed to 2.9 million premature deaths in 2013 globally; among them, 916,000 deaths took place in China (Guan *et al.* 2014; Ma *et al.* 2017). Coal burning was found to be the single largest contributor to ambient $PM_{2.5}$ pollution in China, causing an estimated 366,000 deaths in 2013(GBD MAPS Working Group 2016).

from this area were burned in an open fire simulating the household fire pit (Tian *et al.* 2008b) and the particulate emissions were collected for analysing four groups of suspected carcinogens: PAHs, free radicals (Tian *et al.* 2009a), transition metals including iron, and quartz. Only quartz, estimated by the elemental concentrations of Si and Al, was found to be associated with elevated lung cancer risks (Tian 2005).

Abundant quartz particles in the lung cancer tissues were detected although the quartz deposition is not as serious as to be diagnosed as pneumoconiosis or silicosis. There was also a clear spatial correspondence of high lung cancer rates to areas that produce and utilize coal from the uppermost Permian (C1) coal. It was hypothesized that the availability and usage of the C1 coal, with high concentrations of nanometer-scale quartz grains well mixed within the organic matrix, is responsible for the lung cancer epidemic in Xuan Wei (Tian *et al.* 2008a). The End-Permian C1 coal, precisely adjacent to the Permo-Triassic Boundary, contains unusually high concentrations of authigenic quartz and this quartz is particularly fine grained (Dai *et al.* 2008; Large *et al.* 2009). The high concentrations of authigenic quartz in C1 coal was explained by the interaction of PTB groundwater with the Emeishan Basalts, an interaction highly localized in China.

3.8. Premature deaths

Coal combustion emissions from both industry and households contribute to the serious ambient air pollution in recent years, which is a great public health threat in China. The frequent smog events in large cities and high levels of fine particulate air pollution ($PM_{2.5}$) are part of

3.9. Iodine deficiency disease (IDD), Guizhou province, China – a possible positive impact

Guizhou Province in southwest China is a mountainous province enriched with limestone and coal and is far from the sea, with the area of 174,700 km^2 and the population of 40 million. IDD was prevalent in 57 counties in the southeast of the province. In 1982, there were 2.09 million patients suffered from endemic goitre, and 0.21 million cretinism patients in some endemic areas.

In western Guizhou Province, there were few IDD cases. The reasons for this phenomenon are not clear. Although one explanation may be the difference in the distribution of coal resources and coal use patterns. The coals in the region are exceptionally enriched in iodine – 8 ppm versus a global average of about 1 ppm. Communities using these high-iodine coals appear to have lower incidence of IDD than communities using wood as their main energy source (Wang et al., 2005).

3.9.1. What is being done and what can be done

Improper use of automobiles certain medicines cause widespread and significant health problems but the benefits of automobile use and medications clearly outweigh the negatives. Nevertheless, we strive to minimize the unintended deleterious impacts of automobile use and medicine abuse by making cars and roadways safer, improving signage, producing child-safe bottles, better labelling, etc. The same approach should be taken with regard to coal use. We should make every effort to make coal use and coal mining as safe as possible. Commercial and industrial coal-burning power plants should be required to use the latest pollution control systems to minimize the release of SO_2, $PM_{2.5}$, and potentially harmful trace elements. Domestic coal use should be discouraged but if no alternatives exist, commercial briquettes should be substituted for local raw coal. Stove improvement has played a role in reducing indoor air pollution due to coal combustion (Tian *et al.* 2009b; Lan *et al.* 2002). Nevertheless, lung cancer rates and other disease rates continue to rise (Lin *et al.* 2015), so still more needs to be done. Stove improvement or adding a chimney does not solve the household air pollution problem completely. The chimney has vented the smoke outdoors and contributed to the community air pollution (Liu *et al.* 2016), especially when the air dispersion conditions are poor. The particularly hazardous coal types can be used for certain industries which are capable of special treatment before usage, whereas the more sustainable environmental and public health protection measures in the households would be movement up the energy ladder from coal to cleaner liquid/gaseous fuels, or other renewable energy sources. China is one of the leading nations in developing renewable energy resources (http://www.abc.net.au/news/2017-03-02/china-coal-cuts-and-renewables-transform-climate-change-leader/8316660), thus reducing its dependence on coal. This transition to renewables will undoubtedly have a positive impact on the health problems caused by coal.

4. Concluding remarks

Coal use in China has had a long and unfortunate history of causing widespread and severe health problems. Commercial coal combustion, coal mining, and disposal of coal wastes and coal combustion products contribute to a range of health issues. However, most of the more serious coal-related health problems have resulted from residential coal use. Millions of rural residents have been affected by dental and skeletal fluorosis from residential use of high fluorine briquettes. Many thousands of people suffer from hyperpigmentation to fatal squamous cell carcinoma caused by burning coal enriched in arsenic in their homes. High-selenium stone coal has caused human and animal selenosis. Mercury, quartz, and organic compounds mobilized from coal combustion present dangers to human health as do the widespread uncontrolled coal fires, and exposure of coal miners to pyrite, quartz, and fine particulates. China and many other countries should strive to make coal use and coal mining as safe as possible. For example, (1) industries and power plants using coal should be required to use the latest pollution control systems to minimize the release of potentially harmful effluents; (2) domestic coal use should be discouraged but if no alternatives exist, commercial briquettes should be substituted for local raw coal; (3) stove improvement that has played a role in reducing indoor air pollution should be encouraged; (4) China's emphasis on developing energy resources and thus reducing its dependence on coal should be supported by all. This transition to renewables will undoubtedly have a positive impact on the health problems caused by coal.

Acknowledgements

We wish to acknowledge Baoshan Zheng who was a leading pioneer in researching the links between coal use and health issues in China. We also thank Shifeng Dai and anonymous reviewers for their comments that helped improve this manuscript.

Disclosure statement

No potential conflict of interest was reported by the authors.

ORCID

Robert B. Finkelman http://orcid.org/0000-0002-7295-5952

References

Ando, M., Tadano, M., Asanuma, S., Tamura, K., Matsushima, S., Watanabe, T., Kondo, T., Sakurai, S., Ji, R., Liang, C., and Cao, S., 1998, Health effects of indoor fluoride pollution from coal burning in China: Environmental Health Perspectives, v. 106, p. 239–244. doi:10.1289/ehp.98106239

Baxter, L.L., Mitchell, R.E., and Fletcher, T.H., 1997, Release of inorganic material during coal devolatilization: Combustion and Flame, v. 108, no. 4, p. 494–502. doi:10.1016/0010-2180(95)00120-4

Belkin, H.E., Zheng, B., Zhou, D., and Finkelman, R.B., 1997, Preliminary results on the geochemistry and mineralogy of arsenic in mineralized coals from endemic arsenosis in Guizhou Province, P.R. China, *in* Proceedings of the Fourteenth Annual International Pittsburgh coal conference and workshop: CD-ROM, Pittsburgh, PA: University of Pittsburgh, p. 1–20.

Brauer, M., Freedman, F., Frostad, J., Donkelaar, A., Martin, R.V., Dentener, F., and Van Dingenen, R., 2016, Ambient air pollution exposure estimation for the global burden of disease 2013: Environmental Science & Technology, v. 50, no. 1, p. 79–88. doi:10.1021/acs.est.5b03709

Chen, B., 1994, Identifying the major compounds in coal emission extract of Xuanwei by GC-MS, *in* He, X., and Yang, R., eds, Lung cancer and indoor air pollution from coal burning: Kunming, Yunnan Science and Technology Publishing House, p. 168–190.

Chen, J., Liu, G., Kang, Y., Wu, B., Sun, R., Zhou, C., and Wu, D., 2014, Coal utilization in China: Environmental impacts and human health: Environmental Geochemistry and Health, v. 36, p. 735–753. doi:10.1007/s10653-013-9592-1

Chen, Y., Ebenstein, A., Greenstone, M., and Li, H., 2013, Evidence on the impact of sustained exposure to air pollution on life expectancy from China's Huai River policy:

Proceedings National Academy of Sciences, v. 110, no. 32, p. 12936–12941. doi:10.1073/pnas.1300018110

Chuang, J.C., Wise, S.A., Cao, S., and Mumford, J.L., 1992, Chemical characterization of mutagenic fractions of particles from indoor coal combustion - a study of lung cancer in Xuan Wei, China: Environmental Science & Technology, v. 26, no. 5, p. 999–1004. doi:10.1021/es00029a020

Dai, S., Li, W., Tang, Y., Zhang, Y., and Feng, P., 2007, The sources, pathway, and preventive measures for fluorosis in Zhijin County, Guizhou, China: Applied Geochemistry, v. 22, p. 1017–1024. doi:10.1016/j.apgeochem.2007.02.011

Dai, S., and Ren, D., 2006, Fluorine concentration of coals in China—An estimation considering coal reserves: Fuel, v. 85, p. 929–935. doi:10.1016/j.fuel.2005.10.001

Dai, S., Ren, D., Chou, C.-L., Finkelman, R.B., Seredin, V.V., and Zhou, Y., 2012, Geochemistry of trace elements in Chinese coals: A review of abundances, genetic types, impacts on human health, and industrial utilization: International Journal of Coal Geology, v. 94, p. 3–21. doi:10.1016/j.coal.2011.02.003

Dai, S., Ren, D., and Ma, S., 2004, The cause of endemic fluorosis in western Guizhou Province, Southwest China: Fuel, v. 83, p. 2095–2098. doi:10.1016/j.fuel.2004.03.016

Dai, S., Seredin, V.V., Ward, C.R., Jiang, J., Hower, J.C., Song, X., Jiang, Y., Wang, X., Gornostaeva, T., Li, X., Liu, H., Zhao, L., and Zhao, C., 2014, Composition and modes of occurrence of minerals and elements in coal combustion products derived from high-Ge coals: International Journal of Coal Geology, v. 121, p. 79–97. doi:10.1016/j.coal.2013.11.004

Dai, S., Tian, L., Chou, C.-L., Zhou, Y., Zhang, M., Zhao, L., Wang, J., Yang, Z., Cao, H., and Ren, D., 2008, Mineralogical and compositional characteristics of Late Permian coals from an area of high lung cancer rate in Xuan Wei, Yunnan, China: Occurrence and Origin of Quartz and Chamosite: International Journal of Coal Geology, v. 76, no. 4, p. 318–327.

Dai, S., Zeng, R., and Sun, Y., 2006, Enrichment of arsenic, antimony, mercury, and thallium in a Late Permian anthracite from Xingren, Guizhou, Southwest China: International Journal of Coal Geology, v. 66, p. 217–226. doi:10.1016/j.coal.2005.09.001

Downward, G.S., Hu, W., Large, D., Veld, H., Xu, J., Wu, G., Wei, F., Capman, R.S., Rithman, N., Qing, L., and Vermeulen, R., 2014, Heterogeneity in coal composition and implications for lung cancer risk in Xuanwei and Fuyuan counties, China: Environment International, v. 68, p. 94–104. doi:10.1016/j.envint.2014.03.019

Finkelman, R., 2004, Potential Human Health Impacts of Burning Coal Beds and Waste Piles International Journal of Coal Geology, v. 59, nos. 1–2, p. 19–24. doi:10.1016/j.coal.2003.11.002

Finkelman, R.B., Belkin, H.E., and Zheng, B., 1999, Health impacts of domestic coal use in China: Proceeding of the National Academy of Sciences, v. 96, p. 3427–3431. doi:10.1073/pnas.96.7.3427

Finkelman, R.B., 1993, Trace and minor elements in coal., in Engel, M.H., and Macko, S.A., eds., Organic geochemistry: New York, Plenum Press, p. 593–607.

Finkelman, R.B., Orem, W., Castranova, V., Tatu, C.A., Belkin, H. E., Zheng, B., Lerch, H.E., Maharaj, S.V., and Bates, A.L., 2002, Health impacts of coal and coal use: Possible solutions:

International Journal of Coal Geology, v. 50, p. 425–443. doi:10.1016/S0166-5162(02)00125-8

Fordyce, F., 2005, Selenium deficiency and toxicity in the environment, in Selinus, O., Alloway, B.J., Centeno, J.A., Finkelman, R.B., Fuge, R., Lindh, U., and Smedle, P., eds., Essentials of Medical Geology: New York: Elsevier, p. 373–415.

GBD MAPS Working Group, 2016, HEI Special Report 20: Burden of Disease Attributable to Coal-Burning and Other Major Sources of Air Pollution in China. https://www.healtheffects.org/publication/burden-Disease-Attributable-Coal-Burning-and-Other-Air-Pollution-Sources-China.

Guan, D.B., Su, X., Zhang, Q., Peters, G.P., Liu, Z., Lei, Y.U., and He, K.B., 2014, The socioeconomic drivers of China's primary $PM_{2.5}$ emissions: Environtal Research Letters, v. 9, p. 24010. doi:10.1088/1748-9326/9/2/024010

Huang, R.J., Zhang, Y., Bozzetti, C., Ho, K.F., Cao, J.J., Han, Y., and Daellenbach, K.R., plus 19 additional authors, 2014, High secondary aerosol contribution to particulate pollution during haze events in China: Nature, v. 514, no. 7521, p. 218–222.

Huang, X., Li, W., Attfield, M. D., Nadas, A., Frenkel, K., and Finkelman, R. B., 2004, Mapping and prediction of coal workers' pneumoconiosis with bioavailable iron content in bituminous coals. Environmental Health Perspectives, v. 113, no. 8, p. 964–968.

Jin, Y., Liang, C., He, G., Cao, J., Ma, F., Wang, H., Ying, B., and Ji, R., 2003, A Study on distribution of endemic arsenism in China: Journal of Hygiene Research, v. 32, no. 6, p. 519–540 (in Chinese with English abstract).

Ketris, M.P., and Yudovich, Y.E., 2009, Estimates of Clarkes for carbonaceous bioliths: Word averages for trace element contents in black shales and coals: International Journal of Coal Geology, v. 78, no. 2, p. 135–148. doi:10.1016/j.coal.2009.01.002

Lan, Q., Chapman, R.S., Schreinemachers, D.M., Tian, L.W., and He, X.Z., 2002, Household stove improvement and risk of lung cancer in Xuanwei, China: Journal of the National Cancer Institute, v. 94, no. 11, p. 826–835. doi:10.1093/jnci/94.11.826

Lan, Q., He, X., Shen, M., Tian, L., Liu, L.Z., Chen, W., Berndt, S.I., Hosgood, H.D., Lee, K.-M., Zheng, T., Blair, A., and Chapman, R.S., 2008, Variation in lung cancer risk by smoky coal subtype in Xuanwei, China: International Journal of Cancer, v. 123, no. 9, p. 2164–2169. doi:10.1002/ijc.23748

Large, D.J., Kelly, S., Spiro, B., Tian, L., Shao, L., Finkelman, R., Zheng, M., Somerfield, C., Plint, S., Ali, Y., and Zhou, Y., 2009, Silica-volatile interaction and the geological cause of the Xuan Wei lung cancer epidemic: Environmental Science & Technology, v. 43, no. 23, p. 9016–9021. doi:10.1021/es902033j

Li, R.B., Tan, J.W., and Wang, W.Y., 1982, Discussion on fluorosis induced by foodstuff in Guizhou Province: China Journal of Medicine, v. 62, no. 7, p. 425–428.

Liang, C.K., Quan, N.Y., Cao, S.R., He, X.Z., and Ma, F., 1988, Natural inhalation exposure to coal smoke and wood smoke induces lung cancer in mice and rats: Biomedical and Environmental Sciences, v. 1, no. 1, p. 42–50.

Liang, J., Feng, C., Zeng, G., Gao, X., Zhong, M., Li, X., Li, X., He, X., and Fang, Y., 2017. Spatial distribution and source identification of heavy metals in surface soils in a typical coal mine city, Lianyuan, China. Environmental Pollution, v. 225, p. 681–690. doi:10.1016/j.envpol.2017.03.057

Liang, Y., Liang, H., and Zhu, S., 2014, Mercury emissions from coal seam fires at Wuda Inner Mongolia, China: Atmospheric Environment, v. 83, p. 176–184. doi:10.1016/j.atmosenv.2013.09.001

Liao, Y., Wang, J., Wu, J., Driskell, L., Wang, W., Zhang, T., Xue, G., and Zheng, X., 2010. Spatial analysis of neural tube defects in a rural coal mining area. International Journal of Environmental Health Research, v. 20, p. 439–450. doi:10.1289/ehp.9268

Lin, H.L., Ning, B.F., Li, J.H., Zhao, G.Q., Huang, Y.C., and Tian, L.W., 2015, Temporal trend of mortality from major cancers in Xuanwei, China: Frontiers of Medicine, v. 9, no. 4, p. 487–495. doi:10.1007/s11684-015-0413-z

Liu, J., Liu, G., Zhang, J., Yin, H., and Wang, R., 2012. Occurrence and risk assessment of polycyclic aromatic hydrocarbons in soil from the tiefa coal mine district, Liaoning, China, Journal of Environmental Monitoring, v. 14, no. 10, p. 2634–2642. doi:10.1021/es981203e

Liu, J., Mauzerall, D.L., Chen, Q., Zhang, Q., Song, Y., Peng, W., Zbigniew, K., Qui, X., Zhang, S., Hu, M., Lin, W., Smith, K.R., and Zhu, T., 2016, Air pollutant emissions from Chinese Households: a major and underappreciated ambient pollution source: Proceedings of the National Academy of Sciences, v. 113, no. 28, p. 7756–7761. doi:10.1073/pnas.1604537113

Lockwood, A.H., Welker-Hood, K., Rauch, M., and Gottlieb, B., 2009, Coal's assault on human health: Washington, D.C: Physicians for Social Responsibility, p. 66.

Luo, K., Li, L., and Zhang, S., 2011, Coal-burning roasted corn and chili as the cause of dental fluorosis for children in southwestern China: Journal of Hazardous Materials, v. 185, p. 1340–1347. doi:10.1016/j.jhazmat.2010.10.052

Ma, Q., Cai, S.Y., Wang, S.X., Zhao, B., Martin, R.V., Brauer, M., Cohen, A., Jiang, J.K., Zhou, W., Hao, J.M., Frostad, J., Forouzanfar, M.H., and Burnett, R.T., 2017, Impacts of coal burning on ambient $PM_{2.5}$ pollution in China: Atmospheric Chemistry and Physics, v. 17, p. 4477–4491. doi:10.5194/acp-17-4477-2017

Mou, S.H., Hu, Q.T., and Yan, L., 2007, Progress of researches on endemic selenosis in Enshi District, Hubei Province: Chinese Journal of Public Health, v. 23, no. 1, p. 95–96 (in Chinese).

Mumford, J.L., He, X.Z., Chapman, R.S., Cao, S.R., Harris, D.B., and Li, X.M., 1987, Lung cancer and indoor air pollution in Xuan Wei, China: Science, v. 235, no. 4785, p. 217–220. doi:10.1126/science.3798109

Mumford, J.L., Helmes, C.T., Lee, X., Seidenberg, J., and Nesnow, S., 1990, Mouse skin tumorigenicity studies of indoor coal and wood combustion emissions from homes of residents in Xuan Wei, China with high lung cancer mortality: Carcinogenesis, v. 11, no. 3, p. 397–403. doi:10.1093/carcin/11.3.397

Oller, A.R., Costa, M., and Oberdorster, G., 1997, Carcinogenicity assessment of selected nickel compound:: Toxicology and Applied Pharmacology, v. 143, no. 1, p. 152–166. doi:10.1006/taap.1996.8075

Ren, D.Y., et al., 1999, Distributions of minor and trace elements in Chinese coals: International Journal of Coal Geology, v. 40, no. 2–3, p. 109–118. doi:10.1016/S0166-5162(98)00063-9

Schobert, H.H., 1987, Coal: The energy source of the past and future: Washington, D.C: American Chemical Society, p. 298.

Song, X., Yang, S., Shao, L., Fan, J., and Liu, Y., 2016. Pm10 mass concentration, chemical composition, and sources in the typical coal-dominated industrial city of Pingdingshan, China. Science of the Total Environment, v. 571, p. 1155–1163. doi:10.1016/j.scitotenv.2016.07.115

Stracher, G.B., Kuenzer, C., Hecker, C., Zhang, J., Schroeder, P.A., and McCormack, J.K., 2013, Wuda and Ruqigou coalfield fires in Northern China, in Stracher, G.B., Prakash, A., and Sokol, E.V., eds., Chapter 4 in coal and peat fires: A global perspective, Vol. 2, Amsterdam: Elsevier, p. 554.

Straif, K., Baan, R., Grosse, Y., Secretan, B., and El Ghissassi, F., 2006, Carcinogenicity of household solid fuel combustion and of high-temperature frying: The Lancet Oncology, v. 7, no. 12, p. 977–978. doi:10.1016/S1470-2045(06)70969-X

Sun, D., Zhao, X., and Chen, X., 2005, Investigation on endemic fluorosis in China: Beijing, People's Health Press.

Tan, J.A., and Hou, S., 1989, Environmental selenium and health problems in China, in Tan, J., et al., eds., Environmental selenium and health: Beijing, People Health Press.

Tang, D., Li, T., Liu, J.J., Zhou, Z., Yuan, T., Chen, Y., Rauh, V.A., Xie, J., and Perera, F., 2008. Effects of prenatal exposure to coal-burning pollutants on children's development in China. Environmental Health Perspect, v. 116, p. 674–679. doi:10.1289/ehp.10471

Tian, L.W., 2005. Coal combustion emissions and lung cancer in Xuan Wei, China [PhD dissertation]:, Berkeley, CA: University of California.

Tian, L.W., Dai, S., Wang, J., Huang, Y., Ho, S.C., Zhou, Y., Lucas, D., and Koshland, C.P., 2008a, Nanoquartz in Late Permian C1 coal and the high incidence of female lung cancer in the Pearl River Origin area: A retrospective cohort study: BMC Public Health, v. 8, p. 398. doi:10.1186/1471-2458-8-398

Tian, L.W., Koshland, C.P., and Yano, J., Yachandra, V.K., Ignatius, T.S., Lee, S.C.Y., and Lucas, D., 2009a, Carbon-centered free radicals in particulate matter emissions from wood and coal combustion: Energy & Fuels, v. 23, no. 5, p. 2523–2526. doi:10.1021/ef8010096

Tian, L.W., Lan, Q., Yang, D., He, X., Ignatius, T.S.Y., and Hammond, S.K., 2009b, Effect of chimneys on indoor air concentrations of PM10 and Benzo[a]pyrene in Xuan Wei, China: Atmospheric Environment, v. 43, no. 21, p. 3352–3355. doi:10.1016/j.atmosenv.2009.04.004

Tian, L.W., Lucas, D., and Kashland, C.P., 2008b, Particle and gas emissions from a simulated fire pit: Environmental Science & Technology, v. 42, no. 7, p. 2503–2508. doi:10.1021/es0716610

Wang, Binbin, Jackson, John C., Palmer, C., Zheng, B., and Finkelman, R. B., 2005, Evaluation on determination of iodine in coal by energy dispersive x-ray fluorescence. geochemistry Journal, v. 39, p. 391–394. doi:10.2343/geochemj.39.391

Wang, R., Liu, G., Chou, C.L., Liu, J., and Zhang, J., 2010. Environmental assessment of pahs in soils around the anhui coal district, China, Archives of Environmental Contamination and Toxicology, v. 59, no. 1, p. 62–70. doi:10.1007/s00244-009-9440-6

Wu, D., Zheng, B., and Wang, A., 2004, New understanding on the source of coal combustion type fluorosis in Guizhou Province: China Journal of Endemiology, v. 23, no. 2, p. 135–137.

Xu, Z.Y., Bolt, W.J., Li, G., Fraumeni, J.J.F., Zhao, D.Z., Stone, B.J., Yin, Q., Wu, A., Henderson, B.E., and Guan, B.P., 1991, Environmental determinants of lung cancer in Shenyang, China. IARC Sci Publ: Relevance to Human Cancer of N-Nitroso Compounds, Tobacco Smoking and Mycotoxins, v. 105, p. 460–465.

Yang, G.Q., Wang, S.Z., Zhou, R.H., and Sun, S.Z., 1983, Endemic selenium intoxication of humans in China: American Journal of Clinical Nutrition, v. 37, p. 872–881.

Zheng, B., 1992, Research on endemic fluorosis and industrial fluorine pollution: Beijing, Chinese Environmental Science Pres, p. 151–194.

Zheng, B., Wang, B., Ding, Z., Zhou, D., Zhou, Y., Zhou, C., Chen, C., and Finkelman, R.B., 2005, Endemic arsenosis caused by indoor combustion of high-As coal in Guizhou Province, P.R. China: Environmental Geochemistry and Health, v. 27, p. 521–528. doi:10.1007/s10653-005-8624-x

Zheng, B., Wang, B., and Finkelman, R.B., 2010, Medical geology in China: Then and now. In medical geology: A regional synthesis, *in* Selinus, O., Centeno, J.A., and Finkelman, R.B., eds., Medical geology – A regional synthesis: New York: Springer, p. 303–327.

Zheng, B., Yu, X., Zhang, J., and Zhou, D., 1996, Environmental geochemistry of coal and endemic arsenism in Southwesth Guizhou: China, Abstracts of 30th International Geological Congress Beijing, China, p. 410.

Zheng, B., Zhang, J., Ding, Z., Yu, X., Wang, A., Zhou, D., Mao, D., and Su, H., 1999, Issues of health and disease relating to coal use in southwestern China: International Journal of Coal Geology, v. 40, p. 119–132. doi:10.1016/S0166-5162(98)00064-0

Zheng, X., Pang, L., Wu, J., Pei, L., Tan, L., Yang, C., and Song, X., 2011, Contents of heavy metals in arable soils and birth defect risks in Shanxi, China: A small area level geographic study: Population Environment, v. 33, p. 259–268. doi:10.1007/s11111-011-0138-0

Zhu, J.M., Wang, N., Li, S.H., Li, L., Su, H., and Liu, C., 2008, Distribution and transport of selenium in Yutangba: China: Impact of Human Activities: Science of the Total Environment, v. 392, p. 252–261. doi:10.1016/j. scitotenv.2007.12.019

Addendum

Finkelman, R., 2004, Potential human health impacts of burning coal beds and waste banks: International Journal of Coal Geology, v. 59, nos. 1–2, p. 19–24.

Valuable elements in Chinese coals: a review

Shifeng Dai ⓘ, Xiaoyun Yan ⓘ, Colin R. Ward ⓘ, James C. Hower ⓘ, Lei Zhao ⓘ, Xibo Wang ⓘ, Lixin Zhao ⓘ, Deyi Ren ⓘ and Robert B. Finkelman ⓘ

ABSTRACT

China is, and in the coming decades should continue to be, the largest producer and user of coal in the world. The high volume of coal usage in China has focused attention not only on the toxic trace elements that may be released from coal combustion but also on the valuable elements that may occur in the coal and associated ash. Valuable elements in several coals (or coal ashes) and some coal-bearing strata in China (e.g. Ge, Ga, U, rare earth elements and Y, Nb, Zr, Se, V, Re, Au, and Ag, as well as the base metal Al) occur at concentrations comparable to or even higher than those in conventional economic deposits. Several factors are responsible for these elevated concentrations: (1) injection of exfiltrational solutions during peat accumulation or as part of later epigenetic activity; (2) injection of infiltrational epigenetic solutions; (3) introduction of syngenetic alkali volcanic ashes into the peat-forming environment or into associated non-coal-forming terrestrial environments; (4) input of terrigenous materials into the coal-forming environment; (5) leaching of non-coal partings by groundwater/hydrothermal solutions; and (6) mixed processes involving both hydrothermal solutions and volcanic ash. The valuable elements in Chinese coals may be associated with either the organic matter or mineral matter, or have a mixed organic- and inorganic-affinity. For example, the Ge and U in coal-hosted ore deposits dominantly occur in the organic matter, with only traces of U-bearing minerals being present; gallium mainly occurs in boehmite and kaolinite, and to a lesser extent, in the organic matter. Rare earth elements and Y occur as carbonate-minerals (e.g. florencite, parisite), phosphate-minerals (e.g. rhabdophane, silico-rhabdophane, and xenotime), and in part are associated with the organic matter. Some metals (e.g. Ge, Al, Ga) have been successfully extracted at an industrial scale from Chinese coals, and others have significant potential for such extraction. Major challenges remaining for coal scientists include the development of economic extraction methods from coal ash, and the control of toxic elements released during the metal extraction process to protect human health and to avoid environmental pollution.

1. Introduction

Coal is a precious resource in many countries around the world. China has relatively abundant coal resources, and is and will continue to be the largest coal producer and user in the foreseeable future. From a practical point of view, coal is not only a resource primarily used for power generation; substantial quantities are employed in gasification, in iron and steel production, and as raw materials for activated carbon and other chemicals. Coal can also be considered as a source for valuable metals, such as the base metal Al and rare metals including but not limited to Sc, V, Ga, Ge, Se, Zr, Nb, rare earth elements and Y (REY, or REE if Y is not included), Au, Ag, Re, and U. The economic values of

valuable trace elements in coal have been recognized for more than 100 years. Coals from Wyoming and Utah of the USA had been used for Au and Ag recovery in the late nineteenth and early twentieth centuries (Jenney 1903; Stone 1912). Early in 1933, Goldschmidt and Peters (1933) found highly elevated concentrations of Ge in several coal samples and concluded that there was an opportunity to economically extract Ge from coals. In post–World War II years, U-rich coal was one of the major sources of uranium for the nuclear industry in the Soviet Union and the USA (Kislyakov and Shchetochkin 2000). In the early 1960s, germanium was successfully recovered from coal ashes in the USSR, Czechoslovakia (Jandova and Vu 2001), the

United Kingdom, and Japan (Swaine 1990). Ge-rich coals are currently being utilized as raw sources of Ge in Russia and China (Dai *et al.* 2014b). Germanium provides the most successful example of industrial rare metal extraction from coal ash in more than 50 years (Seredin *et al.* 2013). The Ge extracted from those three coal-hosted Ge ore deposits accounts for more than 50% of the total industrial Ge production in the world (Dai *et al.* 2014b). In recent years, the occurrence of these valuable metals in coal has attracted much attention (e.g. Scott *et al.* 2015; Taggart *et al.* 2015, 2016; Hower *et al.* 2016a), mainly because their concentrations in some coals and/or coal combustion products (e.g. fly ash) are comparable to or even higher than those found in conventional types of corresponding metal ores (Seredin and Finkelman 2008; Seredin and Dai 2012; Seredin *et al.* 2013; Dai *et al.* 2016d).

The terms 'metalliferous coal' or 'coal-hosted ore deposit' have been used to describe coals with practical economic significance for valuable metal production (Seredin and Finkelman 2008; Dai *et al.* 2014d, 2016a). These terms usually refer to coals having valuable element concentrations at least 10 times higher than the respective averages for world coals (Seredin and Finkelman 2008). Recent studies have shown that the valuable metals are enriched not only in the coals themselves, but also in some of the non-coal bands within the coal bed (Hower *et al.* 1999; Dai *et al.* 2014a; Zhao *et al.* 2015; Zhao and Graham 2016) and in the strata above and below certain coal seams (Seredin and Finkelman 2008; Dai *et al.* 2010b, 2014a).

Today, numerous technologies and devices rely on these valuable elements such as the REY, Sc, Ga, Ge, etc. (US Department of Energy 2011; Pecht *et al.* 2012; European Commission 2014). Due to the growing application of these elements in modern technology, many countries are developing strategies to obtain or develop additional sources of rare-metal materials (European Commission 2014; Hower *et al.* 2016a). While traditional mining has typically provided the majority of rare metals (e.g. REY) for industrial use, current limitations on developing new mines have resulted in the search for alternative sources, including coal and coal combustion products (Seredin and Dai 2012; Mayfield and Lewis 2013; Seredin *et al.* 2013; Hower 2016a; Dai *et al.* 2016d).

In China, a number of coal-hosted metalliferous ore deposits have been discovered and have been subjected to various investigations, aimed at evaluating the distribution, origin, modes of occurrence, extraction methods, and economic evaluation of the valuable metals, as well as assessing the potential environmental and human health issues arising from release of toxic elements during coal combustion and metal extraction.

Coal-hosted ore deposits include those containing Ge, Ga, Al, REY, Se, U V, Zr, Nb, Mo, and Re, as well as assemblages of elements such as Al–Ga–REY, REY–Zr (Hf)–Nb(Ta)–Ga, Mo–Se–U–Re–(REY), and Ge–U. Based on published literature combined with new data from our own studies, we have reviewed the distribution, origin, and modes of occurrence of valuable metals in coal-hosted ore deposits, not only as a basis for future studies for Chinese coal scientists but also as a reference for comparative investigations on this subject for worldwide coal and mineral resource researchers.

2. Average contents of valuable metals in common coals

Since the 1930s there have been a number of attempts to calculate average concentrations of metals in common coals (e.g. Goldschmidt 1935; Valkovič 1983a; 1983b; Swaine 1990; Bouška *et al.* 1995; Bragg *et al.* 1998; Ketris and Yudovich 2009) or in coals from certain areas or countries, such as lignite in Czech Republic (Bouška and Pešek 1999), US coals (Finkelman 1993), Chinese coals (Tang and Huang 2004; Bai *et al.* 2007; Dai *et al.* 2012b); UK coals (Spears and Zheng 1999), and Turkish coals (Palmer *et al.* 2004). These concentrations have been referred to as Clarke values (e.g. Ketris and Yudovich 2009), average contents (e.g. Finkelman 1993; Ren *et al.* 2006), abundance (Tang and Huang 2004), or background values (Dai *et al.* 2012b). Table 1 lists the average concentrations of valuable trace elements in common coals or in coals from certain areas.

3. Coal-hosted ore deposits of rare metals in China

Coal-hosted metal ore deposits are mainly distributed in southwestern, northern, northwestern, and northeastern China (Figure 1). Although the concentrations of some metals have been found to be highly elevated in some Chinese coals, as shown in Figure 1, not all the metals in Chinese coals and coals from other areas (such as Cu, Zn, Tl, Pb, Bi, Ni, Co, As, and Sn) could be considered as raw sources for industrial utilization in the foreseeable future (Seredin and Finkelman 2008; Dai *et al.* 2014b). This is because (1) generally they may not be concentrated in coal to a level of economic significance, (2) conventional ore deposits contain substantial reserves of those metals relative to coal deposits, and (3) in a few cases, the elements may occur with sufficient concentrations but the extraction procedures may be so costly that extraction is not currently economic. Some preliminary factors should therefore be considered when the economic significance of metals

Table 1. The average content of selected metals in common coals or in coals from certain areas (ppm unless as indicated).

Elements	World coals			US coals		Chinese coals			UK coals		Turkey coals
	Low-rank coal	Hard coal	Total	AM	GM	C1	C2	C3	AM	GM	
Sc	4.1 ± 0.2	3.7 ± 0.2	3.9	4.2	3	4.38	4	4.4	nd	nd	4.1
V	22 ± 2	28 ± 1	25	22	17	35.1	25	51.18	31.3	25.9	52
Ga	5.5 ± 0.3	6.0 ± 0.2	5.8	5.7	4.5	6.55	9	6.84	3.4	2.9	5.4
Ge	2.0 ± 0.1	2.4 ± 0.2	2.2	5.7	59	2.78	4	2.43	4.4	3.2	1.6
Se	1 ± 0.15	1.6 ± 0.1	1.3	2.8	1.8	2.47	2	2.82	1.7	1.5	1.2
Y	8.6 ± 0.4	8.2 ± 0.5	8.4	8.5	6.6	18.2	9	9.07	5.9	5.37	6.5
Zr	35 ± 2	36 ± 3	36	27	19	89.5	67	112	nd	nd	nd
Nb	3.3 ± 0.3	4 ± 0.4	3.7	2.9	1	9.44	12	nd	nd	nd	3.1
Mo	2.2 ± 0.2	2.1 ± 0.1	2.2	3.3	1.2	3.08	4	2.7	2.1	1.8	5.4
Pd	0.013 ± 0.006	0.001 ± 0.002	0.0074	<0.001	nd	nd	nd	nd	nd	nd	nd
Ag	0.09 ± 0.02	0.1 ± 0.016	0.095	<0.1	0.01	nd	nd	nd	nd	nd	nd
La	10 ± 0.5	11 ± 1	11	12	3.9	22.5	18	nd	nd	nd	nd
Ce	22 ± 1	23 ± 1	23	21	5.1	46.7	35	nd	nd	nd	nd
Pr	3.5 ± 0.3	3.4 ± 0.2	3.5	2.4	nd	6.42	3.8	nd	nd	nd	nd
Nd	11 ± 1	12 ± 1	12	9.5	nd	22.3	15	nd	nd	nd	nd
Sm	1.9 ± 0.1	2.2 ± 0.1	2	1.7	0.35	4.07	3	nd	nd	nd	nd
Eu	0.5 ± 0.02	0.43 ± 0.02	0.47	0.4	0.12	0.84	0.65	nd	nd	nd	nd
Gd	2.6 ± 0.2	2.7 ± 0.2	2.7	1.8	nd	4.65	3.4	nd	nd	nd	nd
Tb	0.32 ± 0.03	0.31 ± 0.02	0.32	0.3	0.09	0.62	0.52	nd	nd	nd	nd
Dy	2.0 ± 0.1	2.1 ± 0.1	2.1	1.9	0.008	3.74	3.1	nd	nd	nd	nd
Ho	0.5 ± 0.05	0.57 ± 0.04	0.54	0.35	nd	0.96	0.73	nd	nd	nd	nd
Er	0.85 ± 0.08	1 ± 0.07	0.93	1	0.002	1.79	2.1	nd	nd	nd	nd
Tm	0.31 ± 0.02	0.3 ± 0.02	0.31	0.15	nd	0.64	0.34	nd	nd	nd	nd
Yb	1 ± 0.05	1 ± 0.06	1	0.95	nd	2.08	2	1.76	nd	nd	nd
Lu	0.19 ± 0.02	0.2 ± 0.01	0.2	0.14	0.06	0.38	0.32	nd	nd	nd	nd
Hf	1.2 ± 0.1	1.2 ± 0.1	1.2	0.73	0.04	3.71	3	nd	nd	nd	nd
Ta	0.26 ± 0.03	0.3 ± 0.02	0.28	0.22	0.02	0.62	0.8	0.4	nd	nd	nd
Re	nd	nd	nd	<0.001	nd	nd	nd	nd	nd	nd	nd
Pt	0.065 ± 0.018	0.005 ± 0.003	0.035	<0.001	nd	nd	nd	nd	nd	nd	nd
Au	3 ± 0.6 ppb	4.4 ± 1.4 ppb	3.7 ppb	<0.05	nd	nd	nd	nd	nd	nd	nd
U	2.9 ± 0.3	1.9 ± 0.1	2.4	2.1	1.1	2.43	3	2.33	nd	nd	6.9

World coals, Ketris and Yudovich (2009). US coals, Finkelman (1993). C1, Dai et al. (2012b); C2, Tang and Huang (2004). C3, Bai et al. (2007). UK coals, Spears and Zheng (1999). Turkey coals, Palmer et al. (2004). nd, no data.

in coal is evaluated: (1) whether the metals are unusually enriched in the coal or its related host rocks, and whether the quantity of those metals is enough to be exploited; (2) the mode or modes of occurrence of the metals in the coal, coal ash, or host rocks; (3) whether the extraction technologies are economical; and (4) any environmental issues that might be caused by release of toxic elements during the metal extraction process.

3.1. Germanium

Exploration and exploitation of germanium in coals around the world has been carried out since the late 1950s. There are three coal-hosted deposits in the world that are currently being mined as raw materials for Ge extraction: Lincang (Yunnan, China; Figure 1), Wulantuga (Inner Mongolia, China; Figure 1), and Spetzugli (Primorye, Russian Far East). The concentration of Ge in the coal-hosted ore deposits is much higher, as described below, than that in Zn–, Ag–, Cu–, Pb–, and Pb–Zn–Cu sulphide ores with concentrations of 100–200 ppm Ge (Seredin et al. 2013), although the average concentrations for world lignite and hard coals are otherwise relatively low, 2.0 and 2.4 ppm, respectively (Table 1; Ketris and Yudovich 2009).

3.1.1. Wulantuga Ge ore deposit

The Wulantuga high-Ge coal deposit is located in the southwestern-most part of the Shengli Coalfield, Inner Mongolia, Northeastern China (Figure 1). The coal-bearing sequences hosting the deposit are part of the Early Cretaceous Shengli Formation (Cui and Li 1991, 1993; Du et al. 2003, 2004, 2009; Huang et al. 2008). The coal seam with elevated concentration of Ge (No. 6 Coal) has a thickness ranging from 0.8 to 36.2 m (with an average of 16.1 m) and covers an area of 2.2 km^2 (Du et al. 2009). However, the part of the same seam with a depleted Ge concentration outside of the Wulantuga deposit covers an area of 342 km^2, and has a thickness of up to 244.7 m at the centre of coal basin (Dai et al. 2012d).

The concentration of Ge greatly varies within the No. 6 Coal. Based on 939 coal samples from 75 boreholes, Du et al. (2009) estimated that the Ge concentration in the Ge-rich part of the No. 6 coal ranges from 32 to 820 ppm (with an average of 137 ppm) and the Ge reserves in the coal (with >100 ppm Ge) are 1700 t. Zhuang et al. (2006) showed that the Ge concentration varies from 22 to 1894 ppm (427 ppm on average; based on 12 bench samples) and that Ge is highly enriched in the middle portion of the seam section. Qi et al. (2007a) showed that the Ge concentration in the

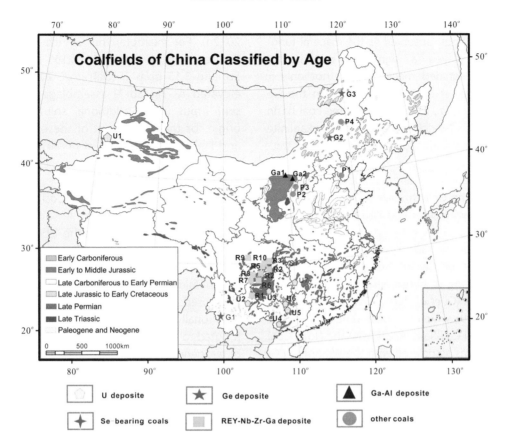

Figure 1. Distribution of coal-hosted ore deposits in China. G1, Lincang. G2, Wulantuga. G3, Yimin. Ga1, Jungar. Ga2, Daqingshan. R1, Xuanwei. R2, Zhongliangshan. R3, Huayingshan. R4, Guxu. R5, Moxinpo. R6, Zhenxiong. R7, Yiliang. R8, Luokan. R9, Xinde. R10, Sujiawan. U1, Yili. U2, Yanshan. U3, Guiding. U4, Fusui. U5, Heshan. U6, Yishan. U7, Chenxi. S1, Enshi. P1, Kailuan. P2, Pingshuo. P3, Datong. P4, Baiyinhua.

No. 6 Coal is from 23.3 to 1424 ppm, with an average of 168 ppm. An investigation on a complete coal section of the No. 6 Coal in the Wulantuga deposit showed that the Ge concentration ranges from 45 to 1170 ppm, with an average value of 273 ppm (Dai *et al.* 2012d; Table 2). The designed capacity for Ge production in the recovery plant at the Wulantuga deposit is 100 t per year. Various Ge compounds, including high-purity zone-refined Ge ingots (from 99.99999% to 99.999999%), are produced from the recovery plant.

The Ge-rich No. 6 Coal is a low-rank ($R_{o,max}$ = 0.45% on average) and low-ash (8.77% on average) coal (Dai *et al.* 2012d). However, the ash yield of the coal may greatly vary laterally within the coal seam. For example, Zhuang *et al.* (2006) reported that the Ge-rich coal has a 21% ash yield. Petrological studies by Hower *et al.* (2013b) and Dai *et al.* (2012d) showed that the total content of inertinite (52.5%) in the Ge-rich coal is slightly higher than that of huminite (46.8%). Textinite (43.9%) dominates the huminite macerals, and the dominant

Table 2. Concentrations of selected elements in Ge-rich coals and their fly ashes (ppm).

Elements	World[a]	Wulantuga[b]		Lincang[b]			Spetzugli[b]	
		Coal	Fly ash	Coal	Fly ash	Yimin[e]	Coal	Ash
Ge	2	273 (427[c])	35170	1294 (852[d])	46580	43	1025	29078
As	7.6	499 (356[c])	21795	104 (47.6[d])	3910	155	65	1250
Hg	0.1	3.16 (4[c])	160	0.24	0.5	0.24	0.51	9.4
Sb	0.84	240 (75[c])	12204	33.7 (32.8[d])	452	0.17	307	8950
Tl	0.68	3.2 (3[c])	137	2.4	70.2	nd	0.5	8.7
Be	1.2	25.7 (33[c])	165	337 (198[d])	340	1.04	67.3	397
Nb	3.3	1.4 (2[c])	5.7	18 (46.8[d])	7.2	1.57	7.4	44.6
W	1.2	115 (507[c])	2370	177 (378[d])	4684	9.84	357	3381
Cs	0.98	5.3 (9[c])	114	28.8 (22.7[e])	279	2.26	15.3	52
U	2.9	0.4 (17[c])	6.1	33.1 (56[d])	62	0.41	3.1	18.4

[a]World low-rank coals, Ketris and Yudovich (2009). [b]Dai *et al.* (2014b). [c]Zhuang *et al.* (2006). [d]Hu *et al.* (2009). [e]Li *et al.* (2014a). Concentrations of trace elements in fly ash are on ash basis (ashed at 815°C). nd, no data.

inertinite macerals are mainly fusinite (33.0%) and semi-fusinite (12.5%). Fungus, observed as the maceral funginite, may have played a significant role in the development of degraded maceral forms not only in the Ge-coals (Hower et al. 2013b) but also in the Ge-depleted coals (Dai et al. 2015a) in the same coal basin. The minerals in the Ge-rich coals are quartz, kaolinite, illite (and/or illite/smectite), gypsum, pyrite, and trace amounts of rutile and anatase. Both detrital quartz of terrigenous origin and cell-filling authigenic quartz have been observed in the Ge-rich coals (Zhuang et al. 2006; Dai et al. 2012d). Pyrite was deposited from sulphur-rich hydrothermal fluids, which were syngenetically injected into the original peat deposit (Dai et al. 2012d).

The enrichment of Ge in the Wulantuga coal is attributed to leaching of hydrothermal fluids from the adjacent granitoids (Qi et al. 2004, 2007b; Zhuang et al. 2006; Du et al. 2009), which are located near the southwest of the Shengli Coalfield. Apart from the No. 6 Coal, the other coal seams associated with the Wulantuga Ge ore deposit are not rich in Ge (Zhuang et al. 2006; Du et al. 2009). The same coal seam (No. 6 Coal) in most of the remaining Shengli Coalfield is also low in Ge (Du et al. 2009; Dai et al. 2015a). The distribution of Ge concentration in the No. 6 Coal follows a fan-shaped trend, decreasing from the SW to the NE of the ore deposit, with a sharp gradient of Ge concentration at the southern margin of the deposit (Du et al. 2009). The spatial distribution of Ge concentration in the ore deposit suggests that the Ge-rich hydrothermal solutions laterally migrated from SW to NE during peat accumulation, and were then adsorbed by the organic matter, leading to the highest concentrations in the southwestern-most part of the ore deposit and lowest concentration in the northeast (Dai et al. 2015a). The isolated association with the leached Ge-rich solutions from the granite was responsible for the isolated enrichment of Ge in the Wulantuga deposit (Dai et al. 2015a).

The elevated-concentration of a Ge–W assemblage (115 ppm W, Dai et al., 2012d; 507 ppm W, Zhuang et al. 2006; 35–627 ppm W, Du et al. 2009) in the Wulantuga ore deposit indicates that an alkaline N_2-bearing hydrothermal solution of amagmatic origin could have played a significant role in the enrichment of Ge and W (Seredin et al. 2006; Dai et al. 2015a). The N_2-bearing hydrothermal solutions may have originally been derived from meteoric waters, which had then been subjected to deep and long-term circulation and had repeatedly and selectively leached the granite, leading to elevated-concentrations of Ge and W in solution. The Wulantuga Ge-rich coals also have elevated-concentrations of As, Hg, Sb, and Tl (Zhuang

et al. 2006; Qi et al. 2007a; Du et al. 2009; Dai et al. 2012d). For example, the Ge-rich part of the No. 6 Coal contains 499 ppm As, 3.165 ppm Hg, 240 ppm Sb, and 3.15 ppm Tl (Dai et al. 2012d; Table 2). This elevated As–Hg–Sb–Tl assemblage appears to represent input of an additional solution of epithermal origin for the enrichment of these elements (Seredin and Finkelman 2008; Dai et al. 2012d).

3.1.2. Lincang Ge ore deposit

The coal-hosted Lincang Ge deposit, located in southwestern Yunnan Province in southwestern China (Figure 1), was discovered in the 1950s during prospecting for U deposits. The Ge-rich coals are in the Miocene Bangmai Formation, deposited on top of a granitic batholith of Middle Triassic age (Zhong 1998), which is mainly composed of biotite- and two-mica granites. Additionally, the upper part of the biotite-granite, distributed to the west of the coal basin, served as a source of terrigenous input to the basin during accumulation of the peat that formed the Ge-rich coals (Lu et al. 2000).

The high-Ge coals are currently mainly being extracted from the Dazhai Mine, the Meiziqing Mine, and the Chaoxiang pit of the Zhongzhai Mine. The estimated Ge reserves in the Dazhai and Meiziqing, as of 31 December, 2009, are 613 and 76 t, respectively; and those in the Zhongzhai Mines are 39 t, as of 31 October, 2010 (Dai et al. 2014b). Production from the Lincang Ge recovery plant is between 39 and 47.6 t Ge per year (Dai et al. 2014b).

The Lincang Ge-rich coals have ash yields, generally in the range of 21.64–28.18%, and are characterized by a medium-sulphur content, generally 1.86–2.33% (Dai et al. 2015c). The huminite random reflectance of the Ge-rich coals varies from 0.33% to 0.48%, indicating a lignite/subbituminous coal rank. The maceral composition of the Lincang Ge-rich coals is quite different from those in the Wulantuga Ge deposit. The former are mainly composed of huminite-group macerals, with more than 88.5% total content of huminite (Dai et al. 2015c), and the latter consist of 52.5% inertinite and 46.8% huminite (Dai et al. 2012d). Ulminite and attrinite generally dominate the huminite macerals in the Ge-rich Lincang coals. Structured inertinite-group macerals are rare, with funginite being the most abundant inertinite form. The minerals in the Lincang coals consist mainly of quartz, and, to a lesser extent, kaolinite, illite, and mica. An unusual mineral in common coals, hydrous beryllium sulphate ($BeSO_4 \cdot 4H_2O$), has been observed in the low-temperature ashes of Ge-rich coals from the Lincang deposit (Dai et al. 2015c).

Qi et al. (2004) reported that the Ge concentration in the Lincang ore deposit ranges from a few tens of ppm to ~2500 ppm, with an average of 850 ppm. Based on three complete coal-seam sections (five benches for the S3 Coal; 13 benches for the Z2 Coal, and 12 benches for the X1 Coal) in the Dazhai Mine, the three Ge-rich coal seams have average Ge concentrations of 1394, 1538, and 1833 ppm, respectively, with respective concentration coefficients (CCs) of 697, 769, and 917 (CC, element concentrations in investigated samples vs. averages for world low-rank coals) (Dai *et al.* 2015c; Table 2). Additionally, Be (up to 2000 ppm, 343 ppm on average) and W (up to 339 ppm, 170 ppm on average) are unusually enriched in the Lincang Ge-rich coals; elements As (156 ppm on average), Sb (38 ppm), Cs (25.2 ppm), and U (52.5 ppm) are significantly enriched; and Nb (28.2 ppm) is enriched (Dai *et al.* 2015c; Table 2). Elements Be and U, as well as precious metals (Seredin and Dai 2014), in the Lincang Ge-rich coals also have potential economic significance.

The highly elevated Ge concentration in the coal was derived from the leaching of biotite- and two-mica granite by hydrothermal solutions (Hu *et al.* 2009; Qi *et al.* 2004). The biotite- and two-mica granites were either hydrothermally altered or -argillized by hydrothermal solutions. The alteration processes appear to have taken place during or shortly after deposition of the peat (Dai *et al.* 2015c). Two types of metasomatites of hydrothermal origin, quartz-carbonate and carbonate, have also been identified (Hu *et al.* 2009; Dai *et al.* 2015c); these occur as partings and as roof and floor strata, and were formed at the syngenetic stage of coal deposition. These rocks have abnormal high Ge contents (Hu *et al.* 2009; Zhuang *et al.* 1998; Qi *et al.* 2004; Dai *et al.* 2015c) and are considered as analogues of hot-spring siliceous sinters or hydrothermal travertines (Hu *et al.* 2009; Qi *et al.* 2004). For example, the carbonate and quartz-carbonate metasomatites respectively contain 410 and 59.7 ppm Ge (Dai *et al.* 2015c). Previous studies have shown that the REY in these metasomatites are characterized by distinct positive Eu anomalies, which were previously attributed to two factors: (1) the ionic radii of Eu^{2+} and Ca^{2+} are similar, and thus Eu^{2+} could be easily incorporated into calcite, leading to fractionation between Eu and other REY (Qi *et al.* 2002); (2) Eu is generally concentrated in the K-feldspar of granite; therefore, intense leaching of this abundant mineral by hydrothermal solutions passing through alkaline basement granites could lead to Eu enrichment in those solutions (Seredin *et al.* 2006). However, as discussed by Dai *et al.* (2016d), the strong positive Eu anomalies could possibly be caused by interference from BaO and/or BaOH during ICP-MS analysis. Eu values from some ICP-MS laboratories are not interpreted if the Ba/Eu ratio is >1000, to exclude

interference from BaO and BaOH (Loges *et al.* 2012). High Ba/Eu values (e.g. >1000) in some coal samples have not been unusually observed, and the Eu anomalies in these coals may thus be incorrect (Dai *et al.* 2016d).

The hydrothermal solutions leaching the batholith granite were a mixture of alkaline N_2-bearing and volcanogenic CO_2-bearing fluids, which led to the enrichment of a range of trace elements, not only including assemblages of Ge–W and Be–Nb–U (both leached from granite and deposited in the peat), but also As–Sb (from volcanogenic solutions), as well as the alteration and argillization of the batholith granites, and the formation of carbonate and quartz-carbonate metasomatites (Dai *et al.* 2015c).

Qi *et al.* (2011) analysed the Ge isotope composition of the Lincang Ge-rich coals using a continuous flow hydride generation system coupled to a multi-collector inductively coupled plasma mass spectrometer. The results indicated that Ge-rich coals show large Ge isotopic fractionation ($\delta^{74}Ge$ values ranging from −2.59‰ to 4.72‰) and that the preferential enrichment of light Ge isotopes in coal in an open system is the main factor in Ge fractionation. The Ge isotope composition of hydrothermal chert (and possibly limestone) might also record the fractionation produced by precipitation of quartz and sorption on coal (Qi *et al.* 2011).

Germanium is mostly enriched in the cross joints of different faults in the Lincang deposit (Hu *et al.* 2009; Qi *et al.* 2004). This is similar to the enrichment in the Spetzugli deposit (Seredin *et al.* 2006; Seredin and Finkelman 2008) but different from that in the Wulantuga deposit, which is mostly enriched in the coal on the margin of the basin (Zhuang *et al.* 2006; Du *et al.* 2009; Dai *et al.* 2012d). The different spatial distributions of the Ge in the Lincang and Wulantuga deposits were caused by the different migration patterns of the hydrothermal solutions. In the Lincang ore deposit, the hydrothermal solutions migrated vertically along fault intersections and then were discharged into the peat swamp, causing a dome-like Ge distribution pattern. Based on the different migration patterns of the hydrothermal solutions, Dai *et al.* (2015a) have proposed formation models for the coal-hosted Ge ore deposits for the Wulantuga and Lincang coal-hosted Ge deposits.

3.1.3. Yimin Ge ore deposit

In addition to the currently mined coal-hosted Ge ore deposits in Lincang and Wulantuga in China, another coal-hosted Ge ore deposit, discovered in 1992 with an estimated resource of 4 kt Ge (Wu *et al.* 2002), is located in the Yimin Coalfield of northeastern Inner Mongolia (Figure 1). The coal-bearing strata of this deposit include the Damoguaihe and Yimin Formations of Early Cretaceous age. The elevated concentrations of Ge occur in the coals and partings of the Damoguaihe Formation. The Yimin coal deposit contains seven coal seams with concentrations of ≥100 ppm Ge (Huang et al. 2007). The Ge concentration in the coals and partings varies greatly, from less than 1 ppm to 470 ppm, and is mostly in the range between 50 and 200 ppm (Li et al. 2014a).

Germanium is mainly enriched in the marginal areas of the Yimin coal deposit. Its concentration decreases from the margin towards the centre of the deposit (Li et al. 2014a), similar to the lateral distribution of Ge in the Wulantuga Ge deposit reported by Zhuang et al. (2006), Du et al. (2009), and Dai et al. (2015a). The coals in some local areas of the coalfield were subjected to shallow felsic intrusions, leading to a high coal rank (low volatile bituminous). Germanium was not enriched in the coals subjected to the igneous intrusions but was enriched in the adjacent subbituminous coals (Liu and Xu 1992). In addition to Ge, elements As (155 ppm) and W (9.84 ppm) are enriched in the coal (Li et al. 2014a), in a similar way to the elemental assemblages in the Wulantuga deposit.

Although the currently-mined Lincang and Wulantuga coal-hosted Ge ore deposits, as well as the Spetzugli deposit in the Russian Far East, have quite different geological settings, different ages of coal formation, and different maceral compositions, and there are great distances (thousands of kilometres) between these ore deposits, they have some common characteristics: (1) The coals with highly elevated Ge concentrations are low in rank. Ge ore deposits have not yet commonly been observed occurring in high-rank coals. One example of a high-rank Ge-rich coal outside China, however, is the Gray Hawk Coal (Early Pennsylvanian, Langsettian) in eastern Kentucky, USA, which has $R_{o,max}$ values of 0.82–0.89% and contains 16.1–569 ppm Ge, with an average of 175 ppm (ash basis) (Hower et al. 2015b). (2) The mode of Ge occurrence in all coal-hosted ore deposits is primarily an organic association (Zhuang et al. 2006; Hu et al. 2009; Li et al. 2011). (3) The highly elevated Ge concentrations were derived from leaching by hydrothermal solutions of nearby Ge-rich granites. (4) The Ge mineralization occurred during peat accumulation or during early diagenesis. (5) All the coal-hosted Ge ore deposits have overall similar elemental assemblages. For example, the Lincang deposit has elevated concentrations of Ge–W, Be–Nb–U, and As–Sb. The Wulantuga and Spetzugli deposits are characterized by highly enriched assemblages of Ge–W and As–Hg–Sb–Tl (Table 2).

3.1.4. Fly ash derived from Ge-rich coals

Although chemical and mineralogical compositions of coal combustion products (CCPs) from power plants have been extensively studied in the last few decades (e. g. Clarke and Sloss 1992; Vassilev and Vassileva 1996; Mardon and Hower 2004; Creelman et al. 2013; Saikia et al. 2015; Liu et al. 2016; Valentim et al. 2016), CCPs derived from Ge-rich coals have seldom been investigated. Qi et al. (2011) determined the concentrations of 14 trace elements in two fly ash and three slag samples from the Lincang deposit. Dai et al. (2014b) investigated the mineralogical and chemical compositions of various CCPs for all the high-Ge coal deposits that are currently being industrially used.

The fly ashes derived from the Lincang and Wulantuga coal-hosted Ge deposits are unique because they have highly elevated concentrations of a number of different elements. For example, the fly ash derived from the Wulangtua Ge-rich coals contains up to 35,170 ppm Ge, 21,795 ppm As, 15,568 ppm F, 5573 ppm W, 12,204 ppm Sb, 137 ppm Tl, and 160 ppm Hg (Dai et al. 2014b; Table 2); and elements including up to 46,580 ppm Ge, 5539 ppm Pb, 1296 ppm Sn, 1172 ppm Ga, 555 ppm Bi, and 279 ppm Cs (Dai et al. 2014b), as well as 7955 ppm Zn, 894 ppm Be, and 106.5 ppm Cd (Qi et al. 2001), are highly enriched in the fly ash derived from the Lincang Ge-rich coal. These high concentrations are attributed both to the elevated levels of the same elements in the feed coals and to the high volatility of those elements during the coal combustion process. Germanium mainly occurs in the fly ash as oxides (e.g. GeO_2). Other Ge-bearing phases have also been observed, such as glass, Ca ferrites (Figure 2), solid solutions of Ge in SiO_2, and probably elemental Ge or Ge (Ge–W) carbide, as well as complex oxides including $(Ge,As)O_x$, $(Ge, As,Sb)O_x$, $(Ge,As,W)O_x$, and $(Ge,W)O_x$ (Figure 2). A portion of the Ge occurs as adsorbed species on unburnt carbon (Dai et al. 2014b). Additionally, the Wulantuga fly ashes have a higher unburned carbon content and a higher loss-on-ignition (LOI) percentage (~57.5%) than the Lincang (~16.3%) fly ashes (Figure 2). This is because of the higher proportion of inertinite macerals in the Wulantuga feed coals than in the Lincang feed coals as described above. Inertinite macerals, e.g. fusinite and secretinite, are more resistant to combustion than vitrinite macerals in pulverized fuel combustion systems (Nandi et al. 1977; Shibaoka 1985; Vleeskens et al. 1993), leading to higher unburnt carbon in the Wulantuga fly ashes.

3.1.5. Potential for impact of toxic elements on human health

Some elements with highly elevated concentrations in Ge-rich coals (e.g. As, Hg, Sb, Cs, Tl, and Be in the Wulantuga coals; As, Be, Zn, Cd, Sb, Cs, Tl, Pb, and U

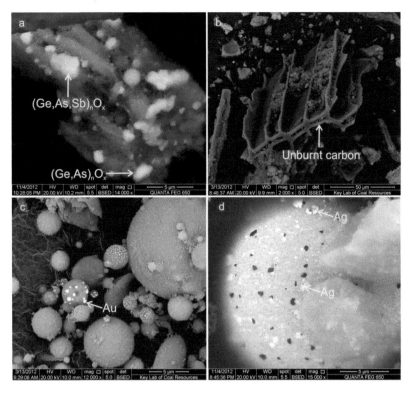

Figure 2. SEM backscattering images of fly ashes derived from Ge-rich coals. A, complex oxides of Ge–As–Sb and Ge–As on the surface of unburnt carbon particle (Dai *et al.* 2014b). B, Unburnt carbon. C and D, drop-like Au and film-like Ag particles on the surface of glass spherules. A, B, and C, Wulantuga fly ash. D, Lincang fly ash.

in the Lincang coals) have a high potential for toxicity and may have adverse influences on human health. The health of workers in power plants using Ge-rich feed coals and in Ge-recovery factories using high-Ge ash as raw material may be endangered by these potentially toxic elements, because: (1) These potentially toxic elements are highly enriched in Ge-rich fly ash as described above, and in bottom ash as well (Dai *et al.* 2014b; Sun *et al.* 2016). For example, the concentrations of some of these elements in a fly ash sample collected from the baghouse filter in the Ge recovery factory at the Wulantuga ore deposit are 21,795 ppm As, 15,568 ppm F, 137 ppm Tl, and 160 ppm Hg (Dai *et al.* 2014b). To the best of our knowledge, these are the highest concentrations that have ever been reported in fly ashes around the world; (2) Very fine grained particles such as fly ash with highly elevated concentrations of toxic trace elements may represent a special danger, because they can be inhaled deep into the lungs of people who come into contact with the fly ash in the course of industrial activity; and (3) The high volatility of many of these elements indicates that they may be concentrated on the surface of the fly ash particles and may be highly soluble in lung or stomach fluids. Caution and preventive measures should be taken in power plants using Ge-rich feed coals and in Ge-recovery factories using high-Ge ash as raw material

in order to control the potential adverse effects of these toxic elements on health. Additionally, the soil around the associated coal mines may have elevated levels of arsenic and other potentially dangerous elements that should also be taken into account (Jia *et al.* 2015; Yang *et al.* 2016).

3.2. Gallium and aluminium

Although Ga has been considered as a valuable element in coal and, thus, has been prospected for more than 50 years in China, a coal-hosted Ga ore deposit was not discovered until 2006. A huge coal-hosted deposit of Al and Ga in the Heidaigou Mine, Jungar Coalfield (Figure 1), Inner Mongolia, was first reported in 2006 (Dai *et al.* 2006a, 2006b) and has been further confirmed by subsequent studies in other mines of the same coalfield (e.g. Haerwusu Mine, Dai *et al.* 2008a; Guanbanwusu Mine, 2012a). Average metal contents have been found to be ~51% Al_2O_3 and ~92 ppm Ga in fly ash derived from this Al–Ga deposit, which are similar to those in conventional bauxite deposits (30–55% Al_2O_3 and 30–80 ppm Ga; Seredin 2012). The confirmed reserves of the metals are 150 Mt of Al_2O_3 and 49,000 t of Ga in the Heidaigou Mine of the Jungar Coalfield. A number of technologies, e.g. acid and alkali methods or salt activation method, have been developed to extract Ga and Al from the Jungar high-Al_2O_3 fly ashes (Guo et al. 2013; Feng et al. 2014; Li et al. 2014c, 2016), and studies have been carried out to evaluate the mineralogical and chemical variations during the Al extraction process (Gong et al. 2016). Based on the original geochemical data on the high-Al and -Ga concentrations in the coal, together with additional work by the Shenhua Group, Zhungeer Energy Corporation Limited (China), a pilot plant with an annual processing capacity of 800,000 t Al_2O_3 and approximately 150 t Ga was built at the beginning of 2011, and Al and Ga are being industrially extracted from fly ashes derived from boehmite-rich coals in the Jungar Coalfield and its surrounding areas. A study by Dai et al. (2012c) showed that Al and Ga also are enriched in the coals from the adjacent Adaohai Mine of the Daqingshan Coalfield, which is located a short distance to the north of the Jungar Coalfield (Figure 1).

The Al–Ga-rich coal seam is located in the uppermost part of the Pennsylvanian Taiyuan Formation and is identified as No. 6 Coal in the Jungar Coalfield and as the CP2 Coal in the Adaohai Mine, Daqingshan Coalfield. The No. 6 coal seam in the Jungar Coalfield, with an average thickness of 30 m (ranging from 2.7 to 50 m; Liu et al. 1991), has reserves of 26.8 Gt and was deposited in a continental-marine transitional environment (Liu et al. 1991; Dai et al. 2006b). The Daqingshan Coalfield is a sub-depression (intermontane basin) in the inner part of the orogenic belt (Zhong et al. 1995; Zhou and Jia 2000; Wang and Ge 2007) and the CP2 Coal has a thickness of 4.72–42.79 m (average 22.58 m). Other coal seams in the Jungar Coalfield (e.g. the minable No. 5 Coal) are not enriched in Al and Ga (Yang et al. 2015).

The No. 6 and CP2 seams in the different coal mines of the two coalfields are different in rank due to heat flow from an igneous intrusion. The coal in the Adaohai Mine of the Daqingshan Coalfield is a low volatile bituminous coal ($R_{o, ran}$ = 1.58%; Dai et al. 2012c). The same coal seam in the Jungar Coalfield, which was not or not strongly affected by the igneous intrusion, is a high volatile bituminous coal ($R_{o,ran}$ generally 0.56–0.58%; Dai et al. 2006b, 2008a, 2012a).

3.2.1. Minerals identified in the coal-hosted Al–Ga ore deposit

With the exception of kaolinite that occurs both in the high and low volatile bituminous coals of the Jungar and Daqingshan Coalfields, the lower-rank (high volatile bituminous) coal is enriched in boehmite and to a lesser extent, goyazite; the relatively high-rank (low volatile bituminous) coals are enriched in diaspore, gorceixite, ammonian illite, and boehmite. The occurrence of boehmite, diaspore, and kaolinite is responsible for the highly

elevated Al and Ga concentrations in the coal. Boehmite, diaspore, and kaolinite occur as cell-fillings and are distributed in collodetrinite (Figure 3). The boehmite in the coals from the Jungar Coalfield, as well as in the Daqingshan Coalfield, was syngenetically formed during peat accumulation, probably from colloidal solutions of gibbsite derived from the exposed bauxite of the Benxi Formation in the sediment source region (Dai et al. 2006b, 2008a, 2012a, 2012c). A critical factor, however, was the absence of silica in the system, which if available would have formed kaolinite. Dehydration of the gibbsite colloid due to compaction of the overlying strata after peat accumulation resulted in boehmite formation (Dai et al. 2006b, 2008a; Zou et al. 2012, 2013). Where the igneous intrusion has increased the rank of the coal (e.g. in the Adaohai Mine), diaspore occurs in the coal. The high-temperature of the igneous intrusion led to the dehydration of gibbsite or boehmite and then to diaspore formation. By contrast, the same coal seam in the adjacent mines of the Jungar Coalfield (e.g. Heidaigou, Haerwusu, and Guanbanwusu), with a relatively low rank contains boehmite rather than diaspore.

Aluminophosphate (goyazite-gorceixite) minerals have been identified in the Ga–Al-rich coals of the two coalfields, having a structure between that of gorceixite (Ba-aluminophosphate), goyazite (Sr-aluminophosphate), and crandallite (Ca-aluminophosphate), and representing a solid solution between these three end members (Ward et al. 1996). These minerals generally occur in the cells of structured macerals (e.g. fusinite, semifusinite, and telinite; Figure 3) and as massive and discrete particles in the collodetrinite. The elements for goyazite-gorceixite formation in the Al–Ga-rich coals were probably derived from the bauxite of the weathered crust (Benxi Formation) in the sediment-source region. Phosphorus was probably released from decomposition of the organic-matter in the peat and then was picked up by those Al-rich solutions to form insoluble aluminophosphates (cf. Ward et al. 1996). The goyazite-gorceixite minerals in the Ga–Al-rich coals are also responsible for elevated concentrations of rare earth elements (Dai et al. 2012a, 2012c).

Ammonian illite only occurs in relatively high rank coals in the Adaohai Mine. It is believed to have been formed by reaction between NH^{4+} released from the organic-matter at high temperature (e.g. >200°C) during hydrothermal alteration and kaolinite already present in the coal (Daniels and Altaner 1993; Ward and Christie 1994; Nieto 2002; Boudou et al. 2008).

3.2.2. Major elements

The oxides of major elements in the Al–Ga-rich coals are dominated by Al_2O_3 and SiO_2 (Table 3), which

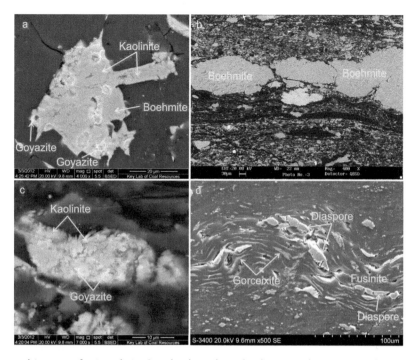

Figure 3. SEM backscattered images of minerals in Ga–Al-rich coals. A, boehmite and goyazite as fusinite-cell fillings in the No. 6 Coal of the Guanbanwusu Mine in the Jungar Coalfield (Dai *et al.* 2012a). B, boehmite in collodetrinite in the No. 6 Coal of the Heidaigou Mine in the Jungar Coalfield. C, kaolinite and goyazite as fusinite-cell fillings in the No. 6 Coal of the Guanbanwusu Mine in the Jungar Coalfield (Dai *et al.* 2012a). D, diaspore and gorceixite as fusinite-cell fillings in the CP2 Coal of the Adaohai Mine in the Daqingshan Coalfield (Dai *et al.* 2012c).

Table 3. Estimation of prospects of metal recovery as by-products from the Jungar and Daqingshan Coalfields (Al_2O_3, Ga, and REO are on ash basis; REO, oxides of rare earth elements and yttrium).

Coal mines	Thickness	A_d (%)	Al_2O_3 (%)	$SiO_2/$ Al_2O_3	Ga (ppm)	REO (ppm)	C_{outl}
Heidaigou	27.81	17.62	62.44	0.76	44.5	1461	0.83
Haerwusu	28.41	18.05	53.43	0.70	135	1404	0.92
Guanbanwusu	11.50	20.25	43.88	0.74	77.8	1121	0.77
Adaohai	16.50	25.00	44.46	0.93	72.9	976	0.93
Cut-off grade	>5		40.0	<1	50	800–900	0.70

have higher and lower concentrations, respectively, than the averages for common Chinese coals reported by Dai *et al.* (2012b). Because quartz is absent in most cases, silicon mainly occurs in clay minerals in the Al–Ga-rich coals. In addition to a small proportion of Al in chlorite and illite in the Guanbangwusu coals, and ammonian illite in the Adaohai coals, the aluminium mainly occurs in the kaolinite and Al-oxyhydroxides (boehmite and diaspore). Due to the abundance of boehmite and/or diaspore, the SiO_2/Al_2O_3 ratio for the coal (0.70–0.93), as might be expected, is lower than the same ratio for kaolinite (1.18), and also lower than the common values for Chinese coals (1.42; Dai *et al.* 2012b). For example, the SiO_2/Al_2O_3 ratios in the coals from the Guanbanwusu, Heidaigou, Haerwusu,

and Adaohai mines are, respectively, 0.74, 0.76, 0.70, and 0.93 (Table 3).

3.2.3. Trace elements

The concentration of Ga is elevated in the coals from the different mines of the Jungar and Daqingshan Coalfields, e.g. Guanbanwusu (12.9 ppm on average), Heidaigou (45 ppm), Haerwusu (18 ppm), and Adaohai coals (16.3 ppm on a whole coal basis; Dai *et al.* 2012a). All of these are much higher than the average value for common Chinese coals (6.6 ppm; Dai *et al.* 2012b) and world hard coals (6 ppm; Ketris and Yudovich 2009). Gallium is generally associated with clay minerals in coal (Finkelman 1981a; Chou 1997); however, it has a mixed (organic and inorganic) affinity in the coals of the Jungar and Daqingshan Coalfields. An inorganic affinity is represented by its occurrence in boehmite, diaspore, goyazite, and kaolinite (Wang *et al.* 2011; Dai *et al.* 2012b).

The concentrations of total REY in the coals from the Jungar and Daqingshan coalfields are all higher than the average for common Chinese coals (136 ppm; Dai *et al.* 2012b) and world hard coals (68.4 ppm; Ketris and Yudovich 2009). For example, the REY concentrations in the coals from Guanbanwusu, Heidaigou, Haerwusu, and Adaohai mines are 154, 257, 253, and 244 ppm, respectively (on a whole coal basis; Dai *et al.* 2012a; Zou et al. 2013). The elevated-concentrations of REY in the coals are due to groundwater leaching of the intra-seam partings in the coal. When the REY-rich leachate infiltrated the underlying benches of organic-matter, three processes of REY incorporation into the coal may have taken place: (1) precipitation of REY-rich leachate as authigenic minerals, e.g. goyazite-gorceixite minerals; (2) incorporation of REY in the leachate into hydrated Al-minerals, such as boehmite; (3) take-up and absorption of REY in the leachate by the organic matter. Thus the Al–Ga-rich coals were noted to have multiple modes of REY occurrence. Similar leaching of REY in the intra-seam partings by groundwater and the consequent REE enrichment in the adjacent coal benches have also been observed by Hower et al. (1999), who showed that REY in a tonstein in the Fire Clay coal bed (Eastern Kentucky) were leached by groundwater, leading to the enrichment of REY in the underlying coal benches. Crowley et al. (1989) also attributed the REE enrichment in coal to the leaching of an overlying intra-seam tonstein by groundwater and subsequent uptake by the underlying organic matter or subsequent incorporation in minerals. REY enrichment in coal benches directly beneath altered volcanic ashes in the Songzao Coalfield (Figure 1), Southwest China, is also due to groundwater leaching (Zhao et al. 2015).

Table 3 summarizes the prospects for metal recovery as by-products from the high-Al_2O_3 coals in the Jungar and Daqingshan Coalfields. It can be seen that Al_2O_3, Ga, and REO concentrations in the ash of all the Jungar and Daqingshan coals are higher than the corresponding suggested cut-off grades. If the thickness of the coal seam is greater than 5 m, the cut-off grades for Al_2O_3, Ga, and REY in coal ash are >40% (or $SiO_2/Al_2O_3 < 1$), >50 µg/g (Dai et al. 2012a), and >800–900 µg/g (Seredin and Dai 2012), respectively.

3.2.4. High-Al_2O_3 fly ash

The higher concentration of Al_2O_3 in the coals leads to unique coal combustion products, e.g. fly ash, which generally have not less than 50% Al_2O_3 (Dai et al. 2010a; Qi and Yuan 2011; Zhao et al. 2012). The fly ash may therefore represent major raw material sources for extraction of Al_2O_3 and Ga (Seredin 2012). The fly ash derived from the high-Al_2O_3 coals mainly consists of amorphous glass and mullite, with traces of corundum, quartz, unburned carbon, K-feldspar, clay minerals, and Fe-bearing minerals (Dai et al. 2010a; Zhao et al. 2008). The mullite content of the fly ash may be as high as 37.4% because of high boehmite and kaolinite contents in the feed coal. Corundum is a characteristic mineral formed during combustion of the boehmite-rich coal (Zhao et al. 2008; Dai et al. 2010a). Zhao et al. (2012)

showed that under temperature treatment aluminium-bearing minerals (boehmite and kaolinite) transformed to γ-Al_2O_3, corundum (α-Al_2O_3), and an amorphous phase. The γ-Al_2O_3 is the main mineral in the high-temperature ash while α-Al_2O_3 and mullite are the main minerals in the fly ash (Zhao et al. 2012).

Dai et al. (2014c) investigated the REY compositions in size-fractionated fly ash (six size fractions from plus-120 mesh to minus-500 mesh) from the Jungar Power Plant, Inner Mongolia. Compared to the upper continental crust, the size-fractionated fly ashes show enrichment in the light rare earths + Y and have negative anomalies of Eu, Ce, and Y. Relative to the raw (unsized) fly ash, Eu and Ce in the size-fractionated ashes, respectively, exhibit positive and negative anomalies, and the finer fly ashes (minus-300 mesh) are enriched in light REY. In contrast to the bottom ash, all the size-fractionated fly ashes show distinctly negative Ce anomalies. Lanthanum, Ce, Pr, and Nd were detected in minerals within the glassy phases in the fly ash (Hower et al. 2013a). Hower et al. (2013a) showed that the ratio of light REE vs. heavy REE decreases with a decrease in flue gas temperature, and the fly ash shows an increase in the concentration of REY from the coarsest to the finest fractions. Hower et al. (2014) used Ce, the most abundant of the REE, as a proxy for other REE in fly ash, and found Ce to be dispersed in the glass.

3.3. Rare earth elements and yttrium

Among the valuable elements discussed in this article, rare earth elements and yttrium (REY) in coal have attracted the most attention in recent years, not only because the demand for these elements has grown rapidly due to their wide applications as metal catalysts, light-emitting diodes, phosphors, permanent magnets, batteries, and various components for renewable energy equipment (Greene 2012; Hatch 2012; Pecht et al. 2012; Hower et al. 2016a. 2016b; Taggart et al. 2016), but also because of the supply crisis of 2010 and the price spike of 2011 (Massari and Ruberti 2013). Although the crisis did not last for long, the short-lived alarm produced by export restrictions from China was sufficient to initiate a treasure hunt by way of exploration for REE deposits all over the world (Barakos et al. 2016).

The estimated average sum of REY concentrations in Chinese coals (138 ppm; Dai et al. 2008a) is around twice the average sum of REY concentrations for world coals (68 µg/g; Ketris and Yudovich 2009). This is because more coal samples of late Permian age from southwestern China, which contain higher REY concentrations (e.g. Zhuang et al. 2012; Dai et al. 2014a, 2015b) than those from northern China (e.g. Huang *et al.* 2000) and northwestern China (e.g. Li *et al.* 2012, 2014b; Jiang *et al.* 2015; Dai *et al.* 2015d), were included in the analysed data set. The cut-off grade for REY in coal ash, expressed as rare earth oxides (REOs), is ≥1000 µg/g, or ≥800–900 µg/g for a coal seam with thickness >5 m (Seredin and Dai 2012). Based on estimations by Ketris and Yudovich (2009), the average concentration of REO for world coal ashes is 485 ppm. If the CC for REO in a coal ash is higher than 2, then the REOs could be an economically viable by-product from the coal combustion residues. Coal is, therefore, becoming a promising alternative raw material for REY (Seredin and Dai 2012), and has attracted much attention in recent years (Blissett *et al.* 2014; Franus *et al.* 2015; Zhang *et al.* 2015; Hower *et al.* 2015a, 2016a, 2016b; Rozelle *et al.* 2016). Rozelle *et al.* (2016) examined two US coal by-products for ion extraction, using ammonium sulphate, an ionic liquid, and a deep eutectic solvent as lixiviants, and showed that extraction of rare earth elements in each case produced high recoveries of rare earth elements to the solution. Rozelle *et al.* (2016)'s data suggest that in at least the cases of the materials examined, US coal by-products may be technically suitable as REE ores.

Five types of elemental assemblage associated with REY enrichment in coal and coal-bearing strata in China have been observed: REY–Zr(Hf)–Nb(Ta)–Ga, REY–U–V–Cr–Se–Re, REY–Nb(Ta)–Zr(Hf)–U, REY–U(Mo,Se), and REY–Al–Ga. The latter two types will be and have been discussed in Sections 3.4 and 3.2, respectively.

Although mineralization of REY–Zr(Hf)–Nb(Ta)–Ga in felsic and alkali intra-seam tonsteins has been known for a long time (e.g. Crowley *et al.* 1989; Seredin 1994; Hower *et al.* 1999; Zhou *et al.* 2000), it has rarely attracted practical interest as a raw material for these rare metals, primarily owing to the small thickness (from 1 to 20 cm, generally 3–7 cm) of these rare metal-bearing intra-seam layers (Seredin and Finkelman 2008). Early in 1999, Zhou (1999) pointed out that 'alkali tonsteins have practical significance in prospecting the rare metal ore deposits related to alkali volcanic activities based on the tonsteins' number of layers, thickness, spatial distribution, and possible location of palaeo-volcanic craters.' The frequent occurrence of these rare metal-bearing tonsteins provides a basis for predicting the possibility of thick horizons of REY–Zr(Hf)–Nb(Ta)–Ga-bearing tuffs outside of coal seam (Seredin 2004). This forecast has been successfully realized in China by discovery of several thick alkali ore beds in Yunnan Province (Dai *et al.* 2010b) and, thus, Hower *et al.* (2016a) called for reconsideration of the previous sceptical views in relation to this mineralization in coal-bearing strata.

The coal-hosted REY–Zr(Hf)–Nb(Ta)–Ga ore deposits in China have only been identified in the southwestern part of the country, covering areas that include eastern Yunnan, Western Guizhou, Chongqing and Southern Sichuan Provinces (Figure 1). The ores are represented both by horizons of argillized (in a few cases, hematized) volcanic-ash (mostly 2–5 m and in some cases, up to 10 m in thickness) and by metalliferous coals, which usually contain alkali tonsteins. The REY–Zr(Hf)–Nb(Ta) deposits contain up to 0.1–0.5% REO, 1–3% $(Zr,Hf)_2O_5$, 0.05–0.1% $(Nb,Ta)_2O_5$ (for metalliferous coals, the rare-metal concentrations are on an ash basis) (Dai *et al.* 2010b, 2014a; Zou *et al.* 2014, 2016; Zhao *et al.* 2015; Table 4). The REY–Zr(Hf)–Nb(Ta)–Ga ore mineralization is associated with three factors: tuffaceous, hydrothermal fluid, and mixed tuffaceous-hydrothermal types (Dai *et al.* 2016a).

3.3.1. Thick argillized tuffs and low-temperature hydrothermal alteration

Thick argillized tuffs containing REY–Zr(Hf)–Nb(Ta)–Ga have been found mainly in eastern Yunnan (e.g. Xuanwei, Dai *et al.* 2010b) and Chongqing (e.g. Zhongliangshan, Zou *et al.* 2016) (Figure 1). The tuffaceous mineralization is mainly associated with mafic alkali and felsic alkali tuffs (Dai *et al.* 2010b, 2014a; Zhao and Graham 2016; Zou *et al.* 2016). Four lithological types are identified in the field, namely clay altered volcanic ash, tuffaceous clay, tuff, and volcanic breccia (Dai *et al.* 2010b), which are different from those of typical sedimentary rocks. The REY–Zr(Hf)–Nb(Ta)–Ga ore beds, as well as the metalliferous coals containing alkali tonsteins, show a significant positive natural gamma-ray anomaly in well logging (generally >250 API, Dai *et al.* 2012b). The gamma-ray peak width in metalliferous coals containing intra-seam alkali tonstein is very narrow because of the small thickness of the tonstein (e.g. mostly 3–7 cm). Because of these anomalies, natural gamma-ray logging has been considered as a useful tool in geophysical prospecting for REY–Zr(Hf)–Nb(Ta)–Ga ores, and thus it has been considered as a geophysical indicator in REY–Zr(Hf)–Nb(Ta)–Ga ore prospecting (Dai *et al.* 2010b, 2012b) (Figure 4).

The rare-metal-enriched argillized tuffs exclusively occur in the lower Xuanwei or Longtan Formations of late Permian age (Figure 4), indicating that they were formed later than the main episode of Emeishan Large Igneous Province (ELIP) magmatism (Dai *et al.* 2010b, 2016a; Zhao and Graham 2016). They are at the ELIP periphery and may thus have resulted from waning activity of the mantle plume (Dai *et al.* 2011b; Zhao and Graham 2016). The highly elevated concentrations of these rare metals in the argillized tuffs probably

Table 4. Concentration of rare metals in REY–Zr(Hf)–Nb(Ta)–Ga ore deposits of southwestern China (μg/g unless indicated as %).

Ore deposits	Sample	Thickness (m)	REO	$(Zr,Hf)_2O_5$	$(Nb,Ta)_2O_5$	Sum-RM (%)
Huayingshan	K1-2p[1]	0.05	1972	2589	433	0.499
	K1-1p2[1]	0.07	2724	2569	448	0.574
	K1-1p1[1]	0.03	2289	2378	618	0.529
	H-Coal[2]	2.03	1710	3617	406	0.573
Baoshan Xuanwei	2009-BX-4[3]	2.5	1358	6011	555	0.792
	2009-YN-5[4]	16.0	1216	3805	302	0.532
	2009-LY-10[5]	3.0	1271	8464	627	1.036
Xinde Xuanwei	301–10[6]	*	1248	3520	558	0.553
	301–13[6]	*	1055	4852	745	0.665
	301–14[6]	*	1227	4703	742	0.667
	301–15[6]	*	1263	4122	673	0.606
	301–16[6]	*	1791	4139	639	0.657
	301–lower part	1.89	1680	3439	625	0.575
	701–1	0.64	1231	1594	252	0.308
	701–2	1.40	1014	3094	509	0.462
	701–3	1.56	1448	2786	493	0.473
	701–4	0.27	518	1678	233	0.243
	701–5	0.45	2348	2999	435	0.578
	701–6	1.35	202	3426	452	0.408
Luokan	802–1	1.15	1494	2679	441	0.461
	802–2	1.20	3375	4642	665	0.868
	802–3	3.47	483	1485	250	0.222
	802–4	2.00	619	1846	315	0.278
Yiliang	1001–1	0.35	1084	2572	421	0.408
	1001–2	1.65	1968	3584	592	0.614
	1001–3	4.50	1941	2169	414	0.452
	1001–4	4.99	1956	3244	484	0.568
	1001–5	0.10	981	1091	224	0.230
	1001–6	3.70	1445	2237	386	0.407
Zhenxiong	1201–1	3.08	1674	2765	455	0.489
	1201–2	0.71	3049	3603	643	0.730
	1201–3	1.66	399	1428	312	0.214
	1201–4	1.90	290	1587	297	0.217
	1201–5	2.00	1022	2073	306	0.340
	1302–1	0.48	1408	2110	407	0.393
	1302–2	2.60	1454	2513	417	0.438
	1302–3	3.23	550	3252	472	0.427
	1908–1	0.05	1817	3599	644	0.606
	1908–2	3.18	1744	2699	525	0.497
	1908–3	3.40	2717	3102	482	0.630
	1908–4	1.74	965	3210	532	0.471
	1908–5	0.50	1453	2105	388	0.395
	1908–6	4.67	2641	4168	561	0.737

Sum-RM, sum of $(Zr,Hf)_2O_5$, $(Nb,Ta)_2O_5$, and REO. [1]alkali tonsteins; [2]Zr(Hf)–Nb(Ta)–REE-rich coal, based on five samples; [3–6]tuffaceous Zr(Hf)–Nb(Ta)–REE ore deposit. [3]based on 14 samples; [4]based on 43 samples; [5]based on five samples; [6]Zr(Hf)–Nb(Ta)–REE-rich layer. *the total thickness of the ore layers including 301-10, −13, −14, −15, and −16, is 2.5 m. With the exception of the samples from the Huayingshan Coalfield, other samples were collected from boreholes, and 2009, 301, 701, 802, 1001, 1201, 1302, and 1908 are the borehole identities.

originated from different mantle sources under various partial melting conditions, and may have been subjected to fluid fractionation and contamination from lithospheric mantle and crustal material (Dai *et al.* 2011b, 2016a). The raw magmas that formed the argillized tuffs had an alkali basalt composition, similar to those of ELIP alkaline Nb–Ta-enriched syenites (Dai *et al.* 2011b, 2016a).

Common REE-, Zr-, and Nb-bearing minerals, such as columbite, samarskite, pyrochlore, or hafnon, have rarely been identified in these ore beds, either by XRD, under the optical microscope, or by SEM-EDS (Dai *et al.* 2010b). The highly elevated concentrations of rare metals in the argillized tuffs mainly occur in minerals such as rhabdophane, silico-rhabdophane, Nb-bearing anatase (or rutile), florencite, REE-bearing carbonates,

zircon, parisite, and xenotime (Figure 5). These minerals occur as finely dispersed grains or coarse-grained particles in clay matrices, microcrystalline aggregates in the pores of clay minerals, films on kaolinite surfaces, cavity-fillings of altered primary minerals of magmatic origin, or along the cleavage planes of clay minerals (Figure 5).

The REE-bearing minerals identified in the tuff commonly contain light rather than heavy REE, although, in a few cases, xenotime, fluor-carbonates, and phosphates (Figure 5) have been found in the tuffs and contain up to 0.2% Y. Ore horizons with a HREE enrichment type ($La_N/Lu_N < 1$; normalized to upper continental crust) have been generally observed in the ore-bearing tuffs. The heavy REY (HREY) in the tuffaceous materials are thus inferred to be mainly associated with the clay

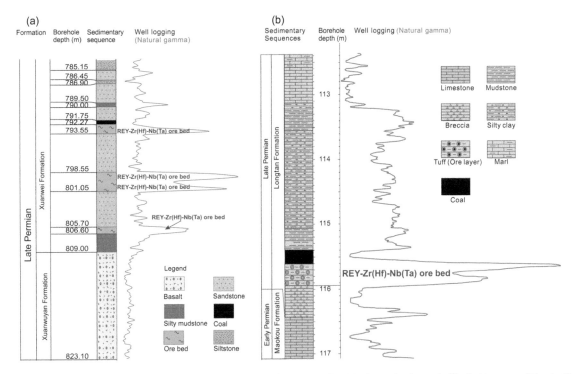

Figure 4. Positive anomalies in natural gamma-ray well logs of REY–Zr(Hf)–Nb(Ta) ore beds. A, Drillhole No. 2009-BX-4 in Xuanwei, eastern Yunnan Province. B, Drillhole No. CK6 in Sujiawan, near Chongqing (Dai *et al.* 2016a).

minerals (e.g. as adsorbed ions, Dai *et al.* 2010b; Seredin and Dai 2012; Zhao *et al.* 2016).

The modes of occurrence of these rare metal-bearing minerals indicate that they are either of authigenic origin or have been hydrothermally altered from primary minerals of magmatic origin. The minerals of authigenic origin were derived from precipitation of rare metals that were already present in the volcanic ash but then leached from the argillized tuffs by hydrothermal fluids (Seredin *et al.* 2011, 2012; Seredin and Chekryzhov 2012). Primary minerals of magmatic origin that have not been subjected to hydrothermal alteration are rarely observed. Chlorite, which has been commonly observed in the argillized tuffs, is associated with hydrothermal solutions and generally does not contain these rare metals (Figure 5).

Particularly, berthierine, which was formed under a low temperature (generally <200°C; cf. Aagaard *et al.*, 2000; Ryan and Hillier 2002; Rivas-Sanchez *et al.* 2006; Rivard *et al.* 2013), is commonly observed the REY–Zr (Hf)–Nb(Ta)–Ga-rich argillized tuffs. REE-bearing minerals (rhabdophane and florencite) are also generally related to low-temperature hydrothermal solutions (Smith *et al.* 2000; Krenn and Finger 2007; Berger *et al.* 2008; Roncal-Herrero *et al.* 2011). The presence of berthierine and REE-bearing minerals in the REY–Zr (Hf)–Nb(Ta)–Ga ore deposits indicate a low-temperature diagenetic environment for the ore formation. Based on

the clay assemblages of kaolinite, mixed layer illite/ smectite, and berthierine, as well as the fact that most of the I/S is R1 ordered with few cases of the R3 I/S, Zhao *et al.* (2016) deduced that the palaeo-diagenetic temperature could be up to 180°C but mostly 100–160°C.

3.3.2. Intra-seam alkali-rhyolite tonsteins and hydrothermal alteration

Some metalliferous coals with highly elevated concentrations of REY–Zr(Hf)–Nb(Ta)–Ga may also contain intra-seam alkali-rhyolite tonsteins. These alkali tonsteins have similar chemical compositions to the thick argillized tuffs discussed above. For example, the K1 Coal in the Huayingshan Coalfield contains 1710 ppm REY, 3617 $(Zr,Hf)_2O_5$, and 406 ppm $(Nb,Ta)_2O_5$ (ash basis; Table 4). The tonsteins with highly enriched REY–Zr(Hf)–Nb(Ta)–Ga (Table 4) have been subjected to leaching by ground water or hydrothermal solutions, with the leached REY–Zr(Hf)–Nb(Ta)–Ga from the tonsteins then being re-deposited as secondary minerals or adsorbed by organic matter in the directly underlying coal bench (Dai *et al.* 2014a). Due to the leaching process, the alkali tonsteins usually have lower Nb/Ta, Zr/Hf, and Yb/La ratios than the underlying coal benches, because of the greater solubility of the first elements (Nb, Zr, and Yb) in each element pair during leaching process

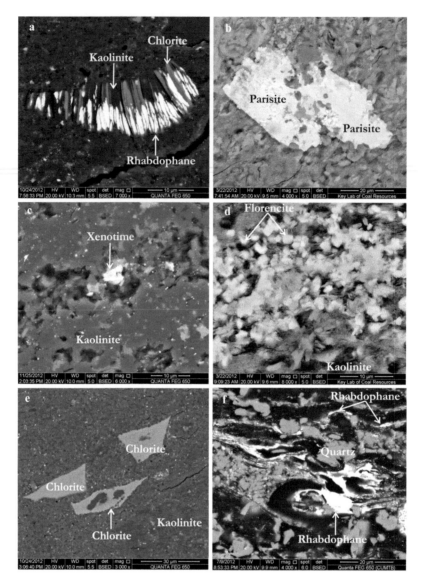

Figure 5. SEM backscattering images for minerals in coal-hosted REY–Zr(Hf)–Nb(Ta)-Ga ore deposits from southwestern China. A, Rhabdophane distributed along cleavage planes of clay minerals and chlorite (Dai *et al.* 2016a). B, Parisite in borehole 1908# in the Zhenxiong Coalfield (Dai *et al.* 2016a). C, xenotime in borehole 1908#. D, florencite in borehole 1908# in the Zhenxiong Coalfield. E, Chlorite derived from vitroclastics by hydrothermal fluids, with authigenic quartz filling in the chlorite cavities. F, Rhabdophane distributed along bedding planes in a coal sample (sample K1-2b) from the Huayingshan Coalfield.

(Figure 6; Dai *et al.* 2014a). Similar leaching processes have been described by Hower *et al.* (1999), Crowley *et al.* (1989), and Zhao *et al.* (2013, 2015).

Zircon is a resistant mineral and, thus, can only be altered with difficulty by ground water or hydrothermal solutions. The leaching properties of Zr, as well as REE in the alkali tonsteins, are related to their modes of occurrence. Zircon is generally observed in felsic tonsteins but rarely occurs in alkali tonsteins (Zhou *et al.* 2000; Dai *et al.* 2011) or in argillized tuffs, although the concentration of Zr is much lower in felsic than in alkali tonsteins. For example, the concentrations of Zr in two felsic tonstein layers at Xinde, eastern Yunnan, are 158 and 329 ppm,

respectively (Dai *et al.* 2014d), and two alkali-rhyolite tonsteins in the Huanyingshan Coalfield (Sichuan Province) contain 1338 and 1232 ppm Zr (Dai *et al.* 2014a). The major carrier of Zr in felsic tonsteins is zircon (Zhou *et al.* 1994; Dai *et al.* 2011), but Zr is probably associated with clay minerals in alkali tonsteins, for example in ion-absorbed form (Zhou *et al.* 2000; Zhao *et al.* 2016). The clay-mineral associated (e.g. ion-absorbed) Zr, as well as rare earth elements, could have been leached out by hydrothermal solutions or ground water and then re-deposited as secondary minerals (e.g. authigenic zircon; Dai *et al.* 2014a) or adsorbed by organic matter (such as heavy rare earth elements).

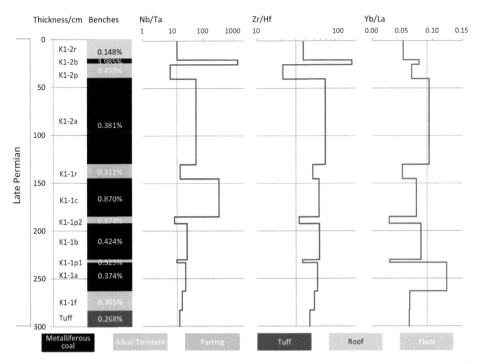

Figure 6. Total concentrations of rare metals and ratios of Nb/Ta, Zr/Hf, and Yb/La throughout the coal-hosted REY–Zr(Hf)–Nb(Ta) ore deposit in the Huayingshan Coalfield (Dai *et al.* 2016a). The number in the bench samples is the total rare metal concentrations. The vertical lines for Nb/Ta, Zr/Hf, and Yb/La represent the values for world hard coals.

The main REE-bearing minerals identified in metalliferous coals and alkali tonsteins both with highly elevated concentrations of REY are light-REE-bearing minerals of secondary origin (e.g. rhabdophane), either distributed along bedding planes (e.g. REE-bearing phosphates and silico-phosphates, which in some cases are also rich in U and Zr), or in collodetrinite and clay minerals. Heavy REE-bearing minerals have not been observed either in metalliferous coals or in alkali tonsteins, and, thus, the heavy REE are deduced to be dominantly associated with the organic matter of the coal and associated with clay minerals in alkali tonsteins. There are a few exceptions represented by monazite that are detrital materials of terrigenous origin (Zhuang *et al.* 2012) in the metalliferous coals.

A stronger organic affinity of H-REE relative to L-REE has been found by some studies (e.g. Goodarzi 1987; Eskenazy 1987a, 1987b, 1999; Palmer *et al.* 1990; Querol *et al.* 1995; Shpirt and Seredin 1999; Arbuzov and Ershov 2007; Arbuzov *et al.* 2012), and also has been confirmed by experimental studies of REE sorption on xylain and humic acid (Eskenazy 1999). In contrast to findings indicating strong HREY–organic associations, however, Finkelman (1981b) estimated that no more than 10% of the total REY in lignite had an organic association and the remaining REE were associated with REY-bearing minerals.

3.3.3. Hydrothermal solutions for enrichment of REY and other associated elements

A recent study by Dai *et al.* (2016c) showed that the K1 Coal (40-cm thick on average) deposited on mafic tuff in the Moxinpo Coalfield, Sichuan Province (Figure 1), is highly enriched in REY oxides (1239 ppm), as well as U (917 ppm), V_2O_5 (13,098 ppm), Cr_2O_3 (8569 ppm), Se (160 ppm), and Re (6.21 ppm) (all on ash basis). These highly elevated concentrations of REY–U–V–Cr–Se–Re were attributed to exfiltrational hydrothermal solutions, based on the mineralogical compositions (e.g. occurrence of roscoelite) and the distribution of U through the coal seam section, floor and roof strata as discussed more fully below.

The late Permian No. 25 Coal (semianthracite) from the Guxu (Gulin-Xuyong) Coalfield, Sichuan Province (Figure 1) has a medium-sulphur content (average 2.73%) and an average ash yield of 20.95%. Geochemical and mineralogical compositions show that mineral matter in the No. 25 Coal was derived from the felsic-intermediate rocks at the top of the Emeishan basalt sequence, rather than from the Emeishan mafic basalts (Dai *et al.* 2016b). The floor strata of the No. 25 Coal can be divided into two sub-sections. The upper sub-section of the sequence immediately below the No. 25 Coal consists of material with a felsic-intermediate composition derived from terrigenous sources and the lower sub-section is composed of mafic tuff. The floor strata have been affected by hydrothermal solutions, leading to enrichment and potentially economic rare metal resources of REY, Nb, Ta, Zr, Hf, and U (Table 4).

3.4. Uranium

Utilization of U from coal provided the first example of industrial metal production from a coal-hosted ore deposit. Uranium production from U-rich coals in the United States and the former USSR accelerated the nuclear industry in those two countries during the post–World War II years (Seredin 2012). However, after later discoveries of richer and larger uranium deposits, utilization of U from coal has attracted less attention. However, prospecting of coal-hosted U deposits has again increased in China in recent years.

Two enrichment types of U (usually accompanied by Re and Se, and in many cases, V, Cr, and Mo, and in a few cases, As and Hg) in coal have been identified (Shao et al. 2003; Zeng et al. 2005; Seredin and Finkelman 2008; Dai et al. 2015b, 2015c, 2015d, 2016c), namely an epigenetic infiltration type (type I) and a syngenetic or early diagenetic exfiltration type (type II).

Type I is connected with sandstone-hosted roll-type uranium deposits. A typical example is the coal-hosted U deposit in the Yili Basin, Xijiang Province, China (Figure 1), where coals with highly elevated U concentrations and associated industrially exploited sandstone roll-type U deposits are located within the Early Jurassic Sangonghe Formation (Min et al. 2001, 2005a, 2005b). The ore in the sandstone-hosted roll-type uranium deposit occurs in medium- to coarse-grained sandstones, which are composed of 40–60% quartz, 8–15% feldspar, 2–3% carbonaceous debris, and 20–40% rock fragments (Min et al. 2001).

The U-rich coals in the Yili deposit have a medium-sulphur content (1.32% on average) and have 0.51–0.59% vitrinite reflectance, indicating a high volatile C/B bituminous rank. The coals are characterized by enrichment assemblages of U–Se–Mo–Re and As–Hg. They contain significantly elevated U (varying greatly from several ppm to 7207 ppm), Se (up to 253 ppm), Mo (up to 1248 ppm), and Re (up to 34 ppm), as well as As (up to 234 ppm) and Hg (up to 3858 ppb). The coals with highly elevated U, Re, Mo, and Se are usually located in the upper part of the seam section and are in direct contact with the sandstone-hosted uranium deposit. The U enrichment in the coals depends on the following conditions (Seredin and Finkelman 2008; Dai et al. 2015d):

(1) U sources. The terrigenous materials for the mineral matter in the U-rich coals, which are mainly composed of Permo-Carboniferous intermediate-felsic igneous rocks and Hercynian granite, are themselves enriched in U (3.0–12.9 and 5.4–20.9 ppm, respectively; Hou et al. 2010) and have provided sources of U not only for the sandstone-hosted deposit (Min et al. 2001, 2005a, 2005b) but also for the associated U-rich coals (Yang et al. 2011a, 2011b; Dai et al. 2015d).

(2) An arid climate. Wang et al. (2006a, 2006b) suggest that aridity may have favoured oxidizing conditions and thus may have favoured leaching of U from the U-rich rocks in the sediment source region as soluble U(VI) (Min 1995; Spirakis 1996; Seredin and Finkelman 2008).

(3) Associated coarse-grained sediments. Coarse-grained sediments (e.g. pebbly or coarse-grained sandstone) immediately overlying the coalbed may serve as channels for migration of U-bearing solutions. The upper portion of the coal bed may thus have access to both oxygen-

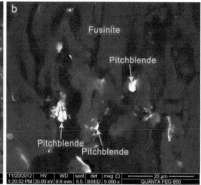

Figure 7. SEM backscattering images of U-bearing minerals (coffinite and pitchblende) in the cavities of fusinite in the Yili coals (Dai et al. 2015d).

and U-rich waters, leading to capture of U by the organic matter.

(4) Porous coal macerals. High proportions of structured fusinite and semifusinite with micro-pores (Figure 7) in the coal may have served not only as channels for migration of the U-bearing solutions but also acted as a reductant to facilitate the redox change from U(VI) to U(IV).

The minerals in the Yili U-rich coals are mainly composed of quartz, kaolinite, illite (illite/smectite), and, to a lesser extent, K-feldspar, chlorite, pyrite, and traces of dolomite, calcite, amphibole, millerite, cattierite, chalcopyrite, ferroselite, siegenite, krutaite, eskebornite, coffinite, pitchblende, zircon, and silicorhabdophane (Dai et al. 2015d). The terrigenous minerals and enriched trace elements in the U-rich coal were derived from a source region mainly composed of felsic and intermediate rocks, and from two different epigenetic solutions (a U–Se–Mo–Re-rich infiltrational and an Hg–As-rich exfiltrational volcanogenic solution). Uranium mainly occurs in the organic matter of the coal (Yang et al. 2011a; b; Dai et al. 2015d), although a small proportion of U-bearing minerals (pitchblende and coffinite; Figure 7) has also been identified, occurring as cavity-fillings in structured inertinite macerals.

Type II U enrichment in coal can be divided into two subtypes (subtypes 1 and 2). Subtype 1 is characterized by an enrichment assemblage of U, Se, Mo, Re, V, and Cr (Table 5), and is usually interlayered between impermeable clays (Seredin and Finkelman 2008; Dai et al. 2016c) or limestones (Dai et al. 2015b).

The U-rich coals of subtype 1 are mostly super-high-organic-sulphur (SHOS) coals interlayered between limestones and preserved within carbonate successions (Lei et al. 1994; Hou et al. 1995; Shao et al. 2003; Zeng et al. 2005; Li et al. 2013; Dai et al. 2015b). Such SHOS coals have been found in Yanshan (Yunnan Province), Guiding (Guizhou Province), Heshan (Guanxi Province), and Chenxi (Hunan Province). The U concentration in these SHOS coals usually ranges from ~100 to >1000 ppm (ash basis; Ren et al. 2006; Dai et al. 2008b; Dang et al. 2016), and the coals have the following characteristics:

(1) The subtype 1 (Type II) U-bearing SHOS coals have elevated concentrations of U–Se–Mo–Re–V–Cr, plus, in some cases, –REY, and are exclusively of late Permian age. By contrast, the coals with Type I U enrichment are of Palaeozoic, Mesozoic, or Cenozoic age (Seredin and Finkelman 2008). However, SHOS coals outside China are not limited to the late Permian age, e.g. coals of Tertiary age along the on-shore margin of the Gippsland Basin, Victoria, Australia (organic-sulphur content is between 5.2% and 7.4%; Smith and Batts 1974); the early Permian Tangorin coal seam of the Cranky Corner Basin, eastern Australia (Marshall and Draycott 1954; Ward et al. 2007); and the upper Palaeocene Raša coal from Istria (Slovenia), which contains up to 11% organic sulphur (Damste et al. 1999).

(2) The highly elevated sulphur and trace element concentrations listed above were derived from exfiltrational hydrothermal solutions during peat accumulation, which were then deposited in a euxinic environment (Dai et al. 2015b). Some studies, however, have variously attributed the highly elevated elements to a marine transgressive environment (Shao et al. 2003; Li et al. 2013) or to the formation of soil horizons prior to peat accumulation (Zeng et al. 2005).

(3) Although, as previously reported by Seredin and Finkelman (2008), the thickness of the U-bearing portion of the coal seams generally varies from 0.1 to 0.5 m and is rarely thicker than 1–2 m, the thickness of the U-bearing SHOS coal seams is generally less than 2 m. For example, the SHOS coals from Yanshan, Heshan, and Guiding coalfields are 1.91 m, <2 m and 0.2 to ~1 m in thickness, respectively.

(4) All the U-bearing SHOS coals are interlayered with marine carbonate rocks. The roof strata of the SHOS coal seams are generally limestones, and in some cases, siliceous or bioclast-rich limestones. The floor strata of the coals are limestones or mudstones.

(5) The trace elements (U, Se, Mo, Re, and V) with elevated-concentrations have a mixed mode of occurrence, but mainly occur in the organic matter. They may also be associated, to a lesser extent, with illite or mixed-layer illite/smectite (Liu et al. 2015), although traces of U-bearing minerals (e.g. coffinite and brannerite) have also been identified in such coals (Dai et al. 2015b).

Table 5. Rare metals in high-U coals and host rocks (ppm, on ash basis).

Coalfield	Ash	U	Se	Mo	Re	REO	V_2O_5	Cr_2O_3
Guiding*	23.1	950	152	1652	1.49	337	7134	2564
Yanshan*	27.51	556	91.6	742	1.1	556	3677	1748
Heshan*	36.69	126	35.5	125	nd	767	680	300
Chenxi*	13.96	539	nd	166	nd	1349	3783	4272
Yishan*	22.65	515	118	837	4.14	1316	3871	796
Fusui*	30.34	21	22.9	27.2	nd	1170	147	107
Moxinpo*	40.97	917	160	16	6.21	1239	13098	8569
Guxu**	82.0	228				1900		
World	nd	15	10	14	nd	485	303	175

*, coal; **, floor strata; nd, no data.

Chromium and Mo are partially distributed in sulphide minerals. A high proportion of the Re is also associated with the carbonate minerals (Liu *et al.* 2015).

(6) In contrast to Type 1, where the U concentration varies significantly through the vertical section and is usually enriched in the upper portions at the contacts with oxidized roof strata (Seredin and Finkelman 2008), the elevated concentrations of these trace elements in the SHOS coals of Type II deposits are consistent throughout the seam section.

As discussed above, the Moxinpo coals in Sichuan Province (Figure 1), which are also enriched in U (917 ppm on average, ash basis), were deposited on mafic tuff and overlain by impermeable clays (Dai *et al.* 2016c). The distribution of U concentration in the floor and roof strata, decreasing from top to bottom in the floor strata but from bottom to top in the roof succession, suggests that the U was derived from hydrothermal solutions during peat accumulation, and then permeated into the adjacent roof and floor strata from the coal bed during early diagenesis. The terrigenous rocks are made up of tuffs that do not contain abundant U and thus could not have provided such highly elevated U concentrations in the coal. Such syngenetic exfiltrational fluids may not only lead to the enrichment of U in coal itself, but also in the adjacent floor and roof strata (e.g. Moxinpo and Guxu; Figure 1). The upper portion of the floor strata of the No. 25 Coal in the Guxu Coalfield have been affected by hydrothermal solutions and is significantly enriched in U (Table 5), as well as REY, Nb, Ta, Zr, and Hf.

Subtype 2 of Type II U enrichment occurs in coal-hosted Ge ore deposits (e.g. Lincang Ge ore deposit) and is accompanied by enrichment of Be–Nb and Ge–W assemblages (Seredin *et al.* 2006; Dai *et al.* 2015c); other elements, including V, Cr, Mo, and Se, are depleted in the coals. The U in the Ge-rich coals of Lincang varies from several ppm to hundreds of ppm, with an average of ~56 ppm (coal basis) (Qi *et al.* 2004; Hu *et al.* 2009; Dai *et al.* 2015c). As with the U, the Ge was derived from leaching of U- and Ge-rich granite by hydrothermal solutions during peat accumulation.

Not all coal-hosted Ge ore deposits, however, are enriched in U. For example, the Wulantuga Ge ore deposit contains 0.36 ppm U on average, much lower than the average for world low-rank coals (2.9 ppm; Ketris and Yudovich 2009); the U concentration in the coals from the Spetzugli Ge ore deposit varies from 0.8 to 10.9 ppm, with an average of 2.6 ppm (Seredin *et al.* 2006).

3.5. Selenium

The average concentration of Se for world low-rank and hard coals is 1.0 and 1.6 ppm, respectively (Ketris and Yudovich 2009). Selenium has a low concentration in world coal ash (8.8 ppm) (Ketris and Yudovich 2009). Chinese coals have a background Se concentration of 2.47 ppm (Dai *et al.* 2012b), close to the arithmetic average for US coals (2.8 ppm; Finkelman 1993). These background values for Se in coal, however, are much higher than the average Se concentration in the upper continental crust (0.09 ppm; Rudnick and Gao 2003). Due to its high volatility (Clarke and Sloss 1992) and its potential for capture from the gas phase on ash particles, the concentration of Se in fly ash may be 20–100 times higher than that in the raw feed coals (Seredin *et al.* 2013), providing an opportunity of its extraction from fly ash, particularly for those coals with highly elevated Se concentrations.

3.5.1. Enrichment origin and mode of occurrence of Se in coal

Two genetic types of coal-hosted Se deposits have been identified in China:

(1) High Se concentrations, up to several hundreds to thousands of ppm, have been reported in U-rich coals connected with sandstone-hosted U deposits (Kislyakov and Shchetochkin 2000; Dai *et al.* 2015d). For example, Se is highly enriched in the Yili coals (up to 253 ppm, coal basis) associated with a sandstone roll-type U deposit (Yang *et al.* 2011a; Dai *et al.* 2015d). The high Se concentrations in the coal, as well as high concentrations of U, Mo, and Re, were derived from passage of infiltrational O_2-rich ground water through the coal basin during epigenesis. The coal itself acted as a reduction or sorption geochemical barrier for Se enrichment (Yudovich and Ketris 2006b).

(2) Some coals preserved within carbonate successions and characterized by very high organic S (4–11%) are also rich in Se (Table 5). The Se enrichment is generally accompanied by high levels of V, Mo, and U. The Se was derived from introduction of exfiltrational hydrothermal solutions with high SO_4 contents during peat accumulation (Dai *et al.* 2008b).

A number of modes of Se occurrence have been identified in high-Se coals in China; these include association with organic matter (Liu *et al.* 2015); pyrite (Dai *et al.* 2015b); marcasite (Dai *et al.* 2013); galena, clausthalite, and selenio-galena (Dai et al. 2006b); ferro-selite ($FeSe_2$), krutaite ($CuSe_2$), eskebornite ($CuFeSe_2$) (Dai et al. 2015d), and elemental form.

3.5.2. Enshi 'stone-coal' hosted Se ore deposit

A successful example of industrial Se utilization from coal-bearing strata is represented by the Enshi 'stone-coal'-hosted Se ore deposit, in Hubei Province of southern China (Figure 1). The carbonaceous siliceous rocks and carbonaceous shales of the middle Permian Gufeng Formation, locally referred to as 'stone-coal' in Enshi County, contain high concentrations of Se. The thicknesses of these carbonaceous siliceous rocks and carbonaceous shales are 5–15 and 3–10 m, respectively (Zheng et al. 2006). In some cases, the stone-coal bearing sequences are included in the upper portion of the early Permian Maokou Formation (Yao et al. 2002). The Se concentration varies greatly, from >10 ppm to several thousands of ppm (Yang et al. 1983; Song 1989; Mao et al. 1990; Zhu et al. 2004a). The Se concentration in two stone coal samples collected by Yang et al. (1981, 1983) and Song (1989) was up to 8290 and 84,123 ppm, respectively. Use of weathered 'stone coal' and stone-coal ashes as fertilizer provided sources for Se enrichment in the soil, which was then taken up by crops and consequently caused endemic selenosis from 1958 to 1987 (Dai et al. 2012b).

From an economic perspective, the stone-coals with highly elevated Se concentrations represent a potential raw material for industrial Se extraction. The stone-coal-hosted ore deposit is now considered as a unique Se deposit, and has been successfully utilized. Two of the major Se ore deposits in Enshi are the Yutangba and Shadi deposits. The Se concentration in the Yutangba deposit varies from 230 to 8590 ppm, with an average of 3638 ppm. The reserves with Se>800 ppm in the deposit are 45.699 t, along with 120.935 t V_2O_5 (concentration 0.38–0.52%) and 12.199 t Mo (concentration 0.036–0.054%)(Peng 2015). The estimated Se resources (Se >100 ppm) in the Shadi deposit are 213 Mt (Peng 2015). It is believed that the Se enrichment in the deposit is due to the influx of exfiltrational hydrothermal solutions (Xia et al. 1995; Zheng et al. 2006; Feng et al. 2009), although biogenic activity might have played a significant role in the Se enrichment (Yao et al. 2002). The elevated-element assemblage (e.g. V–Mo–Se) in the stone-coal hosted Se ore deposit is similar to that in the U-rich coals preserved within carbonate successions as reported by Shao et al. (2003), Zeng et al. (2005), and Dai et al. (2008b, 2015b).

The Se in the stone-coals may occur in pyrite, eskebornite, or mandarinoite (Zheng et al. 1992, 1993; Belkin et al. 2010); in the form of adsorbed Se on the organic matter (Song 1989); or as native Se (Zhu and Zheng 2001; Zhu

et al. 2004a, 2004b, Cu–Fe selenide, krutaite, klockmannite, naumannite, and mandairnoite (Zhu et al. 2004a, 2004b).

3.6. Gold and silver

Although precious metals concentrations up to commercial grade were found in coals of the United States as early as the end of the nineteenth century (Jenney 1903; Stone 1912), and some other coals particularly in Russia have subsequently been found to be anomalous in Au and Ag (e.g. Leonov et al. 1998; Valiev et al. 2002; Bakulin and Cherepanov 2003; Arbuzov et al. 2006; Seredin 2007), only very limited geochemical data on Au and Ag in Chinese coals have been reported.

3.6.1. Gold

The average concentrations of Au for world low-rank and hard coals are 3.0 and 4.4 ppb, respectively (Ketris and Yudovich 2009). Tang and Huang (2004) indicate that the Au concentration in Chinese coals ranges from 0.2 to 6 ppb, with an average of 3 ppb. Zhuang et al. (1998) reported that the late Palaeozoic coals from the Pingshuo and Datong Coalfields (both in Shanxi province; Figure 1) have average Au concentrations of 0.81 ppb (ranging from 0.2 to 2.26 ppb) and 0.46 ppb, respectively. One late Palaeozoic coal sample from the Kailuan Coalfield (Hebei Province; Figure 1) has been found to contain 0.89 ppb Au (Zhuang et al. 1999a,b).

By contrast, elevated Au concentrations have been observed in the late Permian coals of southwestern China. Based on 97 samples, the coals from southwestern China have an average Au concentration of 29.6 ppb, with a maximum value of 366 ppb (Ren et al. 2006). The elevated Au concentration usually occurs in the high-arsenic coals of southwestern China, particularly in western Guizhou Province (Belkin et al. 1997; Nie et al. 2006). Wang et al. (2010) detected Au concentrations of up to 1621 ppb in the host rock of a late Permian coal seam in southwestern China. Gold particles, less than 2 μm in size, have been observed by SEM-EDS (Dai et al. 2016b) in the floor strata of a late Permian coal from the Guxu Coalfield, Sichuan Province (Figure 1). The elevated concentrations of Au in coals from southwestern China are predominantly attributed to epithermal solutions and generally occur near Carlin-type gold deposits (Ding et al. 2001; Wang et al. 2010). Wang et al. (2010) detected ~0.1% Au in the fracture-filling pyrite of a late Permian coal from southwestern China. Not all the late Permian coals of southwestern China, however, have elevated concentrations of Au. For example, the Au concentration in a late Permian coal from the Yanshan Coalfield (Yunnan Province), which is preserved within a carbonate succession, is 2.94 ppb (Ren et al. 2006). A late

Permian coal sample from the Laochang Coalfield (Yunnan Province; Figure 1) contains 1.89 ppb Au (Ren *et al.* 2006).

The coal-hosted Ge ore deposits discussed above are also enriched in Au and Ag. The Au concentration in the Ge-rich coals is from tens of ppb to a few ppm in the Lincang and Wulantuga coal-hosted deposits currently being mined for Ge (Zhuang *et al.* 2006; Dai *et al.* 2012d; Seredin and Dai 2014), as well in the Spetzugli Ge deposit in the Russian Far East (Seredin, 2002, 2007). The Au concentrations in four coal samples from the Wulantuga Ge ore deposit are 26.5, 64.6, 8.71, and 46.5 ppb (Dai *et al.* 2012d). Although Au is also organically associated in the Ge-rich coal (Dai *et al.* 2012d), the Au concentration in the pyrite of the Wulantuga coal is 324.4 ppb, 130 times higher than that of the continental crust (Wedepohl 1995). Gold minerals and organically associated Au have been found in Ge-rich coals from the Pavlovsk Coalfield in the Russian Far East (Seredin 2004b, 2007; Bratskaya *et al.* 2009).

Due to its elevated concentration in the feed coal, Au is also enriched in the coal combustion products (e.g. fly ash) derived from the Ge-rich coal deposit. Gold occurs as drop-like particles with a size of $n \times 0.01$–$0.2 \ \mu m$ on the surface of glass spherules in fly ashes from both the Lincang and Wulantuga high-Ge coals (Seredin and Dai 2014). Gold in the coals, whether occurring as inorganic and organic association, is likely to be volatilized in the high-temperature zone (e.g. 1200–1400°C) of the power plants, and then condensed on the surface of the glass spherules in the cooler zone (<200°C) of the electrostatic and baghouse filters.

Zhang *et al.* (2016) found that the Au concentration in 51 Cretaceous coal samples from the Baiyinhua Coalfield (Figure 1) mostly ranges from 1.49 to 2.99 ppm. However, these reported concentrations of Au are highly questionable, because the authors did not pre-concentrate Au prior to the determinations, and thus Au concentration could have been overestimated due to spectroscopic interference during ICP-MS analysis. Even so, the enriched Au in the coals from the Baiyinhua Coalfield has been reported (e.g. Wen *et al.* 1993) to be enriched in the hydrothermally derived and hydrothermally altered pyrite and to be lower in the pyrite of sedimentary origin. The anthracite produced by an igneous intrusion in the Baiyinhua Coalfield contains up to 39 ppb Au (Wen *et al.* 1993).

3.6.2. Silver

The average concentrations of Ag in world low-rank and hard coals are 0.095 and 0.1 ppm, respectively (Ketris and Yudovich 2009). Swaine (1990) estimated that the concentration of Ag in common coals around the world varies from 0.02 to 0.2 ppm, with an average of less than 0.1 ppm. Based on 1670 lignite samples,

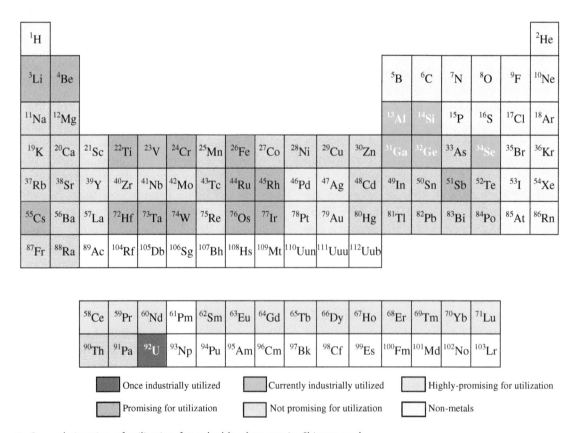

Figure 8. General situation of utilization for valuable elements in Chinese coal.

Bouška and Pešek (1999) estimated that the arithmetic and geometric averages for Ag concentration in lignite are 0.13 and 0.03 ppm respectively, with a maximum value of 11 ppm.

Tang and Huang (2004) estimated the average concentrations of Ag in Chinese coals (based on 76 samples) as ranging from 0.1 to 2 ppm and mostly in the range of 0.2–1 ppm, with an average of 0.5 ppm. Li et al. (1993) reported that the arithmetic and geometric averages of Ag in the Jurassic coals of the Ordos Basin in northern China are 0.3 and 0.1 ppm, respectively. Some coals from the Pingshuo Coalfield and a coal from the Kailuan Coalfield (Figure 1), both in northern China and of late Palaeozoic age, contain 0.94 ppm (ranging from 0.78 to 1.16 ppm) and 1.95 ppm Ag, respectively (Zhuang et al. 1998, 1999a, 1999b).

Elevated concentrations of Ag and Au in Chinese coals are mainly reported from southwestern China and the two coal-hosted Ge deposits (Lincang and Wulangtuga). For example, based on 97 late Permian coal samples from southwestern China, Ren et al. (2006) found that the Ag concentration varies from <0.1 to 3.3 ppm, with an average value of 0.6 ppm. Zeng et al. (1998) reported that the average Ag concentration of three late Permian coals from the Shuicheng Coalfield in Guizhou Province (Figure 1) is 0.36 ppm. Dai et al. (2014b) and Seredin and Dai (2014) found anomalous Ag occurring as thin films on the surface of fly ash particles derived from Ge-rich coals. The anomalous concentration of Ag in both the late Permian coals of southwestern China and Ge-rich coals has been attributed to the influence of epithermal solutions (Seredin and Dai 2014).

4. Conclusions

Based on studies over the past several decades, a number of coal-hosted deposits of valuable elements have been found in China. Uranium had been extracted from coal in the 1950s and Ge is being currently mined and industrially extracted from coal ash. Aluminium and Ga, as well as Si produced during the Al and Ga extraction, are also currently being industrially extracted from coal ash (Figure 8). Selenium and V are currently being industrially extracted from stone coal in southern China (Figure 8). Rare metals including REY, Cr, Ti, Li, Be, Au, Ag, Pt, Pd, etc., have high potential for industrial utilization. The overall features (e.g. locations, concentrations, and ages, and preliminary judgement of origin) of most deposits have become clear (Table 6), although more issues still require further investigation especially the spatial distribution of the ore beds, the modes of occurrence and origin of the rare metals particularly the

Table 6. Summary of rare metals in coal and coal-bearing strata.

Valuable elements	Elemental assemblage	Toxic elements	Host rocks	Age of host rocks	Enrichment origin for valuable elements	Modes of occurrence	Deposits
Ge	Ge–W	As, Hg, Sb, Tl, Be, Cs	Coal	Early Cretaceous	Alkaline N_2-bearing hydrothermal fluids	Organic association	Wulantuga
	Ge–W, Be–Nb–U	Be, As, Sb, Tl, Cs	Coal, metasomatite	Miocene	Mixture of alkaline N_2-bearing and volcanogenic CO_2-bearing fluids		Lincang
Ga and Al	Ge–W	As	Coal	Early Cretaceous	Hydrothermal fluids		Yimin
	Ga–Al–REY	F	Coal	Late carboniferous	Terrigenous	Various minerals and organic association	Jungar, Daqingshan
REY	REY–Zr(Hf)–Nb(Ta)–Ga		Argillized tuffs	Late Permian	Mixed tuffaceous- hydrothermal factors	Various minerals and absorbed ions in clays	Eastern Yunnan, Southern Sichuan, Chongqing
		Se	Coal	Late Permian	Tuffaceous, leaching by hydrothermal solutions on tonsteins	Various minerals and absorbed ions by organic matter	Yanshan, Chenxi, Fusui Moxinpo
	REY–U(Mo,Se)	F (in some cases)	Coal	Late Permian	Exfiltrational hydrothermal fluids	Various minerals and absorbed ions in clays	Guxu
	REY–U–V–Cr–Se–Re	Cd, F, Zn	Coal				
	REY–Nb–Ta–Zr–Hf–U	F	Floor strata				
U	U–Se–Mo–Re	As, Hg	Coal, roof sandstone	Early-Middle Jurassic	Infiltrational solution	Organic association, traces of U-bearing minerals	Yili
	U–Se–Mo–Re–V–Cr	F (in some cases)	Coal	Late Permian	Exfiltrational hydrothermal fluids		Guiding, Yanshan, Heshan, Chenxi
Se	Se–V–Mo		stone-coal	Middle Permian	Exfiltrational hydrothermal fluids	Various minerals and adsorbed by organic matter	Enshi

rare earth elements, the occurrence of toxic elements in the deposits and their adverse effects on human health and the environment, and the methods for economic extraction of the rare metals from the coals and associated non-coal strata.

Because of the wide distribution and huge reserves of coal in China, a number of different coal-forming periods, and the complex environments of coal formation and epigenetic evolution, more coal-hosted ore deposits could still be found in some coalfields, particularly in the southwestern and northeastern parts of the country. Although the study of coal-hosted metal ore deposits is challenging due to the issues discussed above, the extraction and utilization of rare metals from coal deposits are promising because of (1) the relatively low cost of the necessary infrastructure and mining; (2) the potential reduction of adverse environmental impacts due to extraction of otherwise toxic metals from coal ash (Seredin 2012); and (3) current limitations with developing new conventional metalliferous mines (Hower *et al.* 2016a), the decreasing reserves of metals in conventional deposits, and growing application of these valuable metals in modern technology.

Acknowledgements

We would like to thank three anonymous reviewers and Editor-in-Chief Robert J. Stern for their careful reviews and constructive comments, which greatly improved the paper quality.

Disclosure statement

No potential conflict of interest was reported by the authors.

Funding

This work was supported by the National Natural Science Foundation of China [grant number 41272182], [grant number 41420104001]; National Key Basic Research Programme of China [grant number 2014CB238900]; Programme for Changjiang Scholars and Innovative Research Team in University [grant number IRT13099].

ORCID

Shifeng Dai http://orcid.org/0000-0002-9770-1369
Xiaoyun Yan http://orcid.org/0000-0003-1681-3009
Colin R. Ward http://orcid.org/0000-0001-7945-5777
James C. Hower http://orcid.org/0000-0003-4694-2776
Lei Zhao http://orcid.org/0000-0001-9914-0385
Xibo Wang http://orcid.org/0000-0002-8758-2845
Lixin Zhao http://orcid.org/0000-0002-5563-7602
Deyi Ren http://orcid.org/0000-0003-1002-8472
Robert B. Finkelman http://orcid.org/0000-0002-7295-5952

References

Aagaard, P., Jahren, J., Harstad, A., Nilsen, O., and Ramm, M., 2000, Formation of grain-coating chlorite in sandstones: Laboratory Synthesized Vs. Natural Occurrences: Clay Minerals, v. 35, p. 261–269.

Arbuzov, S.I., and Ershov, V.V., 2007, Geochemistry of rare elements in coals of Siberia: Tomsk, D-Print, 468 p. [in Russian].

Arbuzov, S.I., Maslov, S.G., Volostnov, A.V., Il'enok, S.S., and Arkhipov, V.S., 2012, Modes of occurrence of uranium and thorium in coals and peats of Northern Asia: Solid Fuel Chemistry, v. 46, p. 52–66. doi:10.3103/S0361521912010028

Arbuzov, S.I., Rikhvanov, L.P., Maslov, S.G., Arhipov, V.S., and Belyaeva, A.M., 2006, Anomalous gold contents in brown coals and peat in the south-eastern region of the Western-Siberian platform: International Journal of Coal Geology, v. 68, p. 127–134. doi:10.1016/j.coal.2006.01.004

Bai, X., Li, W., Chen, Y., and Jiang, Y., 2007, The general distributions of trace elements in Chinese coals: Coal Quality Technology, no. 1, p. 1–4. [In Chinese with English abstract.]

Bakulin, Y.I., and Cherepanov, A.A., 2003, Gold and platinum in cinder wastes of thermal power station in the City of Khabarovsk: Rudy I Metaly, v. 1, p. 60–67. [in Russian].

Barakos, G., Mischo, H., and Gutzmer, J., 2016, An outlook on the rare earth elements mining industry: https://www.ausimmbulletin.com/feature/an-outlook-on-the-rare-earth-elements-mining-industry/accessed April 2016 24.

Belkin, H.E., Zheng, B., and Zhu, J., 2010, First occurrence of mandarinoite in China: Acta Geologica Sinica - English Edition, v. 77, p. 169–172. doi:10.1111/acgs.2003.77.issue-2

Belkin, H.E., Zheng, B.S., and Zhou, D.X., 1997, Preliminary results on the geochemistry and mineralogy of arsenic in mineralized coals from endemic arsenosis area in Guizhou Province, PR China, *in* 14th International Annual Pittsburgh Coal Conf. and Workshop Proceedings, Taiyuan: p. 1–20. CD-ROM.

Berger, A., Gnos, E., Janots, E., Fernandez, A., and Giese, J., 2008, Formation and composition of rhabdophane, bastnäsite and hydrated thorium minerals during alteration: implications for geochronology and low-temperature processes: Chemical Geology, v. 254, p. 238–248. doi:10.1016/j.chemgeo.2008.03.006

Blissett, R.S., Smalley, N., and Rowson, N.A., 2014, An investigation into six coal fly ashes from the United Kingdom and Poland to evaluate rare earth element content: Fuel, v. 119, p. 236–239. doi:10.1016/j.fuel.2013.11.053

Boudou, J.-P., Schimmelmann, A., Ader, M., Mastalerz, M., Sebilo, M., and Gengembre, L., 2008, Organic nitrogen chemistry during low-grade metamorphism: Geochimica Et Cosmochimica Acta, v. 72, p. 1199–1221. doi:10.1016/j.gca.2007.12.004

Bouška, V., and Pešek, J., 1999, Quality parameters of lignite of the North Bohemian Basin in the Czech Republic in comparison with the world average lignite: International

Journal of Coal Geology, v. 40, p. 211–235. doi:10.1016/S0166-5162(98)00070-6

Bouška, V., Pešek, J., and Peskova, J., 1995, Contents of trace elements of the European and world lignite deposits: Uhlì, Rudy, Geologický Pruůzkum, v. 2, p. 12–19. [Praha (in Czech].

Bragg, L.J., Oman, J.K., Tewalt, S.J., Oman, S.L., Rega, N.H., Washington, P.M., and Finkelman, R.B., 1998, Coal quality (COALQUAL) database: Version 2.0. Open-File Rep: Reston, VA, US Geological Survey, p. 97–134. [unpaginated CD-ROM].

Bratskaya, S.Y., Volk, A.S., Ivanov, V.V., Ustinov, A.Y., Barinov, N.N., and Avramenko, V.A., 2009, A new approach to precious metals recovery from brown coals: correlation of recovery efficacy with the mechanism of metal–humic interactions: Geochimica Et Cosmochimica Acta, v. 73, p. 3301–3310. doi:10.1016/j.gca.2009.03.010

Chou, C.-L., 1997, Abundances of sulfur, chlorine, and trace elements in Illinois Basin coals, USA, in Proceedings of the 14th Annual International Pittsburgh Coal Conference & Workshop, Taiyuan, China, September 1997 23–27, Section 1, p. 76–87.

Clarke, L.B., and Sloss, L.L., 1992, Trace elements—emissions from coal combustion and gasification: London, IEA Coal Research, 111 p.

Creelman, R.A., Ward, C.R., Schumacher, G., and Juniper, L., 2013, Relation between coal mineral matter and deposit mineralogy in pulverized fuel furnaces: Energy and Fuels, v. 27, p. 5714–5724. doi:10.1021/ef400636q

Crowley, S.S., Stanton, R.W., and Ryer, T.A., 1989, The effects of volcanic ash on the maceral and chemical composition of the C coal bed, Emery Coal Field, Utah: Organic Geochemistry, v. 14, p. 315–331. doi:10.1016/0146-6380(89)90059-4

Cui, X., and Li, J., 1991, The Late Mesozoic stratigraphy and palaeontology of the Erlian Basin Group in Inner Mongolia, China: Geoscience, the Journal of the Graduate School, China University of Geosciences, v. 5, p. 397–408. [In Chinese with English Abstract].

Cui, X., Li, J., 1993, Late Mesozoic basin types and their coal accumulation characteristics of Erlian basins in Inner Mongolia: Geoscience, the Journal of the Graduate School, China University of Geosciences, vol. 7, p. 479–484. (In Chinese with English Abstract).

Dai, S., Chekryzhov, I.Y., Seredin, V.V., Nechaev, V.P., Graham, I.T., Hower, J.C., Ward, C.R., Ren, D., and Wang, X., 2016a, Metalliferous coal deposits in East Asia (Primorye of Russia and South China): a review of geodynamic controls and styles of mineralization: Gondwana Research, v. 29, p. 60–82. doi:10.1016/j.gr.2015.07.001

Dai, S., Graham, I.T., and Ward, C.R., 2016d, A review of anomalous rare earth elements and yttrium in coal: International Journal of Coal Geology, v. 159, p. 82–95. doi:10.1016/j.coal.2016.04.005

Dai, S., Jiang, Y., Ward, C.R., Gu, L., Seredin, V.V., Liu, H., Zhou, D., Wang, X., Sun, Y., Zou, J., and Ren, D., 2012a, Mineralogical and geochemical compositions of the coal in the Guanbanwusu Mine, Inner Mongolia, China: further evidence for the existence of an Al (Ga and REE) ore deposit in the Jungar Coalfield: International Journal of Coal Geology, v. 98, p. 10–40. doi:10.1016/j.coal.2012.03.003

Dai, S., Li, D., Chou, C.-L., Zhao, L., Zhang, Y., Ren, D., Ma, Y., and Sun, Y., 2008a, Mineralogy and geochemistry of boehmite-rich coals: new insights from the Haerwusu Surface Mine, Jungar Coalfield, Inner Mongolia, China: International Journal of Coal Geology, v. 74, p. 185–202. doi:10.1016/j.coal.2008.01.001

Dai, S., Liu, J., Ward, C.R., Hower, J.C., French, D., Jia, S., Hood, M.M., and Garrison, T.M., 2016b, Mineralogical and geochemical compositions of Late Permian coals and host rocks from the Guxu Coalfield, Sichuan Province, China, with emphasis on enrichment of rare metals: International Journal of Coal Geology. doi:10.1016/j.coal.2015.12.004

Dai, S., Liu, J., Ward, C.R., Hower, J.C., Xie, P., Jiang, Y., Hood, M.M., O'Keefe, J.M.K., and Song, H., 2015a, Petrological, geochemical, and mineralogical compositions of the low-Ge coals from the Shengli Coalfield, China: a comparative study with Ge-rich coals and a formation model for coal-hosted Ge ore deposit: Ore Geology Reviews, v. 71, p. 318–349. doi:10.1016/j.oregeorev.2015.06.013

Dai, S., Luo, Y., Seredin, V.V., Ward, C.R., Hower, J.C., Zhao, L., Liu, S., Zhao, C., Tian, H., and Zou, J., 2014a, Revisiting the late Permian coal from the Huayingshan, Sichuan, southwestern China: enrichment and occurrence modes of minerals and trace elements: International Journal of Coal Geology, v. 122, p. 110–128. doi:10.1016/j.coal.2013.12.016

Dai, S., Ren, D., Chou, C.-L., Finkelman, R.B., Seredin, V.V., and Zhou, Y., 2012b, Geochemistry of trace elements in Chinese coals: a review of abundances, genetic types, impacts on human health, and industrial utilization: International Journal of Coal Geology, v. 94, p. 3–21. doi:10.1016/j.coal.2011.02.003

Dai, S., Ren, D., Chou, C.-L., Li, S., and Jiang, Y., 2006b, Mineralogy and geochemistry of the No. 6 coal (Pennsylvanian) in the Junger Coalfield, Ordos Basin, China: International Journal of Coal Geology, v. 66, p. 253–270. doi:10.1016/j.coal.2005.08.003

Dai, S., Ren, D., and Li, S., 2006a, Discovery of the superlarge gallium ore deposit in Jungar, Inner Mongolia, North China: Chinese Science Bulletin, v. 51, p. 2243–2252. doi:10.1007/s11434-006-2113-1

Dai, S., Ren, D., Zhou, Y., Chou, C.-L., Wang, X., Zhao, L., and Zhu, X., 2008b, Mineralogy and geochemistry of a super-high-organic-sulfur coal, Yanshan Coalfield, Yunnan, China: evidence for a volcanic ash component and influence by submarine exhalation: Chemical Geology, v. 255, p. 182–194. doi:10.1016/j.chemgeo.2008.06.030

Dai, S., Ren, D., Zhou, Y., Seredin, V.V., Li, D., Zhang, M., Hower, J.C., Ward, C.R., Wang, X., Zhao, L., and Song, X., 2014d, Coal-hosted rare metal deposits: genetic types, modes of occurrence, and utilization evaluation: Journal of China Coal Society, v. 39, no. 8, p. 1707–1715. [in Chinese with English abstract].

Dai, S., Seredin, V.V., Ward, C.R., Hower, J.C., Xing, Y., Zhang, W., Song, W., and Wang, P., 2015b, Enrichment of U–Se–Mo–Re–V in coals preserved within marine carbonate successions: geochemical and mineralogical data from the Late Permian Guiding Coalfield, Guizhou, China: Mineralium Deposita, v. 50, p. 159–186. doi:10.1007/s00126-014-0528-1

Dai, S., Seredin, V.V., Ward, C.R., Jiang, J., Hower, J.C., Song, X., Jiang, Y., Wang, X., Gornostaeva, T., Li, X., Liu, H., Zhao, L., and Zhao, C., 2014b, Composition and modes of occurrence of minerals and elements in coal combustion products derived from high-Ge coals: International Journal of Coal Geology, v. 121, p. 79–97. doi:10.1016/j.coal.2013.11.004

Dai, S., Wang, P., Ward, C.R., Tang, Y., Song, X., Jiang, J., Hower, J.C., Li, T., Seredin, V.V., Wagner, N.J., Jiang, Y., Wang, X., and Liu, J., 2015c, Elemental and mineralogical anomalies in the coal-hosted Ge ore deposit of Lincang, Yunnan, southwestern China: key role of N_2-CO_2-mixed hydrothermal solutions: International Journal of Coal Geology, v. 152, p. 19–46. doi:10.1016/j.coal.2014.11.006

Dai, S., Wang, X., Seredin, V.V., Hower, J.C., Ward, C.R., O'Keefe, J. M.K., Huang, W., Li, T., Li, X., Liu, H., Xue, W., and Zhao, L., 2012d, Petrology, mineralogy, and geochemistry of the Ge-rich coal from the Wulantuga Ge ore deposit, Inner Mongolia, China: new data and genetic implications: International Journal of Coal Geology, v. 90–91, p. 72–99. doi:10.1016/j.coal.2011.10.012

Dai, S., Wang, X., Zhou, Y., Hower, J.C., Li, D., Chen, W., Zhu, X., and Zou, J., 2011, Chemical and mineralogical compositions of silicic, mafic, and alkali tonsteins in the late Permian coals from the Songzao Coalfield, Chongqing, Southwest China: Chemical Geology, v. 282, p. 29–44. doi:10.1016/j.chemgeo.2011.01.006

Dai, S., Xie, P., Jia, S., Ward, C.R., Hower, J.C., Yan, X., and French, D., 2016c, Enrichment of U-Re-V-Cr-Se and rare earth elements in the Late Permian coals of the Moxinpo Coalfield, Chongqing, China: Genetic implications from geochemical and mineralogical data: Ore Geology Review. [under review].

Dai, S., Yang, J., Ward, C.R., Hower, J.C., Liu, H., Garrison, T.M., French, D., and O'Keefe, J.M.K., 2015d, Geochemical and mineralogical evidence for a coal-hosted uranium deposit in the Yili Basin, Xinjiang, northwestern China: Ore Geology Reviews, v. 70, p. 1–30. doi:10.1016/j.oregeorev.2015.03.010

Dai, S., Zhang, W., Ward, C.R., Seredin, V.V., Hower, J.C., Li, X., Song, W., Wang, X., Kang, H., Zheng, L., Wang, P., and Zhou, D., 2013, Mineralogical and geochemical anomalies of late Permian coals from the Fusui Coalfield, Guangxi Province, southern China: influences of terrigenous materials and hydrothermal fluids: International Journal of Coal Geology, v. 105, p. 60–84. doi:10.1016/j.coal.2012.12.003

Dai, S., Zhao, L., Hower, J.C., Johnston, M.N., Song, W., Wang, P., and Zhang, S., 2014c, Petrology, mineralogy, and chemistry of size-fractioned fly ash from the Jungar power plant, Inner Mongolia, China, with emphasis on the distribution of rare earth elements: Energy and Fuels, v. 28, no. 2, p. 1502–1514. doi:10.1021/ef402184t

Dai, S., Zhao, L., Peng, S., Chou, C.-L., Wang, X., Zhang, Y., Li, D., and Sun, Y., 2010a, Abundances and distribution of minerals and elements in high-alumina coal fly ash from the Jungar Power Plant, Inner Mongolia, China: International Journal of Coal Geology, v. 81, p. 320–332. doi:10.1016/j.coal.2009.03.005

Dai, S., Zhou, Y., Zhang, M., Wang, X., Wang, J., Song, X., Jiang, Y., Luo, Y., Song, Z., Yang, Z., and Ren, D., 2010b, A new type of Nb (Ta)-Zr(Hf)-REE-Ga polymetallic deposit in the late Permian coal-bearing strata, eastern Yunnan, southwestern China: possible economic significance and genetic implications: International Journal of Coal Geology, v. 83, p. 55–63. doi:10.1016/j.coal.2010.04.002

Dai, S., Zou, J., Jiang, Y., Ward, C.R., Wang, X., Li, T., Xue, W., Liu, S., Tian, H., Sun, X., and Zhou, D., 2012c, Mineralogical and geochemical compositions of the Pennsylvanian coal in the Adaohai Mine, Daqingshan Coalfield, Inner Mongolia, China: modes of occurrence and origin of diaspore, gorceixite, and ammonian illite: International Journal of Coal Geology, v. 94, p. 250–270. doi:10.1016/j.coal.2011.06.010

Damste, J.A.S., White, C.M., Green, J.B., and De Leeuw, J.W., 1999, Organosulfur compounds in sulfur-rich Raša coal: Energy and Fuels, v. 13, p. 728–738. doi:10.1021/ef980236c

Dang, J., Xie, Q., Liang, D., Wang, X., Dong, H., and Cao, J., 2016, The fate of trace elements in Yanshan coal during fast Pyrolysis: Minerals, v. 6, no. 2, p. 35. doi:10.3390/min6020035

Daniels, E.J., and Altaner, S.P., 1993, Inorganic nitrogen in anthracite from eastern Pennsylvania, U.S.A: International Journal of Coal Geology, v. 22, p. 21–35. doi:10.1016/0166-5162(93)90036-A

Ding, Z., Zheng, B., Zhang, J., Long, J., Belkin, H.E., Finkelman, R.B., Zhao, F., Chen, C., Zhou, D., and Zhou, Y., 2001, Geological and geochemical characteristics of high arsenic coals from endemic arsenosis areas in southwestern Guizhou Province, China: Applied Geochemistry, v. 16, p. 1353–1360. doi:10.1016/S0883-2927(01)00049-X

Du, G., Tang, D.Z., Wu, W., Sun, P.C., Bai, Y.L., Xuan, Y.Q., and Huang, G.J., 2003, Preliminary discussion on genetic geochemistry of paragenetic germanium deposit in Shenli Coalfield, Inner Mongolia: Geoscience, v. 17, p. 453–458. [in Chinese with English abstract].

Du, G., Tang, D.Z., Wu, W., Sun, P.C., Bai, Y.L., Yang, W.B., Xuan, Y.Q., and Zhang, L.C., 2004, Research on grade variation regularity of paragenetic germanium deposit along uprightness in Shengli Coalfield, Inner Mongolia: Coal Geology and Exploration, v. 32, no. 1, p. 1–4. [in Chinese with English abstract].

Du, G., Zhuang, X., Querol, X., Izquierdo, M., Alastuey, A., Moreno, T., and Font, O., 2009, Ge distribution in the Wulantuga high-germanium coal deposit in the Shengli coalfield, Inner Mongolia, northeastern China: International Journal of Coal Geology, v. 78, p. 16–26. doi:10.1016/j.coal.2008.10.004

Eskenazy, G.M., 1987a, Rare earth elements and yttrium in lithotypes of Bulgarian coals: Organic Geochemistry, v. 11, p. 83–89. doi:10.1016/0146-6380(87)90030-1

Eskenazy, G.M., 1987b, Rare earth elements in a sampled coal from the Pirin Deposit, Bulgaria: International Journal of Coal Geology, v. 7, p. 301–314. doi:10.1016/0166-5162(87)90041-3

Eskenazy, G.M., 1999, Aspects of the geochemistry of rare earth elements in coal: an experimental approach: International Journal of Coal Geology, v. 38, p. 285–295. doi:10.1016/S0166-5162(98)00027-5

European Commission, 2014, Report on critical raw materials for the EU. Report of the Ad Hoc Working Group on Defining Critical Raw Materials; European Commission: Bruxelles, Belgium, 41p. http://ec.europa.eu/DocsRoom/documents/10010/attachments/1/translations/en/renditions/native accessed 25 March 2016.

Feng, C., Yao, Y., Li, Y., Liu, X., and Sun, H., 2014, Thermal activation on calcium silicate slag from high-alumina fly ash: a technical report: Clean Technologies and Environmental Policy, v. 16, p. 667–672. doi:10.1007/s10098-013-0648-9

Feng, C.X., Liu, J., Liu, S., Hu, R., and Chi, G.X., 2009, Petrogenesis and sedimentary environment of the cherts from Yutangba, western Hubei Province: evidence from silicon, oxygen, carbon and sulfur isotopic compositions: Acta Petrologica Sinica, v. 25, p. 1253–1259.

Finkelman, R.B., 1981a, Coal geochemistry: practical applications, Volume 81–99: Washington, D.C., US Geol. Surv. Open-File Rep, p. 1–12.

Finkelman, R.B. 1981b, The origin, occurrence, and distribution of the inorganic constituents in low-rank coals. *In* Schobert, H.H., Ed., Proceedings of the Basic Coal ScienceWorkshop, Houston, TX, USA, Grand Forks, ND, December 1981 8–9, Grand Forks Energy Technology Center, p. 70–90.

Finkelman, R.B., 1993, Trace and minor elements in coal, *in* Engel, M., and Macko, S., eds., Organic Geochemistry: New York, Plenum, p. 593–607.

Franus, W., Wiatros-Motyka, M.M., and Wdowin, M., 2015, Coal fly ash as a resource for rare earth elements: Environmental Science and Pollution Research, v. 22, p. 9464–9474. doi:10.1007/s11356-015-4111-9

Goldschmidt, V.M., 1935, Rare elements in coal ashes: Industrial & Engineering Chemistry, v. 27, p. 1100–1102. doi:10.1021/ie50309a032

Goldschmidt, V.M., and Peters, C., 1933, Über die Anreicherung seltener Elemente in Steinkohlen: Nachrichten Von Der Gesellschaft Der Wissenschaften Zu Göttingen, Mathematisch-Physikalische Klasse, v. IV, p. 371–387.

Gong, B., Tian, C., Xiong, Z., Zhao, Y., and Zhang, J., 2016, Mineralogy and trace elements released during extraction of alumina from high aluminum fly ash in Inner Mongolia, China: International Journal of Coal Geology. [under review].

Goodarzi, F., 1987, Concentration of elements in Lacustrine coals from Zone A Hat Creek Deposit No 1, British Columbia, Canada: International Journal of Coal Geology, v. 8, p. 247–268. doi:10.1016/0166-5162(87)90034-6

Greene, J., 2012, Digging for rare earths: the mine where iPhones are born: http://www.cnet.com/news/digging-for-rare-earths-the-mines-where-iphones-are-born/

Guo, C., Zou, J., Wei, C., and Jiang, Y., 2013, Comparative study on extracting alumina from circulating fluidized-bed and pulverized-coal fly ashes through salt activation: Energy and Fuels, v. 27, p. 7868–7875. doi:10.1021/ef401659e

Hatch, G.P., 2012, Dynamics in the global market for rare earths: Elements, v. 8, p. 341–346. doi:10.2113/gselements.8.5.341

Hou, H.Q., Han, S.Y., and Ke, D., 2010, Integrated evaluation of sandstone-hosted uranium metallogenic potential in the southern margin of Yili Basin, Xinjiang, China: Geol. Bull. China, v. 29, p. 1517–1525. [in Chinese with English abstract].

Hou, X., Ren, D., Mao, H., Lei, J., Jin, K., Chu, P.K., Reich, F., and Wayne, D.H., 1995, Application of imaging TOF-SIMS to the study of some coal macerals: International Journal of Coal Geology, v. 27, p. 23–32. doi:10.1016/0166-5162(94)00018-U

Hower, J.C., Dai, S., Seredin, V.V., Zhao, L., Kostova, I.J., Silva, L.F.O., Mardon, S.M., and Gurdal, G., 2013a, A note on the occurrence of yttrium and rare earth elements in coal combustion products: Coal Combustion and Gasification Products, v. 5, p. 39–47.

Hower, J.C., Eble, C.F., Dai, S., and Belkin, H.E., 2016b, Distribution of rare earth elements in eastern Kentucky coals: indicators of multiple modes of enrichment?: International Journal of Coal Geology, v. 160–161, p. 73–81. doi:10.1016/j.coal.2016.04.009

Hower, J.C., Eble, C.F., O'Keefe, J.M.K., Dai, S., Wang, P., Xie, P., Liu, J., Ward, C.R., and French, D., 2015b, Petrology, palynology, and geochemistry of Gray Hawk Coal (Early Pennsylvanian, Langsettian) in eastern Kentucky, USA: Minerals, v. 5, no. 3, p. 592–622. doi:10.3390/min5030511

Hower, J.C., Granite, E.J., Mayfield, D.B., Lewis, A.S., and Finkelman, R.B., 2016a, Notes on contributions to the science of rare earth element enrichment in coal and coal combustion by-products: Minerals, v. 6, no. 2, p. 32. doi:10.3390/min6020032

Hower, J.C., Groppo, J.G., Henke, K.R., Hood, M.M., Eble, C.F., Honaker, R.Q., Zhang, W., and Qian, D., 2015a, Notes on the potential for the concentration of rare earth elements and yttrium in coal combustion fly ash: Minerals, v. 5, no. 2, p. 356–366. doi:10.3390/min5020356

Hower, J.C., Groppo, J.G., Joshi, P., Dai, S., Moecher, D.P., and Johnston, M.N., 2014, Location of cerium in coal-combustion fly ashes: implications for recovery of lanthanides: Coal Combustion and Gasification Products, v. 5, p. 73–78. doi:10.4177/CCGP-D13-00007.1

Hower, J.C., O'Keefe, J.M.K., Wagner, N.J., Dai, S., Wang, X., and Xue, W., 2013b, An investigation of Wulantuga coal (Cretaceous, Inner Mongolia) macerals: paleopathology of faunal and fungal invasions into wood and the recognizable clues for their activity: International Journal of Coal Geology, v. 114, p. 44–53. doi:10.1016/j.coal.2013.04.005

Hower, J.C., Ruppert, L.F., and Eble, C.F., 1999, Lanthanide, yttrium, and zirconium anomalies in the fire Clay coal bed, Eastern Kentucky: International Journal of Coal Geology, v. 39, p. 141–153. doi:10.1016/S0166-5162(98)00043-3

Hu, R., Qi, H., Zhou, M., Su, W., Bi, X., Peng, J., and Zhong, H., 2009, Geological and geochemical constraints on the origin of the giant Lincang coal seam-hosted germanium deposit, Yunnan, SW China: A review: Ore Geology Reviews, vol. 36, p. 221–234.

Huang, W., Sun, L., Ma, Y., Wan, H., Tang, X., Du, G., Wu, W., Qin, S., and Finkelman, R.B., 2007, Distribution and geological feature of the coal-Ge deposit of Shengli coalfield in Inner Mongolia of China: Journal of the China Coal Society, vol. 32, no. 11, p. 1147–1151. [In Chinese with English abstract.]

Huang, W., Wan, H., Du, G., Sun, L., Ma, Y., Tang, X., Wu, W., and Qin, S., 2008, Research on element geochemical characteristics of coal-Ge deposit in Shengli Coalfield, Inner Mongolia, China: Earth Science Frontiers, v. 15, p. 56–64. doi:10.1016/S1872-5791(08)60039-1

Huang, W., Yang, Q., Tang, D., Tang, X., and Zhao, Z., 2000, Rare earth element geochemistry of Late Palaeozoic Coals in North China: Acta Geologica Sinica (English Edition), v. 74, p. 74–83.

Jandova, J., and Vu, H., 2001, Processing of germanium-bearing fly ash: Prague, Metallurgy, Refractory and environment, p. 107–112

Jenney, W.P., 1903, The chemistry of ore deposition: American Institute Miner Engineering Transactions, v. 33, p. 445–498.

Jia, J., Li, X., Wu, P., Liu, Y., Han, C., Zhou, L., and Yang, L., 2015, Human health risk assessment and safety threshold of harmful trace elements in the soil environment of the Wulantuga Open-Cast Coal Mine: Minerals, v. 5, no. 4, p. 837–848. doi:10.3390/min5040528

Jiang, Y., Zhao, L., Zhou, G., Wang, X., Zhao, L., Wei, J., and Song, H., 2015, Petrological, mineralogical, and geochemical compositions of Early Jurassic coals in the Yining Coalfield, Xinjiang, China: International Journal of Coal Geology, v. 152, p. 47–67. doi:10.1016/j.coal.2015.07.011

Ketris, M.P., and Yudovich, Y.E., 2009, Estimations of Clarkes for Carbonaceous biolithes: world averages for trace element contents in black shales and coals: International Journal of Coal Geology, v. 78, p. 135–148. doi:10.1016/j.coal.2009.01.002

Kislyakov, Y.M., and Shchetochkin, V.N., 2000, Hydrogenic ore formation: Moscow, Geoinformmark, 608 p. [In Russian.]

Krenn, E., and Finger, F., 2007, Formation of monazite and rhabdophane at the expense of allanite during Alpine low temperature retrogression of metapelitic basement rocks from Crete, Greece: microprobe data and geochronological implications: Lithos, v. 95, p. 130–147. doi:10.1016/j.lithos.2006.07.007

Lei, J., Ren, D., Tang, Y., Chu, X., and Zhao, R., 1994, Sulfur-accumulating model of superhigh organosulfur coal from Guiding, China: Chinese Science Bulletin, v. 39, p. 1817–1821.

Leonov, S.B., Fedotov, K.V., and Senchenko, A.E., 1998, Economic gold recovery from cinder dumps of thermal power plants: Gorn. Zhurn, v. 5, p. 67–68. [in Russian].

Li, B., Zhuang, X., Li, J., and Zhao, S., 2014b, Geological controls on coal quality of the Yili Basin, Xinjiang, Northwest China: International Journal of Coal Geology, v. 131, p. 186–199. doi:10.1016/j.coal.2014.06.013

Li, H., Hui, J., Wang, C., Bao, W., and Sun, Z., 2014c, Extraction of alumina from coal fly ash by mixed-alkaline hydrothermal method: Hydrometallurgy, v. 147–148, p. 183–187. doi:10.1016/j.hydromet.2014.05.012

Li, H.M., Fei, S.Y., and Wang, S.J., 1993, Inorganic geochemistry of coal in middle Jurassic coal-bearing strata, Ordos basin: Beijing, Geologic Press, 1–52 p. [in Chinese].

Li, J., Zhuang, X., and Querol, X., 2011, Trace element affinities in two high-Ge coals from China: Fuel, v. 90, p. 240–247. doi:10.1016/j.fuel.2010.08.011

Li, J., Zhuang, X., Querol, X., Font, O., Izquierdo, M., and Wang, Z., 2014a, New data on mineralogy and geochemistry of high-Ge coals in the Yimin coalfield, Inner Mongolia, China: International Journal of Coal Geology, v. 125, p. 10–21. doi:10.1016/j.coal.2014.01.006

Li, J., Zhuang, X., Querol, X., Font, O., Moreno, N., Zhou, J., and Lei, G., 2012, High quality of Jurassic coals in the Southern and Eastern Junggar Coalfields, Xinjiang, NW China: geochemical and mineralogical characteristics: International Journal of Coal Geology, v. 99, p. 1–15. doi:10.1016/j.coal.2012.05.003

Li, S.P., Wu, W.F., Li, H.Q., and Hou, X.J., 2016, The direct adsorption of low concentration gallium from fly ash: Separation Science and Technology, v. 51, no. p, p. 395–402. doi:10.1080/01496395.2015.1102282

Li, W., Tang, Y., Deng, X., Yu, X., and Jiang, S., 2013, Geochemistry of the trace elements in the high organic sulfur coals from Chenxi coalfield: Journal of China Coal Society, v. 38, p. 1227–1223. [in Chinese with English abstract].

Liu, H., Sun, Q., Wang, B., Wang, P., and Zou, J., 2016, Morphology and composition of microspheres in fly ash from the Luohuang Power Plant, Chongqing, Southwestern China: Minerals, v. 6, no. 2, p. 30. doi:10.3390/min6020030

Liu, H., Zhang, Y., Wang, H., Jia, R., and Long, Y., 1991, Study on lithofacies paleogeography of coal-bearing formations from the Junger Coalfield: Beijing, Geological Publishing House, 1–35 p. [in Chinese with English abstract].

Liu, J., and Xu, Y., 1992, Distribution of Ge, Ga, As, S in the coal metamorphized by heat of sub-volcanics: Coal Geology & Exploration, v. 20, no. 5, p. 27–32. [in Chinese with English abstract].

Liu, J., Yang, Z., Yan, X., Ji, D., Yang, Y., and Hu, L., 2015, Modes of occurrence of highly-elevated trace elements in super-high-organic-sulfur Coals: Fuel, v. 156, p. 190–197. doi:10.1016/j.fuel.2015.04.034

Loges, A., Wagner, T., Barth, M., Bau, M., GÖb, S., and Markl, G., 2012, Negative Ce anomalies in Mn oxides: The role of Ce4 + mobility during water-mineral interaction: Geochimica et Cosmochimica Acta, vol. 86, p. 296–317.

Lu, J.L., Zhuang, H.P., Fu, J.M., and Liu, J.Z., 2000, Sedimentation, diagenesis, hydrothermal process and mineralization of germanium in the Lincang superlarge germanium deposit in Yunnan Province, China: Geochimica, v. 29, no. 1, p. 36–43. [in Chinese with English abstract].

Mao, D., Su, H., Yan, L., Wang, Y., and Yu, X., 1990, Investigation and analysis of endemic selenois in Exi Autonomous Prefecture of Hunbei province: Chinese Journal of Endemiology, v. 9, p. 311–314. [in Chinese].

Mardon, S.M., and Hower, J.C., 2004, Impact of coal properties on coal combustion by-product quality: examples from a Kentucky power plant: International Journal of Coal Geology, v. 59, p. 153–169. doi:10.1016/j.coal.2004.01.004

Marshall, C.E., and Draycott, A., 1954, Petrographic, chemical and utilization studies of the Tangorin high sulphur seam, Greta coal measures: New South Wales, University of Sydney, Department of Geology and Geophysics Memoir, 1954/1, 66 p.

Massari, S., and Ruberti, M., 2013, Rare earth elements as critical raw materials: focus on international markets and future strategies: Resources Policy, v. 38, p. 36–43. doi:10.1016/j.resourpol.2012.07.001

Mayfield, D.B., and Lewis, A.S., 2013, Environmental review of coal ash as a resource for rare earth and strategic elements. In Proceedings of the 2013 World of Coal Ash (WOCA) Conference, Lexington, KY, USA, April 2013 22–25, p. 22–25. http://www.flyash.info/2013/051-mayfield-2013.pdf (accessed March 2016 25).

Min, M., 1995, Carbonaceous-siliceous-pelitic rock type uranium deposits in southern China: geologic setting and metallogeny: Ore Geology Reviews, v. 10, p. 51–64. doi:10.1016/0169-1368(95)00005-M

Min, M.Z., Chen, J., Wang, J.P., Wei, G.H., and Fayek, M., 2005a, Mineral paragenesis and textures associated with sandstone-hosted roll-front uranium deposits, NW China: Ore Geology Reviews, v. 26, p. 51–69. doi:10.1016/j.oregeorev.2004.10.001

Min, M.-Z., Luo, X.-Z., Mao, S.-L., Wang, Z.-Q., Wang, R.C., Qin, L.-F., and Tan, X.-L., 2001, An excellent fossil wood cell texture with primary uranium minerals at a sandstone-hosted roll-type uranium deposit, NW China: Ore Geology Reviews, v. 17, no. 4, p. 233–239. doi:10.1016/S0169-1368(00)00007-X

Min, M.Z., Xu, H.F., Chen, J., and Fayek, M., 2005b, Evidence of uranium biomineralization in sandstone-hosted roll-front uranium deposits, northwestern China: Ore Geology Reviews, v. 26, p. 198–206. doi:10.1016/j.oregeorev.2004.10.003

Nandi, B.N., Brown, T.D., and Lee, G.K., 1977, Inert coal macerals in combustion: Fuel, v. 56, p. 125–130. doi:10.1016/0016-2361(77)90130-2

Nie, A.G., Huang, Z.Y., and Xie, H., 2006, Research on origin between high-As coal and gold deposit in southwestern area of Guizhou Province: Journal of Hunan University of Science & Technology (Natural Science Edition), v. 21, no. 3, p. 21–25.

Nieto, F., 2002, Characterization of coexisting NH_4- and K-micas in very low-grade metapelites: American Mineralogist, v. 87, p. 205–216. doi:10.2138/am-2002-2-302

Palmer, C.A., Lyons, P.C., Brown, Z.A., and Mee, J.S., 1990, The use of rare earth and trace element concentrations in vitrinite concentrates and companion whole coals (hvA bituminous) to determine organic and inorganic associations, in Chyi, L.L., and Chou, C.-L., eds., Recent advances in coal geochemistry: Colorado, Geological Society of America, Special paper 248, p. 55–63.

Palmer, C.A., Tuncalıb, E., Dennen, K.O., Coburn, T.C., and Finkelman, R.B., 2004, Characterization of Turkish coals: a nationwide perspective: International Journal of Coal Geology, v. 60, p. 85–115. doi:10.1016/j.coal.2004.05.001

Pecht, M.G., Kaczmarek, R.E., Song, X., Hazelwood, D.A., Kavetsky, R.A., and Anand, D.K., 2012, Rare earth materials: insights and concerns: College Park, MD, CALCE EPSC Press, 194 p.

Peng, X.-G., 2015, A preliminary analysis of the potential utilization of Se resources in Enshi, Hubei Province: Resources Environment & Engineering, v. 29, no. 4, p. 436–441. [in Chinese].

Qi, H., Hu, R., Su, W., and Qi, L., 2002, Genesis of carboniferous siliceous limestone in the Lincang germanium deposit and its relation with germanium mineralization: Geochimica, v. 31, p. 161–168. [in Chinese with English abstract].

Qi, H., Hu, R., Su, W., Qi, L., and Feng, J., 2004, Continental hydrothermal sedimentary siliceous rock and genesis of superlarge germanium (Ge) deposit hosted in coal: a study from the Lincang Ge deposit, Yunnan, China: Science in China Series D, v. 47, p. 973–984. doi:10.1360/02yc0141

Qi, H., Hu, R., and Zhang, Q., 2007a, Concentration and distribution of trace elements in lignite from the Shengli coalfield, inner Mongolia, China: implications on origin of the associated Wulantuga germanium deposit: International Journal of Coal Geology, v. 71, p. 129–152. doi:10.1016/j.coal.2006.08.005

Qi, H., Hu, R., and Zhang, Q., 2007b, REE geochemistry of the cretaceous lignite from Wulantuga germanium deposit, inner Mongolia, northeastern China: International Journal of Coal Geology, v. 71, p. 329–344. doi:10.1016/j.coal.2006.12.004

Qi, H., Rouxel, O., Hu, R., Bi, X., and Wen, H., 2011, Germaniumisotopic systematics in Ge-rich coal from the Lincang Ge deposit, Yunnan, Southwestern China: Chemical Geology, vol. 286, p. 252–265.

Qi, L., and Yuan, Y., 2011, Characteristics and the behavior in electrostatic precipitators of high-alumina coal fly ash from the Jungar power plant, Inner Mongolia, China: Journal of Hazardous Materials, v. 192, p. 222–225.

Querol, X., Fernández-Turiel, J.L., and López-Soler, A., 1995, Trace elements in coal and their behaviour during combustion in a large power station: Fuel, v. 74, p. 331–343. doi:10.1016/0016-2361(95)93464-O

Ren, D., Zhao, F., Dai, S., Zhang, J., and Luo, K., 2006, Geochemistry of trace elements in coal: Beijing, Science Press, 556 p. [in Chinese with English abstract].

Rivard, C., Pelletier, M., Michau, N., Razafitianamaharavo, A., Bihannic, I., Abdelmoula, M., Ghanbaja, J., and Villieras, F., 2013, Berthierine-like mineral formation and stability during the interaction of kaolinite with metallic iron at 90°C

under anoxic and oxic conditions: American Mineralogist, v. 98, p. 163–180. doi:10.2138/am.2013.4073

Rivas-Sanchez, M., Alva-Valdivia, L., Arenas-Alatorre, J., Urrutia-Fucugauchi, J., Ruiz-Sandoval, M., and Ramos-Molina, M., 2006, Berthierine and chamosite hydrothermal: genetic guides in the Peña Colorada magnetite-bearing ore deposit, Mexico: Earth, Planets, and Space, v. 58, p. 1389–1400. doi:10.1186/BF03352635

Roncal-Herrero, T., Rodríguez-Blanco, J.D., Oelkers, E.H., and Benning, L.G., 2011, The direct precipitation of rhabdophane ($REEPO_4 \cdot nH_2O$) nano-rods from acidic aqueous solutions at 5–100 °C: Journal of Nanoparticle Research, v. 13, p. 4049–4062. doi:10.1007/s11051-011-0347-6

Rozelle, P.L., Khadilkar, A.B., Pulati, N., Soundarrajan, N., Klima, M.S., Mosser, M.M., Miller, C.E., and Pisupati, S.V., 2016, A study on removal of rare earth elements from U.S. coal byproducts by ion exchange: Metallurgical and Materials Transactions E, v. 3, p. 6–17. doi:10.1007/s40553-015-0064-7

Rudnick, R.L., and Gao, S., 2003, Composition of the continental crust, in Rudnick, R.L., ed., The crust treatise on Geochemistry, Volume 3: (Holland, D., Turekian, K.K., Exec. Eds.). Amsterdam, Elsevier, p. 1–64.

Ryan, P., and Hillier, S., 2002, Berthierine/chamosite, corrensite, and discrete chlorite from evolved verdine and evaporite-associated facies in the Jurassic Sundance Formation, Wyoming: American Mineralogist, v. 87, p. 1607–1615. doi:10.2138/am-2002-11-1210

Saikia, B.K., Ward, C.R., Oliveira, M.L.S., Hower, J.C., Leao, F.D., Johnston, M.N., O'Bryan, A., Sharma, A., Baruah, B.P., and Silva, L.F.O., 2015, Geochemistry and nano-mineralogy of feed coals, mine overburden, and coal-derived fly ashes from Assam (North-east India): a multi-faceted analytical approach: International Journal of Coal Geology, v. 137, p. 19–37. doi:10.1016/j.coal.2014.11.002

Scott, C., Deonarine, A., Kolker, A., Adams, M., and Holland, J., 2015, Size distribution of rare earth elements in coal ash. in Paper Presented at the World of Coal Ash Conference, Nashville, TN May 5–7.

Seredin, V., and Finkelman, R.B., 2008, Metalliferous coals: a review of the main genetic and geochemical types: International Journal of Coal Geology, v. 76, p. 253–289. doi:10.1016/j.coal.2008.07.016

Seredin, V.V., 1994, The first data on abnormal Niobium content in Russian coals: Dokl Earth Science, v. 335, p. 634–636.

Seredin, V.V., 2004, The Au-PGE mineralization at the Pavlovsk Brown Coal Deposit, Primorye: Geology of Ore Deposits, v. 46, p. 36–63.

Seredin, V.V., 2007, Distribution and formation conditions of noble metal mineralization in coal-bearing basins: Geology of Ore Deposits, v. 49, p. 1–30. doi:10.1134/S1075701507010011

Seredin, V.V., 2012, From coal science to metal production and environmental protection: a new story of success: International Journal of Coal Geology, v. 90–91, p. 1–3. doi:10.1016/j.coal.2011.11.006

Seredin, V.V., Chekryzhov, I.Y., and Popov, V.K., 2011, Rare metal-bearing tuffs of Cenozoic coal basins of Primorye formed by transform interaction of lithosphere plates, in Khanchuk, A.I., ed., Geological processes in subduction, collision, and transform environments of lithosphere plates interaction: Vladivostok, Dal'nauka, p. 375–377.

Seredin, V.V., and Dai, S., 2012, Coal deposits as potential alternative sources for lanthanides and yttrium: International Journal of Coal Geology, v. 94, p. 67–93. doi:10.1016/j.coal.2011.11.001

Seredin, V.V., and Dai, S., 2014, The occurrence of gold in fly ash derived from high-Ge coal: Mineralium Deposita, v. 49, p. 1–6. doi:10.1007/s00126-013-0497-9

Seredin, V.V., Dai, S., and Chekryzhov, I.Y., 2012, Rare-metal mineralization in tuffaceous coal basins of Russia and China. Diagnosis of volcanogenic products in sedimentary sequences. In Proceedings of the Russian meeting with the international participations: Geoprint, Syktyvkar, p. 165–167. [In Russian.]

Seredin, V.V., Dai, S., Sun, Y., and Chekryzhov, I.Y., 2013, Coal deposits as promising sources of rare metals for alternative power and energy-efficient technologies: Applied Geochemistry, v. 31, p. 1–11. doi:10.1016/j.apgeochem.2013.01.009

Seredin, V.V., Danilcheva, Y.A., Magazina, L.O., and Sharova, I.G., 2006, Ge-bearing coals of the Luzanovka Graben, Pavlovka brown coal deposit, Southern Primorye: Lithology and Mineral Resources, v. 41, p. 280–301. doi:10.1134/S0024490206030072

Shao, L., Jones, T., Gayer, R., Dai, S., Li, S., Jiang, Y., and Zhang, P., 2003, Petrology and geochemistry of the high-sulphur coals from the Upper Permian carbonate coal measures in the Heshan Coalfield, southern China: International Journal of Coal Geology, v. 55, p. 1–26. doi:10.1016/S0166-5162(03)00031-4

Shibaoka, M., 1985, Microscopic investigation of unburnt char in fly ash: Fuel, v. 64, p. 263–269. doi:10.1016/0016-2361(85)90227-3

Shpirt, M.Y., and Seredin, V.V., 1999, Mode of occurrence REE in coals: Solid Fuel Chemistry, v. 3, p. 91–92.

Smith, J.W., and Batts, B.D., 1974, The distribution and isotopic composition of sulfur in coal: Geochimica Et Cosmochimica Acta, v. 38, p. 121–133. doi:10.1016/0016-7037(74)90198-7

Smith, M., Henderson, P., and Campbell, L., 2000, Fractionation of the REE during hydrothermal processes: constraints from the Bayan Obo Fe-REE-Nb deposit, Inner Mongolia, China: Geochimica Et Cosmochimica Acta, v. 64, p. 3141–3160. doi:10.1016/S0016-7037(00)00416-6

Song, C., 1989, A brief description of the Yutangba sedimentary type Se mineralized area in southwestern Hubei: Mineral Deposits, v. 8, p. 83–88. [in Chinese with English abstract].

Spears, D.A., and Zheng, Y., 1999, Geochemistry and origin of elements in some UK coals: International Journal of Coal Geology, v. 38, p. 161–179. doi:10.1016/S0166-5162(98)00012-3

Spirakis, C.S., 1996, The roles of organic matter in the formation of uranium deposits in sedimentary rocks: Ore Geology Reviews, v. 11, p. 53–69. doi:10.1016/0169-1368(95)00015-1

Stone, R.W., 1912, Coal near the black hills Wyoming-South Dakota: US Geological Survey Bulletin, v. 499, p. 1–66.

Sun, Y., Qi, G., Lei, X., Xu, H., Li, L., Yuan, C., and Wang, Y., 2016, Distribution and mode of occurrence of uranium in bottom ash derived from high-germanium coals: Journal of Environmental Sciences, v. 43, p. 91–98. doi:10.1016/j.jes.2015.07.009

Swaine, D.J., 1990, Trace elements in Coal: London, Butterworths, 278 p.

Taggart, R.K., Hower, J.C., Dwyer, G.S., and Hsu-Kim, H., 2016, Trends in the rare earth element content of U.S.-based coal combustion fly ashes: Environmental Science & Technology, v. 50, p. 5919–5926. doi:10.1021/acs.est.6b00085

Taggart, R.K., Hower, J.C., Dwyer, G.S., Ulug, O., and Hsu-Kim, H., 2015, Comparing extraction methods for rare earth elements in U.S. coal fly ashes. in 2015 World of Coal Ash (WOCA) Conference in Nashville, TN, May 2015 5–7.

Tang, X., and Huang, W., 2004, Trace elements in Chinese coals: Beijing, The Commercial Press, 6–11 24–25 p. [in Chinese].

US Department of Energy, 2011, Critical materials strategy: Washington, DC, U.S. Department of Energy, 197 p. http://energy.gov/sites/prod/files/DOE_CMS2011_FINAL_Full.pdf accessed March 2016 25.

Valentim, B., Shreya, N., Paul, B., Gomes, C.S., Sant'Ovaia, H., Guedes, A., Ribeiro, J., Flores, D., Pinho, S., Suárez-Ruiz, I., and Ward, C.R., 2016, Characteristics of ferrospheres in fly ashes derived from Bokaro and Jharia (Jharkand, India) coals: International Journal of Coal Geology, v. 153, p. 52–74. doi:10.1016/j.coal.2015.11.013

Valiev, Y.Y., Vol'nov, B.A., Pachadzhanov, D.N., and Gofen, G.I., 2002, Gold in Jurassic coals in the mountainous framing of the Tajik Basin and its prospecting significance: Geochemistry International, v. 49, no. p, p. 96–99.

Valkovič, V., 1983a, Trace elements in Coal, Volume 1: Boca Raton, FL, Chem. Rubber Co. Press, 210p.

Valkovič, V., 1983b, Trace elements in Coal, Volume 2: Boca Raton, FL, Chem. Rubber Co. Press, 281p.

Vassilev, S.V., and Vassileva, C.G., 1996, Mineralogy of combustion wastes from coal-fired power stations: Fuel Processing Technology, v. 47, p. 261–280. doi:10.1016/0378-3820(96)01016-8

Vleeskens, J.M., Menendez, R.M., Roos, C.M., and Thomas, C.G., 1993, Combustion in the burnout stage: the fate of inertinite: Fuel Processing Technology, v. 36, p. 91–99. doi:10.1016/0378-3820(93)90014-U

Wang, Q., Pan, J., Cao, S., Guan, T., and Zhang, G., 2006b, Super-enriching mechanism of disperse- elements Re and Se in interlayer oxidation — A case study of the Zhajistan interlayer oxidation zone sandstone-type uranium deposit, Yili Basin, Xijiang: Geological Review, v. 52, p. 358–362. [in Chinese with English abstract].

Wang, S., and Ge, L., 2007, Geochemical characteristics of rare earth elements in kaolin rock from Daqingshan coalfield: Coal Geology & Exploration, v. 35, no. 5, p. 1–5. [in Chinese with English abstract].

Wang, W., Qin, Y., Liu, X., Zhao, J., Zhao, J., Wang, J., Wu, G., and Liu, J., 2011, Distribution, occurrence and enrichment causes of gallium in coals from the Jungar coalfield, inner Mongolia: Science China Earth Sciences, v. 54, p. 1053–1068. doi:10.1007/s11430-010-4147-0

Wang, W.-F., Qin, Y., Sang, S.-X., Wang, J.-Y., and Wang, R., 2010, Advances in geochemical research on gold in coal: Journal of China Coal Society, v. 35, no. 2, p. 236–240. [in Chinese with English abstract].

Wang, Z., Li, Z., Guan, T., and Zhang, G., 2006a, Geological characteristics and metallogenic mechanism of no. 511 sandstone-type uranium ore deposit in Yili Basin, Xinjiang: Mineral Deposits, v. 25, p. 303–311. [in Chinese with English abstract].

Ward, C.R., and Christie, P.J., 1994, Clays and other minerals in coal seams of the Moura-Baralaba area, Bowen Basin,

Australia: International Journal of Coal Geology, v. 25, p. 287–309. doi:10.1016/0166-5162(94)90020-5

Ward, C.R., Corcoran, J.F., Saxby, J.D., and Read, H.W., 1996, Occurrence of phosphorus minerals in Australian coal seams: International Journal of Coal Geology, v. 30, p. 185–210. doi:10.1016/0166-5162(95)00055-0

Ward, C.R., Li, Z., and Gurba, L.W., 2007, Variations in elemental composition of macerals with vitrinite reflectance and organic sulphur in the Greta Coal Measures, New South Wales, Australia: International Journal of Coal Geology, v. 69, p. 205–219. doi:10.1016/j.coal.2006.03.003

Wedepohl, K.H., 1995, The composition of the continental crust: Geochimica Et Cosmochimica Acta, v. 59, p. 1217–1232. doi:10.1016/0016-7037(95)00038-2

Wen, X.D., Li, B.F., and Zheng, Q.W., 1993, Research on gold in coal formation in Pingzhuang Basin, Inner Mongolia: Geoscience - Journal of Graduate School, China University of Geoscience, v. 7, no. 3, p. 346–355. [in Chinese with English abstract].

Wu, W., Mo, R., and Wang, Z., 2002, Occurrence features and geological work of germanium resource in Yimin coal field, Inner Mongolia: Inner Mongolia Geology, v. 1, p. 27–30. [in Chinese].

Xia, B.D., Zhong, L.R., Fang, Z., and Lv, H.B., 1995, The origin of cherts of the early Permian Gufeng Formation in the Lower Yangtze area, eastern China: Acta Geologica Sinica, v. 36, no. 2, p. 125–137. [in Chinese with English abstract].

Yang, G., Wang, S., Zhou, R., and Sun, S., 1983, Endemic selenium intoxication of humans in China: The American Journal of Clinical Nutrition, v. 37, p. 872–881.

Yang, G., Zhou, R., Sun, S., and Wang, S., 1981, Research on the etiology of an endemic disease characterized by loss of nails and hair in Enshi county: Journal of the Chinese Academy Medicine, v. 3, no. Suppl. 2, p. 1–6. [in Chinese].

Yang, J.-Y., Di, Y.-Q., Zhang, W.-G., and Liu, S.-D., 2011b, Geochemistry study of uranium and other elements in brown coal of ZK0161 well in Yili Basin: Journal of China Coal Society, v. 36, no. 6, p. 945–952.

Yang, J.-Y., Wang, G., Shi, Z.-L., Ren, M.-C., Zhang, W.-G., and Liu, S.-D., 2011a, Geochemistry study of uranium and other elements in brown coal of ZK0407 well in Yili basin: Journal of Fuel Chemistry and Technology, v. 39, no. 5, p. 340–346. [in Chinese with English abstract].

Yang, L., Song, J., Bai, X., Song, B., Wang, R., Zhou, T., Jia, J., and Pu, H., 2016, Leaching behavior and potential environmental effects of trace elements in coal gangue of an opencast coal mine area, inner Mongolia, China: Minerals, v. 6, no. 2, p. 50. doi:10.3390/min6020050

Yang, N., Tang, S., Zhang, S., and Chen, Y., 2015, Mineralogical and geochemical compositions of the no. 5 coal in Chuancaogedan mine, Junger Coalfield, China: Minerals, v. 5, no. 4, p. 788–800. doi:10.3390/min5040525

Yao, L., Gao, Z., Yang, Z., and Long, H., 2002, Origin of seleniferous cherts in Yutangba Se deposit, southwest Enshi, Hubei Province: Science in China Series D: Earth Sciences, v. 45, p. 741–754. doi:10.1007/BF02878431

Yudovich, Y.E., and Ketris, M.P., 2006b, Selenium in coal: a review: International Journal of Coal Geology, v. 67, p. 112–126. doi:10.1016/j.coal.2005.09.003

Zeng, R., Zhuang, X., Koukouzas, N., and Xu, W., 2005, Characterization of trace elements in sulphur-rich Late Permian coals in the Heshan coal field, Guangxi, South China: International Journal of Coal Geology, v. 61, p. 87–95. doi:10.1016/j.coal.2004.06.005

Zeng, R.S., Zhao, J.H., and Zhuang, X.G., 1998, Quality of late Permian coal and its controlling factors in Shuicheng Mining District of Liupanshui Area, Guizhou: Acta Petrologica Sinica, v. 14, no. 4, p. 449–558. [in Chinese with English abstract].

Zhang, J.-Q., Zhang, H.-L., Du, J.-L., Huo, C., Pan, H.-Y., and Ding, L., 2016, Distribution characteristics of trace elements in coal of Inner Mongolia Baiyinhua coalfield: Journal of China Coal Society, v. 41, p. 310–315. [in Chinese with English abstract].

Zhang, W., Rezaee, M., Bhagavatula, A., Li, Y., Groppo, J., and Honaker, R., 2015, A review of the occurrence and promising recovery methods of rare earth elements from coal and coal by-products: International Journal of Coal Preparation and Utilization, v. 35, p. 295–330. doi:10.1080/19392699.2015.1033097

Zhao, L., Dai, S., Graham, I., and Wang, P., 2016, Clay mineralogy of coal-hosted Nb-Zr-REE-Ga mineralized beds from late Permian Strata, Eastern Yunnan, SW China: implications for paleotemperature and origin of the micro-quartz: Minerals, v. 6, no. 2, p. 45. doi:10.3390/min6020045

Zhao, L., Dai, S.-F., Zhang, Y., Wang, X.-B., Li, D., Lu, Y.-F., Zhu, X.-W., Sun, -Y.-Y., and Ma, Y.-W., 2008, Mineral abundances of high-alumina fly ash from the Jungar Power Plant, Inner Mongolia, China: Journal of China Coal Society, v. 33, no. 10, p. 1168–1172. [in Chinese with English abstract].

Zhao, L., and Graham, I.T., 2016, Origin of the alkali tonsteins from southwest China: implications for alkaline magmatism associated with the waning stages of the Emeishan large igneous province: Australian Journal of Earth Sciences, v. 63, p. 123–128. doi:10.1080/08120099.2016.1133456

Zhao, L., Ward, C.R., French, D., and Graham, I.T., 2013, Mineralogical composition of Late Permian coal seams in the Songzao Coalfield, southwestern China: International Journal of Coal Geology, v. 116–117, p. 208–226. doi:10.1016/j.coal.2013.01.008

Zhao, L., Ward, C.R., French, D., and Graham, I.T., 2015, Major and trace element geochemistry of coals and intra-seam claystones from the Songzao Coalfield, SW China: Minerals, v. 5, no. 4, p. 870–893. doi:10.3390/min5040531

Zhao, Y., Zhang, J., and Zheng, C., 2012, Transformation of aluminum-rich minerals during combustion of a bauxite-bearing Chinese coal: International Journal of Coal Geology, v. 94, p. 182–190. doi:10.1016/j.coal.2011.04.007

Zheng, B., Hong, Y., Zhao, W., Zhou, H., and Xia, W., 1992, Se-rich carbonaceous-siliceous rocks of West Hubei and local Se poisoning: Chinese Science Bulletin, v. 37, p. 1027–1029.

Zheng, B., Yan, L., Mao, D., and Thornton, I., 1993, The Se resource in southwestern Hubei province, China, and its exploitation strategy: Journal of Natural Resources, v. 8, p. 204–212. [in Chinese].

Zheng, X., Qian, H.-D., and Wu, X.-M., 2006, Geochemical and genetic characteristics of selenium ore deposit in Shuanghe, Enshi, Hubei Province: Geological Journal of China Universities, v. 12, no. 1, p. 83–92. [in Chinese with English abstract].

Zhong, D.L., 1998, Paleo-tethys orogenic belt in Western Sichuan and Yunnan: Beijing, Science Press, 331 p. [in Chinese].

Zhong, R., Sun, S., Chen, F., and Fu, Z., 1995, The discovery of Rhyo-tuffite in the Taiyuan Formation and stratigraphic correlation of the Daqingshan and Datong Coalfield: Acta Geoscience Sinica, v. 3, p. 291–301. [in Chinese with English abstract].

Zhou, A., and Jia, B., 2000, Analysis of Late Paleozoic conglomerates from Daqing Mountain in inner Mongolia: Journal of Taiyuan University Technology, v. 31, p. 498–504. [in Chinese with English abstract].

Zhou, Y., 1999, The alkali pyroclastic tonsteins of early stage of late Permian age in southwestern China: Coal Geology & Exploration, v. 27, p. 5–9. [in Chinese with English abstract].

Zhou, Y., Bohor, B.F., and Ren, Y., 2000, Trace element geochemistry of altered volcanic ash layers (tonsteins) in Late Permian coal-bearing formations of eastern Yunnan and western Guizhou Provinces, China: International Journal of Coal Geology, v. 44, p. 305–324. doi:10.1016/S0166-5162(00)00017-3

Zhou, Y., Ren, Y., Tang, D., and Bohor, B., 1994, Characteristics of zircons from volcanic ash-derived tonsteins in Late Permian coal fields of eastern Yunnan, China: International Journal of Coal Geology, v. 25, p. 243–264. doi:10.1016/0166-5162(94)90018-3

Zhu, J., and Zheng, B., 2001, Distribution of selenium in a mini-landscape of Yutangba, Enshi, Hubei Province, China: Applied Geochemistry, v. 16, p. 1333–1344. doi:10.1016/S0883-2927(01)00047-6

Zhu, J., Zuo, W., Liang, X., Li, S., and Zheng, B., 2004b, Occurrence of native selenium in Yutangba and its environmental implications: Applied Geochemistry, v. 19, p. 461–467. doi:10.1016/j.apgeochem.2003.09.001

Zhu, J.-M., Li, S.H., Zuo, W., Sykorova, I., Su, H.C., Zheng, B.-S., and Pesek, J., 2004a, Mode of occurrence of selenium in black Se-rich rocks of Yutangba: Geochimica, v. 33, no. 6, p. 634–640. [in Chinese with English abstract].

Zhuang, X., Querol, X., Alastuey, A., Juan, R., Plana, F., Lopez-Soler, A., Du, G., and Martynov, V.V., 2006, Geochemistry and mineralogy of the Cretaceous Wulantuga high germanium coal deposit in Shengli coal field, Inner Mongolia, Northeastern China: International Journal of Coal Geology, v. 66, p. 119–136. doi:10.1016/j.coal.2005.06.005

Zhuang, X., Su, S., Xiao, M., Li, J., Alastuey, A., and Querol, X., 2012, Mineralogy and geochemistry of the Late Permian coals in the Huayingshan coal-bearing area, Sichuan Province, China: International Journal of Coal Geology, v. 94, p. 271–282. doi:10.1016/j.coal.2012.01.002

Zhuang, X.G., Xiang, C.F., Zeng, R.S., and Xu, W.D., 1999b, Comparative studies of trace elements in coals from three different types of basins: Acta Petrologica Et Mineralogica, v. 18, no. 3, p. 255–263. [in Chinese with English abstract].

Zhuang, X.G., Yang, S.K., Zeng, R.S., and Xu, W.D., 1999a, Characteristics of trace elements in coals from several main coal districts in China: Geological Science and Technology Information, v. 18, no. 3, p. 63–66. [in Chinese with English abstract].

Zhuang, X.G., Zeng, R.S., and Xu, W.D., 1998, Trace elements in 9 Coal from Antaibao Open Pit Mine, Pingshuo, Shanxi Province: Earth Science - Journal of China University of Geosciences, v. 23, no. 6, p. 583–588. [in Chinese with English abstract].

Zou, J.-H., Li, D.-H., Liu, D., Chen, L.-J., Zhu, C.-S., and Ren, S.-C., 2012, The occurrences and genesis of minerals in Late Paleozoic coal from the Adaohai Mine, Inner Mongolia: Bulletin of Mineralogy, Petrology and Geochemistry, v. 31, no. 2, p. 135–138. [in Chinese with English abstract].

Zou, J.-H., Liu, D., Tian, H.-M., Li, T., Liu, F., and Tan, L., 2014, Anomaly and geochemistry of rare earth elements and yttrium in the late Permian coal from the Moxinpo mine, Chongqing, southwestern China: International Journal of Coal Science & Technology, v. 1, no. 1, p. 23–30. doi:10.1007/s40789-014-0008-3

Zou, J.-H., Liu, D., Tian, H.-M., Liu, F., Li, T., and Yang, H.-Y., 2013, Geochemistry of trace and rare earth elements in the Late Paleozoic Coal from Adaohai Mine, Inner Mongolia: Journal of China Coal Society, v. 38, no. 6, p. 1012–1018. [in Chinese with English abstract].

Zou, J.-H., Tian, H.-M., and Li, T., 2016, Geochemistry and mineralogy of tuff in Zhongliangshan mine, Chongqing, southwestern China: Minerals, v. 6, no. 2, p. 47. doi:10.3390/min6020047

Geochemistry of uranium in Chinese coals and the emission inventory of coal-fired power plants in China

Jian Chen, Ping Chen, Duoxi Yao, Wenhui Huang, Shuheng Tang, Kejian Wang, Wenzhong Liu, Youbiao Hu, Qingguang Li and Ruwei Wang

ABSTRACT

Uranium, which is of chemotoxic and radiotoxic, is connately present in coals. In view of its dual roles in resource recovery and environmental impact, the geochemistry of uranium in Chinese coals and its atmospheric emission by coal-fired plants in China are discussed in this article. The average uranium concentration of the majority of Chinese coals is 2.43 mg/kg, which is comparable to that of world coals. However, in coals from certain local regions of China, i.e. Yunnan, Guizhou, Guangxi, Xinjiang, Inner Mongolia, and Chongqing Provinces, the uranium is significantly enriched; and even abnormally enriched to form coal-hosted uranium deposits, e.g. coals from the Yili and Tarim Basins of Xinjiang Province, Bangmai Basin of Yunnan Province, and Mabugang Basin of Guangdong Province. Uranium in Chinese coals is associated with organic matter, silicates, pores, pelitic components, phosphate minerals, and uranium minerals. The enrichment of uranium in Chinese coals is attributed to the weathering of source rocks, volcanic ashes, magmatic intrusion, marine water influence, groundwater, hydrothermal fluids, organic matter, palaeoclimate, and the geologic conditions of coal-accumulating basins. Uranium is prior to partition into fly ash. About 62.9 tons of uranium were released into atmosphere from Chinese coal-fired power plants in 2014. Some environmental and human health problems were historically related to uranium in coal worldwide.

1. Introduction

Uranium is a member of the actinide series, with 16 radioactive isotopes, dominated by ^{238}U (99.27% by weight), ^{235}U (0.72%), and ^{234}U (0.0055%) (Bird 2012). Uranium is mainly used in two ways, as fuel in nuclear power plants, and as the main ingredient for nuclear weapons (Mudd 2014). The dominant isotope, ^{238}U, with a physical half-life of 4.5×10^9 years, gives a very low specific activity (1.24×10^4 Bq g^{-1} U); thus, uranium is primarily of chemically toxic rather than radiologically toxic (Bird 2012).

Coal contains certain quantities of the naturally occurring primordial ^{238}U and its decay products (Papastefanou 2010). There is a divergence of opinions about the relevance of radioactivity to coal usage (Swaine 1990). Coal combustion is considered as one source of radioactive material in the environment. Radiation doses from airborne emission of a coal-fired plant might be greater than those from a nuclear plant of comparable size (Eisenbud and Petrow 1964; McBride et al. 1978). The radioactive contamination caused by a coal-fired power plant is about 30 times higher than that of atomic power stations even if only 0.9 mg/kg uranium occurs in the coal (Glowiak and Pacyna 1980). Fly ash escaping from stacks contains radionuclides that are concentrated several times in comparison with their contents in coal or surface soil (Dai et al. 2007). The uranium content of the ash from the Ulaanbaatar cogeneration plant No. 4 in Mongolia is 10.6–154 mg/kg (Maslov et al. 2010). The waste disposal site near the Yatagan Thermal Power Plants of Turkey contributes to an enhancement of gross radium isotope activities of groundwater (Baba 2002). The obviously elevated activity concentration of ^{238}U (averaging 116 Bq/kg) in the soil of Ajka town in Hungary is related to the fly ash and slag deposition not only from the coal-fired power plant, but also from industrial activity and domestic use of uraniferous coal (Papp et al. 2002). Radionuclides released together with escaping fly ash from the coal-fired power plants have resulted in a relatively small increase of natural radioactivity levels in the Lodz region of Poland (Bem et al. 2002).

ⓑ Supplemental data for this article can be accessed here.

In contrast, Cujic et al. (2015) concluded that the operation of the largest coal-fired power plant in Serbia has no negative impact on the surrounding environment with regard to the natural radionuclides. The emissions of the Afsin-Elbistan power plants in Turkey have not increased the ^{238}U concentrations in the surface soil (Cayir et al. 2012).

Owing to the growing world population, increased gross domestic product, and reduction of greenhouse gas emissions, more nuclear power plants may be constructed in future (Gabriel et al. 2013), correspondingly increasing the uranium demand. The required uranium supply might rely on some lower grade uranium deposits (Mudd 2014). A large potential for low grade uranium reserves is contained in poor quality coals, including lignite, sub-bituminous coal, and carbonaceous shale (Perricos and Belkas 1969). Historically, uranium production from coals of high uranium content in the United States and the former USSR brought essential acceleration to the establishment of a nuclear industry in both countries during post-WWII years (Seredin 2012).

China is the largest coal producer and consumer in the world (3.87 and 4.12 billion tons in 2014, respectively) (National Bureau of Statistics of China 2016). The uranium in Chinese coals is of great importance because of potential environmental and economic impacts. The abundance and modes of occurrence of uranium in Chinese coals, some typical coal-hosted uranium deposits in China, and genetic factors for uranium enrichment in Chinese coal are comprehensively discussed in this article. Additionally, about 47% of Chinese coal consumption was used by power production and supply during 2010–2014 (National Bureau of Statistics of China 2016). The atmospheric emission inventory of uranium from coal-fired power plants in China is an essential consideration for the local and regional environment. The atmospheric emission of uranium from coal-fired power plants in China is evaluated as well. Finally, some environmental and human health problems related to uranium in coal are introduced.

2. Uranium in Chinese coals

2.1. Abundance of uranium in Chinese coals

Chen et al. (1985, 1989)) presented a pioneering report of the uranium concentration in coal samples collected from 110 coal mines in China; the uranium concentrations of 84% of the samples were lower than 5 mg/kg. Based on 135 coal samples from the main coalfields in China, Ren et al. (1999) reported that uranium contents ranged from 0.16 to 199 mg/kg, with an average of 7.52 mg/kg.

During 2004 and 2007, Tang and Huang (2004), Tang et al. (2006), Ren et al. (2006), and Yang (2007) assigned values of 3.00 mg/kg (1383 samples), 3.84 mg/kg (952 samples), 2.46 mg/kg (1319 samples), and 2.31 mg/kg (1535 samples) for uranium abundance in Chinese coals, respectively.

In 2012, Dai et al. (2012a) indicated an average of 2.43 mg/kg (1383 data) for uranium concentration in common Chinese coals, comparable to that of world coals (2.40 mg/kg) (Ketris and Yudovich 2009).

2.2. Enrichment of uranium in some Chinese coals

The enrichment of trace elements in coal can be classified by concentration coefficients (CC: ratio of element concentration in targeted coal to average element concentration in world (or possibly Chinese) coals of equivalent rank) by Dai et al. (2015a), i.e. abnormal enrichment (CC >100), significant enrichment (100> CC >10), enrichment (10> CC >5), slight enrichment (5> CC >2), and depletion (0.5> CC). Based on this suggestion, if the uranium concentration in coals is higher than 24.3 mg/kg, it can be classified as significant enrichment. Significant uranium enrichment in coal only occurs in local and limited areas in Yunnan, Guizhou, Guangxi, Xinjiang, Inner Mongolia, and Chongqing Provinces of China (see supplementary material).

Almost all the uranium-rich coals occur in southwestern and western China (Guangxi, Yunnan, Guizhou, Tibet, Xinjiang, Inner Mongolia, and Chongqing Provinces) (Figure 1), such as the Heshan coals in Guangxi (73.8 and 111 mg/kg) (Shao et al. 2003, 2006; Zeng et al. 2005; Dai et al. 2013a); the Lincang Dazhai Ge-rich coals (52.5 and 56.0 mg/kg) (Hu et al. 2009; Dai et al. 2015b), the Luquan coals (34.1 mg/kg) (Dai et al. 2006), and the Yanshan coal (153 mg/kg) in Yunnan (Dai et al. 2008a; Liu et al. 2015); the No. 2 coal of Puan Coalfield (133 mg/kg) (Yang 2006), No. 9 coal in the Zhijin Coalfield (49.6 mg/kg) (Dai et al. 2003), the late Permian coals in Qianxi Fault Depression Area (Zhang et al. 2004a), and the Guiding super-high organic sulphur coal in Guizhou (Liu et al. 2015; Dai et al. 2015a); the Late Triassic Tumen coal in Tibet Province (Fu et al. 2013); the Yili coals in Xinjiang (up to 7207 mg/kg) (Dai et al. 2015c); the Wulantuga Ge-rich lignite (Zhuang et al. 2006; Qi et al. 2007), and the boehmite-rich coals in the Haerwusu Surface Mine of Inner Mongolia (Dai et al. 2008b); the Moxinpo coals from the Huayingshan coalfield (Zhuang et al. 2003; Dai et al. 2017), and the No. 12 coal in the Songzao coalfield of Chongqing (18.0 mg/kg) (Dai et al. 2010).

Minor coals from other places in China (Figure 1) are slightly enriched in uranium, i.e. the anthracite in

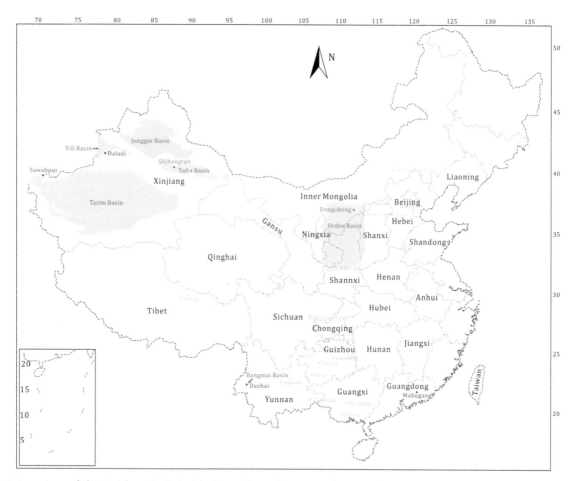

Figure 1. Locations of the U-rich coals (light blue), coal-hosted uranium deposits (dark blue), and some coal accumulating basins (yellow background) in China. The base map was cited from Podwysocki and Lovern (2000).

Shanxi Province (7.70 mg/kg) (Zhang *et al.* 2004b), the late Carboniferous coals in the Weibei coalfield of Shaanxi Province (7.25 mg/kg) (Wang *et al.* 2011), the Chuancaogedan No. 5 coal in Inner Mongolia (7.85 mg/kg) (Yang *et al.* 2015), and the Fusui No. 1 coal in Guangxi (7.34 mg/kg) (Dai *et al.* 2013b).

2.3. Modes of occurrence of uranium in coals

The modes of occurrence of uranium in coal may affect uranium extraction, because most of the uranium associated with minerals might be encapsulated by sintering of the ash (Zhang 1986), which would influence the partition, removal, and fate of the uranium during coal beneficiation and utilization.

In low-rank coals, uranium is generally organically associated (Swaine 1992; Seredin *et al.* 2006; Arbuzov *et al.* 2012; Hasani *et al.* 2014; Hower *et al.* 2016), in particular with humic material, and also partly presents as uranium minerals, e.g. uraninite or coffinite (Ilger *et al.* 1987; Douglas *et al.* 2011), ningyoite, and $CaU(PO_4)_2 \cdot 2H_2O$ (Havelcova *et al.* 2014). Uranium

minerals are always presented in a finely dispersed state, which makes the discrimination between minerals and organic compounds very difficult (Ilger *et al.* 1987). Uranium in the high organic sulphur Indian Palaeogene sub-bituminous coals likely has a clay association (Saikia *et al.* 2015). However, Arbuzov *et al.* (2012) stated that the main ash-forming aluminosilicate, silicate, and carbonate minerals did not play a significant role in the concentration of uranium in coals.

Regarding the Chinese coals, uranium is associated with:

(1) organic matter, such as in the Fengfeng-Handan coal in Hebei Province (Dai and Ren 2007), the Xundian and Zhaotong coals (Wang *et al.* 2015a), Yanshan coal (Liu *et al.* 2015), Luquan coal (Dai *et al.* 2006), and Lincang Ge-rich coal in Yunnan Province (Li *et al.* 2011), the Guiding coal in Guizhou Province (Liu *et al.* 2015; Dai *et al.* 2015a), the Heshan coal in Guangxi Province (Shao *et al.* 2003; Dai *et al.* 2013a), the Chuancaogedan coal in Inner Mongolia

Province (Yang et al. 2016), the No. 11 coal from Antaibao mining district in Shanxi Province (Wang et al. 2005), the Jinshan barkinite coal in Anhui Province (Sun et al. 2007), and even with organic sulphur in the Qinshui coal in Shanxi (Liu et al. 2016);

(2) pores and pelitic components as physical adsorption in lignite from borehole ZK0407 (Yang et al. 2011a), and inherent water in the upper benches of the Nos. 11 and 12 coals from borehole ZK0161 of the Yili Basin in Xinjiang Province (Yang et al. 2011b).

(3) silicates of the Fengfeng-Handan thermally altered coal in Hebei (Dai and Ren 2007), the Leping coal in Jiangxi Province (Querol et al. 2001), and the Jinshan barkinite coal in Anhui (Sun et al. 2007), the Wulantuga coal of Shengli coalfield in Inner Mongolia (Qi et al. 2007; Dai et al. 2012b), the Huayingshan coal in Chongqing Province (Zhuang et al. 2012), the Liupanshui coal in Guizhou (Zhuang et al. 2000), the Haizhou coal from Fuxin Basin of Liaoning Province (Querol et al. 1997); mixed-layer illite/smectite in the Shuoli coal in Anhui (Wang et al. 2015b) and the Yanshan and Guiding coals (Liu et al. 2015); kaolinite in the No. 2 coal of the Puan coalfield in Guizhou (Yang 2006) and the Donglin No. 4 coal of the Nantong coalfield in Chongqing (Chen et al. 2015); clay minerals in the Luquan coal in Yunnan (Dai et al. 2006), the Guanbanwusu No. 6 coal in Inner Mongolia (Xu et al. 2011), and the Wennan No. 15 coal of the Xinwen coalfield in Shandong Province (Querol et al. 1999), handpicking separated clay minerals in the late Permian coals from southwestern Guizhou (Zhang et al. 2002); the veined quartz of low temperature siliceous hydrothermal fluid origin in the No. 30 coal from the Zhijin coalfield (Dai et al. 2004a);

(4) phosphate minerals in the Xinzhouyao Mine of Datong, the Antaibao Surface Mine of Pingshuo, and the Majiata Surface Mine of Shenfu-Dongsheng (Song et al. 2007), the late Permian coals from southeastern Hubei Province (Zhuang et al. 2007); the apatite, xenotime, and monazite in Chongqing coal (Zhuang et al. 2003);

(5) uranium minerals, i.e. pitchblende, coffinite, U-bearing sulphate cavity-filled in structured inertinite macerals of the Yili coal (Dai et al. 2015c), brannerite in the Yanshan coal (Liu et al. 2015) and the Guiding coal (Liu et al. 2015; Dai et al. 2015a); and

(6) both organic and inorganic matter as a mixed affinity, e.g. the Guxu coal in Sichuan Province

(Dai et al. 2016a), the M9 coal of the Yanshan coalfield in Yunnan (Dai et al. 2008a), and the Adaohai coal from the Daqingshan coalfield in Inner Mongolia (Dai et al. 2012c).

Overall, with the exception of minor detectable uranium minerals, the organic matter and silicates are primarily the hosts of uranium in Chinese coals, though the detailed information about speciation and structure is limited.

3. Coal-hosted uranium deposits in china

Coal (with 30–50 mg/kg uranium) can be classified as an unconventional uranium resource along with phosphorite, carbonatite, and black shale (Qi et al. 2011). Uranium production from coal ash is technically feasible, and it was extracted from coal ash in the past in very specific historical situations, mostly due to the Cold War (Monnet et al. 2015).

If a coal had uranium concentration higher than 200 mg/kg, it could be regarded as a resource for industrial extraction (Dai et al. 2004b). However, Huang et al. (2012) suggested that 50 mg/kg was a comparable grade in coal to a low-grade yellowcake deposit. Furthermore, Sun et al. (2014) set the benchmark for uranium recovery as low as 40 mg/kg.

The occurrence of uraniferous coals is globally widespread, but apparently limited to lignites and subbituminous coals on small scale (Breger 1958), as in Chinese coals, with the exception of some bituminous coals preserved within carbonate successions in southern China (Dai et al. 2015a, 2016b).

3.1. Yili basin in Xinjiang

The Yili Basin covers about 40,000 km^2 in Kazakhstan and China, and about 16,000 km^2 in Xinjiang Province of China (Figure 1) (Liu and Jia 2011a). This basin contains a large amount of coal and uranium (coal- and sandstone-hosted).

The coal-hosted uranium deposits, with industrial grades ranging from 0.0503% to 0.5720% (averaging 0.0992%), occur in the Nos. 1 and 10 coals in the eastern part and the No. 12 coal in the western part of the south margin of the Yili Basin (Wang et al. 2015c). As exemplified by the Daladi uranium deposit (Figure 1) in the Daladi syncline, dominated by coal (more than 80% of the economic mineralization is developed at the redox transitional zone in the upper part of the lignite of the M1, M9, and M10 seams), occurs in the Early and Middle Jurassic Shuixigou Group (A and Zhang 2003; Liu and Jia 2011b).

The Yili coal is characterized by high concentrations of U (up to 7207 mg/kg), Se (up to 253 mg/kg), Mo (1248 mg/kg), and Re (up to 34 mg/kg), As (up to 234 mg/kg), and Hg (up to 3.86 mg/kg) (Dai et al. 2015c). Moreover, the concentrations of Se and Mo in the lignite also reach industrial grades (Liu and Jia 2011b).

In addition to the uranium held in the coal, most of the uranium reserves in the Yili Basin occurs as a roll-front deposit in the sandstone from interlayered oxidation zone at the Early and Middle Jurassic coal-bearing measures (Chen et al. 2002). Both the sandstone-hosted uranium deposit and the coal occur in the Shuixigou Group (Liu and Jia 2011a). The enrichment of uranium in the coal is closely related to the overlying sandstone-hosted uranium deposits (Chen et al. 2002; Dai et al. 2015c; Wang et al. 2015c).

The uranium was provided by the Carboniferous–Permian felsic and intermediate volcanic rocks in the source area to the south of the mining area (Liu and Jia 2011b), and partially by the Proterozoic epi-meta-morphic marine clastic and carbonate rocks (Liu and Jia 2011a).

3.2. Tarim basin in Xinjiang

The Sawabuqi uranium deposit in the Tarim Basin of Xinjiang Province (Figure 1) is hosted in the M9 and M1 coalbeds (averaging 360 mg/kg uranium in coal) in the Early and Middle Jurassic Tiemiersu Formation. The coal accumulated at a piedmont braided river channel and delta environment. After diagenesis, three tectonic movements affected this area, of which the Himalaya epoch was most favourable for uranium mineralization in coals. The Aoyibulake coal-hosted uranium minera-lized area was discovered at the western part of the mining area (Liu et al. 2011).

3.3. Bangmai basin in Yunnan

Uranium in the Tengchong coals of Yunnan Province was industrially used a few decades ago (Dai et al. 2012a), although no detailed data are available.

The Dazhai lignite occurs in the Bangmai Basin of Lincang City in Yunnan Province (Figure 1) and is a germanium–uranium deposit. The uranium deposit commonly underlies the Ge orebody (Zhang et al. 1987, 1988; Han et al. 1994). Mean concentrations of 52.5 and 56.0 mg/kg uranium in this coal were reported by Dai et al. (2015b), and Hu et al. (2009), respectively.

3.4. Mabugang basin in Guangdong

The No. 277 uranium deposit is located in the Eocene coal-bearing sequence of the Mabugang Basin in Guangdong Province (Figure 1). A total of 17 ore-bearing layers, which are dominated by coalbeds, occurred in the black strata (Wu 2012). The uranium concentrations in the coals range from 2360 to 15,410 mg/kg, with an average of 8726 mg/kg (Wu 2012).

4. Genetic factors for uranium enrichment in Chinese coals

A variety of factors, including coal rank, chemistry of groundwater, local hydrology, enclosing lithologies, weathering of the surrounding rocks, and depositional environment contributed to the uranium distribution in Canadian coals (Van Der Flier and Fyfe 1985). Based on the origins of uraniferous solutions, two enrichment types of uranium in coals were proposed, i.e. infiltration and exfiltration (Seredin and Finkelman 2008; Dai et al. 2016c, 2017).

With respect to the uranium enrichment in Chinese coals, the weathering of source rocks, volcanic ashes, magmatic intrusion, marine water, hydrothermal fluids, groundwater, organic matter, palaeoclimate, and geologic conditions of coal-accumulating basin, are listed as genetic factors in publications reviewed for this study.

4.1. Weathering of source rocks

The sediment source region located at the margin of a coal basin is the dominant factor for the back-grounds of trace elements in coals (Dai et al. 2012a). The physical and chemical weathering of source rocks, especially for the immediate and felsic igneous rocks with high uranium contents, are the source of uranium for most coals, including uranifer-ous coals.

Examples of source rock influence on the uranium content of Chinese coals include the enriched uranium (20.2 mg/kg) in the Mengtuo coal in Lincang City of Yunnan Province, relating to the weathered Lincang granite around the coal-accumulating basin (Chen et al. 2016). The felsic-intermediate rocks at the top of the Emeishan basalt sequence partially provided the No. 25 coal of Guxu coalfield in Sichuan Province with uranium (Dai et al. 2016a). The source rock is an important geological factor for the enrichment of ura-nium and other trace elements in the late Permian coals from western Guizhou Province that were not influenced by other factors (Dai et al. 2005). The terrigenous detrital materials contributed to the enrichment of uranium (133 mg/kg) in the Puan No. 2 coal in Guizhou (Yang 2006). Moreover, uranium (38.2 mg/kg) in the Datong No. 4 coal in Shanxi Province was provided to the coal basin by terrigenous detrital debris (Wang et al. 2010). The small concentration range of uranium (1.20–3.90 mg/kg) in the late Permian coals from Huainan coalfield in Anhui Province, correlating with ash yields, indicates a terrigenous input (Huang et al. 2001).

4.2. Volcanic ashes and magmatic intrusion

Synsedimentary volcanic ashes, which immediately overlie or are intercalated with the coalbed, and were subsequently altered to tonsteins, are direct sources of uranium for some coals.

High concentration of uranium (49.6 mg/kg) found in the No. 9 coal of the Zhijin Coalfield in Guizhou Province was derived from synsedimentary volcanic ash during peat accumulation (Dai et al. 2003, 2005). Elevated concentration of uranium in the late Permian Huayingshan coal of Sichuan Province arose in part from volcanic material (Zhuang et al. 2012). Additionally, the felsic volcanic debris exerted a significant influence on the enrichment of uranium (7.25 mg/kg) in the late Carboniferous Weibei coals in the southeastern Ordos Basin (Figure 1) (Wang et al. 2011).

The high uranium content (18.0 mg/kg) of the No. 12 coal of the Songzao coalfield in Chongqing Province probably resulted from the transportation of dissolved uranium to the swamp from the mafic tuffs (overlaid by the No. 12 coal) (Dai et al. 2010). Moreover, uranium is enriched in all the samples of the Tonghua No. 10 coal and two coal samples near the alkali tonstein band of the Yuyang No. 11 coal in this coalfield (Zhao et al. 2015).

The higher U/Th ratios in the benches of the Xinde coal from Xuanwei in Yunnan Province (Dai et al. 2014a) and in the Lvshuidong coals from the Huayingshan coalfield (Dai et al. 2014b) relative to the partings and host rocks (roof and floor) were due to the leaching of the uranium from partings of volcanic ash origin. Furthermore, the leaching of the roof altered by felsic volcanic ash makes a significant contribution to slight uranium enrichment (5.80 mg/kg) in the underlying K8 coal of the Changhe Mine in Sichuan (Wang 2009).

The enrichment of uranium in natural coke from the Shuoli Mine in Anhui Province is associated with the veins of mixed-layer I/S precipitated from the intrusive magma (Wang et al. 2015b). Moreover, the magmatic intrusions were the source of uranium in the Fengfeng-Handan thermally altered coals (Dai and Ren 2007).

4.3. Marine water influence

The uranium concentration of marine water (0.0032 mg/L) is generally hundredfold higher than that of stream water (0.00004 mg/L) (Reimann and De Caritat 1998). Thus, some U-rich coals, especially coals preserved within marine carbonate successions, was ascribed to marine water influence (Shao et al. 2003; Zeng et al. 2005). Owing to the invasion of marine water, pH, Eh values, and hydrogen sulphide contents in peat swamp were changed and produced a particular geochemical environment that benefited the enrichment of uranium (Wang et al. 2005, 2008).

The marine water invasion led to the enrichment of uranium (45.0 mg/kg) in the Datong No. 8 coal in Shanxi Province (Wang et al. 2010). A slight uranium enrichment in the Jinshan coal (averaging 8.14 mg/kg) of Guangde City in Anhui Province is also attributable to marine water influence during peat accumulation (Sun et al. 2007). The lower concentration of Th (6.5 mg/kg) than U (14 mg/kg) in western Guizhou Province late Permian coals has been attributed to significant marine influence (Dai et al. 2005), exemplified by the Fenghuangshan No. 27 coal of the Zhina coalfield (Li et al. 2017). The enriched uranium (34.1 mg/kg) in the Luquan coals in Yunnan Province is probably due to seawater influence during coal formation (Dai et al. 2006).

The super high organic sulphur Heshan coal in Guangxi Province, which accumulated in low-lying, marine-influenced palaeo-mires and developed on a confined carbonate platform, has high content of uranium (Shao et al. 2003; Zeng et al. 2005). However, Dai et al. (2013a), Dai et al. (2016c) ascribed the high concentration of uranium in this coal to hydrothermal solutions during peat accumulation or at the early diagenetic stages (syngenetic exfiltration process).

4.4. Hydrothermal fluids

Hydrothermal fluids mainly act as carrier and transporter of either exfiltrated or infiltrated fluids for uranium enrichment in coals, and as uranium sources.

The enriched element assemblage of Ge–Be–Nb–W–U in the Lincang lignite in Yunnan Province has been attributed to the leaching of the associated batholith granite by hydrothermal solutions of alkaline N_2-bearing and volcanogenic CO_2-bearing fluids (Dai et al. 2015b), and in the Ge-rich Wulantuga No. 6–1 coal in the Shengli coalfield of Inner Mongolia Province has been correlated with an epigenetic lateral transfer of Ge-bearing solutions passing through and leaching the granitic source rocks (Zhuang et al. 2006; Qi et al. 2007).

Elements S, U, Se, Mo, Re, and V in the Guiding coal of Guizhou Province were largely derived from exfiltrated hydrothermal solutions during peat accumulation (Dai et al. 2015a), additionally, the highly-elevated concentrations of U–Re–V–Cr–Se assemblage in the Moxinpo K1 coal in Chongqing Province were derived from exfiltrated hydrothermal solutions (Dai et al., 2017). The high concentrations of uranium (up to 7207 mg/kg), Se, and Mo in the Yili coal of Xinjiang Province were related to the later-stage U–Se–Mo–Re rich infiltrated solution (Dai et al. 2015c). The enrichment of uranium at the top of coal seam 15 in west Shandong Province, was probably due to epigenetic input by diagenetic infiltration of U-rich fluids (Querol et al. 1999).

About 8 mg/kg uranium was detected by SEM-EDX in the veined quartz originating from siliceous low-temperature hydrothermal fluids in the No. 30 coal of the Zhijin coalfield in Guizhou (Dai et al. 2004a). The occurrence of albite and dawsonite and strong enrichment of F, S, V, Cr, Ni, Mo, and U in the Yanshan coal of Yunnan resulted from the influence of submarine exhalation which invaded along with marine water into the anoxic peat (Dai et al. 2008a).

4.5. Groundwater

Groundwater, especially oxygenated phreatic water, plays dual roles in carrying oxygen to generate uranyl ions and as a medium for redox reactions.

Huang et al. (2012) stated that uranium in coal was mainly derived from water entering the peat swamp during peat accumulation and from the groundwater interacting with the coalbed after coal formation. The elevated concentrations of uranium in the upper parts rather than the lower parts of the Nos. 21 and 26 coals in the Yining coalfield of Xinjiang Province were attributed to the injection of groundwater or hydrothermal solution (Jiang et al. 2015). The slight enrichment of uranium in coal benches of the Fusui coalfield of Guangxi Province resulted from the leaching of uranium from the overlying parting clays by the S-, F-, and Cl-bearing hydrothermal fluids (i.e. groundwater – personal communication with Professor Shifeng Dai, Dai et al. 2013b).

The uranyl species, which are dominant in oxidizing waters, are strongly adsorbed onto iron (hydro) oxides and preferentially complexed with carbonate in alkaline groundwater. However, uranous species are prone to precipitate or bind to organic matter under reducing conditions in the shallow groundwater of the Datong Basin in Shanxi Province (Wu et al. 2014).

4.6. Organic matter

Organic matter, which has a large amount of active functional groups, e.g. carboxyl, hydroxyl, and phenolic hydroxyl of humic and fulvic acids, is an effective adsorbent and reductant for uranium enrichment in coal.

Coal organic matter may play an important role in the transport, accumulation, and hydrothermal mineralization of uranium in sediments (Havelcova et al. 2014). Organic substances might (1) restore uranium to the tetravalent state with the formation of oxides of uranium, (2) absorb uranium, and (3) form insoluble combinations with uranium (Gerasimovskii 1957). On the basis of material-balance calculations, the organic components carried 98% uranium in a subbituminous uraniferous coal from the Red Desert of Wyoming (Breger et al. 1955). The uranium concentrations are proportional to the organic matter contents in the fine mudstone and siltstone from the No. 277 uranium deposit of the Mabugang Basin of Guangdong Province (Wu 2012).

Humic acids and humates derived from the breakdown of plant material (peat and coal) have multiple roles, such as adsorbents, reactants, and reductants (Pirajno 2009). Humic and fulvic acids are the most powerful adsorbents of uranium at weakly acidic and neutral conditions (Wang 1984), causing the fixation of uranium from very dilute uranium solution (even at a ppb level) (Landais 1996). In the early stages of coal formation, a great portion of uranium in the majority of medium ash peats and brown coals from northern Asia accumulates in the adsorbed form and as strong humate complexes (Arbuzov et al. 2012). Because the humic substances are effective agents in the mobilization of uranium from U(VI) minerals, there is a strong correlation between the humic material and uranium enrichment in a south Texas lignite (Ilger et al. 1987).

The concentration of uranium by organic matter is strongest from the late stage of peatification to early diagenesis (Zhang et al. 1984). The uranyl ion, combined with humic acids to form metallorganics or adsorbed on by organic matter in the syngenetic stage, might damage and redeposit as amorphous uranium minerals during coalification (Peng et al. 1961), which might be the reason for greater uranium enrichment in low rank coal than in high rank coals.

4.7. Palaeoclimate

The weathering of source rocks, activation and migration of uranium, position of the groundwater table, and the growth of plants, which may control the formation of uranium deposits in coal and in coal-bearing strata, are affected by palaeoclimate.

Warm and humid climate during sedimentation, while arid and semiarid climate in mineralization, can provide the hosts with abundant organic matter and oxygenated groundwater for uranium metallogenesis (Chen 2002). Specifically, a warm and humid climate facilitated the deposition of organic mater-rich sediments in coal measures; a subsequent arid climate was beneficial to the oxidation of uranium in minerals to form uranyl complexes by oxidizing meteoric and/or ground waters, and final uranium deposits developed in redox transitional zone (Chen et al. 2002; Liu et al. 2011; Dai et al. 2015c, 2016c; Wang et al. 2015c). For example, the coal-bearing strata contain sandstone-hosted uranium deposits in both the Dongsheng Area of the Ordos Basin in Inner Mongolia Province and the Shihongtan Area of the Tuha Basin in Xinjiang Province (Figure 1) are deposited under warm and arid conditions (Yang et al. 2006).

4.8. Geologic conditions of coal-accumulating basins

Broad-scale geologic conditions, e.g. mantle–crust interaction, plate movement, topography of basin basement, lithological association, and tectonics, may influence the enrichment of uranium in coal and in coal-bearing strata.

(1) Mantle–crust interaction. Metalliferous coal deposits in South China, mainly hosting Zr(Hf)–Nb(Ta)–REE and U(Mo, Se)–REE ores, primarily resulted from the evolution of plumes ascending from deep mantle and/or asthenospheric flows, both of which incorporated some reworking of the continental crust. The mantle–crust interaction not only led to coal-accumulating basin formation but also played a significant role in extensive volcanism and ore-generating hydrothermal activity (Dai et al. 2016b).

(2) Plate movement. The Yili, Tuha, and Junggar uranium-rich intermontane basins in western China (Figure 1) were formed by the collision between the Kazakhstan Plate and Tarim Plate; the Ordos Basin (Figure 1) developed on a depression in craton from the impact of the Tethys Tectonics; and the uranium-bearing basins in eastern China were formed during the stretch stages after the subduction of the Pacific Plate under the Eurasian Plate (Luo and He 2008).

(3) Basement topography. The slopes of the basin basement provided the space for deltas of braided and meandering rivers and alluvial fans, which may be hosts of uranium deposits (Luo and He 2008). Moreover, the topographic relief of basement may control the movement of oxygenated uraniferous fluids.

(4) Lithological association. The lithological associations of mudstone, sandstone, and coal may provide sites for redox, migration, deposition of uranium. Generally, the most favoured lithological association for uranium deposits in coal or coal-bearing strata is sandstone-coal-mudstone, such as in the Daladi uranium deposits of the Yili Basin (Liu and Jia 2011b). Strata with a stable coal-sandstone-mudstone assemble are favourable for coal-hosted uranium deposit (A and Zhang 2003).

(5) Tectonics. Tectonic processes may induce uplift and outcrop of source rocks, infiltrate of oxygenated uraniferous groundwater (Yang et al. 2006), and provide pathways and sites for uranium enrichment in coals. For example, the concentrations of uranium in coals are increased with the intensive tectonic deformation in the Haizi Mine of the Huaibei coalfield in Anhui Province (Li et al. 2014). Moreover, the uranium mineralization in the Sawabuqi coal-hosted deposit is associated with faults (Liu et al. 2011). If the fault plane were paralleled with coalbed, crush zone developed in the coal seam, the coal benches underlain the crush zone were sites from uranium deposit. And if the fault obliquely transected the coalbed, only the intersecting part could develop uranium deposit.

5. Atmospheric emission of uranium from coal-fired power plants in china

5.1. Partition of uranium in coal combustion by-products

When coal is combusted in a coal-fired power plant, the uranium is concentrated in the by-products, i.e. slag, bottom ash, and fly ash. These by-products are commonly enriched in uranium 3–10-fold than that of the feed coals (Bu et al. 1996; Cevik et al. 2008; Krylov and Sidorova 2013; Hasani et al. 2014; Lauer et al. 2015; Hower et al. 2016).

Uranium is preferably partitioned into fine ash particles during coal combustion. Based on the relative enrichment factor, uranium was grouped into Class IIb by Meij and Te Winkel (2007), which is volatile in boiler, but completely condenses in the electrostatic precipitator on the ash particles. During combustion, a considerable fraction of the uranium combined with aluminosilicates enters the bottom ash. The remaining fraction is oxidized to UO_3, volatilized completely and condensed on the fly ash surface (Glowiak and Pacyna 1980; Pacyna 1980). As a result, uranium is primarily concentrated in the fly ash, especially the fine fly ash (Buke 2003; Flues et al. 2006), about 80% in the fly ash, 10% in the bottom ash, and partially in the flue gas desulphurization gypsum (Zhang et al. 2016). The activity of uranium and other radionuclides increases downstream from coal, bottom ash, fly ash, to stack dust (Pacyna 1980).

Oxides and uranates are the dominant modes of uranium occurrence in coal ash. Uranium in coal ash occurs mainly as U_3O_8 and $Ca(UO_2)O_2$ (Zhang 1981). In the bottom ash from the Lincang germanium smelters, it is dominant in the residual and Fe–Mn oxide fractions (Sun et al. 2016). Lithophile uranium is enriched in the magnetospheres (magnetic fractions in fly ashes) from Chinese and Russian power plants (Yang et al. 2014).

5.2. Atmospheric emission of uranium from coal-fired power plants in china

Anthropogenic atmospheric uranium emissions are mainly derived from the burning of wood, peat, coal, and petroleum, the roasting of rock minerals in metal extraction, refining, and cement industries, and the incineration of solid wastes (Bird 2012). A coal-fired power plant of 1000 MW output produces on average, approximately 409 kg uranium annually into the air (Eerkens 2010). Howover, as high as 2975 kg uranium were emitted to atmosphere from one plant in China in 1978 (Zhou 1981).

A simple method (Formula (1)), which was adopted by Chen et al. (2013) to estimate the atmospheric emissions of F, As, Se, Hg, and Sb from coal combustion for power and heat generation in China, is used in this article:

$$E_{g,i} = C_i \times c \times EF, \qquad (1)$$

where C_i is the coal consumption by power plant, c is the average of uranium in common Chinese coals, and EF is the emission factor of uranium. The coal consumption from 2010 to 2014 by power plants in China are presented in Table 1, which are taken from national annual data of China (National Bureau of Statistics of China 2016). The mean of the uranium content in Chinese coals with a value of 2.43 mg/kg issued by Dai et al. (2012a) was incorporated in the calculation.

The emission factor of uranium is rarely informative and shows a large variation from 0.05% to 50%. Tang et al. (2006) conducted simulated combustion tests at 877°C and 1300°C, and found that about 47% and 34% of the uranium in the coals were, respectively, released to the atmosphere. In the Shizuishan Power Plant of China (Song et al. 2011), 12.09% of the uranium in the raw coal was emitted into the atmosphere and in five power plants in western China (Song et al. 2005) an average of 5.40% of the uranium in the raw coal entered in the flue gas. Another coal-fired power plant (with a capacity of 200 MW), which is equipped with a low NO_x burner, a hybrid selective catalytic reduction + selective non-catalytic reduction denitrification system, an electrostatic precipitator, and a wet flue gas desulphurization equipment, only 0.049% of the total uranium input flowed out with the flue gas (Zhang et al. 2016). An emission factor of 1.47%, which is the geometric mean of these reported values in China (12.09%, 5.40%, and 0.049%), was used in this calculation.

The atmospheric emissions of uranium from coal-fired power plants from 2010 to 2014 in China are tabulated in Table 1. About 62.9 tons of uranium were emitted in 2014 due to the huge coal consumption (Table 1). Owing to the rarely informative emission factor of uranium from coal-fired power plants, the calculated atmospheric uranium emissions are of some uncertainty.

6. Environmental and health effects of uranium in coal

6.1. Environmental problems related to uranium in coal and coal-bearing strata

Environmental problems related to uranium in coals might arise from elevated natural radionuclide concentrations in the soil, plants, and water around coal-fired power plants due to the enriched uranium in fly ash. High radioactivity in coal (9–31 Bq/kg [238]U) in the Russian Far East is responsible for the accumulation of radionuclides in the fly (70–370 Bq/kg [238]U) and bottom (56–185 Bq/kg [238]U) ashes, leading to pollution of the atmosphere, lithosphere, biosphere, and hydrosphere

Table 1. Coal consumption and atmospheric uranium emission by production and supply of electric power and heat power during 2010–2014 in China.

Items	2010	2011	2012	2013	2014
Coal consumption (10^4 tons)[a]	151,163	170,744	174,273	189,848	176,098
Uranium emission (ton)	54.0	61.0	62.2	67.8	62.9

[a]Data were cited from the national annual data of China (National Bureau of Statistics of China 2016).

with toxic and radioactive elements (Zvereva and Krupskaya 2013). The activity concentrations of natural radionuclides in fly ash from Greek coal-fired power plants are significantly higher than that of coal and soil (Papastefanou 2010). Additionally, fly ash fallout elevated the concentration of ^{238}U in soils in the inhabited area of Ajka in Hungary (Papp et al. 2002), and ^{210}Pb in soil of the Yatagan Basin at western Turkey (Ugur et al. 2009). And even worse, 'carbon-caged' nanocrystals of uraninite in fly ash, which are protected from immediate oxidation, could lead to an increased mobility of uranium in the environment (Utsunomiya et al. 2002).

The tree bark near a coal-fired power station in United Kingdom shows an increased uranium concentration (0.25–0.38 mg/kg) (Bellis et al. 2001). Enrichments of $^{234,\ 238}U$ and ^{226}Ra in the water used in the four largest Spanish coal-fired power plants' routine procedures were observed (Baeza et al. 2012).

In contrast, the concentrations of uranium in soils were not altered by fly ash deposition from power plant in northwestern New Mexico of United States (Wangen and Williams 1978). Additionally, the gross alpha-radioactivity levels of the fly ashes did not add any risk for public health around the Yatagan coal-fired power plant in Turkey (Buke 2003). The environmental effect of natural radionuclides caused by the Afsin-Elbistan coal-fired power plants in Turkey was found to be negligible as well (Cevik et al. 2008).

The Carboniferous and Permian sedimentary rocks, especially the coal-bearing strata with a higher uranium contents in the west mountain areas, might account for the abnormally high levels of uranium in shallow groundwater of Datong Basin in Shanxi Province of China (Wu et al. 2014).

6.2. Health problems induced by uranium in coals

Uranium accumulating in humans might have a dual effect due to its chemotoxic and radiotoxic properties (Buke 2003; Alloway 2013). The chemical toxicity of uranium is a major hazard to the kidneys (Zhou 1981; Buke 2003). Radiotoxicity arises from the irradiation of bone surfaces and red bone marrow by uranium (Buke 2003).

Enhanced frequencies of chromosomal aberrations are typical symptoms of the radiotoxicity of uranium from coal, such as one Yugoslavia coal-fired power station (burning coal of 14–100 mg/kg uranium) (Bauman and Horvat 1981). The occupationally exposed workers (coal mines and coal-fired plants) to low doses of radiation in Yugoslavia has contributed to the rising percentage of chromatid and chromosome aberrations (Horvat et al. 1980).

The maximum annual effective dose of the natural radionuclides released from coal-fired power plants in China was 0.1 mSv (milli Sievert, a derived unit of ionizing radiation dose) (Li et al. 1987), far below the dose threshold of the Chinese National Standard (GB 18871 2002). The total annual effective dose surrounding lignite-fired power plants in the Megalopolis basin of Greece, both outdoor external irradiation and indoor internal irradiation, was 1.9 mSv for adults, lower than the worldwide average annual effective dose from natural sources (2.4 mSv) (Papaefthymiou et al. 2013).

Swaine (1990) advised that it would be unwise to burn coal with more than about 30 mg/kg uranium, without checking the emissions and solid waste products. The uranium content is 15 mg/kg in coal from the Kuznetsk Basin of Russia, which on that basis would pose no human risk (Iskhakov et al. 2010). The Urtuiskoe coal in Russia with an uranium concentration above 100 mg/kg must be stored in special dumps (Krylov and Sidorova 2013).

Although health problems due to exposure to radioactive materials in coal being mined today is highly unlikely (Huang et al. 2012), in view of the potential health risks from uranium in coal, those coals with high uranium contents should be mined and used with caution.

7. Concluding remarks

Significant uranium enrichment in coal has only occurred in local and limited areas in the Yunnan, Guizhou, Guangxi, Xinjiang, Inner Mongolia, and Chongqing Provinces of China. The uranium concentration of the majority of Chinese coals is comparable to that of world coals. Uranium in Chinese coals is associated with organic matter, silicates, pores, pelitic components, phosphate minerals, and uranium minerals.

Coal-hosted uranium deposits were developed in the Yili and Tarim Basins of Xinjiang Province, the Bangmai Basin of Yunnan Province, and the Mabugang Basin of Guangdong Province. The weathering of source rocks (especially for granite and felsic volcanic rocks), volcanic ashes, magmatic intrusion, marine water influence, groundwater (oxygenated), hydrothermal fluids, organic matter, palaeoclimate, and geologic conditions of coal-accumulating basin contributed to the enrichment of uranium in Chinese coals.

During coal combustion, uranium is partitioned into the coal ash, especially the fine fly ash. About 62.9 tons of uranium were emitted into atmosphere from Chinese coal-fired power plants in 2014. The chemotoxic and radiotoxic properties of uranium in coal have historically posed some harm to both the environment and human health.

Acknowledgements

Special thanks are given to Professor Shifeng Dai, Professor Robert B. Finkelman, Dr Robert J. Stern, Professor Colin R. Ward, Professor Greta Eskenazy, and one anonymous reviewer for their assistance, comments, and suggestions.

Funding

This work was supported by the National Key Basic Research Program of China [grant number 2014CB238901]; the National Key Research and Development Program of China [grant number 2016YFC0201605], and the National Natural Science Foundation of China [grant numbers 41402139 and 41372167].

References

Alloway, B.J., 2013, Uranium, Alloway, B.J., Ed., Heavy metals in soils, Springer, Netherlands, 565–577. doi:10.1007/978-94-007-4470-7_26.

Arbuzov, S.I., Maslov, S.G., Volostnov, A.V., Il'enok, S.S., and Arkhipov, V.S., 2012, Modes of occurrence of uranium and thorium in coals and peats of Northern Asia, Solid Fuel Chemistry, 46, 52–66. doi:10.3103/S0361521912010028.

A, Z. and Zhang, X., 2003, The mechanism of multi-layer ore-forming of Daladi uranium deposit in the Ili Basin, and its inspiration significance for uranium prospecting, Xinjiang Geology, 21, 433–436. in Chinese with an English abstract. doi:10.3969/j.issn.1000-8845.2003.04.010.

Baba, A., 2002, Assessment of radioactive contaminants in by-products from Yatagan (Mugla, Turkey) coal-fired power plant, Environmental Geology, 41, 916–921. doi:10.1007/s00254-001-0469-8.

Baeza, A., Corbacho, J.A., Guillen, J., Salas, A., Mora, J.C., Robles, B., and Cancio, D., 2012, Enhancement of natural radionuclides in the surroundings of the four largest coal-fired power plants in Spain, Journal of Environmental Monitoring, 14, 1064–1072. doi:10.1039/C2EM10991C.

Bauman, A., and Horvat, D., 1981, The impact of natural radio-activity from a coal-fired power plant, The Science of the Total Environment, 17, 75–81. doi:10.1016/0048-9697(81)90109-1.

Bellis, D., Ma, R., Bramall, N., McLeod, C.W., Chapman, N., and Satake, K., 2001, Airborne uranium contamination - as revealed through elemental and isotopic analysis of tree bark, Environmental Pollution, 114, 383–387. doi:10.1016/S0269-7491(00)00236-0.

Bem, H., Wieczorkowski, P., and Budzanowski, M., 2002, Evaluation of technologically enhanced natural radiation near the coal-fired power plants in the Lodz region of Poland, Journal of Environmental Radioactivity, 61, 191–201. doi:10.1016/S0265-931X(01)00126-6.

Bird, G.A., 2012, Uranium in the environment: Behavior and toxicity, Meyers, R.A., Ed., Encyclopedia of Sustainability Science and Technology, Springer, New York, 11220–11262. doi:10.1007/978-1-4419-0851-3_294.

Breger, I.A., 1958, Geochemistry of coal, Economic Geology, 53, 823–841. doi:10.2113/gsecongeo.53.7.823.

Breger, I.A., Deul, M., and Meyrowitz, R., 1955, Geochemistry and mineralogy of a uraniferous subbituminous coal, Economic Geology, 50, 610–624. doi:10.2113/gsecongeo.50.6.610.

Bu, Y., Chen, M., Huang, Z., and Yang, X., 1996, Uranium in coal and its relation to coal mine environment, Coal Mine Environmental Protection, 10, 34–36. in Chinese.

Buke, T., 2003, Dose assessment around the Yatagan coal-fired power plant due to gross alpha-radioactivity levels in flying ash, Journal of Radioanalytical and Nuclear Chemistry, 256, 323–328. doi:10.1023/A:1023905922085.

Cayir, A., Belivermis, M., Kilic, O., Coskun, M., and Coskun, M., 2012, Heavy metal and radionuclide levels in soil around Afsin-Elbistan coal-fired thermal power plants, Turkey, Environmental Earth Sciences, 67, 1183–1190. doi:10.1007/s12665-012-1561-y.

Cevik, U., Damla, N., Koz, B., and Kaya, S., 2008, Radiological characterization around the Afsin-Elbistan coal-fired power plant in Turkey, Energy and Fuels, 22, 428–432. doi:10.1021/ef700374u.

Chen, B., Qian, Q., Yang, Y., and Yang, S., 1985, Contents of trace elements in coals from 110 coal mines in China, Nuclear Techniques, 43–44. in Chinese.

Chen, B., Yang, S., Qian, Q., and Yang, Y., 1989, Content distribution of As, Se, Cr, U and Th elements in Chinese coal samples, Environmental Science, 10, 23–26. in Chinese with an English abstract. doi:10.13227/j.hjkx.1989.06.005.

Chen, F., 2002, Metallogenic geologic prerequisites of sandstone-type uranium deposits and target area selection - Taking Erlian and Ordos basins as examples, Uranium Geology, 18, 138–143. in Chinese with an English abstract. doi:10.3969/j.issn.1000-0658.2002.03.002.

Chen, J., Chen, P., Yao, D., Guo, J., Liu, Z., Lu, J., Li, Q., Liu, W., and Hu, Y., 2016, Geochemistry of trace elements in the Mengtuo Neogene lignite of Lincang, western Yunnan, Earth Science Frontiers, 23, 83–89. in Chinese with an English abstract. doi:10.13745/j.esf.2016.03.011.

Chen, J., Chen, P., Yao, D., Liu, Z., Wu, Y., Liu, W., and Hu, Y., 2015, Mineralogy and geochemistry of Late Permian coals from the Donglin Coal Mine in the Nantong coalfield in Chongqing, southwestern China, International Journal of Coal Geology, 149, 24–40. doi:10.1016/j.coal.2015.06.014.

Chen, J., Liu, G., Kang, Y., Wu, B., Sun, R., Zhou, C., and Wu, D., 2013, Atmospheric emissions of F, As, Se, Hg, and Sb from coal-fired power and heat generation in China, Chemosphere, 90, 1925–1932. doi:10.1016/j.chemosphere.2012.10.032.

Chen, Z., Li, S., Cai, Y., and Chen, D., 2002, Geologic evolution and uranium metallogenic regularity in Yili basin, Mineral Deposits, 21, 849–852. in Chinese. doi:10.16111/j.0258-7106.2002.s1.221.

Cujic, M., Dragovic, S., Dordevic, M., Dragovic, R., Gajic, B., and Miljanic, S., 2015, Radionuclides in the soil around the largest coal-fired power plant in Serbia: Radiological hazard, relationship with soil characteristics and spatial distribution, Environmental Science and Pollution Research, 22, 10317–10330. doi:10.1007/s11356-014-3888-2.

Dai, L., Wei, H., and Wang, L., 2007, Spatial distribution and risk assessment of radionuclides in soils around a coal-fired power plant: A case study from the city of Baoji, China, Environmental Research, 104, 201–208. doi:10.1016/j.envres.2006.11.005.

Dai, S., Chekryzhov, I.Y., Seredin, V.V., Nechaev, V.P., Graham, I. T., Hower, J.C., Ward, C.R., Ren, D., and Wang, X., 2016b, Metalliferous coal deposits in East Asia (Primorye of Russia and South China): A review of geodynamic controls and styles of mineralization, Gondwana Research, 29, 60–82. doi:10.1016/j.gr.2015.07.001.

Dai, S., Han, D., and Chou, C.-L., 2006, Petrography and geochemistry of the Middle Devonian coal from Luquan, Yunnan Province, China, Fuel, 85, 456–464. doi:10.1016/j.fuel.2005.08.017.

Dai, S., Li, D., Chou, C.-L., Zhao, L., Zhang, Y., Ren, D., Ma, Y., and Sun, Y., 2008b, Mineralogy and geochemistry of boehmite-rich coals: New insights from the Haerwusu Surface Mine, Jungar Coalfield, Inner Mongolia, China, International Journal of Coal Geology, 74, 185–202. doi:10.1016/j.coal.2008.01.001.

Dai, S., Li, D., Ren, D., Tang, Y., Shao, L., and Song, H., 2004a, Geochemistry of the late Permian No. 30 coal seam, Zhijin Coalfield of Southwest China: Influence of a siliceous low-temperature hydrothermal fluid, Applied Geochemistry, 19, 1315–1330. doi:10.1016/j.apgeochem.2003.12.008.

Dai, S., Li, T., Seredin, V.V., Ward, C.R., Hower, J.C., Zhou, Y., Zhang, M., Song, X., Song, W., and Zhao, C., 2014a, Origin of minerals and elements in the Late Permian coals, tonsteins, and host rocks of the Xinde Mine, Xuanwei, eastern Yunnan, China, International Journal of Coal Geology, 121, 53–78. doi:10.1016/j.coal.2013.11.001.

Dai, S., Liu, J., Ward, C.R., Hower, J.C., French, D., Jia, S., Hood, M.M., and Garrison, T.M., 2016a, Mineralogical and geochemical compositions of Late Permian coals and host rocks from the Guxu Coalfield, Sichuan Province, China, with emphasis on enrichment of rare metals, International Journal of Coal Geology, 166, 71–95. doi:10.1016/j.coal.2015.1012.1004.

Dai, S., Luo, Y., Seredin, V.V., Ward, C.R., Hower, J.C., Zhao, L., Liu, S., Zhao, C., Tian, H., and Zou, J., 2014b, Revisiting the late Permian coal from the Huayingshan, Sichuan, southwestern China: Enrichment and occurrence modes of minerals and trace elements, International Journal of Coal Geology, 122, 110–128. doi:10.1016/j.coal.2013.12.016.

Dai, S., and Ren, D., 2007, Effects of magmatic intrusion on mineralogy and geochemistry of coals from the Fengfeng-Handan Coalfield, Hebei, China, Energy and Fuels, 21, 1663–1673. doi:10.1021/ef060618f.

Dai, S., Ren, D., Chou, C.-L., Finkelman, R.B., Seredin, V.V., and Zhou, Y., 2012a, Geochemistry of trace elements in Chinese coals: A review of abundances, genetic types, impacts on human health, and industrial utilization, International Journal of Coal Geology, 94, 3–21. doi:10.1016/j.coal.2011.02.003.

Dai, S., Ren, D., Hou, X., and Shao, L., 2003, Geochemical and mineralogical anomalies of the late Permian coal in the Zhijin coalfield of southwest China and their volcanic origin, International Journal of Coal Geology, 55, 117–138. doi:10.1016/S0166-5162(03)00083-1.

Dai, S., Ren, D., Sun, Y., and Tang, Y., 2004b, Concentration and the sequential chemical extraction procedures of U and Th in the Paleozoic coals from the Ordos Basin, Journal of China Coal Society, 29, 56–60. in Chinese with an English abstract.

Dai, S., Ren, D., Tang, Y., Yue, M., and Hao, L., 2005, Concentration and distribution of elements in Late Permian coals from western Guizhou Province, China, International Journal of Coal Geology, 61, 119–137. doi:10.1016/j.coal.2004.07.003.

Dai, S., Ren, D., Zhou, Y., Chou, C.-L., Wang, X., Zhao, L., and Zhu, X., 2008a, Mineralogy and geochemistry of a super-high-organic-sulfur coal, Yanshan Coalfield, Yunnan, China: Evidence for a volcanic ash component and influence by submarine exhalation, Chemical Geology, 255, 182–194. doi:10.1016/j.chemgeo.2008.06.030.

Dai, S., Seredin, V.V., Ward, C.R., Hower, J.C., Xing, Y., Zhang, W., Song, W., and Wang, P., 2015a, Enrichment of U-Se-Mo-Re-V in coals preserved within marine carbonate successions: Geochemical and mineralogical data from the Late Permian Guiding Coalfield, Guizhou, China, Mineralium Deposita, 50, 159–186. doi:10.1007/s00126-014-0528-1.

Dai, S., Wang, P., Ward, C.R., Tang, Y., Song, X., Jiang, J., Hower, J.C., Li, T., Seredin, V.V., Wagner, N.J., Jiang, Y., Wang, X., and Liu, J., 2015b, Elemental and mineralogical anomalies in the coal-hosted Ge ore deposit of Lincang, Yunnan, southwestern China: Key role of N_2-CO_2-mixed hydrothermal solutions, International Journal of Coal Geology, 152, 19–46. doi:10.1016/j.coal.2014.11.006.

Dai, S., Wang, X., Chen, W., Li, D., Chou, C.-L., Zhou, Y., Zhu, C., Li, H., Zhu, X., Xing, Y., Zhang, W., and Zou, J., 2010, A high-pyrite semianthracite of Late Permian age in the Songzao Coalfield, southwestern China: Mineralogical and geochemical relations with underlying mafic tuffs, International Journal of Coal Geology, 83, 430–445. doi:10.1016/j.coal.2010.06.004.

Dai, S., Wang, X., Seredin, V.V., Hower, J.C., Ward, C.R., O'Keefe, J.M.K., Huang, W., Li, T., Li, X., Liu, H., Xue, W., and Zhao, L., 2012b, Petrology, mineralogy, and geochemistry of the Ge-rich coal from the Wulantuga Ge ore deposit, Inner Mongolia, China: New data and genetic implications, International Journal of Coal Geology, 90-91, 72–99. doi:10.1016/j.coal.2011.10.012.

Dai, S., Xie, P., Jia, S., Ward, C.R., Hower, J.C., Yan, X., and French, D., 2017, Enrichment of U-Re-V-Cr-Se and rare earth elements in the Late Permian coals of the Moxinpo Coalfield, Chongqing, China: Genetic implications from geochemical and mineralogical data, Ore Geology Reviews, 80, 1–17. doi:10.1016/j.oregeorev.2016.06.015.

Dai, S., Yan, X., Ward, C.R., Hower, J.C., Zhao, L., Wang, X., Zhao, L., Ren, D., and Finkelman, R.B., 2016c, Valuable elements in Chinese coals: A review, International Geology Review, 1–31. in press. doi:10.1080/00206814.00202016.01197802.

Dai, S., Yang, J., Ward, C.R., Hower, J.C., Liu, H., Garrison, T.M., French, D., and O'keefe, J.M.K., 2015c, Geochemical and mineralogical evidence for a coal-hosted uranium deposit in the Yili Basin, Xinjiang, northwestern China, Ore Geology Reviews, 70, 1–30. doi:10.1016/j.oregeorev.2015.03.010.

Dai, S., Zhang, W., Seredin, V.V., Ward, C.R., Hower, J.C., Song, W., Wang, X., Li, X., Zhao, L., Kang, H., Zheng, L., Wang, P., and Zhou, D., 2013a, Factors controlling geochemical and mineralogical compositions of coals preserved within marine carbonate successions: A case study from the Heshan Coalfield, southern China, International Journal of Coal Geology, 109-110, 77–100. doi:10.1016/j.coal.2013.02.003.

Dai, S., Zhang, W., Ward, C.R., Seredin, V.V., Hower, J.C., Li, X., Song, W., Wang, X., Kang, H., Zheng, L., Wang, P., and Zhou, D., 2013b, Mineralogical and geochemical anomalies of late Permian coals from the Fusui Coalfield, Guangxi Province, southern China: Influences of terrigenous materials and hydrothermal fluids, International Journal of Coal Geology, 105, 60–84. doi:10.1016/j.coal.2012.12.003.

Dai, S., Zou, J., Jiang, Y., Ward, C.R., Wang, X., Li, T., Xue, W., Liu, S., Tian, H., Sun, X., and Zhou, D., 2012c, Mineralogical and geochemical compositions of the Pennsylvanian coal in the Adaohai Mine, Daqingshan Coalfield, Inner Mongolia, China: Modes of occurrence and origin of diaspore, gorceixite, and ammonian illite, International Journal of Coal Geology, 94, 250–270. doi:10.1016/j.coal.2011.06.010.

Douglas, G.B., Butt, C.R.M., and Gray, D.J., 2011, Geology, geochemistry and mineralogy of the lignite-hosted Ambassador palaeochannel uranium and multi-element deposit, Gunbarrel Basin, Western Australia, Mineralium Deposita, 46, 761–787. doi:10.1007/s00126-011-0349-4.

Eerkens, J.W., 2010, Coal and nuclear power generation, Eerkens, J.W., Ed., The Nuclear imperative, Springer, Netherlands, 99–134. doi:10.1007/978-90-481-8667-9_6.

Eisenbud, M., and Petrow, H.G., 1964, Radioactivity in the atmospheric effluents of power plants that use fossil fuels, Science, 144, 288–289. doi:10.1126/science.144.3616.288.

Flues, M., Camargo, I.M.C., Silva, P.S.C., and Mazzilli, B.P., 2006, Radioactivity of coal and ashes from Figueira coal power plant in Brazil, Journal of Radioanalytical and Nuclear Chemistry, 270, 597–602. doi:10.1007/s10967-006-0467-0.

Fu, X., Wang, J., Tan, F., Feng, X., and Zeng, S., 2013, Minerals and potentially hazardous trace elements in the Late Triassic coals from the Qiangtang Basin, China, International Journal of Coal Geology, 116-117, 93–105. doi:10.1016/j.coal.2013.07.013.

Gabriel, S., Baschwitz, A., Mathonniere, G., Eleouet, T., and Fizaine, F., 2013, A critical assessment of global uranium resources, including uranium in phosphate rocks, and the possible impact of uranium shortages on nuclear power fleets, Annals of Nuclear Energy, 58, 213–220. doi:10.1016/j.anucene.2013.03.010.

GB 18871, 2002, Basic standards for protection against ionizing radiation and for the safety of radiation sources: Chinese National Standard, in Chinese, 1–28.

Gerasimovskii, V.I., 1957, On the modes of occurrence of uranium in rocks, The Soviet Journal of Atomic Energy, 3, 1407–1411. doi:10.1007/BF01522506.

Glowiak, B.J., and Pacyna, J.M., 1980, Radiation dose due to atmospheric releases from coal/fired power stations, International Journal of Environmental Studies, 16, 23–28. doi:10.1080/00207238008709843.

Han, Y., Yuan, Q., Li, Y., Zhang, L., and Dai, J., 1994, Dazhai superlarge uranium-bearing germanium deposit in western Yunnan region: Metallogenic geological conditions and prospect, Chinese Nuclear Science and Technology Report, 1–18. in Chinese with an English abstract.

Hasani, F., Shala, F., Xhixha, G., Xhixha, M.K., Hodolli, G., Kadiri, S., Bylyku, E., and Cfarku, F., 2014, Naturally occurring radioactive materials (NORMs) generated from lignite-fired power plants in Kosovo, Journal of Environmental Radioactivity, 138, 156–161. doi:10.1016/j.jenvrad.2014.08.015.

Havelcova, M., Machovic, V., Mizera, J., Sykorova, I., Borecka, L., and Kopecky, L., 2014, A multi-instrumental geochemical study of anomalous uranium enrichment in coal, Journal of Environmental Radioactivity, 137, 52–63. doi:10.1016/j.jenvrad.2014.06.015.

Horvat, D., Bauman, A., and Racic, J., 1980, Genetic effect of low doses of radiation in occupationally exposed workers in coal mines and in coal fired plants, Radiation and Environmental Biophysics, 18, 91–97. doi:10.1007/BF01326048.

Hower, J.C., Dai, S., and Eskenazy, G., 2016, Distribution of uranium and other radionuclides in coal and coal combustion products, with discussion of occurrences of combustion products in Kentucky power plants, Coal Combustion and Gasification Products, 8, 44–53. doi:10.4177/CCGP-D-16-00002.1.

Hu, R.-Z., Qi, H.-W., Zhou, M.-F., Su, W.-C., Bi, X.-W., Peng, J.-T., and Zhong, H., 2009, Geological and geochemical constraints on the origin of the giant Lincang coal seam-hosted germanium deposit, Yunnan, SW China: A review, Ore Geology Reviews, 36, 221–234. doi:10.1016/j.oregeorev.2009.02.007.

Huang, W., Wan, H., Finkelman, R.B., Tang, X., and Zhao, Z., 2012, Distribution of uranium in the main coalfields of China, Energy Exploration and Exploitation, 30, 819–836. doi:10.1260/0144-5987.30.5.819.

Huang, W., Yang, Q., Peng, S., Tang, X., and Zhao, Z., 2001, Geochemistry of Permian coal and its combustion residues from Huainan coalfield, Earth Science - Journal of China University of Geosciences, 26, 501–507. in Chinese with an English abstract. doi:10.3321/j.issn:1000-2383.2001.05.010.

Ilger, J.D., Ilger, W.A., Zingaro, R.A., and Mohan, M.S., 1987, Modes of occurrence of uranium in carbonaceous uranium deposits: Characterization of uranium in a south Texas (U.S.A.) lignite, Chemical Geology, 63, 197–216. doi:10.1016/0009-2541(87)90163-X.

Iskhakov, K.A., Schastlivtsev, E.L., Kondratenko, Y.A., and Lesina, M.L., 2010, Radioactivity of coal and ash, Coke and Chemistry, 53, 198–201. doi:10.3103/S1068364X10050078.

Jiang, Y., Zhao, L., Zhou, G., Wang, X., Zhao, L., Wei, J., and Song, H., 2015, Petrological, mineralogical, and geochemical compositions of Early Jurassic coals in the Yining Coalfield, Xinjiang, China, International Journal of Coal Geology, 152, 47–67. doi:10.1016/j.coal.2015.07.011.

Ketris, M.P., and Yudovich, Y.E., 2009, Estimations of Clarkes for Carbonaceous biolithes: World averages for trace elements in black shales and coals, International Journal of Coal Geology, 78, 135–148. doi:10.1016/j.coal.2009.01.002.

Krylov, D.A., and Sidorova, G.P., 2013, Radioactivity of coal and ash-slag TPP wastes, Atomic Energy, 114, 56–60. doi:10.1007/s10512-013-9670-6.

Landais, P., 1996, Organic geochemistry of sedimentary uranium ore deposits, Ore Geology Reviews, 11, 33–51. doi:10.1016/0169-1368(95)00014-3.

Lauer, N.E., Hower, J.C., Hsu-Kim, H., Taggart, R.K., and Vengosh, A., 2015, Naturally occurring radioactive materials in coals and coal combustion residuals in the United States, Environmental Science and Technology, 49, 11227–11233. doi:10.1021/acs.est.5b01978.

Li, B., Zhuang, X., Li, J., Querol, X., Font, O., and Moreno, N., 2017, Enrichment and distribution of elements in the Late Permian coals from the Zhina Coalfield, Guizhou Province, Southwest China, International Journal of Coal Geology, 171, 111–129. doi:10.1016/j.coal.2017.01.003.

Li, F., Jin, G., Zhang, J., Liu, D., Fang, D., and Liu, Y., 1987, Natural radioactive level in coal and ash from 61 coal-fired power plants in China and its impact on the environment,

Radiation Protection, 7, 260–272. in Chinese with an English abstract.

Li, J., Zhuang, X., and Querol, X., 2011, Trace element affinities in two high-Ge coals from China, Fuel, 90, 240–247. doi:10.1016/j.fuel.2010.08.011.

Li, Y.B., Jiang, B., and Qu, Z.H., 2014, Controls on migration and aggregation for tectonically sensitive elements in tectonically deformed coal: An example from the Haizi mine, Huaibei coalfield, China, Science China Earth Sciences, 57, 1180–1191. doi:10.1007/s11430-014-4857-9.

Liu, B., Huang, W., Ao, W., Yan, D., Xu, Q., and Teng, J., 2016, Geochemistry characteristics of sulfur and its effect on hazardous elements in the Late Paleozoic coal from the Qinshui Basin, Earth Science Frontiers, 23, 59–67. in Chinese with an English abstract. doi:10.13745/j.esf.2016.03.008.

Liu, J., Yang, Z., Yan, X., Ji, D., Yang, Y., and Hu, L., 2015, Modes of occurrence of highly-elevated trace elements in super-high-organic-sulfur coals, Fuel, 156, 190–197. doi:10.1016/j.fuel.2015.04.034.

Liu, W., and Jia, L., 2011a, Sedimentary formation features and it's relationship with sandstone-type uranium ore formation in Yili Basin, World Nuclear Geoscience, 28, 1–5. in Chinese with an English abstract. doi:10.3969/j.issn.1672-0636.2011.01.001.

Liu, W., and Jia, L., 2011b, Metallogenesis of uranium deposit occurring in coals of Daladi, Acta Mineralogica Sinica, 271–272. in Chinese.

Liu, Z., Dong, W., and Liu, H., 2011, Analysis on genesis of uranium-bearing coal in Sawabuqi area, Xinjiang, Uranium Geology, 27, 345–351. in Chinese with an English abstract. doi:10.3969/j.issn.1000-0658.2011.06.005.

Luo, Y., and He, Z., 2008, Metallogeny and technical criteria of evaluation and prediction of uranium deposit occurred in Mesozoic and Cenozoic coal, oil, and natural gas basins in north China, National Seminar on Application of Innovative Remote Sensing Technique in 2008, Nanning, 273–281. in Chinese.

Maslov, O.D., Sh., T., Norov, N., Gustova, M.V., Filippov, M.F., Belov, A.G., Altangerel, M., and Enhbat, N., 2010, Uranium recovery from coal ash dumps of Mongolia, Solid Fuel Chemistry, 44, 433–438. doi:10.3103/S0361521910060133.

McBride, J.P., Moore, R.E., Witherspoon, J.P., and Blanco, R.E., 1978, Radiological impact of airborne effluents of coal and nuclear plants, Science, 202, 1045–1050. doi:10.1126/science.202.4372.1045.

Meij, R., and Te Winkel, H., 2007, The emissions of heavy metals and persistent organic pollutants from modern coal-fired power stations, Atmospheric Environment, 41, 9262–9272. doi:10.1016/j.atmosenv.2007.04.042.

Monnet, A., Percebois, J., and Gabriel, S., 2015, Assessing the potential production of uranium from coal-ash milling in the long term, Resources Policy, 45, 173–182. doi:10.1016/j.resourpol.2015.04.005.

Mudd, G.M., 2014, The future of Yellowcake: A global assessment of uranium resources and mining, Science of the Total Environment, 472, 590–607. doi:10.1016/j.scitotenv.2013.11.070.

National Bureau of Statistics of China, 2016, Energy. http://data.stats.gov.cn/english/easyquery.htm?cn=C01.

Pacyna, J.M., 1980, Radionuclide behavior in coal-fired plants, Ecotoxicology and Environmental Safety, 4, 240–251. doi:10.1016/0147-6513(80)90026-3.

Papaefthymiou, H.V., Manousakas, M., Fouskas, A., and Siavalas, G., 2013, Spatial and vertical distribution and risk assessment of natural radionuclides in soils surrounding the lignite-fired power plants in Megalopolis basin, Greece, Radiation Protection Dosimetry, 156, 49–58. doi:10.1093/rpd/nct037.

Papastefanou, C., 2010, Escaping radioactivity from coal-fired power plants (CPPs) due to coal burning and the associated hazards: A review, Journal of Environmental Radioactivity, 101, 191–200. doi:10.1016/j.jenvrad.2009.11.006.

Papp, Z., Dezso, Z., and Daroczy, S., 2002, Significant radioactive contamination of soil around a coal-fired thermal power plant, Journal of Environmental Radioactivity, 59, 191–205. doi:10.1016/S0265-931X(01)00071-6.

Peng, J., Liao, S., Chen, Y., Chen, G., and Zhang, B., 1961, Genesis of uranium deposit occurred in Jurassic coals, Atomic Energy Science and Technology, 671–684. in Chinese.

Perricos, D.C., and Belkas, E.P., 1969, Determination of uranium in uraniferous coal, Talanta, 16, 745–748. doi:10.1016/0039-9140(69)80106-2.

Pirajno, F., 2009, Uranium hydrothermal mineral systems, Pirajno, F., Ed., Hydrothermal Processes and Mineral Systems, Springer, Netherlands, 1213–1241. doi:10.1007/978-1-4020-8613-7_13.

Podwysocki, S.M., and Lovern, V.S., 2000, Coal-bearing regions and structural sedimentary basins of China and adjacent seas: U.S. Geological Survey Open-File Report 00-047. http://pubs.usgs.gov/of/2000/of00-047/china_coal.pdf.

Qi, F., Zhang, Z., Li, Z., Wang, Z., He, Z., and Wang, W., 2011, Unconventional uranium resources in China, Uranium Geology, 27, 193–199. in Chinese with an English abstract. doi:10.3969/j.issn.1000-0658.2011.04.001.

Qi, H., Hu, R., and Zhang, Q., 2007, Concentration and distribution of trace elements in lignite from the Shengli Coalfield, Inner Mongolia, China: Implication on origin of the associated Wulantuga Germanium Deposit, International Journal of Coal Geology, 71, 129–152. doi:10.1016/j.coal.2006.08.005.

Querol, X., Alastuey, A., Lopez-Soler, A., Plana, F., Fernandez-Turiel, J.L., Zeng, R., Xu, W., Zhuang, X., and Spiro, B., 1997, Geological controls on the mineral matter and trace elements of coals from the Fuxin basin, Liaoning Province, northeast China, International Journal of Coal Geology, 34, 89–109. doi:10.1016/S0166-5162(97)00014-1.

Querol, X., Alastuey, A., Lopez-Soler, A., Plana, F., Zeng, R., Zhao, J., and Zhuang, X., 1999, Geological controls on the quality of coals from the West Shandong mining district, Eastern China, International Journal of Coal Geology, 42, 63–88. doi:10.1016/S0166-5162(99)00030-0.

Querol, X., Alastuey, A., Zhuang, X., Hower, J.C., Lopez-Soler, A., Plana, F., and Zeng, R., 2001, Petrology, mineralogy and geochemistry of the Permian and Triassic coals in the Leping area, Jiangxi Province, southeast China, International Journal of Coal Geology, 48, 23–45. doi:10.1016/S0166-5162(01)00036-2.

Reimann, C., and De Caritat, P., 1998, Chemical elements in the environment, Springer, Berlin Heidelberg, 23–398. doi:10.1007/978-3-642-72016-1.

Ren, D., Zhao, F., Dai, S., Zhang, J., and Luo, K., 2006, Geochemistry of trace elements in coal, Science Press, Beijing, 308–320. in Chinese.

Ren, D., Zhao, F., Wang, Y., and Yang, S., 1999, Distributions of minor and trace elements in Chinese coals, International Journal of Coal Geology, 40, 109–118. doi:10.1016/S0166-5162(98)00063-9.

Saikia, B.K., Wang, P., Saikia, A., Song, H., Liu, J., Wei, J., and Gupta, U.N., 2015, Mineralogical and elemental analysis of some high-sulfur Indian Paleogene coals: A statistical approach, Energy and Fuels, 29, 1407–1420. doi:10.1021/ef502511t.

Seredin, V.V., 2012, From coal science to metal production and environmental protection: A new story of success, International Journal of Coal Geology, 90-91, 1–3. doi:10.1016/j.coal.2011.11.006.

Seredin, V.V., Danilcheva, Y.A., Magazina, L.O., and Sharova, I.G., 2006, Ge-bearing coals of the Luzanovka Graben, Pavlovka brown coal deposit, southern Primorye, Lithology and Mineral Resources, 41, 280–301. doi:10.1134/S0024490206030072.

Seredin, V.V., and Finkelman, R.B., 2008, Metalliferous coals: A review of the main genetic and geochemical types, International Journal of Coal Geology, 76, 253–289. doi:10.1016/j.coal.2008.07.016.

Shao, L., Jones, T., Gayer, R., Dai, S., Li, S., Jiang, Y., and Zhang, P., 2003, Petrology and geochemistry of the high-sulphur coals from the Upper Permian carbonate coal measures in the Heshan Coalfield, southern China, International Journal of Coal Geology, 55, 1–26. doi:10.1016/S0166-5162(03)00031-4.

Shao, L., Lu, J., Jones, T., Gayer, R., Shang, L., Shen, Z., and Zhang, P., 2006, Mineralogy and geochemistry of the high-organic coals from the carbonate coal measures of the Late Permian in central Guangxi, Journal of China Coal Society, 31, 770–775. in Chinese with an English abstract. doi:10.13225/j.cnki.jccs.2006.06.016.

Song, D., Qin, Y., Zhang, J., Wang, W., and Zheng, C., 2005, Distribution of environmentally-sensitive trace elements of coal in combustion, Coal Conversion, 28, 56–60. in Chinese with an English abstract. doi:10.3969/j.issn.1004-4248.2005.02.012.

Song, D., Qin, Y., Zhang, J., Wang, W., and Zheng, C., 2007, Concentration and distribution of trace elements in some coals from Northern China, International Journal of Coal Geology, 69, 179–191. doi:10.1016/j.coal.2006.04.001.

Song, D.-Y., Ma, Y.-J., Qin, Y., Wang, W.-F., and Zheng, C.-G., 2011, Volatility and mobility of some trace elements in coal from Shizuishan Power Plant, Journal of Fuel Chemistry and Technology, 39, 328–332. doi:10.1016/S1872-5813(11)60024-8.

Sun, Y., Lin, M., Qin, P., Zhao, C., and Jin, K., 2007, Geochemistry of the barkinite liptobiolith (Late Permian) from the Jinshan Mine, Anhui Province, China, Environmental Geochemistry and Health, 29, 33–44. doi:10.1007/s10653-006-9059-8.

Sun, Y., Qi, G., Lei, X., Xu, H., Li, L., Yuan, C., and Wang, Y., 2016, Distribution and mode of occurrence of uranium in bottom ash derived from high-germanium coals, Journal of Environmental Sciences, 43, 91–98. doi:10.1016/j.jes.2015.07.009.

Sun, Y., Zhao, C., Li, Y., and Wang, J., 2014, Minimum mining grade of the selected trace elements in Chinese coal, Journal of China Coal Society, 39, 744–748. in Chinese with an English abstract. doi:10.13225/j.cnki.jccs.2013.1718.

Swaine, D.J., 1990, Trace elements in coal, Butterworths, 195.

Swaine, D.J., 1992, The organic association of elements in coals, Organic Geochemistry, 18, 259–261. doi:10.1016/0146-6380(92)90067-8.

Tang, S., Qin, Y., and Jiang, Y., 2006, Geological study on clean coal of China, The Geological Publishing House, Beijing, 34–39. in Chinese.

Tang, X., and Huang, W., 2004, Trace elements in Chinese coals, The Commercial Press, Beijing, 249–257. in Chinese.

Ugur, A., Ozden, B., Yener, G., Sac, M.M., Kurucu, Y., Altinbas, U., and Bolca, M., 2009, Distributions of ^{210}Pb around a uraniferous coal-fired power plant in Western Turkey, Environmental Monitoring and Assessment, 149, 195–200. doi:10.1007/s10661-008-0193-x.

Utsunomiya, S., Jensen, K.A., Keeler, G.J., and Ewing, R.C., 2002, Uraninite and fullerene in atmospheric particulates, Environmental Science and Technology, 36, 4943–4947. doi:10.1021/es025872a.

Van Der Flier, E., and Fyfe, W.S., 1985, Uranium-thorium systematics of two Canadian coals, International Journal of Coal Geology, 4, 335–353. doi:10.1016/0166-5162(85)90019-9.

Wang, J., 1984, Natural organic matter and its implications in uranium mineralization, Geochemistry, 3, 260–271. doi:10.1007/BF03179300.

Wang, J., Wang, W., Li, J., and Qin, Y., 2010, Deposit features of Ge, Ga and U elements in northern part of Datong coalfield, Coal Science and Technology, 38, 117–121. in Chinese with an English abstract. doi:10.13199/j.cst.2010.02.90.wangjy.005.

Wang, M., Li, H., and Qiu, Y., 2015c, Coal-type uranium metallogenic analysis in Honghaigou area, Yili basin, Xinjiang, Coal Geology of China, 27, 12–16. in Chinese with an English abstract. doi:10.3969/j.issn.1674-1803.2015.12.03.

Wang, W., Qin, Y., Sang, S., Zhu, Y., Wang, C., and Weiss, D.J., 2008, Geochemistry of rare earth elements in a marine influenced coal and its organic solvent extracts from the Antaibao mining district, Shanxi, China, International Journal of Coal Geology, 76, 309–317. doi:10.1016/j.coal.2008.08.012.

Wang, W., Qin, Y., Song, D., Sang, S., Jiang, B., Zhu, Y., and Fu, X., 2005, Element geochemistry and cleaning potential of the No. 11 coal seam from Antaibao mining district: Science in China Series D, Earth Sciences, 48, 2142–2154. doi:10.1360/03yd0506.

Wang, X., 2009, Geochemistry of Late Triassic coals in the Changhe Mine, Sichuan Basin, southwestern China: Evidence for authigenic lanthanide enrichment, International Journal of Coal Geology, 80, 167–174. doi:10.1016/j.coal.2009.10.011.

Wang, X., Dai, S., Ren, D., and Yang, J., 2011, Mineralogy and geochemistry of Al-hydroxide/oxyhydroxide mineral-bearing coals of Late Paleozoic age from the Weibei coalfield, southeastern Ordos Basin, North China, Applied Geochemistry, 26, 1086–1096. doi:10.1016/j.apgeochem.2011.03.013.

Wang, X., Feng, Q., Fang, T., Liu, J., and Liu, G., 2015a, Geochemical characteristics of uranium in medium to high sulfur coals from eastern Yunnan, China, Journal of China Coal Society, 40, 2451–2457. in Chinese with an English abstract. doi:10.13225/j.cnki.jccs.2015.0009.

Wang, X., Jiang, Y., Zhou, G., Wang, P., Wang, R., Zhao, L., and Chou, C.-L., 2015b, Behavior of minerals and trace elements during natural coking: A case study of an intruded bituminous coal in the Shuoli Mine, Anhui Province, China, Energy and Fuels, 29, 4100–4113. doi:10.1021/acs.energyfuels.5b00634.

Wangen, L.E., and Williams, M.D., 1978, Elemental deposition downwind of a coal-fired power plant, Water, Air, and Soil Pollution, 10, 33–44. doi:10.1007/BF00161994.

Wu, J., 2012, Mineralization feature and metallogenic model of No. 277 coal-bearing sandstone-type uranium deposit in Mabugang Basin, eastern Guangdong Province, Journal of East China Institute of Technology, 35, 10–16. in Chinese with an English abstract. doi:10.3969/j.issn.1674-3504.2012.01.002.

Wu, Y., Wang, Y., and Xie, X., 2014, Occurrence, behavior and distribution of high levels of uranium in shallow groundwater at Datong basin, northern China, Science of the Total Environment, 472, 809–817. doi:10.1016/j.scitotenv.2013.11.109.

Xu, J., Sun, Y., and Kalkreuth, W., 2011, Characteristics of trace elements of the No. 6 Coal in the Guanbanwusu Mine, Junger Coalfield, Inner Mongolia, Energy Exploration and Exploitation, 29, 827–842. doi:10.1260/0144-5987.29.6.827.

Yang, B., Luo, J., Dai, Y., Liu, X., Lin, T., and Zhang, S., 2006, Petrology of uranium-bearing sandstones and relationship of organic matters, hydrocarbons and coal with uranium: an example from Dongsheng, Ordos Basin and Shihongtan, Tuha Basin, Journal of Northwest University (Natural Science Edition), 36, 982–987. in Chinese with an English abstract. doi:10.16152/j.cnki.xdbzr.2006.06.035.

Yang, J., 2006, Concentrations and modes of occurrence of trace elements in the Late Permian coals from the Puan Coalfield, southwestern Guizhou, China, Environmental Geochemistry and Health, 28, 567–576. doi:10.1007/s10653-006-9055-z.

Yang, J., 2007, Concentration and distribution of uranium in Chinese coals, Energy, 32, 203–212. doi:10.1016/j.energy.2006.04.012.

Yang, J., Di, Y., Zhang, W., and Liu, S., 2011b, Geochemistry study of its uranium and other elements of brown coal of ZK0161 well in Yili Basin, Journal of China Coal Society, 36, 945–952. in Chinese with an English abstract. doi:10.13225/j.cnki.jccs.2011.06.002.

Yang, J., Wang, G., Shi, Z., Ren, M., Zhang, W., and Liu, S., 2011a, Geochemistry study of uranium and other element in brown coal of ZK0407 well in Yili basin, Journal of Fuel Chemistry and Technology, 39, 340–346. in Chinese with an English abstract. doi:10.3969/j.issn.0253-2409.2011.05.004.

Yang, J., Zhao, Y., Zyryanov, V., Zhang, J., and Zheng, C., 2014, Physical-chemical characteristics and elements enrichment of magnetospheres from coal fly ashes, Fuel, 135, 15–26. doi:10.1016/j.fuel.2014.06.033.

Yang, N., Tang, S., Zhang, S., and Chen, Y., 2015, Mineralogical and geochemical compositions of the no. 5 Coal in Chuancaogedan Mine, Junger Coalfield, China, Minerals, 5, 788–800. doi:10.3390/min5040525.

Yang, N., Tang, S., Zhang, S., and Chen, Y., 2016, Geochemistry of trace elements in the no. 5 coal from the Chuancaogedan Mine, Junger Coalfield, Earth Science Frontiers, 23, 74–82. in Chinese with an English abstract. doi:10.13745/j.esf.2016.03.010.

Zeng, R., Zhuang, X., Koukouzas, N., and Xu, W., 2005, Characterization of trace elements in sulphur-rich Late Permian coals in the Heshan coal field, Guangxi, South China, International Journal of Coal Geology, 61, 87–95. doi:10.1016/j.coal.2004.06.005.

Zhang, J., Ren, D., Zheng, C., Zeng, R., Chou, C.-L., and Liu, J., 2002, Trace element abundances in major minerals of Late Permian coals from southwestern Guizhou province, China, International Journal of Coal Geology, 53, 55–64. doi:10.1016/S0166-5162(02)00164-7.

Zhang, J., Ren, D., Zhu, Y., Chou, C.-L., Zeng, R., and Zheng, B., 2004a, Mineral matter and potentially hazardous trace elements in coals from Qianxi Fault Depression Area in southwestern Guizhou, China, International Journal of Coal Geology, 57, 49–61. doi:10.1016/j.coal.2003.07.001.

Zhang, J., Zheng, C., Ren, D., Chou, C.-L., Liu, J., Zeng, R., Wang, Z., Zhao, F., and Ge, Y., 2004b, Distribution of potentially hazardous trace elements in coals from Shanxi province, China, Fuel, 83, 129–135. doi:10.1016/S0016-2361(03)00221-7.

Zhang, R., 1981, The chemical states of uranium in coal ash, Journal of Fuel Chemistry and Technology, 375–383. in Chinese with an English abstract.

Zhang, R., 1986, Sintering of uraniferous coal ash and its inclusion effect on uranium, Uranium Mining and Metallurgy, 5, 7–14. in Chinese with an English abstract.

Zhang, S., Chen, G., and Tang, Y., 1984, Some geochemical characteristics of uranium-bearing coal deposits in China, Acta Sedimentologica Sinica, 2, 77–87. in Chinese with an English abstract.

Zhang, S., Wang, S., and Yin, J., 1987, The study of germanium ore in uranium-bearing coal of the Bangmai Basin, Lincang region, Yunnan Province, Uranium Geology, 3, 267–275. in Chinese with an English abstract.

Zhang, S., Wang, S., and Yin, J., 1988, A study on the forms of existence of germanium in uranium-bearing coals from Bangmai Basin of Yunnan, Chinese Nuclear Science and Technology Report, 1–10. in Chinese with an English abstract.

Zhang, Y., Shi, M., Wang, J., Yao, J., Cao, Y., Romero, C.E., and Pan, W., 2016, Occurrence of uranium in Chinese coals and its emissions from coal-fired power plants, Fuel, 166, 404–409. doi:10.1016/j.fuel.2015.11.014.

Zhao, L., Ward, C.R., French, D., and Graham, I.T., 2015, Major and trace element geochemistry of coals and intra-seam claystones from the Songzao coalfield, SW China, Minerals, 5, 870–893. doi:10.3390/min5040531.

Zhou, Z., 1981, Discussion on the radiation protection during mining and usage of bone coal, Radiation Protection, 74–78. in Chinese.

Zhuang, X., Querol, X., Alastuey, A., Juan, R., Plana, F., Lopez-Soler, A., Du, G., and Martynov, V.V., 2006, Geochemistry and mineralogy of the Cretaceous Wulantuga high-germanium coal deposit in Shengli coal field, Inner Mongolia, Northeastern China, International Journal of Coal Geology, 66, 119–136. doi:10.1016/j.coal.2005.06.005.

Zhuang, X., Querol, X., Alastuey, A., Plana, F., Moreno, N., Andres, J.M., and Wang, J., 2007, Mineralogy and geochemistry of the coals from the Chongqing and Southeast Hubei coal mining districts, South China, International Journal of Coal Geology, 71, 263–275. doi:10.1016/j.coal.2006.09.005.

Zhuang, X., Querol, X., Plana, F., Alastuey, A., Lopez-Soler, A., and Wang, H., 2003, Determination of elemental affinities by density fractionation of bulk coal samples from the Chongqing coal district, Southwestern China, International Journal of Coal Geology, 55, 103–115. doi:10.1016/S0166-5162(03)00081-8.

Zhuang, X., Querol, X., Zeng, R., Xu, W., Alastuey, A., Lopez-Soler, A., and Plana, F., 2000, Mineralogy and geochemistry of coal from the Liupanshui mining district, Guizhou, south China, International Journal of Coal Geology, 45, 21–37. doi:10.1016/S0166-5162(00)00019-7.

Zhuang, X., Su, S., Xiao, M., Li, J., Alastuey, A., and Querol, X., 2012, Mineralogy and geochemistry of the Late Permian coals in the Huayingshan coal-bearing area, Sichuan Province, China, International Journal of Coal Geology, 94, 271–282. doi:10.1016/j.coal.2012.01.002.

Zvereva, V.P., and Krupskaya, L.T., 2013, Environmental assessment of coal ash ponds of thermal power plants in the south of the Russian Far East, Russian Journal of General Chemistry, 83, 2668–2675. doi:10.1134/S1070363213130124.

REVIEW ARTICLE

Emission controls of mercury and other trace elements during coal combustion in China: a review

Yongchun Zhao, Jianping Yang, Siming Ma, Shibo Zhang, Huan Liu, Bengen Gong, Junying Zhang and Chuguang Zheng

ABSTRACT

Trace elements (TEs) in coal result in substantial pollutant emissions and cause serious damage to the ecological environment and human health in China. The emission and control of TEs, especially mercury, during coal combustion are of significant concern, and extensive studies have been performed in China in recent years. This paper reviews the emission characteristics and control strategies of mercury and other TEs during coal combustion in China. The occurrence of TEs in Chinese coals, including the average content of TEs in Chinese coals, the distribution of TEs in Chinese coals from different coal-forming periods, coal ranks, and coal-bearing regions are summarized. The emission characteristics of five specific TEs (Hg, As, F, Se, and Cr) during coal combustion in China are reviewed in detail. Effects of the coal type, combustion temperature, and combustion mode on the partitioning behaviour and emission characteristics of TEs are discussed. The effects of existing air pollution control devices of coal-fired power plants in China on the speciation and emission of TEs are discussed comprehensively. Various sorbents, such as activated carbon, fly ash, calcium-based sorbents, metal oxides, and mineral sorbents, used for TE removal are also summarized. Moreover, the removal performance of different sorbents for capturing certain TEs is compared comprehensively. Finally, future work for TE emission control in China is proposed.

1. Introduction

China is the largest producer and consumer of coal in the world. According to the British Petroleum Statistical Review of World Energy 2015 (BP 2015), coal consumption in China accounts for over half of the coal consumption worldwide. Due to the 'deficient oil, lean gas, rich coal' energy reserves in China, its coal-dominated energy structure will not change in the near future. The total energy consumption in China has increased every year due to rapid economic development, as shown in Figure 1. Coal consumption accounted for over 60% of the total energy consumption each year (National Bureau of Statistics of People's Republic of China 2016).

Elements in coal can be classified into three broad groups based on their concentration: (1) major elements (C, H, O, N, and S), with concentrations above 1000 ppm; (2) minor elements, mainly including coal-mineral matter (Si, Al, Ca, Mg, K, Na, Fe, Mn, and Ti) and halogens (Cl, Br, and I), with concentrations between 100 and 1000 ppm; and (3) trace elements (TEs), with concentrations below 100 ppm (Dai *et al.* 2005; Vejahati

et al. 2010). TEs even in the parts per million level in coal could result in pollutant emissions during coal combustion, owing to the immense coal consumption nationwide. Tian *et al.* (2010) calculated the total atmospheric emissions of Hg, As, and Se from coal combustion in China. They found that the total atmospheric emissions of Hg, As, and Se have rapidly increased from 73.59, 635.57, and 639.69 t in 1980 to 305.95, 2205.50, and 2352.97 t in 2007 at an annually averaged growth rate of 5.4%, 4.7%, and 4.9%, respectively. Gao *et al.* (2013) found a significant correlation between atmospheric emissions of Hg, Pb, and As and coal consumption with squared correlation coefficients of 0.911, 0.971, and 0.996, respectively. It was reported that the atmospheric concentrations of As, Cd, Ni, and Mn in China are 51.0 ± 67.0, 12.9 ± 19.6, 29.0 ± 39.4, and 198.8 ± 364.4 ng/m^3, respectively, which are much higher than the emission limit in the WHO guidelines (Duan and Tan 2013).

Some TEs present in coal are known to be harmful to the environment and human health. TEs such as Cr, Ni,

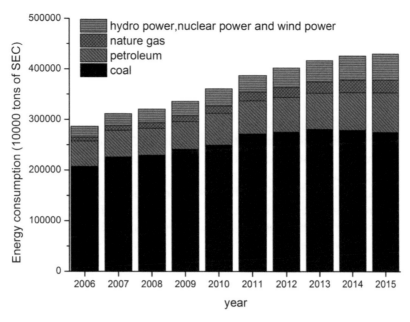

Figure 1. The energy consumption (10,000 t of standard coal equivalent [SCE]) of China from 2006 to 2015 (SEC: coal with 7000 kcal/kg low calorific value as received basis).

Pb, Zn, Cu, V, and Cd are carcinogenic, and As is one of the most common and hazardous carcinogenic elements (Cheng 2003; Ng *et al.* 2003; Duan and Tan 2013). Mercury is a toxic heavy metal that has the characteristics of volatility, bioaccumulation, and persistency (Pavlish *et al.* 2003). Other TEs released from coal combustion can also be harmful. For example, excessive fluoride ingestion can cause skeletal and dental fluorosis (Zheng *et al.* 1999; Finkelman *et al.* 2002; Dai *et al.* 2007). High concentrations of Se may be toxic for organisms, resulting in the loss of hair and nails, as well as nervous system disorders, in humans and animals (Z. Zhu *et al.* 2008). The emission of TEs from coal combustion has caused serious damage to the environment and human health in China. In the southwest region of China, thousands of people are suffering from endemic fluorosis, arsenism, and selenium intoxication. Western Guizhou Province is one of the areas most affected by endemic fluorosis, where approximately 10 million people suffer from dental fluorosis and 1 million people suffer from skeletal fluorosis (Dai *et al.* 2004, 2007). In Enshi Prefecture in Hubei Province, more than 10,000 cases of Se poisoning have occurred in swine due to the combustion of high-Se coal in the past few decades (Mao *et al.* 1990). Thus, it is important to study the emission characteristics of TEs and develop effective control technologies for TE emission. On the other hand, the concentration of some valuable elements (e.g. Ge, Ga, U, rare earth elements, Y, Nb, Zr, Se, V, Re, and Al) in some coals or coal ashes is near or even higher than those in conventional economic

deposits. Thus, these coals or coal ashes can be used as raw sources of these valuable elements (Seredin 2012; Seredin and Finkelman; Seredin and Dai 2012; Seredin *et al.* 2013; Dai *et al.* 2016a, 2016b; Hower and Dai 2016; Hower *et al.* 2014, 2015, 2016). Additionally, the noble metals Au and Ag have also attracted attention due to their highly elevated concentration in fly ash (Seredin and Dai 2014). Understanding the behaviour of these elements and controlling their emission will provide basic information for the development of extraction methods.

The objective of this paper is to provide an overview of the emission characteristics and control strategies of TEs during coal combustion in China. The first section briefly summarizes the distribution of TEs in Chinese coals, including the average content of TEs in Chinese coals and the distribution of TEs in Chinese coals from different coal-forming periods, coal ranks, and coal-bearing regions. The second section summarizes the TE emissions during coal combustion in China. Then, the simultaneous control of TEs by existing air pollution control devices (APCDs) in power plants is described. Finally, the sorbents used for TE removal are discussed.

2. Distribution of TEs in Chinese coals

2.1. Average content of TEs in Chinese coals

There are four statistical methods that have been published in the literature to calculate the average concentration of TEs in coals: arithmetic mean, geometric mean, weighted mean, and bootstrap mean. The weighted mean of TEs in Chinese coals was proposed by Ren et al. (2006a). Ren classified coals based on the coal-forming periods and calculated the weighted mean of TEs according to the percentage of coal reserves from different coal-forming periods in national overall coal reserves. Tian et al. (2013) evaluated the variability and uncertainty of the average concentration of TEs in Chinese coals by the bootstrap simulation method. Bootstrap simulation is a type of Monte Carlo simulation, which is a statistical processing method involving resampling and replacement, and details of the bootstrap simulation method can be found in the literature (Tian et al. 2013). Among the four statistical methods, the weighted mean and bootstrap mean are more accurate because the influence of very low or high data with low frequency is not significant. Table 1 shows the average concentration of TEs in Chinese coals calculated by different researchers and different statistical methods. The average concentration of Hg, Pb, F, and Se in Chinese coals is higher than that in global coals reported by Ketris and Yudovich (2009); the concentrations of the elements Sb, Be, Ni, and Cr in Chinese coals are near those of the world average, while As in Chinese coals is lower than the world average.

2.2. Distribution of TEs in Chinese coals from different coal-forming periods and coal ranks

The main coal-forming periods in China can be classified into six periods, late Carboniferous and early Permian (C_2–P_1), late Permian (P_2), Late Triassic (T_3), Early to Middle Jurassic (J_1–J_2), Late Jurassic and Early Cretaceous (J_3–K_1), and Palaeogene and Neogene

(E–N), which are essentially consistent with the global main coal-forming periods (Ren et al. 2006a). In addition, there is a minor but significant amount of early Carboniferous (C1) coal in Southern China (Dai et al. 2012a). The reserves of coal formed in different coal-forming periods are quite different. The total recoverable coal reserves are made up of 38.1% C_2–P_1, 7.5% P_2, 0.4% T_3, 39.6% J_1–J_2, 12.1% J_3–K_1, and 2.3% E–N coals (China Coal Geology Bureau 1996). Different coal-forming periods, which vary in botanical composition, sedimentary environment, climate, and epigenetic geological factors that affect coalification, can significantly influence the distribution of TEs. Thus, it is important to investigate the distribution of TEs in Chinese coals from different coal-forming periods. Table 2 summarizes the average concentration of TEs in Chinese raw coals from different coal-forming periods. It can be clearly seen that most TEs are enriched in P_2 and T_3 coals, especially in T_3 coals, while the enrichment of TEs in J_1–J_2 and J_3–K_1 coals is relatively low. Elements of Ni, Co, As, F, and Cr are concentrated in E–N coals. Based on the metamorphic stage, coals can be divided into three main coal ranks: lignite, bituminous coal, and anthracite. The distribution of TEs in coals of different ranks also varies. Zhang et al. (2004) systematically analysed the distribution of TEs in different-rank coals from Shanxi Province and found that Shanxi bituminous coals are highly enriched in Hg and slightly enriched in Cd, F, and Th compared to the average values for world coals. Tian et al. (2013) investigated the distribution of eight TEs in coals of different ranks on a nationwide level and concluded that the average concentration of Hg, Se, Pb, and Cr increases gradually with an increase in coal rank. Elements of Ni, As, and Sb are more enriched in lignite than in bituminous coal and minimally enriched in anthracite coal.

Table 1. The average concentration (µg/g) of TEs in Chinese coals.

	Chinese coals						Continental crust abundance	World coals
	Ren et al. (2006a)		Bai et al. (2007)	Dai et al. (2012a)	Tian et al. (2013)		Taylor and McLennan (1985)	Ketris and Yudovich (2009)
TEs	Arithmetic mean	Weighted mean	Weighted mean	Weighted mean	Geometric mean	Bootstrap mean	–	Arithmetic mean
Hg	0.195	0.188	0.154	0.163	0.15	0.20	0.089	0.10
As	4.05	3.796	4.09	3.79	3.96	5.78	2.2	8.30
Be	2.13	2.127	1.75	2.11	–	–	1.3	1.6
Sb	0.95	0.83	0.71	0.84	0.89	2.01	0.62	0.92
Cd	0.25	0.2415	0.81	0.25	0.22	0.61	0.15	0.22
Cr	16.12	15.334	16.94	15.4	15.01	30.37	110	16
F	167[a]	–	157	130	136[a]	–	450	88
Co	6.88	7.052	10.62	7.08	–	–	25	5.1
Pb	16.91	15.549	16.64	15.1	13.98	23.04	12	7.80
Mo	3.72	3.113	2.70	3.08	–	–	1.3	2.2
Ni	14.04	13.711	14.44	13.7	15.31	17.44	89	13.00
Se	2.78	2.472	2.82	2.47	1.92	3.66	18	1.30
Th	7.00	5.811	5.88	5.84	–	–	5.8	–
U	2.46	2.405	2.33	2.43	–	–	1.7	–

[a]From Wu et al. (2004). (Reprinted with permission from Wu et al. (2004). Copyright 2004 International Society for Fluoride Research)
TE: Trace elements.

Table 2. The average concentration of TEs in Chinese coals of different coal-forming periods [Arithmetic mean, μg/g, origin from Ren *et al.* (2006a)].

	C–P$_1$	P$_2$	T$_3$	J$_1$–J$_2$	J$_3$–K$_1$	E–N
Hg	0.222 (0.29)	0.418 (0.39)	0.478 (0.50)	0.113 (0.24)	0.189 (0.27)	0.071
As	2.59 (3.71)	6.42 (6.99)	9.16 (10.37)	3.23 (3.94)	6.04 (6.49)	12.16
Be	1.92	2.42	2.67	2.41	1.88	0.93
Sb	0.68 (0.78)	1.67 (1.65)	3.62 (3.90)	0.61 (0.90)	1.38 (1.17)	1.03
Cd	0.30 (0.37)	0.62 (0.69)	0.99 (0.82)	0.13 (0.26)	0.125 (0.18)	0.44
F[a]	149.4	200.9	84.1	84.46	134.2	352.8
Cr	15.81 (19.36)	29.02 (31.24)	50.40 (33.43)	11.35 (12.52)	11.89 (22.80)	43.33
Co	4.24	10.07	7.72	8.59	8.09	11.75
Pb	20.70 (22.34)	26.95 (24.07)	20.38 (21.81)	8.76 (15.00)	12.29 (15.80)	26.21
Mo	3.47	8.40	3.14	1.92	2.55	3.44
Ni	11.75 (15.66)	24.05 (26.21)	28.1 4 (20.30)	11.9 7 (10.80)	10.56 (13.03)	56.46
Se	4.84 (3.35)	4.71 (4.33)	1.47 (2.76)	0.43 (0.98)	0.63 (0.91)	1.01
Th	8.71	7.81	9.07	3.01	4.91	3.72
U	2.6	4.29	3.9	1.24	1.72	3.93

[a]Weighted mean, origin from Dai and Ren (2006).
() Bootstrap mean, origin from Tian *et al.* (2013). (Reprinted with permission from Den and Ren *et al.* (2006). Copyright 2006 Elsevier)

2.3. Distribution of TEs in Chinese coals from different coal-bearing regions

The geographic distribution of coal reserves in China is uneven. The main coal-bearing areas in China can be divided into five regions: Northern China, Northeastern China, Northwestern China, Southern China, and the Tibet-western Yunnan coal-bearing area (Dai *et al.* 2012a), as shown in Figure 2. More than 90% of the coal reserves are in the Northern China, Northeastern China, and Northwestern China coal-bearing regions. Furthermore, the coal reserves in Shanxi, Ningxia, Xinjiang, Shaanxi, Gansu, Guizhou, and Inner Mongolia Provinces account for almost 80% of the national coal reserves. The concentration of TEs in Chinese coals from different provinces and coal-bearing regions also varies. Table 3 gives the arithmetic mean concentrations of TEs in Chinese coals from different coal-bearing regions.

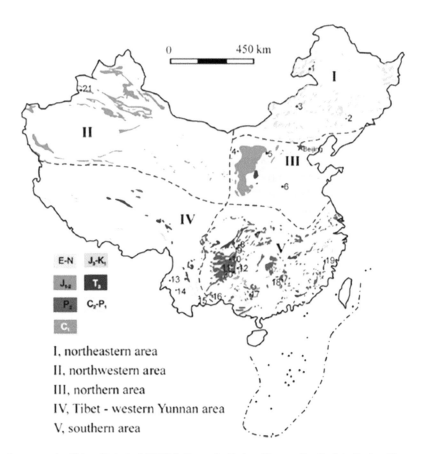

I, northeastern area
II, northwestern area
III, northern area
IV, Tibet - western Yunnan area
V, southern area

Figure 2. Coal distribution areas in China (Dai *et al.* 2012a). C$_1$: early Carboniferous; C$_2$–P$_1$: late Carboniferous, and early Permian; P$_2$: late Permian; T$_3$: late Triassic; J$_1$–$_2$: Early to Middle Jurassic; J$_3$–K$_1$: Late Jurassic and Early Cretaceous; E–N: Palaeogene and Neogene. (Reprinted with permission from Dai *et al.* (2012a). Copyright 2012 Elsevier).

Table 3. The arithmetic mean concentration of TEs in Chinese coals of different coal-bearing regions (µg/g) (Bai 2003).

Coal-bearing region	Hg	As	Se	Pb	Cd	F	Cr	Ni	Sb
Southern China	0.24	7.39	5.32	19.70	1.15	323.26	32.52	21.71	2.06
Northeastern China	0.10	5.40	0.72	15.91	0.94	198.63	17.89	12.68	0.77
Northern China C_2–P_1	0.18	1.75	5.30	23.40	1.01	144.13	13.67	13.45	0.44
Northern China J_1–J_2	0.17	5.54	0.60	14.28	0.87	128.23	15.87	15.00	0.51
Northwestern China	0.10	4.26	0.36	7.47	0.46	135.14	19.24	16.30	0.78

The enrichment of TEs results from geologic processes, i.e. the emplacement of elements by hydrothermal solutions generated by tectonic activity. Different coal-bearing regions in China experienced different geologic processes. Thus, the concentrations of TEs in Chinese coals from different provinces and coal-bearing regions vary. The coal-bearing regions of Northern China, Northeastern China, Northwestern China usually have low Hg, As, Pb, Cd, and Cr concentrations, while coals distributed in Southern China, especially Southwestern China, are enriched in Hg, As, Pb, Cd, and Cr. In the regional distribution of Chinese coals, Northern China, Northeastern China, Northwestern China, and Southern China are rich in late Carboniferous to early Permian coals, Late Jurassic to Early Cretaceous coals, Early to Middle Jurassic coals, and late Permian coals, respectively (Wang et al. 2006). As mentioned in Section 2.2, most TEs are enriched in P_2 and T_3 coals and are relatively low in J_1–J_2 and J_3–K_1 coals. Therefore, the average concentration of Hg, As, Pb, Cd, and Cr in Chinese coals gradually increases from northern China to southern China. High-Cr coals in China are found in Heshan coalfield (Shao et al. 2003; Zeng et al. 2005; Dai et al. 2013), Guiding coalfield (Dai et al. 2015; Liu et al. 2015), Yanshan coalfield (Dai et al. 2008), Shenbei (SB) coalfield, Huaibei coalfield, Beipiao coalfield, and Weibei coalfield. The Cr concentration in these coal fields ranges from 55.8 to 180 µg/g (Wu et al. 2005; Dai et al. 2015; Liu et al. 2015). Coal fields with high As concentrations (up to 35,000 ppm) are located in Southern China, especially in Southwestern China. It has been reported that the As concentration in coals in Southwestern China is 267 µg/g (Tian et al. 2016). However, arsenic in a few coals from Northern China (e.g. the Wulantuga coal mine in the Shengli coalfield in Inner Mongolia) is highly enriched, reaching up to 500 µg/g (Zhuang et al. 2006; Dai et al. 2012b). With the exception of Inner Mongolia (e.g. coals from the Wulantuga coal mine) (Zhuang et al. 2006; Dai et al. 2012b), the concentration of Sb in coals from Northern China is lower than that in coals from Southern China (Qi et al. 2008). In the regional distribution of Se in Chinese coals, coal from South-central China, which includes Anhui, Jiangsu, Henan, Chongqing, and Hubei province, has a higher Se concentration compared to that of other provinces (Wang et al. 2010a). Enshi Prefecture in Hubei Province has the most serious Se poisoning situation, and the Se concentration of carbonaceous–siliceous rock samples collected in Yutangba, Enshi Prefecture, reaches up to 5371–8290 µg/g (Song 1989; Zheng et al. 1992, 1993). The coal-bearing regions of Northeastern China and Southern China have high F concentrations, while the F content in Northern and Northwestern China coals are relatively low. Guizhou province suffers from endemic fluorosis caused by coal burning. However, the arithmetic mean concentration of fluorine in coals sampled from Guizhou Province is not high. The main fluorine source in these villages is not from coal but from the clay mixed with the coal to make coal–clay briquettes (Dai et al. 2004, 2007). Highly elevated concentrations of F have been reported in the late Permian coals from Heshan coalfield, reaching up to 3362 µg/g, and fluorine mainly occurs in fluorite (Dai et al. 2013).

3. TE emissions during coal combustion

The TEs were divided into three groups according to their partitioning behaviour during coal combustion (Clarke and Sloss 1992), as shown in Figure 3:

Group I includes non-volatile elements; they almost always remain in the combustion residue due to their high boiling point.

Group II includes semi-volatile elements. Part or all of these elements can be released at the high temperatures present in the boiler, but they are present at homogeneous nucleation sites and in heterogeneous condensate in the low-temperature downstream of flue gas cooled in the upper furnace and heat recovery section.

Group III includes Hg, F, Se, and Cl, which are volatile elements. These elements are present in the gaseous phase in the combustion process and are released to the atmosphere.

The emission characteristics of TEs have been discussed by many researchers, and mechanisms of TEs transformation have been proposed. TEs vaporized at high temperature will nucleate or condense when the flue gas is cooled downstream. Thus, TEs will be released in the gaseous phase or remain in the solid

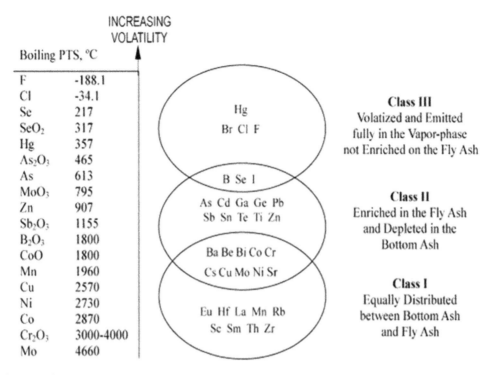

Figure 3. Classification of TEs during coal combustion based on their volatility (Clarke and Sloss 1992).

products in power plants, such as bottom ash, coal fly ash, and even the flue gas desulfurization (FGD) gypsum (Tian *et al.* 2013). Figure 4 shows the transformation of TEs during coal combustion, including the processes of evaporation, nucleation, condensation, and coagulation (Lockwood and Yousif 2000). On the other hand, rare earth elements were hardly fractionated between fly and bottom ash and were almost completely retained in these ashes, rather than being evaporated in the gaseous phase (Hower *et al.* 2010; Dai

et al. 2014b). In this section, the emission and control of five elements (Hg, F, Se, As, and Cr) have been chosen for detailed discussion.

3.1. *Hg emissions during coal combustion in China*

China is one of the largest atmospheric Hg emission sources in the world, and nearly 50% of its Hg emissions result from stationary combustion, especially coal combustion (Wang *et al.* 2012; Tian *et al.* 2013). Hg

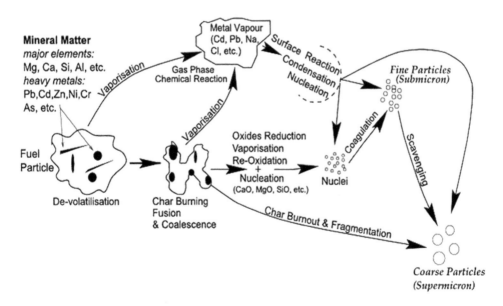

Figure 4. Transformation and distribution of TEs during coal combustion (Lockwood and Yousif 2000).

emissions cause serious harm to the environment and to human health. Therefore, it is essential to study the Hg release characteristics and the influencing factors of Hg volatility during coal combustion.

Zheng et al. (2010) conducted a combustion experiment with two coals from southwestern Guizhou province in China in a tube furnace to determine the occurrence of Hg in the coals. The volatilization rate of Hg at 150°C was in the range of 49.33–51.16%, with an average of 50.25%, indicating that Hg volatized at low temperature. Water-soluble and ion-exchangeable Hg is easy to volatilize at low temperature, and the Hg adsorbed on clay minerals will also volatilize under low temperature (Zheng et al. 2010). The volatilization rate of Hg will rise sharply with an increase in temperature below 400°C, and when the temperature increased to 550°C, the Hg contained in sulphide and carbonate minerals that gradually oxidized and decomposed was released (Zheng et al. 2010). Wang et al. (2003) conducted a coal combustion experiment in a one-dimensional furnace, which is a laboratory-scale combustion instrument, using three Chinese coals at 1200°C. The results showed that 85% of the Hg in the coals was volatilized. The largest release rate of Hg from one coal was 95.8%, and only a small amount of Hg remained in the bottom ash. Finkelman et al. (1990) analysed the volatilization rate of Hg during the Argonne coal combustion, and 40–75% volatilized below 550°C. Rizeq et al. (1994a) determined that Hg will almost completely vaporize at temperatures higher than 800°C. The results of Hg emission in Chinese coals are consistent with those of Finkelman et al. (1990) and Rizeq et al. (1994a). Hg, which could be adsorbed on fly ash carbons (Hower et al. 2010), was present in higher concentrations in the relatively carbon-rich coarse fraction than in the C-depleted fine fraction of fly ash from the Jungar Power Plant in Inner Monglia (Dai et al. 2010, 2014b).

The speciation of gaseous Hg in flue gas was investigated by laboratory-scale experiments, as shown in Table 4. The studies of Wang et al. (2003) showed that the total Hg concentration in the flue gas ranged from 48 to 80 µg/N m³. In their experiment, most of the Hg was particulate Hg (Hg^P), while gaseous Hg ($Hg_{(g)}$) accounted for 11–48% of the total Hg in the flue gas. This content of Hg^P is higher than that determined by other researchers (Ren et al. 2002; Wu et al. 2009). This is possibly because Hg was cooled and adsorbed in the test bench. Moreover, in their study, Hg in flue gas was mainly in the form of Hg^0. The percentage of Hg^0 in the flue gas ranged from 52% to 83%, while Hg^{2+} only accounted for 17–48%. Zhou et al. (2013) chose two Chinese coals, including Changguang coal and Xinwen coal, to study the speciation of Hg during combustion, and the combustion temperatures were 700, 900, and 1200°C. They found that $Hg_{(g)}$ was the dominant species in the flue gas, while approximately 20% of the Hg remained in the ash of the two coals (Table 4). Hg^0 was the dominant species in the flue gas at 700°C, while the percentage of Hg^{2+} gradually increased with increasing temperature. When the temperature increased from 700 to 1200°C, the Hg^0 content reduced from 83% to 45% in the flue gas (Table 4). This observation was significantly different with the studies of Wang et al. (2003). The proportion of Hg^{2+} to Hg^0 depends on coal composition and chemical kinetics.

Mercury in the coal will decompose to gaseous Hg^0 in the high temperature combustion zone of boiler (Senior et al., 2000). Due to the decrease of flue gas temperature in the post-combustion section, Hg^0 oxidation reaction with halogen species originally present in the coal occurs. Thermodynamic calculations predict that Hg oxidation occurs at temperatures below 700°C and that Hg will be completely oxidized at or below 450°C (Senior et al., 2000). However, the extent of Hg oxidation is highly dependent on the coal rank and composition, halogen concentration in coal, and the cooling rate of the flue gas (Pavlish et al. 2003; Wilcox et al. 2012). Generally, the gas-phase oxidation reactions do not reach equilibrium under the conditions of

Table 4. Gaseous Hg speciation in flue gas of laboratory scale experiments.

Coal types	Temperature (°C)	Combustion mode	$Hg_{(g)}$ (µg/m³)	Hg^0 (%)	Hg^{2+} (%)	Reference
Lignite	1200	One-dimension furnace	17.32	77.3	22.7	Wang et al. (2003)
Bituminous coal	1200	One-dimension furnace	23.06	83.3	16.7	
Anthracite	1200	One-dimension furnace	8.89	52.1	47.9	
Bituminous coal	700	Tube furnace	–	83	17	Zhou et al. (2013)
Bituminous coal	1200	Tube furnace	–	45	55	
Mixed coals	–	Circulating fluidized-bed	6.8–9.3	59–71	29–41	
Mixed coals	1200	One-dimension furnace	–	45.2	54.8	Wu et al. (2009)
Bituminous coal	1100–1300	Bench-scale pulverized coal furnace	10–15	>40	>50	Ren et al. (2002)
Three coal samples	600	Tube furnace	–	55.9–89.2	6.1–22.4	Yao et al. (2006)
Bituminous coal	600–1000	Tube furnace	–	15.5–73.3	26.3–84.5	Liu et al. (2008)
Four UK coal samples	–	1 MW combustion test facility	8.0–11.7	12–34	66–88	Gibb et al. (2000)

–: No data; $Hg_{(g)}$: gaseous Hg content in flue gas; Hg^0: partitioning of elemental Hg in flue gas; Hg^{2+}: partitioning of divalent Hg in flue gas.

rapid quenching. The studies of Sliger et al. (1998) showed that the 50% equilibrium conversion of Hg^0 to $HgCl_2$ occurred at 675 and 550°C in the presence of 500 and 50 ppm HCl, respectively. The experimental studies also demonstrated that not all of the Hg is oxidized regardless of the content of halogen present in the coal. Thus, the Hg oxidation reaction is kinetically controlled, rather than thermodynamically limited (Senior et al., 2000).

Extensive studies about the homogeneous and heterogeneous reaction mechanisms and kinetics of Hg oxidation in coal-fired flue gas have been preformed. Wilcox (2014) carried out a comprehensive study on the homogeneous Hg oxidation kinetics of Hg–Cl and Hg–Br reactions. It was found that the increasing HCl concentration can enhance the oxidation reaction. The direct elementary oxidation pathway of Hg by HCl is difficult to occur because of the high energy barrier of the $Hg + HCl \rightarrow HgCl + H$ reaction; however, it can occur via an intermediate Cl species from HCl. A similar result was also obtained from Hg–Br reactions. Yang et al. (2016d, 2017b) studied the homogeneous and heterogeneous Hg oxidation by HCl and HBr, and the catalytic effects of unburned carbon (UBC) and Fe_2O_3 in fly ash were included as well. The results showed that the dominant pathway of heterogeneous Hg oxidation by Cl and Br on fly ash surface is $Hg^0 \rightarrow StHgCl(s) \rightarrow HgCl_2$ and $Hg^0 \rightarrow StHgBr(s) \rightarrow HgBr_2$, respectively, whereby a chlorinated and brominated carbon site partially oxidizes Hg^0 into $StHgCl(s)$ and $StHgBr(s)$ which is subsequently oxidized into $HgCl_2$ and $HgBr_2$. UBC and Fe_2O_3 can promote the Hg oxidation, and UBC exhibits much higher catalytic oxidation activity for Hg oxidation than Fe_2O_3. In addition, SO_2 weakly promotes heterogeneous Hg oxidation on fly ash, and H_2O played an inhibitive role in heterogeneous Hg oxidation through the elimination of chlorinated sites.

After Hg^0 oxidation reaction, the formed $Hg^{2+}(g)$ may remain in flue gas or adsorb onto fly ash surface. The increasing flue gas temperature alters the equilibrium partitioning of Hg^{2+}: at lower temperatures, more Hg^{2+} can be adsorbed on surfaces, while at higher temperatures, equilibrium shifts to the vapour phase. It is well known that much larger surface-to-volume ratio of lab-scale experiments likely leads to greater loss of condensable Hg^{2+} than Hg^0, resulting in an artificial enrichment of Hg^0 in lab measurements as compared to full-scale electric generation units.

To understand and control the TE emission from coal-fired power plants, it is very important to study the partitioning behaviour of gaseous and solid TEs and the factors influencing TE emission during coal combustion. Some relevant research results are summarized in Table 5. The residual carbon in fly ash can retain some of the Hg, and Hg is more likely to present in the fly ash than in bottom ash (Gibb et al. 2000; Dai et al. 2010). Huang et al. (2004) investigated the factors influencing TE partitioning in a Chinese thermoelectric power plant and found that the Hg percentage retained in the bottom ash and fly ash was 0% and 15%, respectively, while more than 80% of the Hg was distributed in the flue gas. An extremely high Hg concentration (up to 67.5 µg/g on a whole sample basis and 160 µg/g on an ash basis) has been observed in the Ge-rich fly ash from the Wulantugan power plant (Dai et al. 2014a). Zhou et al. (2008) studied the emission of Hg in a 300-MW unit. Both before and after the electrostatic precipitator (ESP), the Hg in flue gas accounted for the largest proportion of total Hg. The percentage of Hg in slag is minor (0.5%) before ESP and that in fly ash after ESP accounted for 13.1%. Before the ESP, Hg^0 and Hg^{2+} accounted for 25% and 75%, respectively, while the proportions after the ESP were 46.4% and 53.6%. Many TEs are more likely to be enriched in fine particles, especially in submicron particles. The TEs vaporized during combustion will condense on the surface of particles at low temperature, and the fine particles of fly ash have a larger specific surface area than bottom ash and coarse fly ash, which is beneficial for Hg adsorption (Martinez-Tarazona et al. 1996; Yi et al. 2008). Huang et al. (2004) found that the Hg content slightly increased in fine fly ash. The Hg content in fly ash was 0.066 ppm from the third field of ESP, which can capture smaller particles than the preceding fields of ESP, while that from the first field of ESP was only 0.019 ppm. Wang et al. (1998) demonstrated that fine fly ash (<38 µm) from a pulverized coal combustion power plant contained a higher Hg concentration than the coarse ash.

Huang et al. (2004) demonstrated that the Hg concentration in fly ash can increase with the addition of reducing oxygen during coal combustion, and Yan et al. (2001) also obtained similar conclusions. Zhu et al. (2001) investigated the distribution of Hg at different boiler loads in a coal-fired power plant in China. The partitioning of Hg in the slag and bottom ash was minor, which were 0–0.9% and 0.25–0.65%, respectively, while that in the fly ash was 24.5–30.9%, which was much higher than that in the slag and bottom ash. The emission of Hg in the flue gas will be reduced with decreasing boiler load, and the Hg enrichment in fly ash will also be reduced. Yang et al. (2008) studied the distribution of Hg in several typical Chinese boiler combustion modes and showed that Hg will be enriched in fly ash during circulating fluidized bed combustion (CFBC) compared to that in a pulverized coal-fired

Table 5. Partitioning of Hg during coal combustion.

Column groups: "Hg partitioning in combustion products (%)" covers Bottom ash / Fly ash / Flue gas / FGD. "Hg concentration and speciation in flue gas" covers the Excluding Hg^P columns ($Hg_{(g)}$, Hg^0, Hg^{2+}) and the Including Hg^P columns (Total Hg, $Hg_{(g)}$, Hg^0, Hg^{2+}, Hg^P).

Boiler type	Measure point	Bottom ash	Fly ash	Flue gas	FGD	$Hg_{(g)}$ (μg/m³) excl. Hg^P	Hg^0 (%) excl.	Hg^{2+} (%) excl.	Total Hg (μg/m³) incl. Hg^P	$Hg_{(g)}$ (μg/m³) incl.	Hg^0 (%) incl.	Hg^{2+} (%) incl.	Hg^P (%)	Reference
PCB	–	<0.2	15–29	69–84	–									Zhou et al. (2006)
CFB	–	6.7	59.3	24	–									Zhou et al. (2004)
PCB	Before ESP	0.5	–	99.5	–									Zhou et al. (2008)
PCB	After ESP	0.4	13.4	86.1	–									Guo et al. (2004b)
PCB	–	1	13	86	–									Zhang and Xie (2014)
PCB	–	0.7	14.6	6.74	69.77									Zhang and Xie (2014)
PCB	–	0.1	10.38	89.52	–									Yu et al. (2015)
PCB	–	<1	5–43	19–27	13–63									Wang et al. (2010b)
CCFF	–					Average 6.3	Average 23	Average 67						Prestbo et al. (1995)
PCB	After ESP					14.14	51.0	35.5						Zhou et al. (2013)
PCB	After WFGD					5.29	92.1	7.9						Zhou et al. (2013)
PCB	After ESP					15.06	8.7	91.3						Zhou et al. (2013)
PCB	After ESP					13.7–21.2	31–35	55–56						Guo et al. (2004b)
PCB	Stack (after ESP)					12.2–14.1	77.7	22.3						Lu et al. (2009b)
PCB	After ESP					–	47.6–66.2	33.8–52.4						Zhou et al. (2006)
CFB	Before ESP					17.44	88.15	11.85						Zhou et al. (2004)
PCB	After ESP					11.24	85.3	14.7						Zhou et al. (2008)
PCB	Before ESP					–	25	75						Zhou et al. (2013)
PCB	After ESP					–	46.4	53.6						Zhou et al. (2013)
PCB	Before ESP								9.5	8.21	51.0	35.5	13.5	Hu et al. (2008)
PCB	After ESP								15.5	13.57	48.3	39.2	12.5	Hu et al. (2008)
PCB	Before ESP								13.06	12.19	43.4	49.3	7.3	Hu et al. (2008)
PCB	After ESP								2.84	2.812	15.7	83.2	1.1	Hu et al. (2008)
PCB	Before SCR								16.7–21.7	–	10–40	30–45	15–55	Hu et al. (2008)
CFB	Before ESP								13.11	–	About 50	About 40	About 15	Duan et al. (2008)
CFB	After ESP								0.062	–	About 55	0	97.5	Duan et al. (2008)
PCB	Before ESP								2.6–3.0	1.2–2.0	27.0–47.9	0.41–0.66	24.3–58.8	Lee et al. (2006a)
PCB	After ESP								1.2–2.5	1.2–2.2	36.6–50	48.4–59.6	0.9–3.8	Lee et al. (2006a)
PCB	Stack								1.03–2.4	1.0–2.4	63.1–84.9	14.6–34	0.4–3.3	Lee et al. (2006a)
PCB	Before ESP								–	–	48.6	33.5	17.9	Lu et al. (2009b)
PCB	After ESP								–	–	55.8	44.2	–	Lu et al. (2009b)
CFB	Before ESP								–	–	–	–	41.1	Zhou et al. (2004)
CFB	After ESP								–	–	–	–	3.5	Zhou et al. (2004)
PCB	Before ESP								0.3575	0.345	63.2	33.3	3.5	Yang et al. (2006)
PCB	After ESP								0.548	0.5468	90.3	9.5	0.2	Yang et al. (2006)
PCB	Before ESP								29.16–33.92	27.36–32.15	38.49–45.01	48.84–56.28	5.23–6.15	Li et al. (2013)
PCB	After ESP								23.47–26.16	24.3–26.14	32.03–47.89	51.82–67.9	0.8–2.9	Li et al. (2013)
PCB	After WFGD								12.2–18.73	11.87–18.73	7.12–6.42	92.89–93.57	0	Li et al. (2013)

–: No data; CCFF: commercial coal fired facility; PCB: pulverized-coal boiler; CFB: circulating fluidized bed boiler; $Hg_{(g)}$: gaseous Hg content in flue gas; Hg^0: partitioning of elemental Hg in flue gas; Hg^{2+}: partitioning of divalent Hg in flue gas; Hg^P: particle Hg.

boiler. Tian *et al.* (2011) summarized the Hg release rate in different boiler types, showing that the average Hg release rate in a pulverized-coal boiler, fluidized-bed furnace, and stoker fired boiler was 99.42%, 98.92%, and 83.15%, respectively.

3.2. Arsenic emissions during coal combustion in China

To understand the characteristics of As emissions during coal combustion, Guo *et al.* (2003) chose a Chinese lignite to carry out combustion experiments in a tube furnace with long residence time. As the coal only partially vaporized during the combustion process, the volatilization rate of As increased with temperature from 400 to 700°C. Only 30–40% of the As was released between 850 and 1150°C, while the rest remained in the ash. Xu *et al.* (2004) studied the volatility of some TEs in three Chinese coals of different ranks via a tube furnace experiment. For bituminous coal and lignite, the trend of the As release rate increasing with temperature was similar for each sample during combustion, while the release rate of As in anthracite was lower than the other two coals. The release rate of all the samples increased sharply between the temperature of 1000 and 1100°C, and most of the As had volatilized at 1100°C. Monahan-Pendergast *et al.* (2008) studied the thermochemical properties of As species during coal combustion using *ab initio* methods. The following thermodynamically favoured reactions of gaseous As oxidation reactions involving OH and HO₂ were found:

$$As + OH \rightarrow AsOH \tag{1}$$

$$As + HO_2 \rightarrow AsO + OH \tag{2}$$

Seames and Wendt (2000) studied the partitioning of As during the combustion of Pittsburgh and Illinois No. 6 coals in a self-sustained combustor. The volatilization and subsequent heterogeneous transformation of As to submicron particles occurred for Illinois No. 6 fly ash in the post-combustion zone but not for Pittsburgh fly ash. Most of the As in the submicron fly ash enters the post-combustion zone in the solid phase. The formation of As–Ca products was an important factor in the heterogeneous transformation of As from the vapour phase to supermicron particles. This was also consistent with the studies of Mahuli *et al.* (1997), where it has been proposed that formation of calcium arsenate complexes is important for the phase transformation of As. The studies of Sakulpitakphon *et al.* (2003) also showed that most of As was captured in the fly ash at a Kentucky power plant. Although many researchers have studied the distribution and enrichment of arsenic

in the PM, there isn't a quantifiable description about As emission during coal combustion for the typical high As coal in China. Zhao *et al.* (2008) investigated the As emission characteristics of a Chinese high-As coal (LT-K2) using a bench-scale drop tube furnace (DTF). Different temperatures and combustion atmospheres were considered in the experiment. The As emission of particulate matter less than 1.0 μm (PM₁) increased from 0.07 to 0.25 mg/N m³, while that of PM₁₀ increased from 0.18 to 1.03 mg/N m³ as the temperature increased from 1100 to 1400°C (Figure 5). This is because the total PM increased with increasing temperature, as more As vapour condensed and reacted on the surface of the PM at higher temperature.

Compared to the results of As emission with different oxygen contents (Figure 6), almost each particle size contained less As in 20% oxygen content than in 50%. The total As emission of PM₁₀ was 0.1 mg/N m³ in 20% oxygen, which is only a tenth of the total As emission of PM₁₀ in 50% oxygen content. This is due to the higher temperature in 50% oxygen content. The effects of different kinds of coals were also considered by Zhao *et al.* (2008). Two coals from southwest Guizhou province (LT-K2) and central Henan province (Pingdingshan [PDS]) in China were chosen. The As behaviour of the two coals is quite different, due to the different As occurrence modes in coal. The As in LT-K2 was in pyrite, which undergoes explosive volatilization. However, the As in PDS is in an oxidized form as arsenate minerals, most of which will be assimilated as arsenate in the particles of fly ash.

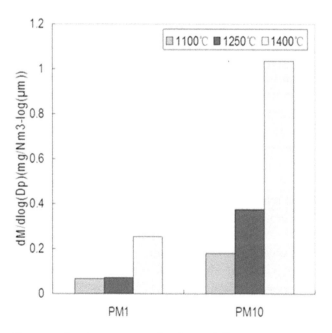

Figure 5. The As emission within PM in different temperature (Zhao *et al.* 2008).

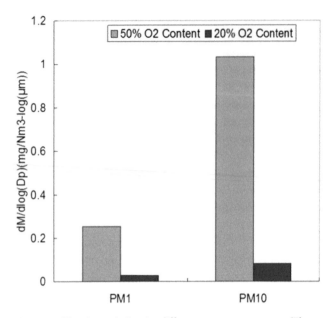

Figure 6. The As emission in different oxygen contents (Zhao *et al.* 2008).

As is regarded as a semi-volatile element in coal combustion and is enriched fly ash and depleted in bottom ash as the flue gas cools in coal-fired power plants. Guo *et al.* (2004c) investigated the As concentrations in feed coal, slag, and fly ash from a Chinese coal-fired power plant. The percentage of As retained in the slag, gaseous phase and fly ash is 0.53%, 2.16%, and 84.6%, respectively. Linak *et al.* (2000) and Guo *et al.* (2004a) found that there was a significant negative correlation between As and the particle sizes of fly ash, indicating that As was concentrated in the finer particles. It was considered that the mechanism of As enrichment in fly ash was not physical adsorption but probably an outer surface reaction, such as As reacting with CaO in the fly ash. Wang *et al.* (2013a) analysed the TE distribution in coal and different particle sizes of bottom ash and fly ash from a coal-fired power plant in China. The results showed that the As concentration increased with decreasing particle size. Stanislav *et al.* (1994) concluded that 30–40% of As will be emitted to the atmosphere in various ways in Bulgarian thermo-electric power stations. However, only a few per cent of As is released to the atmosphere due to the effect of modern pollution control systems for As partitioning in Chinese power plants. Rizeq *et al.* (1994b) suggested that there was little As retention in the bottom ash and gaseous phase, and most of the As was distributed in the fly ash (>90%). Kang *et al.* (2011) summarized that the worldwide average ratio of As released to the atmosphere (solid + gas) was 7.28% in coal-fired power plants. Clemens *et al.* (1999) considered that CaO in coal will inhibit the emission of As during coal

combustion. China is dominated by pulverized coal boilers, fluidized bed furnaces, and stoker fired boilers. The release rate of As during coal combustion in these kinds of boilers is 98.46%, 75.6%, and 77.18%, respectively. The overall removal efficiency of downstream pollutant control devices (ESP and Wet Flue Gas Desulfurization, WFGD) for As can be as high as 97.3% (Tian *et al.* 2011).

3.3. Selenium emissions during coal combustion in China

Selenium emission from coal-fired power plants is of particular concern due to its hazardous effect to the environment and human health. Guo *et al.* (2003) studied the volatilization rate of Se coal at temperatures of 400–1150°C. Se evaporated rapidly as the temperature increased, and 98% of the Se was transferred from coal to the gaseous phase at 850°C, indicating that almost all of the Se will be vaporized in the gaseous phase during the actual combustion event.

Zhang *et al.* (2007) conducted an experiment in a laboratory-scale DTF to determine the release characteristics of TEs during coal combustion. Se is the most volatile element in coal (Xu *et al.* 2013), and its volatilization rate increases with increasing combustion temperature. More than 80% of the Se is released at 550°C, and the volatilization rate reached up to 97% at 815°C. When the temperature was increased to 950°C, almost all of the Se in coal was volatilized to the gaseous phase. A coal combustion experiment carried out by other researchers also indicated that the volatilization rate of Se was more than 75% at 750°C (Zhang *et al.* 2003). Monahan-Pendergast *et al.* (2008) studied the thermochemical properties of Se species during coal combustion using *ab initio* methods. Through a comparison of equilibrium constants for a series of gaseous Se oxidation reactions involving OH and HO_2, the following thermodynamically favoured reactions were found:

$$Se + OH \rightarrow SeOH \qquad (3)$$

$$Se + HO_2 \rightarrow SeO + OH \qquad (4)$$

$$SeO + HO_2 \rightarrow SeO_2 + OH \qquad (5)$$

SeO_2 is water-soluble and the reaction between SeO and HO_2 may present a way for Se from coal combustion to end up in drinking water. Thus, the thermodynamically and kinetically favoured species of Se should be highlighted. The studies of Seames and Wendt (2000) showed that, a large fraction of Se are present in the fly ash, which may be complexed with Ca. The formation of Se–Ca complexes is consistent with the results from Ca-based sorbent experiments reported by Agnihotri *et al.* (1998), indicating that $CaSeO_3$ will be formed at the post-combustion zone temperature conditions.

Tang *et al.* (2012) investigated the emission of Se in four Chinese coal-fired power plants. It was found that the Se content in bottom ash was lower than that in fly ash, and the relative enrichment (RE) of Se in fly ash was more than 0.7, which indicated that Se was easily volatilized during combustion and will condense on the surface of fly ash when the flue gas is cooled downstream. FGD gypsum is also an important by-product of coal combustion. Senior *et al.* (2011) found that the largest proportion of Se was distributed in the fly ash. There were also some portions of Se distributed in the flue gas and gypsum. Xu *et al.* (2005) found that 16.5% of the Se was emitted to the atmosphere; the fly ash accounted for 71.3% of the total Se in coal, and 46.1% of the Se was concentrated in particles with sizes lower than 19 μm.

The Se release rates could be affected by the boiler types and combustion parameters. The pulverized coal boiler and fluidized bed furnace showed higher Se release rates, which were 96.22% and 98.05%, respectively, and the rate was 80.95% for the stoker fired boiler (Tian *et al.* 2011). It was found by some researchers that the concentration of Se in fly ash increased with an increase in the excessive air coefficient (Huang *et al.* 2004). Tian *et al.* (2009) estimated the inventory of the Se discharge amount in China in 2005 using the emission factor method, according to the provincial coal consumption and average Se concentration. China's discharge amount of Se was 1331.1 t in 2005, and the province that had largest discharge amount was Shandong (175.90 t), followed by Hebei (111.79 t), Henan (105.36 t), Jiangsu (97.32 t), Shanxi (84.66 t), etc. The regions of lowest Se emission were Qinghai (0.73 t), followed by Xinjiang (1.80 t), Hainan (2.43 t), and Gansu (5.04 t).

3.4. Fluorine emissions during coal combustion in China

Qi *et al.* (2005b) discussed the emission characteristics of gaseous F and its influencing factors during coal combustion in a tube furnace. The coal samples include the typical Chinese lignite, anthracite, bituminous coal, and stone coal. The volatilization rate of F varied in different stages as the temperature increased. F can be released from coal at temperatures lower than the ignition temperature of coal. This indicated that a certain amount of F was vaporized before the coal was fired, but the volatilization rate is lower than 20% at temperatures under 400°C, with an average of 7.88% (Qi *et al.* 2002a). For Heshan coal, the release rate of F was 12.5% (Qi *et al.* 2005a). The volatilization rate and volatile speed significantly increase at 400–1000°C, and more than 90% of F is released at 900°C. Other influencing factors, such as the steam content in the combustion atmosphere, SiO_2 content in the coal and reaction time, were also investigated by Qi *et al.* (2005a, 2005b). If the reaction time is increased from 5 to 40 min, the release rate of F will increase from approximately 40% to 75%. When the combustion atmosphere contained steam, the release rate of F can evidently increase by an average of 11.43%, compared to the condition without steam. SiO_2 in coal can promote the release of F, and the release rate could be increased by 18%, which is due to SiO_2 reacting with basic oxides, such as CaO and MgO, which weakens the F retention of coal.

Yu *et al.* (2004) studied 33 coal samples from Shanxi province and Guizhou province, China. It was found that the release rate of F in coal had little relationship with the content of F in coal during combustion below 1000°C. The correlation coefficient between the release rate and content of F was only 0.022. The release rate of F gradually increased at 300°C but was not higher than 31% before 700°C. The rate increased rapidly with increasing temperature at 700–1000°C and approached 100% at 1200°C.

Many Chinese cities are polluted by F in the atmosphere (Feng *et al.* 2003), and coal-fired power plants are considered to be one of the largest sources of F emissions in China. Xu *et al.* studied the F concentration in flue gas of the Nos. 1 and 2 thermoelectric power plants in Baotou, China. The average concentrations of F in the 1 power plant were 2.61 mg/m³ in 1999 and 1.80 mg/m³ in 2000, and those in the 2 power plant were 1.12 and 2.71 mg/m³, respectively (Zheng *et al.* 2002). Li *et al.* (2012) studied the geochemical characteristics of the coal and combustion by-products from two coal-fired power plants in Xinjiang, China. Their results showed that F was depleted in the fly ash and slag of both power plants, and the volatile proportion of F at the ESP was 64% and 79%, respectively. Liu *et al.* (2004) studied the enrichment of TEs in three power plants in Shandong province, China. The F concentration in fly ash was evidently lower than that in raw coal, which demonstrated that F was not concentrated in the fly ash. During coal combustion in a power plant, most of the F is discharged into the atmosphere. There was a positive correlation between F and the particle size of the ashes, which indicated that the F content will increase with increasing ash particles. The bottom ash contained a higher F content than the fly ash, due to its larger size and higher content of residual carbon.

However, the emission and enrichment of F is complex because of the different feed coals, combustion conditions, boiler structure, etc. Dai et al. (2014a) investigated the distribution of TEs in coal combustion products derived from Ge-rich coals, which are also highly enriched in other TEs, including Hg, As, and F. The results showed that the fly ashes are significantly enriched in a number of TEs, including Hg, As, F, Sb, and Pb. F is enriched in the fine-grained fly ash of Lincang Ge-rich coals, with an enrichment factor >10. Wang et al. (2015b) studied the occurrence modes of F in coal and coal combustion by-products (slag, FGD gypsum, and fly ash) from Inner Mongolia, China. The results indicated that 3.02% of the F was retained in slag, the proportion in ashes and FGD gypsum was 3% and 72.74%, respectively, and only a small amount of F was emitted to the air. The emission of F from actual coal-fired power plants still needs be investigated broadly with respect to their various equipment and complex conditions.

3.5. Chromium emissions during coal combustion in China

The release characteristics of Cr during coal combustion have been described in the literature. Lu et al. (2009a) studied the migration features of Cr in the tube furnace at different temperatures. They found that Cr is a non-volatile element, which is mainly retained in the bottom ash and slag. The Cr content underwent a certain reduction when the temperature reached 900°C. Zhao et al. (2008) investigated the Cr in PM during coal combustion in a bench-scale DTF. Lignite from SB coalfield and mixed bituminous coal of PDS were chosen for the experiment. The results showed that the Cr emissions of the two coals were obviously different. The Cr in SB coal was mainly in the organic form and undergoes experiences explosive volatilization, most of which will be released during the combustion process. The PM of SB coal showed a higher Cr emission than that of mixed PDS coal. Most of the Cr in PDS coal was enriched in particles larger than 10 μm, and the emission concentrations of Cr in PM_1 of the SB and PDS coals were 4.8 and 2.3 mg/N m^3, respectively. Zhang et al. (2003) conducted a coal combustion experiment in a one-dimensional furnace. They observed little change in the Cr content in the ashes, and the release rate was seldom affected by the coal type at low temperature (□ 950°C). Xu et al. found that the release rate of Cr at 800°C was very low for all the anthracite, lignite and bituminous coal samples from China, which then increased at 800–1100°C. The Cr emission of anthracite was slightly higher than that of lignite and bituminous coal. Huang et al. (2004) found that fine fly ash contained a slightly

higher concentration of Cr than the larger particles. When the excessive air coefficient was increased, the Cr content in fly ash slightly decreased, which was not consistent with the behaviour of Hg and Se. The effects of reducing conditions on the volatility of TEs depended on the element and coal type, and reducing conditions only reduced Cr volatilization in low-ash coal (Yan et al. 2001).

The inventory of the Chinese anthropogenic atmospheric emission of Cr was estimated by Tian et al. (2012), and the total Cr emission from Chinese coal rapidly increased from 1019.07 t in 1980 to 8593.35 t in 2008. Li et al. (2012) studied two coal-fired power plants in China, and the results suggested that Cr exhibited different distributions in slag and fly ash. In one power plant, Cr was not evidently enriched or depleted in slag or fly ash, while in the other one, Cr was enriched in fly ash and depleted in slag. Wang et al. (2013a) found that the RE of Cr in bottom ash was only 0.26, while the RE was higher than 4 and 10 in fly ashes with different particle sizes of 1–0.5 and 0.5–0.2 μm, respectively. This indicated that Cr was enriched in the fine particles, especially the micron particles. Wang et al. (1998) divided fly ash into five classes according the particle size. They also found that the smallest fly ash (<38 μm) had the maximum Cr content. Zhao et al. (2016) studied Cr specification in a coal-fired power plant in China and found that the oxidation state of Cr was the main form in flue gas, i.e. CrO_3 and Cr_2O_3. Zhu et al. (2003) studied the distribution of total Cr and Cr^{6+} in coal, bottom ash, and three ESP fields. The results showed that the Cr^{6+} content in the coal combustion products was higher than that in coal, and both Cr and Cr^{6+} were concentrated in fine particles.

Tian et al. (2012) studied the release rate of Cr during coal combustion in different boiler types. The pulverized-coal boiler and fluidized-bed furnace had higher Cr release rates of 84.50% and 81.33%, respectively. The Cr release rate of the stoker fired boiler was only 26.74%. Li et al. (2005) suggested that the vast majority of Cr was partitioned in bottom ash (>90%), and only a small amount of Cr was partitioned in fly ash during coal combustion in a stoker fired combustion unit. Zhang et al. (2015a) studied the partitioning behaviour of Cr in a coal gangue-fired CFB plant in Shanxi Province, China. It was found that the content of Cr in bottom ash was higher than that in fly ash, which indicated that Cr was relatively concentrated in the bottom ash of the coal gangue-fired CFB plant.

4. Simultaneous control for TEs by APCDs

In the past 10 years, coal-fired power plants have been equipped with energy-saving and emission-reducing devices in China. The devices mainly include selective catalytic reduction (SCR), lower temperature economizer (LTE), ESP, FGD, and wet electrostatic precipitator (WESP). Among these devices, SCR and FGD are used to control the emission of NO_x and SO_2, respectively. ESP and WESP are used to capture particulate matter. LTE is added to collect the existing energy from flue gas. In addition to the intrinsic functions, these devices have been found to have the ability to alter the speciation of TEs in flue gas and control their emission. In this section, the roles of each of the five devices on the speciation and emission control of TEs are summarized.

4.1. Role of SCR system

The technology of SCR denitration is applied to remove NO_x from the flue gas of coal-fired power plants (Tronconi et al. 1992; Shelef 1995). The process is to reduce NO_x to N_2 using NH_3 as the reducing agent at approximately 350°C (Bosch et al. 1988). The removal of TEs across the SCR system focuses on the control of Hg^0. Studies have shown that SCR systems have important effects on Hg speciation and emission in flue gas (Senior 2006; Zhuang et al. 2007; Cao et al. 2008; Wilcox et al. 2012; Negreira et al. 2013, 2014; Lee et al. 2008). Although many references that have been published on Hg oxidation across SCR catalysts, there are few references to report the Hg oxidation within an SCR in China comprehensively. Xu et al. (2014) tested the Hg speciation and concentration before and after passing through the SCR systems of 20 typical coal-fired power plants in China. The capacity of the units ranged from 150 to 1000 MW. The kind of coal included bituminous coal, sub-bituminous coal, and lignite. Thus, the selected power plants covered most types of coal-fired power plants in China. The results of the tests indicated that SCR had no obvious effect on the total Hg (Hg^T) emission. When the flue gas passed through the SCR, the Hg^P concentration showed no significant change, while the Hg^0 concentration decreased and the Hg^{2+} concentration increased. This suggested that SCR could oxidize Hg^0 to Hg^{2+}, which could be captured by WFGD because of its solubility. However, the Hg^0 oxidation efficiency within SCR at different power plants were significantly different, which were determined by the effects of flue gas composition, SCR operating conditions, catalyst ageing, and boiler load. The most significant variation in Chinese power plants is the coal characteristics. The chlorine content in Chinese coals are generally low, resulting in the low Cl concentration in flue gas, which played an important role in Hg oxidation over SCR catalysts (Senior 2006; Cao et al. 2007; Zhuang et al. 2007, Lee et al. 2008).

4.1.1. Impacts of the flue gas components on Hg^0 conversion across the SCR catalyst

Many studies have shown that the flue gas components have a significant effect on Hg^0 oxidation over the SCR catalyst (Eswaran et al. 2005, 2008; Presto et al. 2006; Senior 2006). First, Hg^0 oxidation on SCR is closely related to the HCl concentration (Senior 2006; Cao et al. 2007; Zhuang et al. 2007, Lee et al. 2008). The experimental data of Zhuang et al. (2007) showed that 5–50 ppm HCl in flue gas enhanced mercury oxidation within the SCR catalyst. The Hg^0 oxidation of SCR catalyst can reach to 45% with the addition of 5 ppm HCl in the simulated flue gas, which is much higher than the 4% of Hg^0 oxidation when 1 ppm HCl is in the flue gas. As HCl concentration increased to 50 ppm, 63% of Hg^0 oxidation was reached. Similar results were also obtained in the studies of Senior (2006) and Lee et al. (2008). A possible mechanism includes HCl adsorbing on the active sites to form active Cl species. Then, Hg^0 can be oxidized by active Cl to generate $HgCl_2$. Hence, the existence of HCl promotes Hg^0 oxidation over the SCR catalyst. The HCl concentration in flue gas is positively correlated with the chlorine content in coal (Eswaran et al. 2008). When chlorine content in burning coal was increased from 109 to 876 mg/kg, the Hg^0 oxidation rate of SCR increased from 2% to 89% (Xu et al. 2014). The chlorine content in bituminous coal is higher than that in sub-bituminous coal. Thus, the Hg^0 oxidation rate over the SCR catalyst of power plants burning bituminous coal is usually higher than that burning sub-bituminous coal (Cao et al. 2008).

The effect of NO on Hg^0 oxidation over the SCR catalyst is not completely clear. NO could be oxidized by O_2 or lattice oxygen, generating oxidizing materials such as NO^+ and NO_2 (Li et al. 2008). These materials can oxidize Hg^0 to $Hg(NO_3)_2$, suggesting that NO is helpful for Hg^0 oxidation (Li et al. 2011). However, if NO and O_2 coexist in the flue gas, some materials, such as nitrite, which have no Hg^0 oxidation capacity, are also easily formed (Busca et al. 1998). The active sites can be blocked by these materials, inhibiting the activity of the SCR catalyst for Hg^0 oxidation.

As NH_3 is the reducing agent for SCR de-NO_x, the effect of NH_3 on Hg^0 oxidation over the SCR catalyst should also be given more attention. The Hg^0 oxidation rate declines as NH_3/NO increases. NH_3 exhibits strong inhibition towards Hg^0 oxidation across the SCR. Wang et al. (2013b) tested gaseous Hg species before and after passing through the SCR system of a 300-MW unit burning two kinds of coal. The results showed that NH_3 inhibited Hg^0 oxidation, which may be due to the following reasons: (1) NH_3 competed with Hg^0 for the active sites, inhibiting the adsorption of Hg^0 on the catalyst surface (Ding *et al.* 1998; Qi *et al.* 2004); (2) the strong adsorption capacity leads to a decrease in adsorbed HCl on the catalyst, which is an important intermediate for Hg^0 oxidation; (3) NH_3 directly reacts with and consumes HCl, which participates in Hg^0 oxidation reactions; and (4) Hg^{2+} can be reduced to Hg^0 by NH_3 (Dranga *et al.* 2012). Overall, the inhibition of NH_3 results from competitive adsorption, the consumption of Hg^0 oxidation reactants, and Hg^{2+} reduction. The inhibitory effect becomes obvious as NH_3/NO increases.

4.1.2. Impacts of catalyst ageing and load rate on Hg^0 conversion across the SCR catalyst

In addition to the component of flue gas, catalyst service time was an important factor for determining the SCR oxidation of Hg^0. Eswaran *et al.* (2008) found that the activity of SCR catalyst decreased as the increase of catalyst service time. The results of Kamata *et al.* (2008) showed that the Hg^0 oxidation efficiency of fresh catalyst was about 40%, while that decreased to 15% as the severice time of SCR catalyst increased to 71,000 h in the presence of NH_3. The study of Zhao *et al.* (2015) showed that a commercial SCR catalyst equipped in a coal-fired power plant has lost its Hg^0 oxidation capacity when the service time reached for 35,000 h. Similar results were also obtained by Zhou *et al.* (2015). This is mainly because that the SCR catalyst may be saturated with the adsorbed Hg after servicing for a long time, and Hg can not be adsorbed on the catalyst.

Zhou *et al.* (2015) found that Hg^0 conversion across the SCR system was determined by the load conditions. At load ratios of 60%, 75%, and 100%, the Hg^0 conversion was 72.12%, 65.71%, and 61.78%, respectively. Therefore, Hg^0 conversion is inversely proportional to the load ratio. A decrease in the load ratio means lower coal consumption and lower flue gas flow rate, resulting in a lower GHSV, lower temperature, and higher O_2 concentration, which all benefit Hg^0 adsorption and oxidation over the catalyst. As a consequence, Hg^0 conversion is enhanced as the load rate decreases from 100% to 60%.

4.2. Role of particulate matter control device (ESP/FF)

As the flue gas temperature decreases, some TEs attach to fly ash. Thus, a particulate matter control device can control the TE emission by capturing the fly ash. By 2007, more than 90% of Chinese coal-fired power plants were equipped with ESPs. The collection efficiency of current ESPs is generally above 99%. Therefore, ESP has a significant impact on the emission of TEs adsorbed on fly ash.

4.2.1. Effects of ESP/FF on Hg emission

Gao *et al.* (2007) tested the effects of ESP on the Hg emission characteristics using a 35-t/h CFB power plant boiler. They found that the ESP could control Hg emission to a certain extent. The Hg^T concentration at the outlet of the ESP was significantly lower than that at the inlet. Hg speciation changed significantly across the ESP. After passing through the ESP, Hg in the flue gas is mainly in the form of Hg^0, while the content of Hg^{2+} is greatly reduced. This phenomenon is due to the lower gas flow rate and longer pace time of the flue gas in the ESP, which increased the contact time of Hg and the flue gas. Thus, the adsorption of Hg, especially Hg^{2+}, on fly ash was increased.

The enrichment of Hg in fly ash significantly determined the Hg removal efficiency of ESP/fabric filter (FF), which was affected by the combustion mode, boiler capacity, boiler load, particulate control device type, and others. The study of Yang *et al.* (2008) showed that the Hg^0 removal capacity of ESP/FF decreased as the raise of boiler capacity. This was mainly because that the bigger boiler capacity resulted in a longer retention time of UBC particles in boiler, enhancing the burning of coal particles. As a result, the content of UBC in fly ash decreased with the raise of boiler capacity, which would decrease the adsorption of Hg in fly ash. The Hg removal efficiency of different type of particulate control devices was varied significantly (James *et al.* 2002). Generally, the FF presents a more superior Hg removal capacity than cool-side ESP (CS-ESP) and hot-side ESP (HS-ESP), due to the longer contact time between Hg and fly ash in FF. Compared to CS-ESP, the Hg removal capacity of HS-ESP is generally poor because of the higher operating temperature of HS-ESP, which would decrease the adsorption of Hg in fly ash. The combustion mode also played an important role in Hg removal by ESP/FF. The CFBC can promote the Hg removal by ESP/FF (Yang *et al.* 2008). The combustion temperature in CFBC is lower than that in PCC, which is beneficial to the formation of pores in fly ash. The content of UBC in CFBC fly ash is generally higher than that in PCC fly ash. These two factors can promote the adsorption of Hg in CFBC fly ash and thus promote the Hg removal by ESP/FF in CFBC.

Studies by Xu *et al.* (2014) showed that the removal efficiencies of different forms of Hg in ESP were different. The reduction of Hg^T emission by ESP was mainly reflected in the Hg^P removal. The average Hg^P proportion in flue gas decreased from 28.4% at the inlet of the ESP to 5.3% at the outlet. Thus, the Hg^P removal efficiency in the ESP can be more than 90%. Arrangements of other APCDs have a significant impact on Hg removal in the ESP. For example, after installing SCR, the Hg removal capability of ESP increased obviously. For power plants equipped with an SCR system, the Hg^T removal efficiency in the ESP can reach as high as 80.6%. This is because the SCR catalyst oxidizes Hg^0 to Hg^{2+}, while Hg^{2+} is easily adsorbed on the fly ash surface and then synergistically removed in the ESP. In addition, the study of Lei *et al.* (2007) demonstrated that continuous Hg oxidation occurred when the flue gas passed through the ESP/FF, suggesting that the reaction between Hg and oxidants in flue gas is kinetically controlled. Coal characteristics like Cl concentration and ash compositions in coal played more important roles in the speciation of gaseous Hg than boiler capacity and particulate control device type. Chlorine in flue gas from coal is the major oxidant for Hg oxidation. Ash compositions are also the key factor that affects Hg oxidation, which might react with Cl to compete against Hg.

4.2.2. Effect of ESP/FF on the release of other TEs

Yi *et al.* (2008) tested the control capacity of FF on TE (As, Se, Cd, Cr, etc.) emissions in a 220-MW power plant burning anthracite. The fine particles were divided into 12 sizes from 30 nm to 10 μm, and the enrichment factors of each element in particles with different sizes were analysed. According to the results, TEs can be grouped into three categories. The first category includes Fe, Al, and Mn, of which the relative enrichment factor is approximately 1. The factor contrasts the trend of changing particle size, indicating that there are no obvious enrichments and losses of the three elements in PM_{10}. The second category includes Hg, Se, and Cr. The contents of these elements increase as the

particle diameter decreases. The relative enrichment factor in submicron particles is greater than 1, while in the coarse particles, it is less than 1, indicating the enrichment of these three elements in submicron particles and their loss from coarse particles. The third category includes Cd, Cu, V, and Zn. The relative enrichment factors of these four elements in particles of all sizes are greater than 1, which decreases as the particle size increases. The degree of enrichment increases with increasing elemental volatility. The trend of each TE after FF is similar to that before FF. The difference is the enrichment of elements in submicron particles. Their relative enrichment factors increase after flowing across FF. When the flue gas passes through PMCD, these elements condense on the surface of fine particles.

Figure 7 shows the capture efficiencies of TEs in particles with different sizes. The capture efficiency of PM_{total} and PM_1 is the highest and lowest, respectively. The enrichment and loss extent of each TE in particles with different size levels is related to its migration and transformation behaviour during the combustion and emission processes. Therefore, the capture efficiencies of non-volatile elements, such as Al and Fe, are close to that of PM_{total}. However, the capture efficiencies of volatile elements, such as Hg, Zn, and Se, of PM_1 are lower than that of PM_{total}.

4.3. Role of LTE

The role of LTEs on the control of TEs in flue gas has been investigated in recent years. The operation parameters of LTE influence the temperature of the flue gas

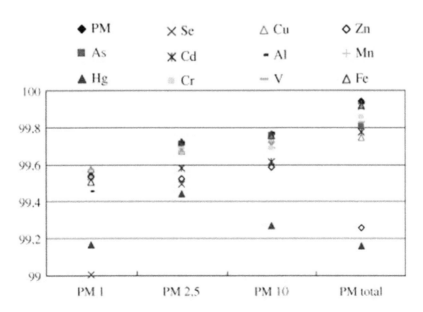

Figure 7. Capture efficiencies of TEs in particles with different size levels (Yi *et al.* 2008).

in the ESP. As a result, the removal efficiencies of TEs in the ESP are different at different temperatures. Zhou et al. (2015) compared the change in Hg speciation and Hg removal efficiency across the ESP under different LTE operation conditions, as shown in Figure 8. When the LTE was off, only approximately 10% of the total Hg was removed across the ESP, the majority of which was in the form of Hg^P. The concentration of Hg^0 was decreased by 46%, and Hg^{2+} was increased by 51%, suggesting that most of the Hg^0 reduction may be due to conversion on fly ash rather than adsorption. Hg^{2+} mainly presented in the flue gas instead of being captured by the fly ash. The ESP could not control the gaseous Hg emission efficiently when the LTE was off.

When the LTE was on, the flue gas temperature in the ESP decreased, benefiting gaseous Hg (Hg^0 and Hg^{2+}) adsorption by the UBC and inorganic minerals in the fly ash, which were eventually removed by the ESP. As a result, 46.07% of the gaseous Hg was captured, and the total and elemental Hg removal efficiency across the ESP was increased by 42.87% and 18.85%, respectively. Therefore, LTE can promote Hg removal across the ESP, especially for gaseous Hg.

Wang et al. (2015a) studied the TE (As, Cd, Cr, Pb, and Mn) emission from the ESP under different LTE operation conditions. More than 90% of the TEs were collected alongside fly ash by the ESP, while few TEs were emitted from the ESP. The amount of TEs at the

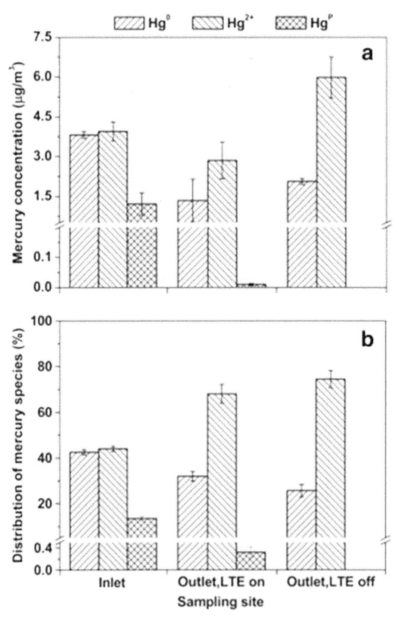

Figure 8. Hg concentration and mass distribution of the Hg measured at ESP under different operation states of the LTE (Zhou et al. 2015).

outlet of the ESP with the LTE on was less than that with the LTE off, which suggests the benefit of LTE in reducing TE emission. The reason was that the flue gas temperature in the ESP was decreased when the LTE was on. The decrease in the flue gas temperature reduced the enrichment of As, Cd, Cr, and Pb in submicron particle matter, which are difficult to be captured by ESP. As a consequence, the LTE decreased the mass of the TEs escaping from the ESP.

4.4. Role of FGD

4.4.1. Effects of FGD on Hg emission control

Because of its high desulfurization efficiency and extensive adaptability, WFGD is the main desulfurization method used inmost Chinese coal-fired power plants. Currently, the effects of WFGD on controlling Hg emissions have gained widespread attention. This is mainly due to the solubility of Hg^{2+} compounds, such as $HgCl_2$, in flue gas. WFGD can remove approximately 90% of Hg^{2+}. Researchers have meticulously studied the factors that affect Hg removal in WFGD.

Li et al. (2013) carried out field tests on Hg emission in a 300-MW coal-fired power plant with installed ESP and WFGD. The results showed that 79.93–90.35% of Hg^{2+} can be removed by WFGD. However, WFGD has no removal effects on Hg^0, and the Hg^0 content exhibits a small increase instead. This is consistent with the earlier work of Chang and Ghorishi (2003), where the reduction of Hg^{2+} to Hg^0 by aqueous S(IV) (sulphite and/or bisulphite) species can occur within FGD units. The extent of Hg^{2+} reduction across a WFGD scrubber can be reduced by decreasing the pH, increasing S(IV) concentration and lowering the temperature. Wang et al. (2008b) found that the Hg^T removal efficiency in WFGD is 9.68–29.36%. The combination of ESP + WFGD can remove all of the Hg^p and most of the Hg^{2+}, and the Hg^T removal efficiency is in the range of 25.38–38.38%. This is related to the low Cl content of the coal. The proportion of Hg^{2+} in Hg^T is vital to the Hg control capacity in WFGD. Therefore, it is important to develop Hg emission control technologies in WFGD to improve Hg^0 removal and suppress Hg^{2+} reduction in the slurry.

Hu et al. (2009) examined the Hg concentration and speciation before and after WFGD in six typical coal-fired power plants in China and studied the effect of WFGD capacity on Hg control and transformation. The results showed that after WFGD, Hg is mainly in the form of Hg^0, and Hg^{2+} is completely captured. The Hg removal efficiency of WFGD is related to the proportion of Hg^{2+} in Hg^T before WFGD. A larger ratio of Hg^{2+} in the flue gas entering the WFGD system will enhance the Hg removal efficiency across the WFGD.

4.4.2. Effect of FGD on the emission of TEs

Córdoba et al. (2012a, 2012b) investigated the partitioning of TEs in a coal-fired power plant equipped with a WFGD system. They found that the WFGD had significant effects on controlling the emission of TEs, besides Hg. The retention of elements in gypsum sludge and their dissolution in the water of the gypsum slurry can be regarded as the main outputs of the FGD system. S, F, Se, and As are mainly retained in the gypsum sludge, whereas Cl and B are retained in the water of the gypsum slurry. A high gaseous retention capacity of 92–100% for S, Cl, F, Se, As, and B can be attained in the WFGD, with a subsequent reduction in their gaseous emissions. In addition, the emission control of TEs in WFGD is influenced by factors such as water recirculation to the scrubber and the co-combustion of coal with petroleum coke.

4.5. Role of the WESP

To attain near-zero emissions, a few coal-fired power plants in China are equipped with WESPs. The combination of APCDs in these power plants was SCR + ESP + WFGD + WESP. Zhao et al. (2016) investigated the TE emission control of these plants. The results showed that TEs can be removed by each of these devices, as shown in Figure 9. Most of the TEs exist in the ash, which can be captured efficiently after the flue gas passes the ESP. WESP, together with WFGD, removes the residual TEs in the flue gas, further reducing TE emissions to the atmosphere. The TE removal rates across ESP + WFGD are shown in Figure 10. The capture efficiencies of Mn, Co, As, Ag, Sb, Ba, and Pb in WESP can reach over 60%, which verifies that some gaseous TEs are further removed under an increased water vapour and a lower temperature of approximately 50°C. The overall TE removal rate across ESP + WFGD + WESP is over 99.9%, which demonstrates that coal-fired power plants implementing extra equipment with WESPs have excellent capture capacity of TEs.

5. Sorbents for capturing TEs in flue gas

The existing APCDs can alter the speciation of TEs in flue gas and control TE emission to some degree. However, the existing APCDs cannot efficiently remove certain species of TEs, such as Hg^0. Sorbent injection and blending sorbents with coal are considered promising methods to control TE emission in coal-fired power plants. In recent years, extensive studies have carried out to develop various sorbents for TE capture, including carbon-based sorbents, calcium-based sorbents, metal oxides, mineral sorbents, and others. In this section,

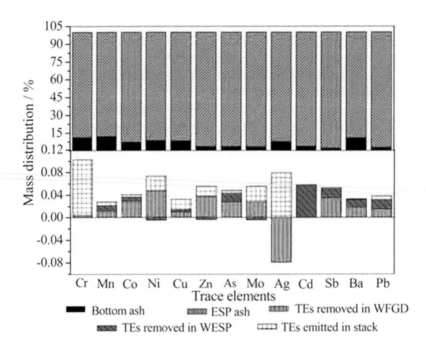

Figure 9. Mass distribution of TEs in the coal-fired power plant (Zhao *et al.* 2016).

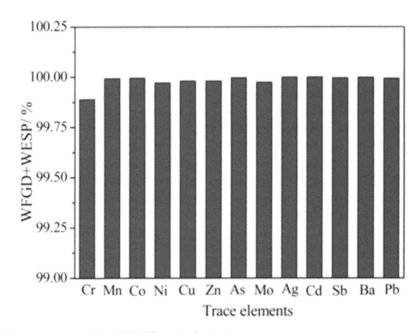

Figure 10. TEs removal rate across WFGD + WESP (Zhao *et al.* 2016).

sorbents for capturing five typical TEs (Hg, As, F, Se, and Cr) and their removal performance are summarized.

5.1. Sorbents for capturing Hg

5.1.1. Adsorbents for capturing Hg in coal combustion flue gas

5.1.1.1. Hg removal by activated carbon. Among the various adsorbents, activated carbon (AC), with large specific surface area and strong adsorption ability, is widely used for Hg removal from flue gas (Scala *et al.* 2011; De *et al.* 2013; Rupp *et al.* 2014). Fuente-Cuesta *et al.* (2012) studied the Hg adsorption capacity of nine adsorbents, and the results showed that the adsorption capacity of commercial AC impregnated with sulphur could reach as high as 226.66 μg/g. De *et al.* (2013) reported that halide salts, especially KI, could significantly improve the Hg removal efficiency of AC (to nearly 100%) through a modified process. Diamantopoulou *et al.* (2010) showed that the Hg uptake capacity of

Calgon F400 AC was greatly increased in presence of HCl. The Hg concentration at the outlet was only 20% of that at the inlet after a 2500-min adsorption experiment. Rupp et al. (2014) studied the Hg^0 removal performance over brominated AC. The packed bed was exposed to 25 ppbv Hg^0 for 120 min and the Hg^0 removal efficiency was >97%. However, the interaction of SO_2 and NO_x in flue gas with surface Br leads to sorbent deactivation.

Scala et al. (2011) conducted Hg capture experiments using powdered AC in a fluidized bed reactor. The experimental results showed that the Hg^0 removal efficiency increased when (a) the AC feed rate increased, (b) the gas superficial velocity decreased, (c) the bed particle size decreased, or (d) the bed height increased. In addition, a fluidized bed was better than an entrained bed for Hg capture. However, Lee et al. (2006b) found that the Hg removal efficiency will not further increase after a certain C/Hg ratio, giving the highest removal efficiency of 60%.

Due to the good Hg removal performance, activated carbon injection (ACI) has been commercially applied to capture Hg in partial coal-fired power plants. In March 2009, the Institute of Clean Air Companies (ICAC) reported that Hg control systems were installed in 135 power plants in the US and Canada, representing more than 55 GW of generation, 54 GW (or more than 98%) of which are ACI systems (Sjostrom et al. 2010). A continuous 30-day, 24-h-per-day Hg removal experiment using brominated powdered activated carbon injection was conducted in the St. Clair power plant in America. The results showed that with a 48-g/m^3 injection rate, the total Hg removal across the ESP averaged 94% over the 30-day test period (Liu et al. 2009). However, ACI technologies are limited due to their high operating cost and negative effect on fly ash quality.

5.1.1.2 Hg removal by fly ash.

Studies on Hg removal by fly ash have been conducted by many researchers in recent years (Table 6). Experiments conducted in fixed-bed reactors showed that no Hg capture was observed without accompanying oxidation. Magnetite in the fly ash and the surface area are important factors for the oxidation and adsorption of Hg (Dunham et al. 2003; Yang et al. 2017a; Yang et al. 2014). However, iron species do not affect the retention of Hg or heterogeneous Hg oxidation (Abad-Valle et al. 2011a). UBC in fly ash also played important role in the capture of Hg, while Hg capture depended on the nature and type of anisotropic particles (López-Antón et al. 2009; Hower et al. 2017). Generally, with the presence of N_2, CO_2, and O_2, the ability of fly ash to capture Hg is proportional to the UBC content; however, this relation is invalid when SO_2, HCl, and H_2O are also present (Abad-Valle et al. 2011b).

Zhao et al. (2010a) studied the reaction mechanism of Hg capture by fly ash. The results indicated that the active sites on the fly ash surface could be classified into four types: low-binding-energy catalytic oxidation active sites, catalytic oxidation active sites, adsorption active sites, and high-binding-energy adsorption active sites. Wang et al. (2016) conducted full-scale modified fly ash injection experiments for Hg pollution control at a 300-MW power plant, and the results showed that the Hg concentration in the flue gas decreased by 30% due to adsorption by the modified fly ash.

5.1.1.3. Hg removal by calcium-based adsorbents.

Ghorishi et al. (2002) researched various calcium-based adsorbents for Hg removal. The results revealed that oxidant-enriched Ca-based adsorbents could remove both Hg^0 and Hg^{2+} as effectively as AC in fixed-bed

Table 6. Hg removal experiments by fly ash in published papers.

Sorbents	Flue gas	Hg removal efficiency (%)		Adsorption capacity	Temperature (°C)	References
		Hg^T	Hg^{2+}	(μg/g)		
AA	N_2, O_2, CO_2,		85		121	Dunham et al. (2003)
MA	H_2O, SO_2, NO,		80		121	
Absaloka	NO_2, HCl		70		177	
ESP1-fly ash	N_2, O_2, CO_2, SO_2	39.10	12.21		135	Meng et al. (2011)
ESP2-fly ash		35.26	17.11			
FF-fly ash		23.08	8.17			
CTL-EC	N_2, O_2, CO_2, SO_2, H_2O, HCl	97.6	69.3		120	Abad-Valle et al. (2011b)
CTE-EC		39.1	13.3			
FA2 > 80	Air			10.3	120	Zhao et al. (2010a; 2010b)
FA3 > 100				9.36		
CAT > 150	N_2, O_2, CO_2, SO_2, H_2O			15.19×10^3	120	López-Antón et al. (2009)
CTSR > 80				43.8×10^3		
CTES > 200				16.3×10^3		
CTP				5.76×10^3		
CTL-EC	N_2, O_2			20	120	Abad-Valle et al. (2011a)
CTE-EC				8.5		
CTSR-NM				11		
Co−MF	N_2, O_2, CO_2, HCl	95.5			150	Yang et al. (2014; 2015;
CuCl$_2$-MF		90.6				2016a; 2016b; 2016c)

simulations of coal-fired flue gas at 80°C. Moreover, these Ca-based adsorbents were proven to be far superior to AC with respect to SO_2 uptake. Ren et al. (2006b) found that SO_2 could increase the Hg removal efficiency of $Ca(OH)_2$ to 15–20% and improve the Hg adsorption quantity by more than 50%. In addition, an appropriate increased temperature contributed to the capture of Hg, which was due to the formation of active sites between SO_2 and Ca-based adsorbents. Huang et al. (2009, 2011) studied the ability of several calcium-based adsorbents in Hg removal from simulated flue gas. The results demonstrated that the adsorption of $Ca(OH)_2$ was mainly controlled by a physical mechanism, while that of CaO was the outcome of the combined action of physical and chemical adsorption. The roles of SO_2 and HCl in Hg removal were similar to the conclusion of Ren et al. (2006b). $KMnO_4$-modified Ca$(OH)_2$ can oxidize a large amount of Hg^0 into Hg^{2+} and adsorb more than 50% of the total Hg in the presence of SO_2 or HCl, while $AgNO_3$-modified $Ca(OH)_2$ can obtain a 90% Hg removal efficiency. Wang et al. (2011) found that Hg removal by $Mn^{x+}/Ca(OH)_2$ was higher than that by Ca$(OH)_2$. From the XPS results, Hg^0 could be effectively oxidized by MnO_2.

5.1.1.4. Hg removal by metal and metal oxide adsorbents.

Metal and metal oxides, including some transition metals and noble metals, are also effective in removing Hg (Aboud et al. 2008; Li et al. 2010; Lim, et al. 2013; Mei et al. 2014; Cimino et al. 2016). Although AC is one of the most widely used sorbents for Hg removal from flue gas, the increase of the amount of AC can hinder the recycling of fly ash for concrete manufacturing. The use of non-carbon sorbent materials which can capture Hg efficiently but is also concrete-friendly would be promising. Aboud et al. (2008) and Lim et al. (2013) studied the Hg adsorption over noble metals like Pd, Au, and Ag by density functional theory (DFT) calculations. The results showed that the noble metals and the alloys are reactive for Hg^0 adsorption, which are potential candidates for concrete-friendly sorbents. Cimino et al. (2016) studied the Hg removal performance of MnO_x catalysts supported on TiO_2 or Al_2O_3. Mei et al. (2014) studied the detailed Hg oxidation/chemical adsorption mechanisms on the surface of N-doped $CuCo_2O_4$(1 1 0) through DFT calculations. N doping could greatly increase the activity of O_w and decrease the activity of O_s, which greatly affects the Hg oxidation/chemical adsorption abilities on the $CuCo_2O_4$(1 1 0) surface. In addition, N doping can decrease the adsorption energy of Hg and mercuric oxide. Li et al. (2010) employed several metal elements to modify the Mn/α-Al_2O_3 catalyst, among which the best performance was

achieved with molybdenum (Mo) as the dopant, which showed even better Hg^0 removal capacity than the noble metal catalyst Pd/α-Al_2O_3 and had a strong ability to resist SO_2. When the content of SO_2 reached 500 ppm, the oxidation efficiency of Mo(0.03)–MnO_x/α-Al_2O_3 could also reach 95%.

5.1.1.5. Hg removal by mineral adsorbents.

Natural mineral materials possess good adsorption characteristics, which can be applied to control Hg in flue gas in coal-fired power plants. Jurng et al. (2002) investigated the capability of natural zeolite and bentonite in Hg removal from flue gas. The results showed that both the natural materials and the sulphur-impregnated materials had low Hg removal efficiencies of no more than 50%. Eswaran et al. (2007) tested the Hg removal performance of mordenite. In the presence of acidic gases, the adsorption rate of mordenite was accelerated, while the absence of acidic gases enhanced its oxidation ability. Ding et al. (2012) studied the Hg removal performances of bentonite, mordenite, and attapulgite. Among the materials modified by $CuCl_2$, $NaClO_3$, KBr, and KI, the first three modifiers were successful, reaching Hg^0 removal efficiencies of more than 90%. Zhang et al. (2012) used a variety of active substances to modify six adsorbent carriers (kaolinite, bentonite, zeolite, calcium silicate, alumina, and column-layer chromatographic silica gel) and studied their Hg removal performance at 140°C in a fixed-bed reactor. The results indicated that these carriers had poor Hg adsorption performance, though some modified carriers had better performance than commercial AC.

5.1.2. Hg capture during coal combustion

Adding additives, such as halogen compounds, into pulverized coal to control Hg emission is also recognized as an effective method and has been studied by many scholars. Table 7 summarizes the published work on Hg removal by blending additives with coal. Zhuang et al. (2007) added $CaCl_2$ to fire coal and available APCDs and found that the Hg removal efficiency could reach 95%. Pan et al. (2009) conducted a series of experiments on adding NaCl and NH_4Cl into a pulverized coal firing trial furnace. The results showed that with an increase in the additive amount, the percentage of Hg^{2+} and Hg^0 decreased, while the percentage of Hg^P increased. Yu et al. (2012) also noted that additives in pulverized coal could promote the transformation of Hg. $CaBr_2$ was added in the boiler fronts of 600 MW coal-fired units to control Hg emission, and 12 ppm of the additive could increase the Hg efficiency to 88% (Zhao et al. 2013a; Liu et al. 2009). Zhong et al. (2016) also added halogen compounds to fired coal and

Table 7. Hg removal by blending additives with coal.

Coal	Hg concentration (μg/N m^3)	Cl concentration (ppm)	Boiler	Additives	Dosage (ppm)	Hg removal efficiency (%)	Author
Subbituminous coal	14.2	3.3	Pulverized coal	CaCl$_2$	800	87	Zhuang *et al.* (2007)
Mixed coal	–	0.08	Pulverized coal	NH$_4$Cl	28,000	95	Pan *et al.* (2009)
Powder River Basin (PRB) coal	13	–	Pulverized coal	CaBr$_2$	12	88	Liu *et al.* (2009)
Bituminous coal	19.38	90–180	Pulverized coal	Halide	355	90	Zhong *et al.* (2016)

improved the Hg removal efficiency from 33% to 90% with only a small increase in commissioning cost. The results indicated that the best molar ratio of Br and Hg was 3670.

5.2. Adsorbents for capturing as in coal combustion flue gas

Similar to Hg, As is difficult to control because of its high volatility. Elemental arsenic (As^0) and arsenic trioxide (As$_2$O$_3$) are the forms of As most likely to be present in an oxidizing flue gas environment. In this section, reports about the control of As from flue gas in coal-fired power plants are summarized. Studies have shown that fly ash has good As adsorption capacity and that the removal performance can be extensively affected by the physicochemical characteristics and reaction conditions (López-Antón *et al.* 2006). In addition, AC, metal oxide, and Ca-based adsorbents can also be used to adsorb As (López-Antón *et al.* 2007b; Fierro *et al.* 2009; Chen *et al.* 2012; Chen 2013). López-Antón *et al.* (2006) studied the effect of fly ashes from a pulverized coal-fired boiler (CTA) and fluidized-bed combustion boiler (CTP) on As removal. The UBC in CTA fly ash is responsible for As capture, while the calcium and iron content in CTP fly ash were positively associated with the amount of As adsorption. López-Antón *et al.* (2007b) also studied the capture of As by AC, which contained approximately 30% ash, a small amount of Fe$_2$O$_3$ and CaO. The desorption experiments showed that little As was released, which was ascribed to the presence of Fe$_2$O$_3$ and CaO in the thermodynamic calculation. Chen *et al.* (2012) conducted experiments on the removal of As and S by calcium-based materials. The results showed that, with an increase in n(Ca)/n(S) to 3.0, the As removal efficiency could increase to 92% at 1050°C.

Zhang *et al.* (2015b) studied the adsorption of gas-phase As$_2$O$_3$ by CaO, Fe$_2$O$_3$, and Al$_2$O$_3$ using a fixed-bed reactor at a temperature range of 600–900°C. The results indicated that Fe$_2$O$_3$ showed the best adsorption capacity, followed by CaO and Al$_2$O$_3$. The amount of As captured by Fe$_2$O$_3$ was 1.38 mg/g at 600°C, while CaO captured 0.63 mg/g of As and Al$_2$O$_3$ captured 0.17 mg/g of As. In the temperature range of 600–900°C, the adsorption capacity increases in the order of Al$_2$O$_3$ < CaO < Fe$_2$O$_3$. Several studies have combined different metal oxides to explore their common ability of As uptake. Chen (2013) prepared different CaO/γ-Al$_2$O$_3$ ratios to capture As. The results showed that the optimum CaO/γ-Al$_2$O$_3$ ratio could achieve 43.25% removal efficiency for As in flue gas. In addition, FTIR and XRD analysis results verified the existence of As-O and Ca$_3$(AsO$_4$)$_2$ in dearsenic products. Baltrus *et al.* (2010, 2011), Poulston *et al.* (2011) and Rupp *et al.* (2013) focused on the development of palladium-alumina sorbents for As capture from flue gas. The experimental results indicated that a higher temperature, within the temperature range of 204–371°C, and good palladium dispersion in the pores of the support are favourable for As adsorption. In fact, 20 wt% Pd/γ-Al$_2$O$_3$ could remove 100% of the target contaminant after 1500 min, while Pd in the form of 5 wt% Pd on 1-mm alumina beads was also shown to be an effective sorbent for the removal of AsH$_3$ from simulated fuel gas at temperatures of 204 and 288°C. Zhao *et al.* (2008) investigated the retention capacity of a calcium-based adsorbent towards As through bench-scale DTF experiments, finding that the calcium-based sorbent is an effective additive to control the emission of As during coal combustion. With the addition of the calcium-based sorbent, the As concentration in PM$_1$ sharply decreased from 0.25 to 0.11 mg/N m^3, which agrees with the above conclusions.

5.3. Sorbents for F capture during coal combustion

Unlike the Hg removal process, F is mostly removed from coal by blending sorbents with coal during the coal combustion process. Carbonates and hydroxides of Group II$_A$ alkaline-earth metals have different retention actions on F emission during coal combustion. Among them, calcium-based carbonates and hydroxides are suitable sorbents due to their good F retention performance, low cost, and natural abundance (Qi *et al.* 2006). Deng *et al.* (2014) studied F emissions from six coal-fired boiler units in China and found that over 96.63% of the F contained in coal was released during coal combustion, and F compounds in the flue gas were mainly in the form of HF. Calcium-based sorbents were decomposed to CaO under high-temperature conditions, and then CaO reacted with HF to form solid products (CaF_2).

The decomposition rate of calcium-based carbonates or hydroxides is much higher than the CaO–HF fluorine retention rate. The CaO–HF reaction is a key step during the calcium-based F retention process. Wang *et al.* (2014) established an unreacted shrinking core dynamic model of CaO–HF in the F retention reaction and found that the initial CaO particle size had a significant impact on the F retention rate. The conversion rate significantly improved with a decrease in the initial CaO particle size.

CaO was used to reduce the emission of F during coal combustion in a fixed-bed reactor (Qi *et al.* 2002b). The results indicated that the efficiency of F removal by CaO at 1173 K is between 13.62% and 80.44%, with a mean value of 47.35%. The Ca/F molar ratio and particle size of lime significantly affect the F removal efficiency. The F removal efficiency can reach 66.7–70.0% with a Ca/F ratio of 60–70 and a lime particle size at 0.2–1.0 mm. The F retention performance of CaO could be improved by promoting the reactivity of CaO and the high-temperature stability of the F retention reaction products. Under high-temperature conditions, the addition of alkali metal compounds could change the lattice structure, and improve the pore structure characteristics of CaO, thus promoting the reactivity of CaO.

5.4. Adsorbents for capturing Se in coal combustion flue gas

Many solid sorbents have been used to capture Se during coal combustion, including calcium-based adsorbents (Li *et al.* 2006a, 2006b, 2007; Xu *et al.* 2013), AC (López-Antón *et al.*, 2007a; Jadhav *et al.* 2000), fly ash (López-Antón *et al.* 2006, 2007a), and metal oxide sorbents(Baltrus *et al.* 2011; Rupp *et al.* 2013). Among these sorbents, calcium-based adsorbents have exhibited enormous potential for the removal of Se from coal combustion flue gas.

López-Antón *et al.* (2006, 2007b) studied the removal of selenium (Se) by fly ash and the related reaction mechanism. CTA contained UBC, which concentrated on coarse particles, while CTP had a high calcium content, which contributed to the capture of Se. UBC had no obvious influence on the adsorption of Se. Similarly, the amount of Se adsorbed by fly ash showed no obvious correlation with the calcium content. López-Antón *et al.* (2007b) also studied the adsorption of three kinds of AC, including commercial ACs (Norit RB3 and RBHG3) and high-sulphur coal-activated carbon (CA). The sulphur content of RB3, RBHG3 and CA was 0.43%, 6.07%, and 5.02%, respectively. Se mainly

exists in the form of Se^{4+} in AC. In addition, the adsorption of Se by CA was more stable than that by RB3 and RBHG3.

Xu *et al.* (2013) combined the nanosized metal oxides ZnO, Al_2O_3, and Fe_3O_4 with CaO and used them as adsorbents for Se removal from flue gas generated by high-Se coal combustion over a temperature range of 700–1000°C. The experimental results showed that the maximum adsorption efficiency of composite adsorbents was obtained at 800°C due to a trade-off between the relatively weak adsorption reaction between Se and the sorbents at low temperature and the desorption of Se at high temperature. The composite adsorbents, especially those containing ZnO and Fe_3O_4, exhibited higher adsorption efficiencies than pure CaO. The highest adsorption efficiency of 95.46% was obtained over the CaO–ZnO composite sorbent under optimized experimental conditions.

Li *et al.* (2007) studied the effect of CO_2 on the adsorption of SeO_2 by CaO in flue gas and confirmed that SeO_2 could be captured by CaO in high concentrations of CO_2 (volume fraction was 6000 times than that of SeO_2) in the range of 700–760°C. However, due to the competition of CO_2, the adsorption ability of CaO was reduced by 12–35% compared to that of CaO without CO_2.

5.5. Adsorbents for capturing Cr in coal combustion flue gas

Various sorbents have been developed to capture Cr during the combustion process. Table 8 summarizes previously reported Cr adsorption experiments using different sorbents. Most of the studies on Cr capture focussed on the solid fuel incineration process (Chen *et al.*, 1996, 1998; Ho *et al.* 2001; Liu 2007; Kuo *et al.* 2011). Alumina, limestone, kaolinite, AC, and rice husk coke showed good Cr adsorption capacity during coal combustion in a laboratory-scale fluidized-bed combustor (Lu *et al.* 2004; Li *et al.* 2016). CaO and calcite are also effective for Cr removal (Cheng *et al.*, 2001b). Zhao *et al.* (2013b) evaluated the Cr retention performance of six sorbents (Al_2O_3, CaO, Fe_2O_3, zeolite, bentonite, and bauxite) in a fixed-bed reactor. The experimental results showed that bauxite had the highest retention rate for SB high-Cr lignite due to its highest surface area. The Cr retention rate of the six sorbents for SB high-Cr lignite follows the order of bauxite > zeolite > CaO > Fe_2O_3 > bentonite > Al_2O_3. The interaction mechanisms of Cr compounds and sorbents were also discussed. Both physical adsorption and chemical reactions are involved in the Cr capture process. Physical adsorption is the main

Table 8. Cr adsorption by sorbents in published papers.

Fuel	Reactor	Sorbents	Cr removal efficiency (%)	Adsorption capacity (µg/g)	Temperature (°C)	Author
Shenbei high-Cr lignite	Fixed-bed reactor	CaO	33		1400	Zhao *et al.* (2013a)
		Zeolite	36			
		bauxite	92			
Qingshan coal, Pingdingshan coal	Electrically heated drop-tube furnace	Bauxite CaO	~0		1400	Cheng *et al.* (2001b)
Plastics, sawdust, heavy metal solution	Fluidized bed	Kaolinite + NaCl	85.94		700	Chen *et al.* (1998)
		Water + NaCl	90.28			
Plastics, sawdust, heavy metal solution	Fluidized bed	Limestone	87		800	Chen *et al.* (1996)
Wood pellets	Fluidized bed	Bauxite + limestone + zeolite	83		900	Ho *et al.* (2001)
Sawdust, polyethlene, heavy metal solution	Fluidized bed	CaO + Na	61.8		800	Kuo *et al.* (2011)
		Al_2O_3 + Na	49			
Sawdust, polypropylene, heavy metal solution	Fluidized bed incinerator	ACFs	>90	2×10^4	150	Liu (2007)
Yima bituminous coal	One-dimension pulverized coal furnace	CaO		99.80	1100	Cheng *et al.* (2001a)
		Calcite		101.29		
		kaolinite		104.02		
Pingdingshan bituminous coal, Jiaozuo anthracite coal, Liupanshui lean coal	Fluidized bed combustor	Alumina		64.25	950	Lu *et al.* (2004)
		Limestone		66.36		
		kaolinite		51.55		
Guizhou bituminous coal	Fluidized bed	Activated carbon, rice husk coke		186	850	Li *et al.* (2016)

ACF: Activated carbon fibres.

mechanism for the capture of organically bound Cr, whereas the formation of new Cr compounds of Al_3CrO_6, $FeCr_2O_4$, and $Ca_5Cr_3O_{13}H$ indicates that the capture process of Cr involved a chemical reaction. Bauxite and CaO are not effective for Cr removal from Qingshan coal and PDS coal in a DTF (Cheng *et al.* 2001b), which is related to the occurrence of Cr in coal and the reaction conditions.

6. Conclusions and future work

China is the world's largest producer and consumer of coal. Coal consumption in China accounts for over half of the coal consumption worldwide. TEs in coal result in substantial pollutant emissions and cause serious damage to the ecological environment and human health in China, owing to the huge amount of coal consumption nationwide. The concentrations of TEs in Chinese coals vary depending on their coal-forming period and the coal-bearing area. Most of the TEs in Chinese coals are enriched in P_2 and T_3 coals, especially in T_3 coals, while J_1–J_2 and J_3–K_1 coals have relatively low enrichment. The coal from coal-bearing regions of southern China, especially Southwestern China, has high Hg, As, Pb, Cd, and Cr concentrations.

TEs exhibit different emission characteristics during coal combustion. A large amount of Hg is volatilized at low temperatures and is almost completely released at 800°C. As is a semi-volatile element; only some of the As content is released at low temperature, and its release rate increases sharply at high temperature. Almost all of

the Se will be released from coal at temperatures higher than 800°C. F can be volatized at low temperature, and its release rate increases with an increase in the temperature and combustion time. Cr is a non-volatile element, but the form in which Cr occurs can affect its volatility. During coal combustion in a power plant, TEs can volatilize at high temperature. A portion of the TEs was released to the atmosphere from the stack, while the other part was redistributed in the solid combustion products, such as the bottom ash, fly ash, and FGD gypsum. In the field test results, the partitioning of TEs may vary over a certain range, which is determined by several factors, such as the volatile characteristics of the TEs, the coal type, the combustion conditions, the APCD equipment, etc.

The existing APCDs equipped in coal-fired power plants, such as SCR, LTE, ESP/FF, FGD, and WESP, can alter the speciation of TEs in flue gas and control TE emission, in addition to removing other common pollutants (NO_x, SO_2, and PM). The SCR catalyst can promote Hg^0 oxidation but has no obvious effect on the total Hg emission. When the LTE was on, the flue gas temperature in the ESP decreased, benefiting gaseous Hg adsorption by the UBC and inorganic minerals in the fly ash and reducing the enrichments of TEs in submicron particle matter. Thus, LTE enhances TE removal by ESP. ESP and FF could control the TE emission by capturing the fly ash. WFGD had significant effects on controlling the emission of TEs. Most of the Hg^{2+} could be removed by WFGD due to its solubility. A high gaseous retention capacity for F, Se, and As is attained in the WFGD. WESP benefits TE removal. The overall TE removal rate across ESP + WFGD + WESP is larger than 99.9%.

Sorbent injection or blending sorbents with coal are considered to be promising methods to efficiently control TE emission in coal-fired power plants. Various sorbents have been developed to capture TEs, especially Hg. Sorbents, such as AC, fly ash, calcium-based sorbents, metal oxides, and mineral sorbents, have been used to capture TEs in flue gas. ACI exhibits good Hg removal performance and had been commercially applied in many coal-fired power plants. Fly ash has good adsorption properties towards As and Hg. The removal performance can be extensively affected by the physicochemical characteristics and reaction conditions. F and Hg can be removed by blending calcium-based sorbents and halogen compounds, respectively, with coal during the coal combustion process. Metal oxides, such as CaO, Fe_2O_3, and Al_2O_3, can also capture As, Se, and Cr. Among these mineral sorbents, bauxite, kaolinite, and zeolite are usually used to capture Cr.

However, much work needs to be done to control TE emissions in China. First, Hg speciation, emission, and removal from power plants have been reported extensively, while information on other TEs is very limited. Therefore, comprehensive studies on the transformation mechanism for other easily vaporized TEs are imperative. A more effective and accurate chemical kinetics model should be developed to describe the transformation and mobility characteristics of easily vaporized TEs during coal combustion. Second, the co-beneficial removal of TEs by APCDs is an effective control technology. With the 'Transformation and Upgrading Action Plan for Coal Energy Conservation and Emission Reduction' in China, many power plants have been equipped with ultra-clean emission technology. After equipping power plants with ultra-clean emission technology, the emissions of SO_2, NO_x, and PM are lower than the emissions limitation of a gas turbine ($SO_2 < 35$ mg/m^3, $NO_x < 50$ mg/m^3, PM < 5 mg/m^3). However, the migration and emission characteristics of TEs in power plants with ultralow emission APCDs is not very clear. Third, more focus is needed to develop simultaneous control technologies for TEs and other pollutants with high efficiency and low investment. Sorbent injection is a promising technology for Hg removal, but further research is needed to understand the competitive adsorption mechanism between Hg and other TEs, with the intention to develop sorbents for the simultaneous removal of Hg and other TEs. In addition, fundamental research on the regeneration and enhancement of deactivated sorbents by adding chemical agents during the regeneration process are also necessary.

Acknowledgements

This research was supported by the National Key Basic Research Program (973) of China: [Grant Number 2014CB238904], the National Key Technologies R&D Program: [Grant Number 2016YFB0600604], and the National Natural Science Foundation of China (NSFC): [Grant Numbers 51376074 and 41672148]. We appreciate Prof. Shifeng Dai and Dr. Robert B. Finkelman for careful review and detailed comments to improve the early version of the manuscript. Author would like to thank anonymous reviewers for their critical comments.

Disclosure statement

No potential conflict of interest was reported by the authors.

Funding

This work was supported by the National Key Basic Research Program (973) of China: [Grant Number 2014CB238904], the National Key Technologies R&D Program: [Grant Number 2016YFB0600604], and the National Natural Science Foundation of China: [Grant Number 51376074].

References

Abad-Valle, P., Lopez-Anton, M.A., Diaz-Somoano, M., Juan, R., Rubio, B., Garcia, J.R., Khainakov, S.A., and Martinez-Tarazona, M.R., 2011a, Influence of iron species present in fly ashes on mercury retention and oxidation: Fuel, v. 90, p. 2808–2811. doi:10.1016/j.fuel.2011.04.031

Abad-Valle, P., Lopez-Anton, M.A., Diaz-Somoano, M., and Martinez-Tarazona, M.R., 2011b, The role of unburned carbon concentrates from fly ashes in the oxidation and retention of mercury: Chemical Engineering Journal, v. 174, p. 86–92. doi:10.1016/j.cej.2011.08.053

Aboud, S., Sasmaz, E., and Wilcox, J., 2008, Mercury adsorption on PdAu, PdAg and PdCu alloys: Main Group Chemistry, v. 7, p. 205–215. doi:10.1080/10241220802465213

Agnihotri, R., Chauk, S., Mahuli, S., and Fan, L.S., 1998, Selenium capture using sorbent powders: Mechanism of sorption by hydrated lime: Environmental Science & Technology, v. 32, p. 1841–1846. doi:10.1021/es971119j

Bai, X.F., 2003, The distribution, modes of occurrence and volatility of trace elements in coals of China [Ph.D.]: Beijing, China Coal Research Institute, 26 p.

Bai, X.F., Li, W.H., Chen, Y.F., and Yang, Y., 2007, The general distributions of trace elements in Chinese coals: Coal Quality Technology, v. 1, p. 1–4. in Chinese with English abstract.

Baltrus, J.P., Granite, E.J., Pennline, H.W., Stanko, D., Hamilton, H., Rowsell, L., Poulston, S., Smith, A., and Chu, W., 2010, Surface characterization of palladium–alumina sorbents for high-temperature capture of mercury and arsenic from fuel gas: Fuel, v. 89, p. 1323–1325. doi:10.1016/j.fuel.2009.09.030

Baltrus, J.P., Granite, E.J., Rupp, E.C., Stanko, D.C., Howard, B., and Pennline, H.W., 2011, Effect of palladium dispersion on the capture of toxic components from fuel gas by palladium-alumina sorbents: Fuel, v. 90, p. 1992–1998. doi:10.1016/j.fuel.2011.01.001

Bosch, H., and Janssen, F., 1988, Catalytic reduction of nitrogen oxides: Catalysis Reviews, v. 2, p. 369–531.

BP, 2015, Statistical review of world energy, June, London, UK: British Petroleum.

Busca, G., Lietti, L., and Ramis, G., 1998, Chemical and mechanistic aspects of the selective catalytic reduction of NO_x, by ammonia over oxide catalysts: A review: Applied Catalysis B: Environmental, v. 18, p. 1–36.

Cao, Y., Chen, B., Wu, J., Cui, H., Smith, J., Chen, C.K., Chu, P., and Pan, W.P., 2007, Study of mercury oxidation by a selective catalytic reduction catalyst in a pilot-scale slipstream reactor at a utility boiler burning bituminous coal: Energy & Fuels, v. 21, p. 145–156. doi:10.1021/ef0602426

Cao, Y., Gao, Z.Y., Zhu, J.S., Wang, Q.H., Huang, Y.J., Chiu, C.C., Parker, B., Chu, P., and Pan, W.P., 2008, Impacts of halogen additions on mercury oxidation, in a slipstream selective catalyst reduction (SCR), reactor when burning sub-bituminous coal: Environmental Science & Technology, v. 42, p. 256–261. doi:10.1021/es071281e

Chang, J.C.S., and Ghorishi, S.B., 2003, Simulation and evaluation of elemental mercury concentration increase in flue gas across a wet scrubber: Environmental Science & Technology, v. 37, p. 5763–5766. doi:10.1021/es034352s

Chen, J.C., and Wey, M.Y., 1996, The effect of operating conditions on the capture of metals with limestone during incineration: Environment International, v. 22, p. 743–752. doi:10.1016/S0160-4120(96)00066-9

Chen, J.C., Wey, M.Y., and Lin, Y.C., 1998, The adsorption of heavy metals by different sorbents under various incineration conditions: Chemosphere, v. 37, p. 2262–2617. doi:10.1016/S0045-6535(98)00161-1

Chen, J.F., 2013, Dry dearsenic by Cao/γ-Al$_2$O$_3$ adsorbent during coal combustion: Journal of Huazhong Normal University (Natural Sciences), 47, p. 519–522.

Chen, J.F., and Shuai, Q., 2012, Study of simultaneous dearsenic and desulfurization by calcium-based materials during coal combustion: Journal of Hefei University of Technology (Natural Science), v. 35, p. 112–115.

Cheng, J.F., Han, J., Liu, Y.H., Zeng, H.C., Xu, M.H., and Luo, K.L., 2001a, Effect of air staging with absorbents on trace metal during coal combustion: Environmental Science, v. 22, p. 34–38. in Chinese with English abstract

Cheng, J.F., Zeng, H.C., Zhang, Z.H., and Xu, M.H., 2001b, The effects of solid absorbents on the emission of trace elements, SO_2, and NO_x during coal combustion: International Journal of Energy Research, v. 25, p. 1043–1052. doi:10.1002/(ISSN)1099-114X

Cheng, S.P., 2003, Heavy metal pollution in China: Origin, pattern, and control: Environmental Science and Pollution Research, v. 10, p. 192–198. doi:10.1065/espr2002.11.141.1

China Coal Geology Bureau, 1996, Atlas of coal resources in major coal mines of China: Internal data of China Coal Geology Bureau, Beijing, China.

Cimino, S., and Scala, F., 2016, Removal of elemental mercury by MnO_x catalysts supported on TiO_2 or Al_2O_3: Industrial & Engineering Chemistry Research, v. 55, p. 5133–5138. doi:10.1021/acs.iecr.5b04147

Clarke, L.B., and Sloss, L.L., 1992, Trace elements - emissions from coal combustion and gasification: IEA Coal Research, London.

Clemens, A.H., Damiano, L.F., Gong, D., and Matheson, T.W., 1999, Partirioning behaviour of some toxic volatile elements during stoker and fluidised bed combustion of alkaline sub-bituminous: Fuel, v. 78, p. 1379–1385. doi:10.1016/S0016-2361(99)00066-6

Córdoba, P., Font, O., Lzquierdo, M., Querol, X., Leiva, C., Lopez-Anton, M.A., Diaz-Somoano, M., Martinez-Tarazona, M.R., Ochoa-Gonzalez, R., Fernandez, C., and Gomez, P., 2012b, The retention capacity for trace elements by the flue gas desulphurisation system under operational conditions of a co-combustion power plant: Fuel, v. 102, p. 773–788. doi:10.1016/j.fuel.2012.06.059

Córdoba, P., Ochoa-Gonzalez, R., Font, O., Lzquierdo, M., Querol, X., Leiva, C., Lopez-Anton, M.A., Diaz-Somoano, M., Martinez-Tarazona, M., Fernandez, C., and Tomas, A., 2012a, Partitioning of trace inorganic elements in a coal-fired power plant equipped with a wet flue gas desulphurisation system: Fuel, v. 92, p. 145–157. doi:10.1016/j.fuel.2011.07.025

Dai, S., Graham, I.T., and Ward, C.R., 2016b, A review of anomalous rare earth elements and yttrium in coal: International Journal of Coal Geology, v. 159, p. 82–95. doi:10.1016/j.coal.2016.04.005

Dai, S., Li, W., Tang, Y., Zhang, Y., and Feng, P., 2007, The sources, pathway, and preventive measures for fluorosis in Zhijin County, Guizhou, China: Applied Geochemistry, v. 22, p. 1017–1024. doi:10.1016/j.apgeochem.2007.02.011

Dai, S., and Ren, D., 2006, Fluorine concentration of coals in China—An estimation considering coal reserves: Fuel, v. 85, p. 929–935. doi:10.1016/j.fuel.2005.10.001

Dai, S., Ren, D., Chou, C.L., Finkelman, R.B., Seredin, V.V., and Zhou, Y.P., 2012a, Geochemistry of trace elements in Chinese coals: A review of abundances, genetic types, impacts on human health, and industrial utilization: International Journal of Coal Geology, v. 94, p. 3–21. doi:10.1016/j.coal.2011.02.003

Dai, S., Ren, D., and Ma, S., 2004, The cause of endemic fluorosis in western Guizhou Province, Southwest China: Fuel, v. 83, p. 2095–2098. doi:10.1016/j.fuel.2004.03.016

Dai, S., Ren, D., and Tang, Y., 2005, Modes of occurrence of major elements in coal and their study significance: Coal Geoglogy & Exploration, v. 33, p. 1–5. in Chinese with English abstract

Dai, S., Ren, D., Zhou, Y., Chou, C.L., Wang, X., Zhao, L., and Zhu, X., 2008, Mineralogy and geochemistry of a superhigh-organic-sulfur coal, Yanshan Coalfield, Yunnan, China: Evidence for a volcanic ash component and influence by submarine exhalation: Chemical Geology, v. 255, p. 182–194. doi:10.1016/j.chemgeo.2008.06.030

Dai, S., Seredin, V.V., Ward, C.R., Hower, J.C., Xing, Y., Zhang, W., Song, W., and Wang, P., 2015, Enrichment of U–Se–Mo–Re–V in coals preserved within marine carbonate successions: Geochemical and mineralogical data from the Late Permian Guiding coalfield, Guizhou, China: Mineralium Deposita, v. 50, p. 159–186. doi:10.1007/s00126-014-0528-1

Dai, S., Seredin, V.V., Ward, C.R., Jiang, J., Hower, J.C., Song, X., Jiang, Y., Wang, X., Gornostaeva, T., Li, X., Liu, H., Zhao, L., and Zhao, C., 2014a, Composition and modes of occurrence of minerals and elements in coal combustion products derived from high-Ge coals: International Journal of Coal Geology, v. 121, p. 79–97. doi:10.1016/j.coal.2013.11.004

Dai, S., Wang, X., Seredin, V.V., Hower, J.C., Ward, C.R., O'Keefe, J.M.K., Huang, W., Li, T., Li, X., Liu, H., Xue, W., and Zhao, L., 2012b, Petrology, mineralogy, and geochemistry of the Ge-rich coal from the Wulantuga Ge ore deposit, Inner Mongolia, China: New data and genetic implications: International Journal of Coal Geology, v. 90–91, p. 72–99. doi:10.1016/j.coal.2011.10.012

Dai, S., Yan, X., Ward, C.R., Hower, J.C., Zhao, L., Wang, X., Zhao, L., Ren, D., and Finkelman, R.B., 2016a, Valuable elements in Chinese coals: A review: International Geology Review, p. 1–31. doi:10.1080/00206814.2016.1197802

Dai, S., Zhang, W., Seredin, V.V., Ward, C.R., Hower, J.C., Wang, X., Li, X., Song, W., Zhao, L., Kang, H., Zheng, L., and Zhou, D., 2013, Factors controlling geochemical and mineralogical compositions of coals preserved within marine carbonate successions: A case study from the Heshan coalfield, southern China: International Journal of Coal Geology, v. 109–110, p. 77–100. doi:10.1016/j.coal.2013.02.003

Dai, S., Zhao, L., Hower, J.C., Johnston, M.N., Song, W., Wang, P., and Zhang, S., 2014b, Petrology, mineralogy, and chemistry of size-fractioned fly ash from the Jungar power plant, Inner Mongolia, China, with emphasis on the distribution of rare earth elements: Energy and Fuels, v. 28, p. 1502–1514. doi:10.1021/ef402184t

Dai, S., Zhao, L., Peng, S., Chou, C.-L., Wang, X., Zhang, Y., Li, D., and Sun, Y., 2010, Abundances and distribution of minerals and elements in high-alumina coal fly ash from the Jungar power plant, inner Mongolia, China: International Journal of Coal Geology, v. 81, p. 320–332. doi:10.1016/j.coal.2009.03.005

De, M., Azargohar, R., Dalai, A.K., and Shewchuk, S.R., 2013, Mercury removal by bio-char based modified activated carbons: Fuel, v. 103, p. 570–578. doi:10.1016/j.fuel.2012.08.011

Deng, S., Liu, Y., Zhang, C., Wang, X.F., Cao, Q., Wang, H.M., and Zhang, F., 2014, Fluorine emission of pulverized coal-fired power plants in China: Research of Environmental Sciences, v. 27, p. 225–231.

Diamantopoulou, I., Skodras, G., and Sakellaropoulos, G.P., 2010, Sorption of mercury by activated carbon in the presence of flue gas components: Fuel Processing Technology, v. 91, p. 158–163. doi:10.1016/j.fuproc.2009.09.005

Ding, F., Zhao, Y.C., Mi, L.L., Li, H.L., Li, Y., and Zhang, J.Y., 2012, Removal of gas-phase elemental mercury in flue gas by inorganic chemically promoted natural mineral sorbents: Industrial & Engineering Chemistry Research, v. 51, p. 3039–3047. doi:10.1021/ie202231r

Ding, Z.Y., Li, L., Wade, A.D., and Gloyna, E.F., 1998, Supercritical water oxidation of NH_3 over a MnO_2/CeO_2 catalyst: Industrial & Engineering Chemistry Research, v. 37, p. 1707–1716. doi:10.1021/ie9709345

Dranga, B.A., Li, L., and Koeser, H., 2012, Oxidation catalysts for elemental mercury in flue gases—A review: Focus on Catalysts, v. 2, p. 139–170. doi:10.3390/catal2010139

Duan, J., and Tan, J., 2013, Atmospheric heavy metals and arsenic in China: Situation, sources and control policies: Atmospheric Environment, v. 74, p. 93–101. doi:10.1016/j.atmosenv.2013.03.031

Duan, Y., Jiang, Y., Yang, L., and Wang, Y., 2008, Experimental study on mercury emission and adsorption in circulating fluidized bed boiler: Proceedings of the CSEE, v. 28, p. 1–5.

Dunham, G.E., DeWall, R.A., and Senior, C.L., 2003, Fixed-bed studies of the interactions between mercury and coal combustion fly ash: Fuel Processing Technology, v. 82, p. 197–213. doi:10.1016/S0378-3820(03)00070-5

Eswaran, S., and Stenger, H.G., 2005, Understanding mercury conversion in selective catalytic reduction (SCR) catalysts: Energy & Fuels, v. 19, p. 2328–2334. doi:10.1021/ef050087f

Eswaran, S., and Stenger, H.G., 2008, Effect of halogens on mercury conversion in SCR catalysts: Fuel Processing Technology, v. 89, p. 1153–1159. doi:10.1016/j.fuproc.2008.05.007

Eswaran, S., Stenger, H.G., and Fan, Z., 2007, Gas-phase mercury adsorption rate studies: Energy & Fuels, v. 21, p. 852–857. doi:10.1021/ef060276d

Feng, Y.W., Ogura, N., Feng, Z.W., Zhang, F.Z., and Shimizu, H., 2003, The concentrations and sources of fluoride in atmospheric depositions in Beijing, China: Waste Air & Soil Pollution, v. 145, p. 95–107. doi:10.1023/A:1023680112474

Fierro, V., Muñiz, G., Gonzalez-Sánchez, G., Ballinas, M.L., and Celzard, A., 2009, Arsenic removal by iron-doped activated carbons prepared by ferric chloride forced hydrolysis: Journal of Hazardous Materials, v. 168, p. 430–437. doi:10.1016/j.jhazmat.2009.02.055

Finkelman, R.B., Orem, W., Castranova, V., Tatu, C.A., Belkin, H. E., Zheng, B., Lerch, H.E., Maharaj, S.V., and Bates, A.L., 2002, Health impacts of coal and coal use: Possiblesolutions: International Journal of Coal Geology, v. 50, p. 425–443. doi:10.1016/S0166-5162(02)00125-8

Finkelman, R.B., Palmer, C.A., Krasnow, M.R., Aruscavage, P.J., Sellers, G.A., and Dulong, F.T., 1990, Combustion and leaching behavior of elements in the argonne premium coal samples: Energy & Fuels, v. 4, p. 755–766. doi:10.1021/ef00024a024

Fuente-Cuesta, A., Diaz-Somoano, M., Lopez-Anton, M.A., Cieplik, M., Fierro, J.L.G., and Martinez-Tarazona, M.R., 2012, Biomass gasification chars for mercury capture from a simulated flue gas of coal combustion: Journal of Environmental Management, v. 98, p. 23–28. doi:10.1016/j.jenvman.2011.12.013

Gao, H.L., Wang, X.Y., Zhou, J.S., and Luo, Z.Y., 2007, The influence of ESP on mercury emission from coal-fired power plant: Boiler Technology, v. 38, p. 63–67.

Gao, W., Zhi, G.R., and Xue, Z.G., 2013, Analysis of atmospheric emission trends of mercury, lead and arsenic from coal combustion in china from 1980-2007: Research of Environmental Sciences, v. 26, p. 822–828.

Ghorishi, S.B., Singer, C.F., Jozewicz, W.S., and Srivastava, R.K., 2002, Simultaneous control of $Hg°$, SO_2, and NO_x by novel oxidized calcium-based sorbents: Journal of Air & Waste Management Association, v. 52, p. 273–278. doi:10.1080/10473289.2002.10470786

Gibb, W.H., Clarke, F., and Mehta, A.K., 2000, The fate of coal mercury during combustion: Fuel Processing Technology, v. 65-66, p. 365–377. doi:10.1016/S0378-3820(99)00104-6

Guo, X., Zheng, C., Jia, X., and Sun, T., 2003, The behaviour of mercury, arsenic, selenium during coal combustion: Journal of Engineering Thermophysics, v. 24, p. 703–706.

Guo, X., Zheng, C., and Jia, X., 2004a, Studies on characteristics of mercury and arsenic distribution in combustion products at the coal-fired utility boiler: Journal of Engineering Thermophysics, v. 25, p. 714–716.

Guo, X., Zheng, C., Jia, X., Lin, Z., and Liu, Y., 2004b, Study on mercury speciation in pulverized coal fired flue gas: Proceedings of CSEE, v. 24, p. 185–188.

Guo, X., Zheng, C., and Xu, M., 2004c, Characterization of arsenic emissions from a coal-fired power plant: Energy & Fuels, v. 18, p. 1822–1826. doi:10.1021/ef049921b

Ho, T.C., Chuang, T.C., Chelluri, S., Lee, Y., and Hopper, J.R., 2001, Simultaneous capture of metal, sulfur and chlorine by sorbents during fluidized bed incineration: Waste Management, v. 21, p. 435–441. doi:10.1016/S0956-053X(00)00135-5

Hower, J.C., Senior, C.L., Suuberg, E.M., Hurt, R.H., Wilcox, J.L., and Olson, E.S., 2010, Mercury capture by native fly ash carbons in coal-fired power plants: Progress Energy Combustion Science, v. 36, p. 510–529. doi:10.1016/j.pecs.2009.12.003

Hower, J.C., Groppo, J.G., Joshi, P., Dai, S., Moecher, D.P., and Johnston, M.N., 2014, Location of cerium in coal-combustion fly ashes: Implications for recovery of lanthanides: Coal Combustion and Gasification Products, v. 5, p. 73–78. doi:10.4177/CCGP-D13-00007.1

Hower, J.C., Groppo, J.G., Henke, K.R., Hood, M.M., Eble, C.F., Honaker, R.Q., Zhang, W., and Qian, D., 2015, Notes on the potential for the concentration of rare earth elements and yttrium in coal combustion fly ash: Minerals, v. 5, no. 2, p. 356–366. doi:10.3390/min5020356

Hower, J.C., and Dai, S., 2016, Petrology and chemistry of sized Pennsylvania anthracite, with emphasis on the distribution of rare earth elements: Fuel, v. 185, p. 305–315. doi:10.1016/j.fuel.2016.07.055

Hower, J.C., Granite, E.J., Mayfield, D.B., Lewis, A.S., and Finkelman, R.B., 2016, Notes on contributions to the science of rare earth element enrichment in coal and coal combustion by-products: Minerals, v. 6, no. 2, p. 32. doi:10.3390/min6020032

Hower, J.C., Groppo, J.G., Graham, U.M., Ward, C.R., Kostova, I. J., Maroto-Valer, M.M., and Dai, S., 2017, Coal-derived unburned carbons in fly ash: A review: International Journal of Coal Geology, v. 179, p. 11–27. doi:10.1016/j.coal.2017.05.007

Hu, C.X., Zhou, J.S., He, S., Zhang, L., Zheng, J.M., Luo, Z.Y., and Cen, K.F., 2009, Influence and control of electrostatic precipitators and wet flue gas desulfurization systems on the speciation of mercury in flue gas: Journal of Power Engineering, v. 29, p. 400–404.

Hu, C.X., Zhou, J.S., He, S., Zheng, J., Luo, Z.Y., and Cen, K.F., 2008, Influence of chlorine and ash on flue gas mercury speciation of large scale coal-fired boilers: Journal of Power Engineering, v. 28, p. 946–948.

Huang, Y., Jin, B., Zhong, Z., Xiao, R., Tang, Z., and Ren, H., 2004, Trace elements (Mn, Cr, Pb, Se, Zn, Cd and Hg) in emissions from a pulverized coal boiler: Fuel Processing Technology, v. 86, p. 23–32. doi:10.1016/j.fuproc.2003.10.022

Huang, Z.J., Duan, Y.F., Wang, Y.J., Meng, S.L., and Jiao, Y.G., 2009, Experimental investigation on absorption of Hg in simulated flue gas by modified $Ca(OH)_2$: Proceedings of CSEE, v. 17, p. 56–62.

Huang, Z.J., Duan, Y.F., Wang, Y.J., Meng, S.L., and Jiao, Y.G., 2011, Simulation of mercury emission control by calcium-based sorbent under fixed-bed operations: Boiler Technology, v. 42, p. 65–69.

Jadhav, R.A., Agnihotri, R., Gupta, H., and Fan, L.S., 2000, Mechanism of selenium sorption by activated carbon: Canadian Journal of Chemical Engineering, v. 78, p. 168–174. doi:10.1002/cjce.5450780122

James, D., Kilgroe, C.B., and Sedman, R.K., 2002, Control of mercury emissions from coal-fired electric utility boilers. Interim Report. EPA-600.

Jurng, J., Lee, T.G., Lee, G.W., Lee, S.J., Kim, B.H., and Seier, J., 2002, Mercury removal from incineration flue gas by organic and inorganic adsorbents: Chemosphere, v. 47, p. 907–913. doi:10.1016/S0045-6535(01)00329-0

Kamata, H., Ueno, S.-I., Naito, T., and Yukimura, A., 2008, Mercury oxidation over the $V_2O_5(WO_3)/TiO_2$ commercial SCR catalyst: Industrial & Engineering Chemistry Research, v. 47, p. 8136–8141. doi:10.1021/ie800363g

Kang, C.M., Gupta, T., Ruiz, P.A., Wolfson, J.M., Ferguson, S.T., Lawrence, J.E., Rohr, A.C., Godleski, J., and Koutrakis, P., 2011, Aged particles derived from emissions of coal-fired power plants: the teresa field results: Inhalation Toxicology, v. 23, p. 11–30.

Ketris, M.P., and Yudovich, Y.E., 2009, Estimations of clarkes for carbonaceous biolithes: World averages for trace element contents in black shales and coals: International Journal of Coal Geology, v. 78, p. 135–148. doi:10.1016/j.coal.2009.01.002

Kuo, J.H., Lin, C.L., and Wey, M.Y., 2011, Effect of particle agglomeration on heavy metals adsorption by Al- and Ca-based sorbents during fluidized bed in cineration: Fuel Processing Technology, v. 92, p. 2089–2098. doi:10.1016/j.fuproc.2011.06.014

Lee, S., Seo, Y., Jang, H., Park, K., Baek, J., An, H., and Song, K., 2006a, Speciation and mass distribution of mercury in a bituminous coal-fired power plant: Atmospheric Environment, v. 40, p. 2215–2224. doi:10.1016/j.atmosenv.2005.12.013

Lee, S.H., Rhim, Y.J., Cho, S.P., and Baek, J.I., 2006b, Carbon-based novel sorbent for removing gas-phase mercury: Fuel, v. 85, p. 219–226. doi:10.1016/j.fuel.2005.02.030

Lei, C., Duan, Y., Zhuo, Y., Yang, L., Zhang, L., Yang, X., Yao, Q., Jiang, Y., and Xu, X., 2007, Mercury transformation across particulate control devices in six power plants of china: the co-effect of chlorine and ash composition: Fuel, v. 86, p. 603–610.

Lee, C.W., Serre, S.D., Zhao, Y., Lee, S.J., and Hastings, T.W., 2008, Mercury oxidation promoted by a selective catalytic reduction catalyst under simulated powder river basin coal combustion conditions: Journal of the Air & Waste Management Association, v. 58, p. 484-493.

Li, H., Li, Y., Wu, C.Y., and Zhang, J., 2011, Oxidation and capture of elemental mercury over $SiO_2–TiO_2–V_2O_5$ catalysts in simulated low-rank coal combustion flue gas: Chemical Engineering Journal, v. 169, p. 186–193. doi:10.1016/j.cej.2011.03.003

Li, J., Zhuang, X., Querol, X., Font, O., Moreno, N., and Zhou, J., 2012, Environmental geochemistry of the feed coals and their combustion by-products from two coal-fired power plants in Xinjiang Province, Northwest China: Fuel, v. 95, p. 446–456. doi:10.1016/j.fuel.2011.10.025

Li, J.F., Yan, N.Q., Qu, Z., Qiao, S.H., Yang, S.J., Guo, Y.F., Liu, P., and Jia, J.P., 2010, Catalytic oxidation of elemental mercury over the modified catalyst $Mn/\alpha-Al_2O_3$ at lower temperature: Environmental Science & Technology, v. 44, p. 426–431. doi:10.1021/es9021206

Li, X.L., Sun, H.C., Duan, L.B., and Zhao, C.S., 2016, Influence of different additives/adsorbents on migration of trace elements in circulating fluidized bed combustion: Journal of

Combustion Science and Technology (Chinese, Abstract in English), v. 22, p. 45–49

Li, Y., Murphy, P.D., Wu, C.Y., Powers, K.W., and Bonzongo, J.J., 2008, Development of silica-vanadia-titania catalysts for removal of elemental mercury from coal-combustion flue gas: Environmental Science & Technology, v. 42, p. 5304–5309. doi:10.1021/es8000272

Li, Y., Tong, H., Zhuo, Y., Chen, C., and Xu, X., 2006a, Simultaneous removal of SO_2 and trace SeO_2 from flue gas: Effect of product layer on mass transfer: Environmental Science & Technology, v. 40, p. 4306–4311.

Li, Y., Tong, H., Zhuo, Y., Wang, S., and Xu, X., 2006b, Simultaneous removal of SO_2 and trace SeO_2 from flue gas: Effect of SO_2 on selenium capture and kinetics study: Environmental Science & Technology, v. 40, p. 7919–7924.

Li, Y.Z., Tong, H.L., Li, Y., Gao, Y.X., and Xu, X.C., 2007, Effect of CO_2 on trace selenium adsorption by CaO from flue gas: Journal of Tsinghua University (Science and Technology), v. 47, p. 699–702.

Li, Z., Clemens, A.H., Moore, T.A., Gong, D., Weaver, S.D., and Eby, N., 2005, Partitioning behaviour of trace elements in a stoker-fired combustion unit: An example using bituminous coals from the Greymouth coalfield (Cretaceous), New Zealand: International Journal of Coal Geology, v. 63, p. 98–116. doi:10.1016/j.coal.2005.02.007

Li, Z.C., Duan, Y.F., Wang, Y.J., Huang, Z.J., Meng, S.L., and Shen, X.Z., 2013, Mercury removal by ESP and WESP in a 300 MW coal-fired power plant: Journal of Fuel Chemistry & Technology, v. 41, p. 491–498.

Lim, D.H., and Wilcox, J., 2013, Heterogeneous mercury oxidation on Au (111) from first principles: Environmental Science & Technology, v. 47, p. 8515–8522. doi:10.1021/es400876e

Linak, W.P., and Miller, C.A., 2000, Comparison of particle size distributions and elemental partitioning from the combustion of pulverized coal and residual fuel oil, Journal Of The Air & Waste Management Association, v. 50, p. 1532–1544.

Liu, G.J., Zhang, H.Y., Gao, L.F., Zheng, L.G., and Peng, Z.C., 2004, Petrological and mineralogical characterizations and chemical composition of coal ashes from power plants in Yanzhou mining district, China: Fuel Processing Technology, v. 85, p. 1635–1646. doi:10.1016/j.fuproc.2003.10.028

Liu, J., Yang, Z., Yan, X., Ji, D., Yang, Y., and Hu, L., 2015, Modes of occurrence of highly-elevated trace elements in super-high-organic-sulfur coals: Fuel, v. 156, p. 190–197. doi:10.1016/j.fuel.2015.04.034

Liu, Z.S., 2007, Control of heavy metals during incineration using activated carbon fibers: Journal of Hazardous Materials, v. 142, p. 506–511. doi:10.1016/j.jhazmat.2006.08.055

Liu, Y., Wei, H.M., Xu, J.R., Zhou, J.H., and Cen, K.F., 2008, Effect Of O_2/co_2 And Air On Mercury Speciation In Coal Fired Flue Gases: Proceedings Of Csee, v. 28, p. 48–53. [in Chinese with English abstract.]

Liu, X., and Jiang, Y., 2009, Development and status of mercury control technologies in American coal-fired power plants: High-Technology & Industrialization, v. 5, p. 92–95.

Lockwood, F.C., and Yousif, S., 2000, A model for the particulate matter enrichment with toxic metals in solid fuel flames: Fuel Processing Technology, v. 65-66, p. 439–457. doi:10.1016/S0378-3820(99)00109-5

López-Antón, M.A., Abad-Valle, P., Díaz-Somoano, M., Suarez-Ruiz, I., and Martinez-Tarazona, M.R., 2009, The influence of carbon particle type in fly ashes on mercury adsorption: Fuel, v. 88, p. 1194–1200. doi:10.1016/j.fuel.2007.07.029

López-Antón, M.A., Díaz-Somoano, M., Abad-Valle, P., and Martinez-Tarazona, M.R., 2007a, Mercury and selenium retention in fly ashes: Influence of unburned particle content: Fuel, v. 86, p. 2064–2070. doi:10.1016/j.fuel.2007.03.031

López-Antón, M.A., Díaz-Somoano, M., Fierro, J.L.G., and Martinez-Tarazona, M.R., 2007b, Retention of arsenic and selenium compounds present in coal combustion and gasification flue gases using activated carbons: Fuel Processing Technology, v. 88, p. 799–805. doi:10.1016/j.fuproc.2007.03.005

Lu, J., Chen, X., Duan, L., and Zhou, W., 2009a, Experimental study of trace element migration characteristics in an O_2/CO_2 atmosphere: Journal of Engineering for Thermal Energy & Power, v. 24, p. 648–651.

Lu, P., Wu, J., and Pan, W., 2009b, Mercury emission and its speciation from flue gas of a 860 MW pulverized coal-fired boiler: Journal of Power Engineering, v. 29, p. 1067–1072.

Lu, X.D., Yu, L.Y., and Zhang, J., 2004, Study on control with absorbents on trace elements during fluidized bed combustion: Proceedings of the CSEE, v. 24, p. 192–197.

Mahuli, S., Agnihotri, R., Chauk, S., Ghosh-Dastidar, A., and Fan, L.-S., 1997, Mechanism of arsenic sorption by hydrated lime: Environmental Science & Technology, v. 31, p. 3226-3231.

Mao, D.J., Su, H.C., and Yan, L.R., 1990, An epidemiologic investigation on selenium poisoning in southwestern Hubei Province: Chinese Journal of Epidemiology, v. 9, p. 311–314

Martinez-Tarazona, M.R., and Spears, D.A., 1996, The fate of trace elements and bulk minerals in pulverized coal combustion in a power station: Fuel Processing Technology, v. 47, p. 79–92. doi:10.1016/0378-3820(96)01001-6

Mei, Z.J., Fan, M.H., Zhang, R.Q., Shen, Z.M., and Wang, W.H., 2014, The effect of nitrogen doping on mercury oxidation/chemical adsorption on the $CuCo_2O_4$ (110) surface: A molecular-level description: Physical Chemistry Chemical Physics, v. 16, p. 13508–13516. doi:10.1039/C4CP01362J

Meng, S.L., Duan, Y.F., Huang, Z.J., Wang, Y.J., and Yang, L.G., 2011, Experimental study on characteristics of mercury removal by ESP and FF fly ash: Boiler Technology, v. 42, p. 70–74.

Monahan-Pendergast, M.T., Przybylek, M., Lindblad, M., and Wilcoxet, J., 2008, Theoretical predictions of arsenic and selenium species under atmospheric conditions: Atmospheric Environment, v. 42, p. 2349–2357. doi:10.1016/j.atmosenv.2007.12.028

National Bureau of Statistics of People's Republic of China, 2015, China statistical yearbook, March 2015, China Statistics Press: Beijing, China.

Negreira, A.S., and Wilcox, J., 2013, Role of WO_3 in the Hg oxidation across the V_2O_5–WO_3–TiO_2 SCR catalyst: A DFT study: The Journal of Physical Chemistry C, v. 117, p. 24397–24406. doi:10.1021/jp407794g

Negreira, A.S., and Wilcox, J., 2014, Uncertainty analysis of the mercury oxidation over a standard SCR catalyst through a lab-scale kinetic study: Energy & Fuels, v. 29, p. 369–376. doi:10.1021/ef502096r

Ng, J.C., Wang, J.P., and Shraim, A., 2003, A global health problem caused by arsenic from natural sources: Chemosphere, v. 52, p. 1353–1359. doi:10.1016/S0045-6535(03)00470-3

Pan, W.G., Wu, J., Wang, W.H., He, P., Zhang, Y.D., Leng, X.F., and Shen, M.Q., 2009, Study on the effect of NH$_4$Cl addition on Hg and NO produced by coal combustion: Proceedings of the CSEE, v. 29, p. 41–46.

Pavlish, J.H., Sondreal, E.A., Mann, M.D., Olson, E.S., Galbreath, K.C., Laudal, D.L., and Benson, S.A., 2003, Status review of mercury control options for coal-fired power plants: Fuel Processing Technology, v. 82, p. 89–165. doi:10.1016/S0378-3820(03)00059-6

Poulston, S., Granite, E.J., Pennline, H.W., Hamilton, H., and Smith, A.W.J., 2011, Palladium based sorbents for high temperature arsine removal from fuel gas: Fuel, v. 90, p. 3118–3121. doi:10.1016/j.fuel.2011.05.012

Prestbo, E.M., and Bloom, N.S., 1995, Mercury speciation adsorption (MESA) method for combustion flue gas: Methodology, artifacts, intercomparison, and atmospheric implications: Water Air & Soil Pollution, v. 80, p. 145–158. doi:10.1007/BF01189663

Presto, A.A., and Granite, E.J., 2006, Survey of catalysts for oxidation of mercury in flue gas: Environmental Science & Technology, v. 40, p. 5601–5609. doi:10.1021/es060504i

Qi, C.C., Liu, G.J., Chou, C.L., and Zheng, L.G., 2008, Environmental geochemistry of antimony in Chinese coals: Science of the Total Environment, v. 389, p. 225–234. doi:10.1016/j.scitotenv.2007.09.007

Qi, G., Yang, R.T., and Chang, R., 2004, MnO$_x$-CeO$_2$ mixed oxides prepared by co-precipitation for selective catalytic reduction of NO with NH$_3$ at low temperatures: Applied Catalysis B: Environmental, v. 51, p. 93–106. doi:10.1016/j.apcatb.2004.01.023

Qi, Q.J., Liu, J.Z., Cao, X.Y., Zhou, J.H., Zhang, S.X., and Cen, K.F., 2002b, Restraining of fluorine emission by blending CaO or lime with coal during coal combustion: Journal of Fuel Chemistry and Technology, v. 30, p. 204–208.

Qi, Q.J., Liu, J.Z., Cao, X.Z., Zhou, J.H., and Cen, K.F., 2002a, Fluorine distribution characteristics in coal and behavior of luorine during coal combustion: Journal of Chemical Industry and Engineering (China), v. 53, p. 572–577.

Qi, Q.J., Wu, X., Liu, J.Z., Yu, H., Zhou, J.H., and Cen, K.F., 2005a, Experimental on fluorine emission characteristics and influence factors during coal combustion (□): Journal of Liaoning Technical University, v. 24, p. 465–468.

Qi, Q.J., Wu, X., Liu, J.Z., Yu, H., Zhou, J.H., and Cen, K.F., 2005b, Experimental research on fluorine emission characteristics and influence factors during coal combustion (□): Journal of Liaoning Technical University, v. 24, p. 625–628.

Qi, Q.J., Yu, G.S., Zhang, H., Liu, J.Z., and Cen, K.F., 2006, Experimental research on fluorine retention of alkaline-earth metal compounds during coal combustion: Journal of Liaoning Technical University, v. 25, p. 801–803.

Ren, D.Y., Zhao, F.H., Dai, S.F., Zhang, J.Y., and Luo, K.L., 2006a, Geochemistry of trace elements in coals, Science Press: Beijing, China.

Ren, J., Zhou, J., Luo, Z., and Cen, K., 2002, Study of mercury emission during coal combustion: Acta Scientiae Circumstantiae, v. 22, p. 289–293.

Ren, J.L., Zhou, J.S., Luo, Z.Y., Xu, Z., and Zhang, X.M., 2006b, Ca-based sorbents for mercury vapor removal from flue gas: Journal of Fuel Chemistry & Technology, v. 34, p. 557–561.

Rizeq, R.G., Hansell, D.W., and Seeker, R.W., 1994a, Predictions of metals emissions and partitioning in coal-fired combustion systems: Fuel, v. 39, p. 219–233.

Rizeq, R.G., Hansell, D.W., and Seeker, W.R., 1994b, Predictions of metal emissions and partitioning in coal-fired combustion systems: Fuel Processing Technology, v. 39, p. 219–236. doi:10.1016/0378-3820(94)90181-3

Rupp, E.C., Granite, E.J., and Stanko, D.C., 2013, Laboratory scale studies of Pd/γ-Al$_2$O$_3$ sorbents for the removal of trace contaminants from coal-derived fuel gas at elevated temperatures: Fuel, v. 108, p. 131–136. doi:10.1016/j.fuel.2010.12.013

Rupp, E.C., and Wilcox, J., 2014, Mercury chemistry of brominated activated carbons–packed-bed breakthrough experiments: Fuel, v. 117, p. 351–353. doi:10.1016/j.fuel.2013.09.017

Sakulpitakphon, T., Hower, J.C., Trimble, A.S., Schram, W.H., and Thomas, G.A., 2003, Arsenic and mercury partitioning in fly ash at a kentucky power plant: Energy & Fuels, v. 17, p. 1028–1033.

Scala, F., Chirone, R., and Lancia, A., 2011, Elemental mercury vapor capture by powdered activated carbon in a fluidized bed reactor: Fuel, v. 90, p. 2077–2082. doi:10.1016/j.fuel.2011.02.042

Seames, W.S., and Wendt, J.O.L., 2000, Partitioning of arsenic, selenium, and cadmium during the combustion of pittsburgh and illinois #6 coals in a self-sustained combustor: Fuel Processing Technology, v. 63, p. 179–196.

Senior, C., Blythe, G., and Chu, P., 2011, Multi-media emissions of selenium from coal-fired electric utility bolilers: ADA-ES Inc Report , Arlington.

Senior, C.L., 2006, Oxidation of mercury across selective catalytic reduction catalysts in coal-fired power plants: Journal of the Air & Waste Management Association, v. 56, p. 23–31. doi:10.1080/10473289.2006.10464437

Senior, C.L., Sarofim, A.F., Zeng, T., Helble, J.J., and Mamani-Paco, R., 2000, Gas-phase transformations of mercury in coal-fired power plants: Fuel Processing Technology, v. 63, p. 197–213.

Seredin, V.V., 2012, From coal science to metal production and environmental protection: A new story of success: International Journal of Coal Geology, v. 90–91, p. 1–3. doi:10.1016/j.coal.2011.11.006

Seredin, V.V., and Dai, S., 2012, Coal deposits as potential alternative sources for lanthanides and yttrium: International Journal of Coal Geology, v. 94, p. 67–93. doi:10.1016/j.coal.2011.11.001

Seredin, V.V., and Dai, S., 2014, The occurrence of gold in fly ash derived from high-Ge coal: Mineralium Deposita, v. 49, p. 1–6. doi:10.1007/s00126-013-0497-9

Seredin, V.V., Dai, S., Sun, Y., and Chekryzhov, I.Y., 2013, Coal deposits as promising sources of rare metals for alternative power and energy-efficient technologies: Applied Geochemistry, v. 31, p. 1–11. doi:10.1016/j.apgeochem.2013.01.009

Shao, L., Jones, T., Gayer, R., Dai, S., Li, S., Jiang, Y., and Zhang, P., 2003, Petrology and geochemistry of the high-sulphur coals from the upper Permian carbonate coal measures in the Heshan coalfield, southern China: International Journal of Coal Geology, v. 55, p. 1–26. doi:10.1016/S0166-5162(03)00031-4

Shelef, M., 1995, Selective catalytic reduction of NO_x with N-Free reductants: Chemical Reviews, v. 95, p. 209–225. doi:10.1021/cr00033a008

Sjostrom, S., Durham, M., Bustard, C.J., and Martin, C., 2010, Activated carbon injection for mercury control: Overview: Fuel, v. 89, p. 1320–1322. doi:10.1016/j.fuel.2009.11.016

Sliger, R.N., Going, D.J., and Kramlich, J.C., 1998, Kinetic investigation of the high-temperature oxidation of mercury by chlorine species: Proceedings of Western States Section, Seattle, WA.

Song, C., 1989, A brief description of the Yutangba sedimentary type selenium mineralized area in southwestern Hubei: Mineral Deposits, v. 3, p. 83–89.

Stanislav, V.V., 1994, Trace elements in solid waste products from coal burning at some Bulgarian thermoelectric power: Fuel, v. 73, p. 367–374. doi:10.1016/0016-2361(94)90089-2

Tang, Q., Liu, G., Yan, Z., and Sun, R., 2012, Distribution and fate of environmentally sensitive elements (arsenic, mercury, stibium and selenium) in coal-fired power plants at Huainan, Anhui, China: Fuel, v. 95, p. 334–339. doi:10.1016/j.fuel.2011.12.052

Taylor, S.R., and McLennan, S.M., 1985, The continental crust: Its composition and evolution, Blackwell, Oxford.

Tian, C., Zhang, J.Y., Gupta, R., and Zhao, Y.C., 2016, Emissions of arsenic in fine particles generated from a typical high arsenic coal in high temperature: Energy & Fuel, v. 30, p. 6201–6209. doi:10.1021/acs.energyfuels.6b00279

Tian, H.Z., Cheng, K., Wang, Y., Zhao, D., Lu, L., Jia, W.X., and Hao, J.M., 2012, Temporal and spatial variation characteristics of atmospheric emissions of Cd, Cr, and Pb from coal in China: Atmospheric Environment, v. 50, p. 157–163. doi:10.1016/j.atmosenv.2011.12.045

Tian, H.Z., Lu, L., Hao, J.M., Gao, J.J., Cheng, K., Liu, K.Y., Qiu, P.P., and Zhu, C.Y., 2013, A review of key hazardous trace elements in Chinese coals: Abundance, occurrence, behavior during coal combustion and their environmental impacts: Energy & Fuels, v. 27, p. 601–614. doi:10.1021/ef3017305

Tian, H.Z., Qu, Y.P., Wang, Y., Pan, D., and Wang, X.C., 2009, Atmospheric selenium emission inventories from coal combustion in China in 2005: China Environmental Science, v. 29, p. 1011–1015.

Tian, H.Z., Wang, Y., Xue, Z.G., and Cheng, K., 2010, Trend and characteristics of atmospheric emissions of Hg, As, and Se from coal combustion in China, 1980–2007: Atmospheric Chemistry & Physics, v. 10, p. 11905–11919. doi:10.5194/acp-10-11905-2010

Tian, H.Z., Wang, Y., Xue, Z.G., Qu, Y., Chai, F.H., and Hao, J., 2011, Atmospheric emissions estimation of Hg, As, and Se from coal-fired power plants in China, 2007: Science of the Total Environment, v. 409, p. 3078–3081. doi:10.1016/j.scitotenv.2011.04.039

Tronconi, E., Forzatti, P., Martin, J.P.G., and Mallogi, S., 1992, Selective catalytic removal of NO_x: A mathematical model for design of catalyst and reactor: Chemical Engineering Science, v. 47, p. 2401–2406. doi:10.1016/0009-2509(92)87067-Z

Vejahati, F., Xu, Z., and Gupta, R., 2010, Trace elements in coal: Associations with coal and minerals and their behavior during coal utilization – A review: Fuel, v. 89, p. 904–911. doi:10.1016/j.fuel.2009.06.013

Wang, C., Liu, X., Xu, Y., Wu, J., Wang, J., Xu, M., Lin, X., Li, H., and Xia, Y., 2013a, Distribution characteristics of minor and trace elements in fine particulate matters from a 660MW coal-fired boiler: CIESC Journal, v. 64, p. 2975–2981.

Wang, C., Liu, X.W., Li, D., Si, J.P., Zhao, B., and Xu, M.H., 2015a, Measurement of particulate matter and trace elements from a coal-fired power plant with electrostatic precipitators equipped the low temperature economizer: Proceedings of the Combustion Institute, v. 35, p. 2793–2800. doi:10.1016/j.proci.2014.07.004

Wang, G., Luo, Z., Zhang, J., and Zhao, Y., 2015b, Modes of occurrence of fluorine by extraction and SEM method in a coal-fired power plant from Inner Mongolia, China: Minerals, v. 5, p. 863–869. doi:10.3390/min5040530

Wang, H.W., Qi, Q.J., Wang, J.R., Liu, J.Z., Zhou, J.H., and Cen, K.F., 2014, Kinetic mechanics of thereaction between CaO and HF during coal combustion: Journal of China Coal Society, v. 39, p. 161–165.

Wang, L., Ju, Y.W., Liu, G.J., Chou, C.L., Zheng, L.G., and Qi, C.C., 2010a, Selenium in Chinese coals: Distribution, occurrence, and health impact: Environmental Earth Sciences, v. 60, p. 1641–1651. doi:10.1007/s12665-009-0298-8

Wang, M., Yang, N., Zhu, J., and Zheng, B., 2008a, Estimation of arsenic emission from coal combustion in China: Coal Conversion, v. 31, p. 1–5.

Wang, M.S., Zheng, B.S., Wang, B.B., Li, S.H., Wu, D.S., and Hu, J., 2006, Arsenic concentrations in Chinese coals: Science of the Total Environment, v. 357, p. 96–102. doi:10.1016/j.scitotenv.2005.04.045

Wang, Q., Duan, Y.F., Wu, C.J., Yang, L.G., Wang, Y.J., and Jiang, Y.M., 2008b, Demercurization property of flue gas desulphurization system in coal fired power plants: Boiler Technology, v. 39, p. 69–74.

Wang, Q., Liu, Y., Jia, X., Liu, J., Zhang, J., and Qiu, J., 2003, Experimental study on mercury partitioning behavior during coal comobustion: Coal Conversion, v. 26, p. 67–70.

Wang, Q., Shao, Q., and Zhou, C., 1998, Grain size distribution of 16 trace elements in fly ash of burning coal: Environmental Pollution & Control, v. 5, p. 37–41.

Wang, S., Zhang, L., Zhao, B., Meng, Y., and Hao, J., 2012, Mitigation potential of mercury emissions from coal-fired power plants in China: Energy & Fuels, v. 26, p. 4635–4642. doi:10.1021/ef201990x

Wang, S.M., Zhang, Y.S., Gu, Y.Z., Wang, J.W., Liu, Z., Zhang, Y., Cao, Y., Romero, C.E., and Pan, W.P., 2016, Using modified fly ash for mercury emissions control for coal-fired power plant applications in China: Fuel, v. 181, p. 1230–1237. doi:10.1016/j.fuel.2016.02.043

Wang, S.X., Zhang, L., Li, G.H., Hao, J.M., Pirrone, N., Sprovieri, F., and Ancora, M.P., 2010b, Mercury emission and speciation of coal-fired power plants in China: Atmospheric Chemistry and Physics, v. 10, p. 1183–1192. doi:10.5194/acp-10-1183-2010

Wang, Y.J., and Duan, Y.F., 2011, Effect of manganese ions on the structure of $Ca(OH)_2$ and mercury adsorption performance of $Mn^{x+}/Ca(OH)_2$ composites: Energy & Fuels, v. 25, p. 1553–1558. doi:10.1021/ef200113t

Wang, Z., Xue, J.M., Xu, Y.Y., Wang, H.L., and Liu, J., 2013b, Research on influencing factors of SCR's cooperative control in mercury emissions from coal-fired flue: Proceedings of the CSEE, v. 33, p. 32–37.

Wilcox, J., Rupp, E., Ying, S.C., Lim, D.-H., Negreira, A.S., Kirchofer, A., Feng, F., and Lee, K., 2012, Mercury adsorption and oxidation in coal combustion and gasification processes: International Journal of Coal Geology, v. 90, p. 4–20. doi:10.1016/j.coal.2011.12.003

Wilcox, J., 2014, Atomistic-level models, *in* Senior, c. and Granite E. J., Eds., Mercury control: For coal-derived gas streams, p. 389–412, Wiley-VCH Verlag GmbH & Co. KGaA, Weinheim, Germany.

Wu, D.S., Zheng, B.S., Tang, X.Y., Li, S.H., Wang, B.B., and Wang, M.S., 2004, Fluorine in Chinese coals: Fluorine, v. 37, p. 125–132.

Wu, J., Pan, W., Ren, J., He, P., Wang, W., Shen, M., Leng, X., Du, Y., and Jin, Y., 2009, Mercury speciation distribution in flue gas and the influence of chloride additive on it: Journal of Power Engineering, v. 29, p. 405–408.

Wu, J.P., Yan, J., Liu, G.J., and Zheng, L.G., 2005, Advance of research on the distribution, mode of occurrence, and enrichment factors of chromium in Chinese coals: Bulletin of Mineralogy, Petrology and Geochemistry, v. 24, p. 239–244.

Xu, L., Cheng, J., and Zeng, H., 2004, Experimental investigation of the release characteristics of trace elements As, Cd and Cr during the combustion of coal: Journal of Engineering for Thermal Energy and Power, v. 19, p. 478–482.

Xu, S., Qin, S., Huang, Y., Bao, Z., and Hu, S., 2013, Se capture by a Cao-Zno composite sorbent during the combustion of se-rich stone coal: Energy & Fuels, v. 27, p. 6880–6886. doi:10.1021/ef4013449

Xu, W., Zeng, R., Ye, D., and Querol, X., 2005, Distributions and environmental impacts of selenium in wastes of coal from a power plant: Environmental Science, v. 26, p. 64–68.

Xu, Y.Y., Xue, J.M., Wang, H.L., Li, B., Guan, Y.M., and Liu, J., 2014, Research on mercury collaborative control by conventional pollutants purification facilities of coal-fired power plants: Proceedings of the CSEE, v. 34, p. 3924–3931.

Yan, R., Gauthier, D., and Flamant, G., 2001, Volatility and chemistry of trace elements in a coal: Fuel, v. 80, p. 2217–2226. doi:10.1016/S0016-2361(01)00105-3

Yang, J., Zhao, Y., Zhang, S., Liu, H., Chang, L., Ma, S., Zhang, J., and Zheng, C., 2017a, Mercury removal from flue gas by magnetospheres present in fly ash: Role of iron species and modification by HF: Fuel Processing Technology, v. 167, p. 263–270. doi:10.1016/j.fuproc.2017.07.016

Yang, J.P., Zhao, Y.C., Chang, L., Zhang, J.Y., and Zheng, C.G., 2015, Mercury adsorption and oxidation over cobalt oxide loaded magnetospheres catalyst from fly ash in oxyfuel combustion flue gas, Environmental Science & Technology, v. 49, p. 8210–8218. doi:10.1021/acs.est.5b01029

Yang, J.P., Zhao, Y.C., Zhang, J.Y., and Zheng, C.G., 2014, Regenerable cobalt oxide loaded magnetosphere catalyst from fly ash for mercury removal in coal combustion flue gas: Environmental Science & Technology, v. 48, p. 14837–14843. doi:10.1021/es504419v

Yang, J.P., Zhao, Y.C., Zhang, J.Y., and Zheng, C.G., 2016a, Removal of elemental mercury from flue gas by recyclable $CuCl_2$ modified magnetospheres catalyst from fly ash. Part 1. Catalyst characterization and performance evaluation: Fuel, v. 164, p. 419–428. doi:10.1016/j.fuel.2015.08.012

Yang, J.P., Zhao, Y.C., Zhang, J.Y., and Zheng, C.G., 2016b, Removal of elemental mercury from flue gas by recyclable $CuCl_2$ modified magnetospheres catalyst from fly ash. Part 2. Identification of involved reaction mechanism: Fuel, v. 167, p. 366–374. doi:10.1016/j.fuel.2015.11.003

Yang, J.P., Zhao, Y.C., Zhang, J.Y., and Zheng, C.G., 2016c, Removal of elemental mercury from flue gas by recyclable $CuCl_2$ modified magnetospheres catalyst from fly ash. Part 3. Regeneration performance in realistic flue gas atmosphere: Fuel, v. 173, p. 1–7. doi:10.1016/j.fuel.2015.12.077

Yang, L., Duan, Y., Wang, Y., Jiang, Y., Yang, X., and Zhao, C., 2008, Influence of boiler capacities on enrichment law of mercury: Journal of Power Engineering, v. 28, p. 302–307.

Yang, X., Duan, Y., Zhuo, Y., Yang, L., Chen, L., Li, Y., Zhang, L., Yang, L., Shen, X., and Xu, X., 2006, The characteristic of mercury sepciation around ESP in a coal fired power plant: Journal of Hohai University, v. 20, p. 39–42.

Yang, Y., Liu, J., Shen, F., Zhao, L., Wang, Z., and Long, Y., 2016d, Kinetic study of heterogeneous mercury oxidation by HCl on fly ash surface in coal-fired flue gas: Combustion and Flame, v. 168, p. 1–9. doi:10.1016/j.combustflame.2016.03.022

Yang, Y., Liu, J., Wang, Z., and Zhang, Z., 2017b, Homogeneous and heterogeneous reaction mechanisms and kinetics of mercury oxidation in coal-fired flue gas with bromine addition: Proceedings of the Combustion Institute, v. 36, p. 4039–4049. doi:10.1016/j.proci.2016.08.068

Yao, H., Luo, G., and Xu, M., 2006, Mercury emissions and species during combustion of coal and waste: Energy & Fuels, v. 20, p. 1946–1950. doi:10.1021/ef060100b

Yi, H.H., Hao, J.M., Duan, L., Tang, X.L., Ning, P., and Li, X.H., 2008, Fine particle and trace element emissions from an anthracite coal-fired power plant equipped with a baghouse in China: Fuel, v. 87, p. 2050–2057. doi:10.1016/j.fuel.2007.10.009

Yu, J., Feng, F., Wang, W., Luo, K., Chen, D., Bai, G., Li, Y., Zheng, L., Bai, A., and Li, Y., 2004, Regularity of flourine release from fuorine-rich coal combustion in the fluorine poisoning area: Environmental Science, v. 25, p. 43–46.

Yu, L., Li, C., Han, Z., Liu, J., and Jiang, L., 2015, Mercury emission characteristics from coal-fird power plants based on field tests: Environmental Engeineering, v. 33, p. 136–139.

Yu, M., Dong, Y., Wang, P., and Ma, C.Y., 2012, Progress of effects of chloride on mercury removal for coal-fired power plants: Chemical Industry and Engineering Progress, v. 31, p. 1610–1614.

Zeng, R., Zhuang, X., Koukouzas, N., and Xu, W., 2005, Characterization of trace elements in sulphur-rich late permian coals in the Heshan coal field, Guangxi, south China: International Journal of Coal Geology, v. 61, p. 87–95. doi:10.1016/j.coal.2004.06.005

Zhang, D., and Xie, X., 2014, Distribution features and emission characteristics of mercuryn in a Nanjing coal-fired power plant: The Administration and Technique of Environmental Monitoring, v. 26, p. 64–67.

Zhang, J., Lu, J., Yu, L., Wang, S., and Zhang, B., 2003, Distribution of trace elements in coal combustion with low temperature: Journal of Engineering Thermophysics, v. 24, p. 531–533.

Zhang, J.Y., Zhao, Y.C., Ding, F., Zeng, H.C., and Zheng, C.G., 2007, Preliminary study of trace element emissions and control during coal combustion: Frontiers of Energy and Power Engineering in China, v. 1, p. 273–279. doi:10.1007/s11708-007-0038-2

Zhang, J.Y., Zheng, C.G., Ren, D.Y., Chou, C.L., Liu, J., Zeng, R.S., Wang, Z.P., Zhao, F.H., and Ge, Y.T., 2004, Distribution of potentially hazardous trace elements in coals from Shanxi province, China: Fuel, v. 83, p. 129–135. doi:10.1016/S0016-2361(03)00221-7

Zhang, L., Zhuo, Y.Q., Du, W., Tao, Y., Chen, C.H., and Xu, X.C., 2012, Hg removal characteristics of noncarbon sorbents in a fixed-bed reactor: Industrial & Engineering Chemistry Research, v. 51, p. 5292–5298. doi:10.1021/ie202750c

Zhang, Y., Nakano, J., Liu, L., Wang, X., and Zhang, Z., 2015a, Trace element partitioning behavior of coal gangue-fired CFB plant: Experimental and equilibrium calculation: Environmental Science and Pollution Research International, v. 22, p. 15469–15478. doi:10.1007/s11356-015-4738-6

Zhang, Y., Wang, C.B., Li, W.H., Liu, H.M., Zhang, Y.S., Hack, P., and Pan, W.P., 2015b, Removal of gas-phase As2O3 by metal oxide adsorbents: Effects of experimental conditions and evaluation of adsorption mechanism: Energy & Fuels, v. 29, p. 6578–6585. doi:10.1021/acs.energyfuels.5b00948

Zhao, B., Liu, X.-W., Zhou, Z.-J., Shao, H.-Z., Wang, C., and Xu, M.-H., 2015, Mercury oxidized by V2O5–MoO3/TiO2 under multiple components flue gas: An actual coal-fired power plant test and a laboratory experiment: Fuel Processing Technology, v. 134, p. 198–204. doi:10.1016/j.fuproc.2015.01.034

Zhao, S.L., Duan, Y.F., Tan, H.Z., Liu, M., Wang, X.B., Wu, L.T., Wang, C.P., Lu, J.H., Yao, T., She, M., and Tang, H.J., 2016, Migration and emission characteristics of trace elements in a 660 MW coal-fired power plant of China: Energy & Fuels, v. 30, p. 5937–5944. doi:10.1021/acs.energyfuels.6b00450

Zhao, Y., Xue, F.M., Dong, L.Y., and Shao, Y., 2013a, Flue gas mercury removal technology for coal fired boiler: Thermal Power Generation, v. 42, p. 59–62.

Zhao, Y., Zhang, J., Huang, W., Wang, Z., Li, Y., Song, D., Zhao, F., and Zheng, C., 2008, Arsenic emission during combustion of high arsenic coals from southwestern Guizhou, China: Energy Conversion and Management, v. 49, p. 615–624. doi:10.1016/j.enconman.2007.07.044

Zhao, Y.C., Zhang, J.Y., Liu, J., Diaz-Somoano, M., Martinez-Tarazona, M.R., and Zheng, C.G., 2010a, Study on mechanism of mercury oxidation by fly ash from coal combustion: Chinese Science Bulletin, v. 55, p. 163–167. doi:10.1007/s11434-009-0567-7

Zhao, Y.C., Zhang, J.Y., Liu, J., Diaz-Somoano, M., Martinez-Tarazona, M.R., and Zheng, C.G., 2010b, Experimental study on fly ash capture mercury in flue gas: Science China Technological Sciences, v. 53, p. 976–983. doi:10.1007/s11431-009-0367-y

Zhao, Y.C., Zhang, J.Y., and Zheng, C.G., 2013b, Release and removal using sorbents of chromium from a high-Cr lignite in Shenbei coalfield, China: Fuel, v. 109, p. 86–93. doi:10.1016/j.fuel.2012.09.049

Zheng, B., Ding, Z.H., Huang, R.G., Zhu, J.M., Yu, X.Y., Wang, A.M., Zhou, D.X., Mao, D.J., and Su, H.C., 1999, Issues of health and disease relating to coal use in southwestern China: International Journal of Coal Geoglogy, v. 40, p. 119–132. doi:10.1016/S0166-5162(98)00064-0

Zheng, B., Hong, Y., Zhao, W., Zhou, H., and Xia, W., 1992, Se-rich carbonaceous-siliceous rocks of west Hubei and local Se poisoning: Chinese Science Bulletin, v. 37, p. 1027–1029.

Zheng, B., Yan, L., Mao, D., and Thornton, I., 1993, The Se resource in southwestern Hubei province, China, and its exploitation strategy: Journal of Natural Resource, v. 8, p. 204–212.

Zheng, C., Xu, M., Zhang, J., and Liu, J., 2002, Emission and control of trace elements during coal combustion: Hubei Science and Technology Press, p. 63, Wuhan, China.

Zheng, C., Zhang, J., Zhao, Y., Liu, J., and Guo, X., 2010, Emission and control of mercury from coal combustion: Science Press: Beijing, China.

Zhong, L., Xiao, P., Jiang, J.Z., Guo, T., and Mei, Z.F., 2016, Experimental study on mercury removal by oxidation in a coal-fired boiler: Thermal Power Generation, v. 45, p. 52–58.

Zhou, J., Luo, Z., Zhu, Y., and Fang, M., 2013, Mercury emission and its control in Chinese coal-fired power plants: Zhejiang University Press, p. 30–31, Hangzhou, China.

Zhou, J., Wang, G., Luo, Z., and Cen, K., 2006, An experimental study of mercury emission from a 600 MW pulverized coal-fired boiler: Journal of Engineering for Thermal Energy and Power, v. 21, p. 569–572.

Zhou, J., Wu, X., Gao, H., Luo, Z., and Cen, K., 2004, Experimental study on mercury emission and control for CFB boilers: Thermal Power Generation, v. 33, p. 72–75.

Zhou, J., Zhang, L., Luo, Z., Hu, C., He, S., Zheng, J., and Cen, K., 2008, Study on mercury emission and its control for boiler of 300MW unit: Thermal Power Generation, v. 37, p. 22–27.

Zhou, Z.J., Liu, X.W., Zhao, B., Chen, Z.G., Shao, H.Z., Wang, L.L., and Xu, M.H., 2015, Effects of existing energy saving and air pollution control devices on mercury removal in coal-fired power plants: Fuel Processing Technology, v. 131, p. 99–108. doi:10.1016/j.fuproc.2014.11.014

Zhu, J.M., Wang, N., Li, S.H., Li, L., Su, H.C., and Liu, C.X., 2008, Distribution and transport of selenium in Yutangba, China: Impact of human activities: Science of the Total Environment, v. 392, p. 252–261. doi:10.1016/j.scitotenv.2007.12.019

Zhu, Z., Tan, Y., Zheng, J., Zhang, C., Li, Y., Zhang, D., and Wang, Q., 2003, Study on chromium existence and distribution in a 300MW pulverized coal fired utility boiler: Proceedings of CSEE, v. 23, p. 167–171.

Zhu, Z., Xue, L., Tan, Y., Zhang, C., Li, Y., Zhang, D., Wang, Q., Pan, L., and Ke, J., 2001, Studies characteristics of mercury distribution in combustion products at various loads of a P.C. fired utility boiler: Proceedings of CSEE, v. 21, p. 87–90.

Zhuang, X., Querol, X., Alastuey, A., Juan, R., Plana, F., Lopez-Soler, A., Du, G., and Martynov, V.V., 2006, Geochemistry and mineralogy of the Cretaceous Wulantuga high germanium coal deposit in Shengli coal field, Inner Mongolia, northeastern China: International Journal of Coal Geology, v. 66, p. 119–136. doi:10.1016/j.coal.2005.06.005

Zhuang, Y., Thompson, J.S., Zygarlicke, C.J., and Pavlish, J.H., 2007, Impact of calcium chloride addition on mercury transformations and control in coal flue gas: Fuel, v. 86, p. 2351–2359. doi:10.1016/j.fuel.2007.02.016

ARTICLE

A review on the applications of coal combustion products in China

Jing Li, Xinguo Zhuang, Xavier Querol, Oriol Font and Natalia Moreno

ABSTRACT

Owing to the continuous high demand for energy, huge amounts of coals are consumed every year in China, giving rise to large volumes of coal combustion products (CCPs) from coal-fired power plants. Fly ash (FA), bottom ash (BA), boiler slag (BS), and flue gas desulphurization (FGD) gypsum are the primary CCPs generated. The disposal of these CCPs may occupy large areas of useful cultivated land, and cause serious environmental problems. However, these products might be also utilized as recoverable resources. Therefore, it is of economic and environmental significance to carry out research on the utilization of these CCPs. The present review first describes the physicochemical, mineralogical, and environmental geochemical properties of FA, slag, and FGD gypsum. Then the authors focus on the current and potential high value-added applications for these products in China. The utilization of FA for concrete and cement, soil amendment and fertilizer, in the ceramic industry, for catalysis, as adsorbents for the removal of flue gas, heavy metals, dyes and organic compounds, for zeolite and geopolymer synthesis, for recovery of valuable metals, and for recovery of unburned carbon and cenospheres, is discussed. The utilization of slag, such as reclamation of the burnable carbon, use in concrete, cement, and building materials, for roadway pavement and waste-water treatment, and for the production of acoustic barriers is reviewed as well. The current utilization of FGD gypsum includes use as a cement retarder, for the production of building plaster (β-hemihydrated gypsum) and calcium sulphate whiskers (α-hemihydrated gypsum), the production of fire-resistant panels, and use as a fertilizer and soil amendment agent. Furthermore, the possible influence of CCP properties on their utilization, and the advantages and disadvantages of various applications are discussed in this review. Finally, new directions for the future utilization prospects of CCPs in China are proposed.

1. Introduction

When coal is used as fuel in most modern coal-fired power stations, it is crushed (or pulverized in pulverized-fuel power stations), and blown into a combustion chamber where it immediately ignites and burns to heat boiler tubes (Figure 1). The inorganic components of the coal either remain in the combustion chamber or are carried away by the flue gas stream and retained in emission control systems to form materials that are commonly known as coal combustion products (CCPs) (ADAA 2016). Owing to the mineral components of coal and the combustion technique, the CCPs produced are represented by fly ash (FA), bottom ash (BA), boiler slag (BS), and fluidized-bed combustion (FBC) ash, as well as the products from dry or wet flue gas desulphurization (FGD), especially semi-dry absorption (SDA) product and FGD gypsum (ECOBA 2016). As combustion technologies, such as fluidized-bed or fixed-bed combustion processes, are not widely used in China, this article focuses only on CCPs from pulverized-fuel power stations, and does not consider products of other combustion technologies. The major CCPs from pulverized-coal power stations mainly include FA, BA, BS, and FGD gypsum.

FA is retained in the fabric filters or electrostatic/mechanical precipitators as dust-like particles from the flue gases of furnaces fired with coal at 1100–1400°C (Figure 1, ECOBA 2016). FA is a fine powder, which is mainly composed of spherical glassy particles (Figure 2 (a)). As the most important CCP, FA usually accounts for the largest proportion (67–90%) of the total CCPs (Ahmaruzzaman 2010).

BA and BS are the heavier and coarser CCPs collected from the bottom of the furnace (Figures 1 and 2(b)). BA is mainly composed of agglomerated ash particles formed in pulverized-coal furnaces that are too large to be carried in the flue gases and impinge on the furnace walls or fall through open grates to an ash

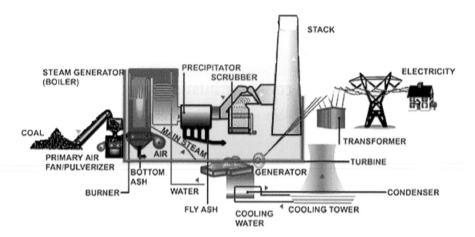

Figure 1. Schematic diagram of coal combustion products (CCPs) generation from typical coal-fired power plants.

a.   b.   c.

Figure 2. Main components of CCPs produced from coal-fired power plants. (a) Fly ash; (b) slag; (c) FGD gypsum.

hopper at the bottom of the furnace (ACAA Glossary of Terms 2003). BA is typically grey to black in colour, is quite angular, and has a porous surface structure. BS is a molten ash collected at the base of the slag tap and cyclone furnaces that is quenched with water and shatters into black, angular particles having a smooth, glassy appearance (ACAA Glossary of Terms 2003).

FGD gypsum is a gypsum-like product arising from the interaction of SO_2 with the $CaCO_3$ added in the wet scrubbers, where such units are fitted (Figure 1). It is obtained by the wet desulphurization of flue gas in emission control systems (scrubbers) that remove sulphur oxides from power plant flue gas streams (Figures 1 and 2(c), ECOBA 2016). It is estimated that the wet limestone (or lime) FGD system accounts for about 80% of all of the FGD systems that have been extensively installed in coal-fired power plants in China and elsewhere (Wang and Wu 2004). Because the wet FGD systems are designed to introduce primarily lime or limestone as the reagent sorbents and are generally combined with FA removal facilities, the obtained end product is usually a mixture of gypsum, calcium sulphite ($CaSO_3$), FA, and unreacted lime (CaO), portlandite ($Ca(OH)_2$), or calcite ($CaCO_3$) (TFHRC 2010).

The worldwide annual CCP production was approximately 780 Mt in 2011, and the largest CCP-producing countries or regions were China (395 Mt), followed by the US (118 Mt), India (105 Mt), Europe (EU15) (52.6 Mt), Africa (31.1 Mt), and the Middle East as a minor contributor (Heidrich et al. 2013). Based on the statistics and forecasts, the worldwide CCPs production has been and will keep increasing steadily over the next several decades, due to the continued energy demands from coal combustion.

The utilization of coal resources for energy production promotes economic development, but large volumes of CCPs produced from pulverized-coal combustion (PCC) power plants and the emissions of gaseous pollutants and particulate matter to the atmosphere may cause serious environmental and human health problems. Many of these CCPs, particularly FA, may contain relatively high concentrations of elements of environmental concern. A number of these elements may be present in highly leachable species, increasing their potential environmental and health impacts, especially when large amounts of CCPs are disposed of in landfills. Furthermore, the emplacement of CCPs also involves significant costs and environmental impacts associated with landfill disposal. As a consequence, investigations into the potential utilization and applications of CCPs is environmentally beneficial and of high economic and social significance.

There are several environmental and economic benefits associated with the use of CCPs, such as saving natural resources, saving energy, saving emissions of pollutants, saving greenhouse gas (GHG) emissions, reducing exploitation costs for resources, and saving useful land (ACAA 2016). Nowadays, a large proportion of CCP production has been used for different purposes worldwide (Koukouzas *et al.* 2006; Cox *et al.* 2008; CIRCA 2010; Heidrich *et al.* 2013; ECOBA 2016; among others). However, a substantial proportion of the world's CCPs is still discharged into ash ponds, landfills, and/or lagoons. Thus, according to Heidrich *et al.* (2013), of the 780 Mt of CCPs produced, some 415 Mt or 53% were reported as utilized; however, the CCPs utilization varies widely in different countries. Japan ranked first, with the highest effective CCPs utilization rate of 96.4%, followed by the European Union (EU15) with 90.9%, China with 67.1%, other Asia with 66.5%, Australia with 45.8%, US with 42.1%, and Africa/Middle East with the lowest rate of 10.6% (Heidrich *et al.* 2013).

In the USA, CCPs have been used in a number of applications, with the leading proportion in the construction and civil engineering industries, followed by mining applications and other uses. FA is used as a major component in concrete structures, such as roads and bridges, dams, and buildings. FA is also used in fills (flowable, structural), road base for structure fill, agriculture, mine backfilling, cenospheres for paints/coatings/adhesives, soil stabilization, and Portland cement (Kentucky Ash Education Site 2016). BA is usually used for construction and drainage purposes, and used to build embankments, fill trenches, and backfill foundations. BS is used for sand blasting grit, roofing granules, and landscaping, for roofing tiles/shingles, for surface coatings of asphalt in road paving, as well as for traction control material for icy roads in cold climates. The most common utilization of FGD gypsum is in wallboard manufacture and agriculture.

In the EU 15, CCPs are utilized in a wide range of applications in the building and construction industry (ECOBA 2016), including their use as an additive in concrete as a cement replacement material and as an aggregate or binder in road construction. They can also be utilized as mineral fillers and as fertilizers.

Within the EU, in 2010, the utilization of FA in the construction industry was around 43% and for BA around 46%, whereas the utilization rate for BS was 100% (ECOBA 2016).

In Canada, on average, 31% of the FA, 15% of the BA, and 95% of the FGD gypsum generated were directed to beneficial use over 2004–2006, amounting to an overall 30% use rate (P&U Statistics 2004-2006). As of December 2013, the use rate of FA had climbed only slightly to 32% and FGD gypsum to 100%, but the use rate of BA had decreased to 12%, and the total use rate of all CCPs dropped to 28% (P&U Statistics 2011-2013). FA is used in Canada in cement, concrete, and mining applications and in a growing range of 'non-traditional' applications, including paints and plastics. BA and BS are commonly used in cement and as a road base/sub-base. FGD gypsum is used principally in the manufacture of wallboard.

In Australia, the beneficial use of CCPs during 2015 resulted in 4.8 million tonnes or 40% being effectively utilized in various value-added products or to some beneficial end (ADAA 2016). The main utilization of FA is in cement and concrete or as mineral fillers, although 21% of the effectively utilized coal ash was used in non-cementitious applications such as flowable fills, structural fills, road bases, coarse/fine aggregates, and mine site remediation (ADAA 2016).

As the largest coal producer and consumer, China contributed 47.7% of the global coal production and 50.0% of the global coal consumption in 2015 (BP, 2016). Accordingly, China is the largest CCP-producing country in the world (Heidrich *et al.* 2013). A large portion of the FA produced in China has been used for low value-added applications, such as in cement, grout, brick manufacture, and road/dam construction, as well as for road base pavement, structural fill, backfill and agriculture fertilizer, concrete, and mineral extraction (Wang and Wu 2004; Cao *et al.* 2008; Lu *et al.* 2011). Only a very minor proportion of the total FA produced was used in the manufacture of novel materials (Hui *et al.* 2005; Peng *et al.* 2005; Sheng *et al.* 2003; Hsu *et al.* 2008; Wu *et al.* 2008). Despite these uses, a large amount of the Chinese FA is still dumped into ponds or piled on land (Lu *et al.* 2011). With respect to FGD gypsum, the production is rising rapidly with increasing requirements for SO_2 emission control; however, the use rate is barely reaching 30% (Tian *et al.* 2006). From both environmental and economic points of view, it is therefore imperative to investigate the full and optimal utilization for the CCPs generated in China.

There are several reviews on FA utilization domestically and overseas (Iyer and Scott 2001; Cao *et al.* 2008; Cox *et al.* 2008; Jha *et al.* 2008; González *et al.* 2009; Ahmaruzzaman 2010; Blissett and Rowson 2012; Tharaniyil 2013; Yao *et al.* 2015; among others). Cox *et al.* (2008), Ahmaruzzaman (2010), and Tharaniyil (2013) presented details on the properties of coal FA and its various possible applications. Song *et al.* (2006), Cao *et al.* (2008), and Yao *et al.* (2015) reported the characteristics and comprehensive applications of FA in China. However, the properties and possible utilization of other CCPs, such as BS and FGD gypsum, have not been summarized.

Hence, this review will first summarize the physico-chemical, mineralogical, geochemical, and environmental characteristics of FA, slag, and FGD gypsum in China, and then focus on reviewing and evaluating the various applications of these CCPs in China, including common/traditional utilization as well as novel/advanced applications with high added value. Based on this, this article will finally investigate the future research and prospects for integrated utilization of Chinese CCPs.

2. General characteristics of CCPs

The characteristics of CCPs in terms of chemical composition, mineralogy, geochemistry, and environmental features are of fundamental significance in developing the various applications of CCPs. The specific properties depend on the type of coal used, the percentage of incombustible matter in the coal, the nature of the mineral matter in the coal, sulphur content, the pulverization process, furnace types, the combustion conditions, and the collector setup, among other factors (Creelman et al. 2013; Tharaniyil 2013).

2.1. Characterization of FA

2.1.1. Physical characteristics

FA consists of fine, powdery particles that are predominantly spherical in shape, either solid or hollow, and mostly amorphous in nature (Figure 3(a)). A small proportion of unburned carbon usually also occurs in the FA. The proportions of unburned carbon and iron influence the apparent colour of the FA, which varies from white-grey to grey to yellow to black, or from orange to deep red (Fisher et al. 1978).

The particle size distribution of most coal FAs is generally similar to that of silt, ranging from 1 μm to several hundred μm, but typically less than 75 μm (Song et al. 2006; Cao et al. 2008; Liu 2009; Ahmaruzzaman 2010). FA particles are usually characterized by a lognormal grain size distribution, with polymodal coarse (micron particle size) and/or fine (submicron particle size) modes (Li et al. 2012). Yao et al. (2015) reported an average particle size of <20μm for Chinese FA. The majority of the FA particles are smaller than those in cement. The specific surface area and specific gravity of FA worldwide generally varies from 170 to 1000 m²/kg and from 2.1 to 3.0 g/cm³, respectively (Ahmaruzzaman 2010). By contrast, Chinese FA is usually characterized by a high surface area of 300–500 m²/kg, and the average specific gravity

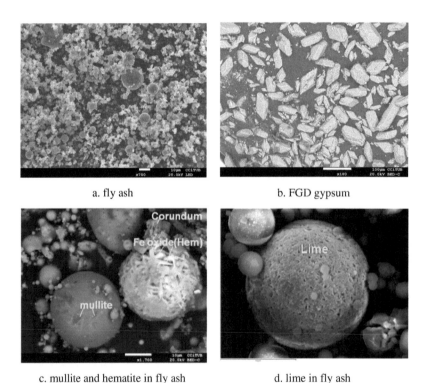

a. fly ash

b. FGD gypsum

c. mullite and hematite in fly ash

d. lime in fly ash

Figure 3. Morphology of fly ash and FGD gypsum and typical mineral phases in fly ash identified by SEM-EDX, BED, SEM back-scattered image. (a) Fly ash; (b) FGD gypsum; (c) mullite and hematite in fly ash; (d) lime in fly ash.

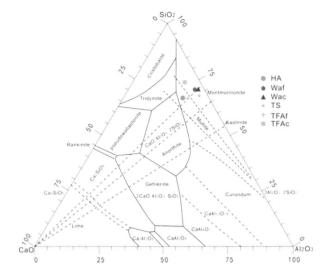

Figure 4. Chemical composition of the glass matrix in five fly ashes from Chinese power plants (modified after Moreno *et al.* 2005).

of Chinese FA is 2.1 g/cm³ (Song *et al.* 2006; Yao *et al.* 2015).

2.1.2. Chemical characteristics

With respect to the chemical composition, the main components of FA are silica, alumina, ferrous, and other metal oxides, as well as calcium, potassium, and sodium oxides, with varying proportions of carbon (Blissett and Rowson 2012). The chemical composition of FAs generated from power plants in China has been reported by several researchers (Ma *et al.* 1999; Liu *et al.* 2004; Koukouzas *et al.* 2006; Song *et al.* 2006; Cao *et al.* 2008; Zhang *et al.* 2010; Qi and Yuan 2011; Li *et al.* 2012; Yan *et al.* 2012; among others). The range for the main chemical components of Chinese FAs is given in Table 1. It is obvious that the content of each metal

oxide varies greatly in different FAs, due to the difference in the source and combustion conditions (Ma *et al.* 1999). However, the contents of metal oxides in FAs are commonly in the decreasing order of SiO_2 >Al_2O_3>Fe_2O_3>CaO>MgO>K_2O>Na_2O>TiO_2.

The American Society for Testing and Materials (ASTMs) classifies FA into C and F Classes based on the combined contents of SiO_2, Al_2O_3, and Fe_2O_3 as well as the loss on ignition (LOI), SO_3, and moisture content (Table 2). The main difference between Class F and Class C FA is in the amount of CaO and the SiO_2, Al_2O_3, and Fe_xO_y in the ash (Ahmaruzzaman 2010). In Class F FA, the total CaO contents typically range from 1 to 12%, whereas those in Class C FA is reported to be as high as 30–40%. Vassilev and Vassileva (2007) present a new classification for FAs and create four tier classification systems (Table 2), which could help in assessing a particular use of FAs in applications other than cement. In China, according to the fineness, water requirement ratio, LOI, moisture, and SO_3 content (GB1596-1991 1991), FA can be classified into three grades for its utilization for cement and concrete (Table 2).

Generally, the chemical properties of FA are influenced to a large extent by the properties of the coal burned and the techniques used for the handling and storage of coals. Concerning the coal rank, there are basically four ranks of coal: anthracite, bituminous, sub-bituminous, and lignite, with different calorific values, moisture contents, and volatile matter yields (Ahmaruzzaman 2010). As shown in Table 1, lignite and sub-bituminous coal FAs usually have higher CaO, MgO, and SO_3 contents, but lower SiO_2 and Al_2O_3 contents and lower LOI than FAs from bituminous and anthracite coals (Ahmaruzzaman 2010; Blissett and Rowson 2012). This may be ascribed to the different mineral matter characteristics and the significant proportions of non-mineral inorganic elements (Ca, Al, Fe,

Table 1. Content range of chemical composition for CCPs generated from power plants in China.

wt, %	Fly ash[a]	Slag[b]	FGD gypsum[c]	Common range of fly ash by coal type[d]		
				Bituminous	Sub-bituminous	Lignite
SiO_2	31.4–60.0	31.8–64.1	0.8–7.2	20–60	40–60	15–45
Al_2O_3	11.2–55.0	1.4–42.5	0.3–3.7	5–35	20–30	10–25
Fe_2O_3	1.5–19.3	0.08–8.5	0.1–0.9	10–40	4–10	4–15
CaO	0.8–31.0	1.4–20.1	25.0–50.0	1–12	5–30	15–40
K_2O	0.7–2.9	0.4–2.5	0.1–0.3	0–3	0–4	0–4
Na_2O	0.2–2.4	0.2–1.1	0.3	0–4	0–2	0–4
SO_3	0.1–3.5	0.1–0.2	24.0–53.0	0–4	0–2	0–6
MgO	0.4–4.8	0.2–2.2	0.1–1.8	0–5	1–6	3–10
P_2O_5	0.3–1.5	0.1–0.2	<dl	–	–	–
MnO	0.05–0.06	0.07–0.2	0.01	–	–	–
TiO_2	0.2–1.5	0.3–0.8	0.07	–	–	–
LOI	2.4–18.3	1.7–7.8	19.2–23.4	0–15	0–3	0–5

[a]Data from Ma *et al.* (1999), Liu *et al.* (2004), Song *et al.* (2006), Cao *et al.* (2008), Qi and Yuan (2011), Li *et al.* (2012) and Yan *et al.* (2012).
[b]Data from Zhang *et al.* (2010), Dai *et al.* (2010a, 2014), Li *et al.* (2012) and Sun *et al.* (2016b).
[c]Data from Yang *et al.* (2011), Li *et al.* (2015) and Zhang *et al.* (2015).
[d]Data from Ahmaruzzaman (2010).

Table 2. Classification for fly ashes based on ASTM C618, chemical composition, and GB1569.

	Classification for fly ashes based on ASTM C618[a]			
Class	$SiO_2+Al_2O_3$ $+Fe_2O_3$ (%)	SO_3 (%)	Moisture (%)	LOI (%)
C	>50, <70	<5	<3	<6
F	>70			<12
	Classification of fly ash based on chemical composition[b]			
Class	$SiO_2+Al_2O_3+K_2O$ $+TiO_2+P_2O_5$ (%)	$CaO+MgO+SO_3$ $+Na_2O+MnO$ (%)		Fe_2O_3 (%)
Sialic	>77	<11.5		<11.5
Calsialic	<89	>11.5		<11.5
Ferrisialic	<89	<11.5		>11.5
Ferricalcsialic	<77	>11.5		>11.5
	Classification for fly ashes based on GB1596-1991[c]			
Class	Fineness (amount retained on 0.045 mm sieve, %)≤	Water requirement ratio, %≤	LOI, %≤	Moisture, %≤	SO$_3$, %≤
I	12	95	5	1	3
II	20	105	8	1	3
III	45	115	15	-	3

[a]ASTM C 618-05 (2005).
[b]Vassilev and Vassileva (2007).
[c]GB1596-1991.

Mg, Ti) of these lower-rank coals, such as brown coals, lignites, and sub-bituminous materials (Ward 2002; Li et al. 2010). Mineral matter in coal encompasses dissolved salts in the pore water, inorganic elements associated with the organic compounds, as well as discrete crystalline and non-crystalline mineral particles. The first two forms of mineral matter are also described as non-mineral inorganics, which are usually prominent constituents in the mineral matter of lower-rank coals. As indicated by Li et al. (2010), significant proportions (>0.05% and up to around 1.5%) of non-mineral inorganic elements (Ca, Mg, Fe, Ti, and Al) occur as minor constituents of the macerals, especially vitrinites, of lower-rank coals, and the proportions of such non-mineral inorganic elements in the vitrinites decrease with coal rank. Unlike the same elements in crystalline mineral particles, which form oxides and similar compounds during combustion, these non-mineral inorganic elements are typically released in a reactive atomic form, contributing significantly to ash formation in lower-rank coals (Ward 2002).

Given the aforementioned main difference, high-Ca Class C FA is normally produced from the burning of low-rank lignites or sub-bituminous coals, whereas low-Ca Class F FA is commonly produced from the burning of higher-rank bituminous coals or anthracites (Wang 2008). Class C FA has cementitious properties (self-hardening when reacted with water), whereas Class F FA is characterized by pozzolanic properties (hardening when reacted with $Ca(OH)_2$ and water), which are siliceous, or siliceous and aluminous, and has no intrinsic cementitious property (Ahmaruzzaman 2010). The different LOI values of FAs depend on different combustion conditions and efficiencies, and the high LOI is indicative of the high occurrence of unburned char particles (Li et al. 2012). Volatilization or decomposition of inorganic species (probably sulphate and carbonate species) may also have an impact on the LOI values. With increase in boiler capacity, coal combustion tends to be more complete, which means that the LOI tends to be lower. Therefore, the coal ash may be regarded as a kind of pure building material that has been burned at a high temperature (Song et al. 2006). However, higher LOI is acceptable or even welcomed when the FA is used in brick manufacturing, since the carbon in the ash could be burnt during the calcination process, thereby saving energy.

The chemical composition of the glass matrix of the FA could be obtained through a mass balance method, combining the chemical and mineralogical data (Moreno et al. 2005; Ward and French 2006). According to Li et al. (2012) and some unpublished data, the chemical composition of glass matrix in five FAs from power plants in Xinjiang and Inner Mongolia is commonly SiO_2 and Al_2O_3, with variable proportions of Ca, Fe, K, S, Ti, and V oxides. The $SiO_2/Al_2O_3/CaO$ ratio in the glass matrix is critical to favour mullite ($3Al_2O_3.2SiO_2$) crystallization (Figure 5). The composition of the FA glass matrix indicates the suitability of the FA for potential applications, such as zeolite synthesis, fire-resistance panel, and ceramics (Cheng and Chen 2004; Leiva et al. 2008; Arenas et al. 2011). High glass content and high SiO_2/Al_2O_3 ratios in the glass make FA suitable for zeolite synthesis (Moreno et al. 2005). The relatively high CaO content in some FAs may be a limitation for zeolite synthesis, but it is regarded as

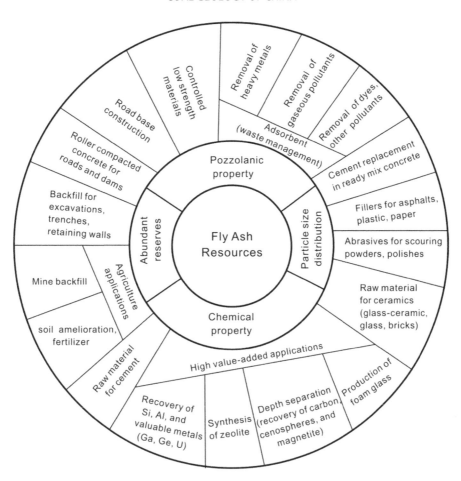

Figure 5. Possible applications of fly ash in China based on the properties of fly ash (modified after Wang and Wu 2006).

an advantage in the production of fire-resistant panels (Leiva *et al.* 2008).

2.1.3. Mineralogical characteristics

In the order of decreasing abundance, FA is made of Si-Al-rich amorphous glass and crystalline mineral matter, as well as organic constituents represented by unburned carbon. With respect to the mineralogical composition, the major mineral phases in most FAs are mullite ($3Al_2O_3.2SiO_2$), quartz (SiO_2), and haematite (Fe_2O_3) (Wang 2008). The less-predominant minerals in FAs are magnetite (Fe_3O_4), anhydrite-gypsum ($CaSO_4$-$CaSO_4.2H_2O$), feldspars ($KAlSi_3O_8$-$NaAlSi_3O_8$-$CaAl_2Si_2O_8$), lime-portlandite (CaO-$Ca(OH)_2$), cristobalite-tridymite, calcite-ankerite, corundum, and some Ca and Ca-Mg silicates (Vassilev and Vassileva 2007). The morphology of typical mineral phases in the FA is shown in Figure 3. Blissett and Rowson (2012) divided FA into four types, namely Pozzolanic (P), Inert (I), Active (A), and Mixed (M), based on the distinct behaviour of glass, quartz +mullite, and the sum of any other mineral-bearing phases such as Fe-Ca-Mg-K-Na-Ti-Mn oxyhydroxides, sulphates, carbonates, and silicates, which would help

simplify the choice of application for each unique composition of FA.

The mineralogical composition of FAs depends on the geological factors related to the formation and deposition of the coal, on the composition of the inorganic fraction of the feed coal as well as on combustion technology and the operational conditions of the power plants (Chinchón *et al.* 1991; Van Dyk *et al.* 2009; Dai *et al.* 2010a, 2014a; Li *et al.* 2012). For instance, the occurrence of mullite is attributed to the melting of clay and other Al-Si minerals when the coal combustion temperature exceeds 1200°C. Quartz may be a relict mineral from the feed coal or partially crystallized from the silicate melt. Haematite in FA is predominantly produced by the decomposition of pyrite, and the Fe-carbonate species present in coal. The occurrence of lime is attributed to the decomposition of feed coal $CaCO_3$ into CaO and CO_2 at around 900°C. It is worth mentioning that a significant proportion of the organically associated Ca may also occur in coal, especially in lower-rank materials. The occurrence of $CaCO_3$ and dolomite may be due to the carbonation of lime Ca(Mg)O or to the relict $CaCO_3$ and dolomite from the feed coal. Anorthite can be formed from the crystallization of the Ca-Si-Al melt, and can also be formed by solid-state reaction processes without melting (Creelman *et al.* 2013). Even so, it is more common in BAs and furnace deposits than in FAs.

2.1.4. Geochemical and environmental characteristics

FA also contains many trace elements, some of which are of high added value (Li, Ga, Ge, V, and U), whereas others, such as As, Ba, Cd, Cr, Cu, Mo, Ni, Pb, Sb, Se, Zn, Hg, Cl, and F, are of potential environmental concern. As a consequence, mobility and leaching studies have also been conducted to assess the probability of these elements migrating to the environment (Jankowski *et al.* 2006; Ram *et al.* 2007; Sarode *et al.* 2010; Izquierdo and Querol 2012; Li *et al.* 2012).

The trace element concentrations in several FAs from different power plants in China have been reported by Liu *et al.* (2005), Koukouzas *et al.* (2006), Lu *et al.* (2009), Dai *et al.* (2010a), Li *et al.* (2012), and Sun *et al.* (2016b), among others. The concentrations of trace elements in such FAs are highly dependent on their bulk concentrations in the coal burned. The trace element concentrations in FA are sometimes 4–10 times higher than those in the parent coal (Yao *et al.* 2015). For instance, the FA generated from the combustion of high-Ge coals from Lincang and Ge-coals in Yunnan, China, are characterized by high Ge concentration (374 mg/kg, Dai *et al.* 2014a; Sun *et al.* 2016b). The FAs from the high-Al-Li Jungar coals are characterized by high Li concentration (453 mg/kg, Dai *et al.* 2010a). Secondarily, the particle size of FA has a strong influence on the concentration and partition of trace elements in FAs. Most of the trace elements are more enriched in the fine fraction of FA rather than the coarse fraction. This may be attributable to the larger surface area of the finer particles of FA, which allows for the condensation/absorption of semi-volatile elements on the surface of fine FA (Querol *et al.* 1995; Dai *et al.* 2010a).

From an environmental point of view, the leachability of trace elements in Chinese FAs has been widely investigated by means of different leaching tests (Wang *et al.* 1999; Ke *et al.* 2005; Wang *et al.* 2008d; Lu *et al.* 2009; Li *et al.* 2012; Yao *et al.* 2015; Wang *et al.* 2016a; among others). Hung *et al.* (2009) indicated that Pb and Cr leaching from large-scale FA landfills may induce serious human health risks. Li *et al.* (2012) obtained the leachable concentration and pH range for some environmentally significant trace elements in several Chinese FAs (Table 3), following the European Standard leaching test EN-12457 (water/solid ratio of 10 L/kg and stirring time of 24 h). The pH values of FAs vary from 1.2 to 12.5 (Kolbe *et al.* 2011), but most Chinese FAs tend towards alkalinity (Li *et al.* 2012). The pH value of the ash-water system depends mainly on the Ca/S molar ratio in the ash, although other minor alkalis or alkaline earth cations may also contribute to the balance (Ward *et al.* 2009; Izquierdo and Querol 2012). Based on the Ca/S ratio and pH value, FA can be classified into three main groups: strongly alkaline ash (pH 11–13), mildly alkaline ash (pH 8–9), and acidic ash (Yao *et al.* 2015).

It is found that some environmental-concerned trace elements in FAs from several Chinese power plants are characterized by low leaching potential (Table 3). These FAs are considered as non-hazardous-inert landfill materials according to the European Council Decision 2003/33/EC (EULFD). In China, the Chinese Standard for Pollution Control on Security Landfill Site for Hazardous Wastes also regulates the maximum concentration limit

Table 3. Leachable concentration range of environmental-concerned elements in CCPs from Chinese power plants.

mg/kg	Fly ash[a]	Slag[a]	FGD gypsum[a]	Chinese hazardous limit[b]	Limit values 2003/33/CE[c]		
					Inert	Non-hazardous	Hazardous
pH	11.5–12.5	9.3–9.7	6.5	–			
As	<dl-0.02	0.04–0.05	<dl	25	0.5	2	25
Ba	3–7	0.8–1	0.4	1500	2	100	300
Cd	<dl-0.02	<dl	<dl	5	0.04	1	5
Cr	1.6–3.4	0.01–0.2	<dl	120	0.5	10	70
Cu	<dl-0.01	<dl-0.03	<dl	750	2	50	100
Mo	2.2–7.2	0.2–0.5	0.2	–	0.5	10	30
Ni	0.02–0.1	<dl-0.1	<dl	150	0.4	10	40
Pb	<dl	<dl	<dl	50	0.5	10	50
Sb	<dl-0.04	0.01	0.1	–	0.06	0.7	5
Se	0.02–0.4	<dl	<dl	–	0.1	0.5	7
Zn	0.03–0.04	0.03–0.04	<dl	750	4	50	200
Hg	0.004–0.005	0.004–0.008	0.009	2.5	0.01	0.2	2
F-	8.5–220	<dl	<dl	1000	10	150	500
Cl-	162–287	218–341	3028	–	800	15000	25000

[a]Data from Li *et al.* (2012) and unpublished data.
[b]Data from GB 18,598–2001.
[c]Data from European Council Decision 2003/33/CE.

of some potential hazardous wastes (GB 18598-2001). The concentrations of potential hazardous elements in these FAs are far below the Chinese Standard hazardous limit (Table 3).

The leachability of the elements is closely related to the phase in which they occur and the leaching conditions such as pH of the extraction solution, leaching time, and the solid/liquid ratio (Jankowski et al. 2006). Other factors, such as the particle size and initial concentration of trace elements, also affect the mobility of the trace elements in aqueous environments (Saikia et al. 2006).

Radiation exposure from FA is another important parameter that may have a serious impact on human beings and the surroundings. During coal combustion, most of the U, Th, and their decay products are released from the coal and are distributed between the gas phase and the solid combustion products. Some late Permian coals in South and North China, for example, Guiding in Guizhou Province, Yanshan in Yunnan Province, Heshan and Fusui in Guangxi Province, and Yili in Xinjiang Province, are highly enriched in U (Dai et al. 2008, 2015a, 2015b). However, the radioactivity of their combustion residues has not been yet studied (Sun et al. 2016b). As the trace element concentrations may be 4–10 times higher in FA than in the parent coal, FAs generated from the combustion of high-U coals in China may pose a high risk of radiation exposure from the FA. Assessment of the radiation exposure from FA is critically dependent on the concentration of the radioactive elements in the parent coal and in the FA that remains after combustion (Yao et al. 2015).

2.2. Characterization of BS

BS is the molten BA discharged into a water-filled pit where it is quenched and then collected at the bottom of the wet-bottom boilers as vitreous particles. The BS in the wet-bottom boilers is kept in a molten state and tapped off as a liquid, which is cooled by quenching water contained in the ash hopper in a wet-bottom furnace. Thus, the molten slag is fractured instantaneously, and the resulting product is often a coarse, hard, black, angular, porous, glassy material, with a rough surface texture (NCASI 2003). The particle size of slag is much coarser than that of FA, predominantly consisting of coarse to medium sand-sized particles, usually with 90–100% of particles passing a 4.75 mm (No. 4) sieve, 40–60% a 2.0 mm sieve, ≤10% passing a 0.42 mm sieve, and ≤5% passing a 0.075 mm sieve (Cox et al. 2008).

Several studies have also reported the chemical composition of slag in China (Dai et al. 2010a; Li et al. 2012;

Sun et al. 2016b; among others). As shown in Table 1, the major components of slag are also SiO_2 and Al_2O_3, with various amounts of other metal oxides, as well as CaO.

With respect to the mineral components, slag is also made up of a predominant Al-Si amorphous glass (up to >90%), with different proportions of quartz, anorthite, and mullite as the major crystalline constituents. Haematite may be present in much lower proportions as well. The mineral composition in the slag is dependent on the original minerals in the feed coals, on the combustion technology, and the operational conditions of the power plants (Chinchón et al. 1991; Li et al. 2012).

As another product generated from coal combustion, it is certain that there are a number of trace elements with various concentrations existing in the slag, some of which, especially those of environmental concern, may even migrate to the surroundings during disposal on land. Li et al. (2012) reported the environmental geochemical characteristics of slag from two Chinese power plants and found that the trace element concentration in slag mainly depends on the concentration in feed coals. Furthermore, with the exception of Ca, Fe, Mn, Zr, Ni, and Mg, most of the other elements are more enriched in the FA than in the slag. Segregation of Fe, Mn, Ni, and Zr in slag instead of FA is due to the lower volatility and higher density of the Fe-Mn-Ni-Zr-bearing species. Calcium, Mg, and Na species are flux-like elements favouring the slagging of the aforementioned elements (Meij 1994; Li et al. 2012).

Slag usually has a slightly alkaline pH, lower than that of the corresponding FA (Table 3). Similar to FA, the leachability of the elements in slag is influenced by the particle size of the slag and the initial concentration of trace elements in the slag, and the leaching conditions such as pH of the extraction solution, leaching time, and the solid–liquid ratio (Jankowski et al. 2006; Saikia et al. 2006). Nevertheless, experience suggests that elements are less likely to be leached from (slag and) BA, because of the coarser particle size and the lesser abundance of condensed coatings on the particle surfaces.

2.3. Characterization of FGD gypsum

FGD gypsum, which is chemically known as calcium sulphate dihydrate (CaSO4.2H2O), is a gypsum-like solid residue generated from the wet desulphurization of flue gas from coal-fired power plants. FGD gypsum is generally characterized by a yellowy colour due to limestone impurities and due to the abrasion of iron balls used for limestone milling, and consists of small, fine particles with 10%–15% moisture content.

The wet product from the limestone-based wet scrubbing processes is predominantly $CaSO_3$. Generally, $CaSO_3$ is converted to calcium sulphate ($CaSO_4$) by forced oxidation and the moisture and other impurities in the FGD gypsum are reduced by appropriate measures, such as vacuum dewatering and physical processing, to satisfy the requirements for industrial applications (Kalyoncu 2001). Generally, the gypsum content of FGD gypsum can be >95%, with some $CaSO_3$, $CaCO_3$ (unreacted lime or limestone), and $MgCO_3$ as the main impurities (TFHRC 2010; Tharaniyil 2013). Trace amounts of different impurities, such as quartz, dolomite, and clinochlore, have been detected in FGD gypsum from some Chinese power plants (Li *et al.* 2012). FGD gypsum is characterized by a moderate grain size, usually with >95% of particles passing a 200-mesh sieve (Tharaniyil 2013; Li *et al.* 2015).

Trace amounts of Fe_2O_3, K_2O, Na_2O, MgO, MnO, and TiO_2 in FGD gypsum from Chinese power plants have also been identified by several researchers (Yang *et al.* 2011; Li *et al.* 2015; Zhang *et al.* 2015) (Table 1). The geochemical characteristics of FGD gypsum from several power plants in China have been reported by Li *et al.* (2015). As substantial amounts of FGD gypsum are sometimes disposed of in unlined landfills, water from rainfall can potentially dissolve toxic elements in the FGD gypsum and then pollute the surrounding water systems (Li *et al.* 2015). It has been found that the leachable F^- content of FGD gypsum could be very close to or even surpass the limit required by the legislation of the 2003/33/CE decision on waste disposal for this waste to be characterized as acceptable at landfills for non-hazardous wastes, which could impose consequent serious environmental and economic implications (Álvarez-Ayuso *et al.* 2006; Álvarez-Ayuso and Querol 2008). Therefore, investigation on the environmental geochemical characteristics of FGD gypsum is also of high significance (Li *et al.* 2015).

3. Applications of coal FA

Since it represents a CCP with the largest proportion of total CCP production, utilization of coal FA can be a good alternative to disposal, and could also achieve significant economic and environmental benefits. FA in China has been traditionally utilized in construction, engineering, and agriculture. Among the sectors of utilization, the top three were cement (41%), bricks and tiles (26%), and concrete (19%) (Yao *et al.* 2015). The remainder includes, but is not limited to, roads and embankments, agriculture, and resource recovery (e.g. alumina, magnetite, carbon, cenospheres, and gallium).

Wang and Wu (2006) showed the possibility of multiple FA applications in different sectors based on the natural advantages of this resource (Figure 5). The utilization of Chinese FAs for various industrial applications or pilot- and laboratory-scale experimental studies is described in the following sections.

3.1. Applications of FA in construction and engineering

3.1.1. Use in cement production

Among the most common applications of CCPs is the substitution of FA for Portland cement in concrete or the use of FA as a mineral additive in cement (Song *et al.* 2006), because of the advantages in particular geotechnical properties, for example, pozzolanic character, specific gravity, permeability, internal angular friction, and consolidation characteristics (Ahmaruzzaman 2010). In China, approximately 41% and 19% of the total ash production were, respectively, applied to cement and concrete manufacture in 2011 (Yao *et al.* 2015).

In the cement production industry, due to the similarity of chemical composition of coal ash to that of clay, coal ash can replace natural clay-rich resources to be used as a raw material of cement. A large number of studies have been carried out on the application of FA in the cement industry (Song *et al.* 2006; Cao *et al.* 2008; Sun *et al.* 2011; Yao *et al.* 2015; among others).

As stated above, FA can be referred to as either cementitious or pozzolanic depending mainly on the CaO content. Owing to the pozzolanic properties of FA (high capacity of the material to react with lime to form a hydraulic agglomerate), the cements produced from FA are characterized by high water retention potential. Furthermore, replacing clay by coal ash as raw material can increase the yield of cement kiln. Furthermore, this replacement might also benefit from the unburned carbon of the FA, which can lower fuel consumption by 16–17% (Song *et al.* 2006). According to the statistics, it can save 120 thousand tonnes of fuel when 1 million tonne of coal ash is added (Song *et al.* 2006).

3.1.2. Use as concrete admixture

Concrete is primarily a mixture of aggregates (rock and sand), cement, and water. In the manufacture of concrete, FA may be used as an additive, substituting for a given proportion of cement and consequently reducing the amount of cement, or as an aggregate improving the grain size distribution produced by the sand. In general, the proportion of FA added for concrete ranges from 15 to 35 wt.%, and can be up to 70 wt.% for pavements, walls, and parking lots, and even as much as 80 wt.% in autoclaved aerated concrete (Dilmore and Neufeld 2001). Recently, the use of FA as a concrete admixture has been investigated by several researchers in China (Zhang *et al.* 1997; Yang and Qin 1999; Gao *et al.* 2007; Cao *et al.* 2008; Sun *et al.* 2011; among others). Sun *et al.* (2011) have initiated the research on the optimization and application of coal ash of different qualities in the Three Gorges Dam construction project. The utilization of FA in the construction of concrete dams was also investigated by Gao *et al.* (2007). It was shown that the compressive strengths of dam concrete in 90 days with 50% of FA are higher than those with 30% of FA or without FA. Moreover, dam concrete with 50% of FA is also highly resistant to deformation and the shrinkage and expansion strain may be about 33% and 40% less than for concrete without FA, respectively.

When FA is added to the concrete mix, not only a given proportion of the cement can be eliminated, but also it decreases the water demand and the cost, improves the workability, reduces the permeability and damage from the heat of hydration, and improves the durability, the ultimate strength of the concrete, as well as the resistance to sulphate and other chemical attacks of the concrete (Gu and Hu 2002; Li 2012). Inclusion of FA in the binder also improves the workability and reduces the bleeding of freshly mixed concrete (Yao *et al.* 2015). Furthermore, the use of FA in concrete reduces not only the cost, but also the CO_2 emissions by reducing the energy demand needed to produce the fraction of substituted cement. In addition, chemical activation of FA reactivity is an effective method to increase the use of FA in concrete. For instance, addition of chemical activators such as Na_2SO_4 or $CaCl_2$, along with FA as raw materials for concrete, is believed to accelerate the pozzolanic reactions, and change the pozzolanic reaction mechanisms between FA and lime (Shi and Qian 2003), which may result in decreased setting time, accelerated strength development, and increased strength of materials containing FA, especially with a high percentage of cement substitution. From the environmental point of view, Zhang *et al.* (1999) studied the leachability of hazardous trace metal elements from FA concrete, and found that none of the trace metals analysed (As, Cd, Cr, Cu, Pb, Se, and Zn) in the leachates from FA concrete exceeded the regulated concentration levels specified in the toxicity characteristic leaching procedure (TCLP).

When used as a cementitious material in concrete, different FA properties show various effects. For instance, SiO_2 and Al_2O_3 in FA are regarded as effective ingredients

because they participate in the secondary hydration reaction, which is beneficial to strength development and the improvement of durability. However, CaO and MgO are regarded as useless ingredients, because the free-CaO and free-MgO in FA have densely compacted crystalline structures due to the high-temperature combustion process. These kinds of structures lead to low hydration activity in fresh concrete and volume expansion at later stages (Ma *et al.* 1999). The unburned carbon is also categorized as a harmful ingredient due to its porous structure (Ma *et al.* 1999). High carbon content can also lead to the discolouration of concrete and mixture segregation (Freeman *et al.* 1997). Nevertheless, both Class C and Class F FAs have their own specific advantages when used as a cementitious material in concrete. It may be ideal to have an FA with a low to moderate LOI and that can be used to prepare a concrete with high resistance to alkali sulphate reactions, sulphate attack, and at the same time have high early compressive strengths. One way to achieve that is to blend Class C and Class F FA. The exact ratio of the blend will depend upon the specific FA and the desired behaviour in the concrete (Ahmaruzzaman 2010).

3.1.3. Use in roadway construction

Large benefits have recently been demonstrated in China when recycling FA in pavement and roadway construction applications (Song *et al.* 2006; Sun *et al.* 2011). Utilization of FA in roadway construction and associated projects, such as use in embankment soil stabilization, subgrade base coarse material, as aggregate filler, as a bituminous pavement additive, and as a mineral filler for bituminous concrete, has been a significant outlet for FA applications (Ahmaruzzaman 2010). Coal ash is also used in the construction of highways, as well as in the building of bridge, slope protection, major roads and airport runways, including the Shanghai Cuisong and Hujia highways, the fifth ring highway around Beijing city, the Yellow Sea Bridge in Xiamen, the Huangpu River tunnel, etc. The technique is becoming more and more mature and the economic benefits are high (Song *et al.* 2006).

Easy availability combined with positive physical properties can make FA soil stabilization cost effective. Hydrated ash may have stiffness equal to or greater than that of lime rock when used in both subgrade stabilization and base coarse applications (Yao *et al.* 2015). From the environmental point of view, chances of pollution due to use of FA in road works are negligible (Ahmaruzzaman 2010). When using FA in construction of road embankments, leaching of heavy metals is prevented since there is no seepage of rain water into the FA core. When FA is used in stabilization work, it chemically reacts with cement, which binds FA particles and reduces any leaching effect. Furthermore, use of FA in road works leads to reduction in construction cost by about 10–20% (Ahmaruzzaman 2010).

3.1.4. Use in building construction

Coal FA has been widely applied in the building industry in China. It involves the use of FA in the production of various building bricks, slab foundations, frame and concrete construction, mineral cotton, and microlite glass, among others (Qu *et al.* 2005; Lin 2006; Song *et al.* 2006; Li 2012).

Considering the large quantities of raw materials needed for brick production, as well as the previously mentioned features of FA (high glass content, fine grain size, among others), it can be used as one of the major raw materials for brick production (Li 2012). FA bricks are commonly produced by the traditional extrusion method using a blend of FA and clay as the predominant raw materials, and sintered at different temperatures. The technologies are eco-friendly, reducing solid waste and dust emissions. The cost is reasonable compared with that of hollow bricks and clay bricks. Moreover, the abundance of low-cost FA available to make the bricks leads to ultimate financial gains from the increased physical and chemical properties. It has been shown that FA may improve the compressive strength of bricks, make them more resistant to frost, and increase the mullite content of the brick, which markedly improves the properties of the stoneware products (Xu *et al.* 2005). The use of FA bricks provides a stronger, more durable construction that is better protected from efflorescence and salinity, with meaningful savings in construction costs (Xu *et al.* 2005).

Commonly, an amount of up to 20–30 %wt FA is used in the raw materials for brick production (Xu *et al.* 2005). Li (2012) has succeeded in producing bricks using up to 50% FA in the starting material, with good physicochemical and mechanical properties. However, increasing the FA dose will lead to deteriorating green strength of the ceramic body. The amount of ash that can be added to clay to form green bodies depends on both the nature of the clay and the particle size distribution of the FA (Xu *et al.* 2005). In addition, the properties of the FA brick are also influenced by the sintering temperature, and the nature of the clay and FA used. Apart from FA bricks, low-quality FA in a wet state was also used as a raw material to replace clay to make fired bricks (Xu *et al.* 2005).

Coal ash slab foundation is another application technology for FA in the building industry, which is a kind of unreinforced slab foundation with large stiffness. As a replacement of reinforced concrete slab and pile foundation, it is applicable on uniform soft ground (Song *et al.* 2006). Compared with reinforced concrete slab foundations, coal ash slab foundations are characterized by lower cost, lower density, and more enhanced bearing capacity, solidification, strength, and durability.

Another application of FA in building construction is the production of a frame concrete structure. The frame concrete construction is a composite structure in which the pillar frame is the main bearing structure, and the walls are built with small-sized hollow bricks made partly of FA (Song *et al.* 2006). Frame concrete construction may work as a support against the force of earthquake due to the ductility of the frame. This frame concrete construction has been used in building construction for several years in China, and problems such as leakage and peeling off of paints have never been found.

3.2. Applications of FA in agriculture

The spreading of FA in various agriculture applications (cultivation, fertilizer, soil conditioner, among others) has been carried out in the UK since the 1940s (Cox *et al.* 2008), and is well developed in other countries (Kikuchi 1999; Mittra *et al.* 2005). The application of coal FA in agriculture requires low investments, has the capacity to use large quantities of ash, and thus has a great potential. It is a comprehensive application that is important and appropriate in China under contemporary conditions (Song *et al.* 2006). Studies of the application of coal ash in agriculture mainly focus on its effect on improving and fertilizing the soil as well as on reclaiming underground mines□Application of coal ash in agriculture also includes, but is not limited to, fishery (Liu *et al.* 2007a). FA and caustic sludge have been used in place of cement for building the artificial fish reef, which has positive effects on growing fishery resources and improving the oceanic ecological environment (Liu *et al.* 2007a).

3.2.1. Use as a soil amelioration agent and/or fertilizer

Use of FA as a soil amelioration agent may provide various benefits. By adding coal ash, the quality of the soil can be improved, and its physical and chemical properties, such as unit weight, density, porosity, air permeability, penetration coefficient, and pH value can also be improved, which may in turn lead to an increase in the yield of particular crops (Song *et al.* 2006). The effect is obvious if coal ash is used to improve cohesive soil and acid soil. The addition of alkaline FA to acidic soil can raise the pH of the soil.

Some alkaline FA has been found to be chemically equivalent to approximately 20% reagent-grade $CaCO_3$ in increasing the soil pH (Yao et al. 2015). Class C FA (generally having ≥15% CaO) has been shown to have the ability to raise the pH and hence has been applied at relatively high rates. Most Class F ashes with low CaO content (<15%), on the other hand, have limited potential to ameliorate soil acidity (Yao et al. 2015).

In addition, coal FA itself contains P, K, Mg, B, Mn, Ca, Fe, Si, and other elements that are essential for plant growth. Therefore, the addition of FA to soil can, to some extent, serve as a fertilizer (Gu and Hu 2002). Gu and Hu (2002) have shown that the yield and quality of peanuts and beans were both significantly improved on a soil amended by FA. Furthermore, due to the structure of the FA particles, the water held in the added coal ash is easier to be absorbed by crops than that in the soil (Song et al. 2006).

The use of FA as an agricultural amendment can be enhanced by blending it with humus and other potentially acid-forming organic by-products, which may enhance nutrient availability, decrease the bioavailability of toxic metals, buffer the soil pH, and enhance the soil organic matter content, indirectly stimulating microbial activity and overall improving the soil health and increasing the crop yield (Gu and Hu 2002).

It is worth noting that, when used in agriculture, some elements of environmental concern, such as F^-, Cl^-, As, Se, Sb, Hg, Cd, Cr, Cu, Ni, Pb, Zn, Ba, and Mo, may be leached in significant amounts from the FA and migrate into the surrounding water systems (Hung et al. 2009). With respect to the possible harmful effects of these hazardous components on the crops, it is indicated that the utilization of FA at an appropriate proportion does not cause contamination of soil (Xu 1999). It was shown that the contents of those hazardous components in crops grown on fly-ash-amended soil are similar to their corresponding contents in those grown on regular soil (Xu 1999).

3.2.2. Use in mine backfill

Bulk quantities of coal FA have been used to replace the conventionally used sand for reclaiming underground mines (Ahmaruzzaman 2010). Backfilling of underground mines especially holds good potential for those areas where sand is scarce. Open cast mine filling can again be considered as land reclamation. In China, coal ash was initially used to fill mines in the 1920s to prevent possible underground fires (Song et al. 2006). Mixing coal ash with cobblestones to fill the void has been shown to increase the hardness and to lead to improved water-proofing and sound-proofing effects. Coal ash was successfully used to fill the collapsed areas of coal mines, and can be used to fill the hollow left in brickfields after the brick clay is removed. FA grout injection is currently being considered for use at a closed underground mining site (Gu and Hu 2002). The injection process would reduce acid mine drainage, which occurs in areas that have previously been mined for coal and contain pyritic materials in spoil piles or in mine shafts. FA mine void filling has been carried out both under controlled circumstances and in actual field applications. The potential application of coal FA in reclaiming abandoned coal mines is of great practical significance. Research is ongoing for the commercial use of such large volumes of FA as a mine-filling material (Song et al. 2006). The technical feasibility of disposing coal combustion by-products in underground mines has been proven and the selection of this disposal alternative will be decided based primarily on cost and regulatory compliance issues (Ahmaruzzaman 2010). The way to add the coal ash and the quantity needed are determined by the objective and the requirement (Song et al. 2006). The use of FA for mine backfilling requires a range of technical information with the potential for adverse environmental impacts, such as the chemistry and mineralogy of the ash, impact on groundwater, impact on the community, likely contamination from the ash, etc. (National Research Council 2006). More research into the chemical aspect and the interaction of the CCPs, mine water, local geology, and groundwater is needed to assess the environmental impact of coal FA injection. Ward et al. (2010) have shown that the concentration of at least some potentially significant trace elements releases into percolating groundwater from a mine-site ash deposit may be attenuated by contact of the ash leachate with the adjacent rock strata, especially those with a high proportion of clay minerals or Fe oxy-hydroxides.

3.3. Applications of FA as an adsorbent

FA is commonly alkaline, and its surface is negatively charged at high pH, so it can be expected to remove metal ions from solutions by precipitation or electrostatic adsorption. When alkaline coal ash is mixed with water, the Al and Fe ions will flocculate to form suspended particles in the water and settle together as a combination. As a result, the ash completes the separation of the pollutant and the suspended matter from the water, which makes the water become clear and transparent in the end (Song et al. 2006). Besides, FA contains a certain volume of unburned carbon, which has a high adsorption capacity (Ahmaruzzaman 2010). It was found that unburned carbon remaining in the FA particles contribute to the surface area of the FA, and that this carbon can be activated to further improve the adsorption performance of FA (Ahmaruzzaman 2010). Probably, the silicate and aluminate groups (or aluminosilicates) at the surface of the FA particles are also involved in these adsorption processes (Lieberman et al. 2014).

The low cost and huge resources of FA could make it an economically viable alternative for the removal of pollutants from waste water, although it is an inferior adsorbent relative to activated carbon. In China, FA has been used experimentally for treatment of domestic sewage, waste water from paper mills and tanneries as well as in the adsorption of SOx, NOx, Hg, and other hazardous gases (Song et al. 2006; Sun et al. 2011).

3.3.1. Use for the removal of gaseous pollutants

Fly has been widely used on a laboratory scale for the removal of SO_x, NO_x, and Hg from flue gases and organic gases in China (Wei et al. 2004). Coal FA is a cheap and environmentally friendly absorbent for dry-type FGD and this process has been successfully commercialized (Wang and Wu 2006). FA recycling in the FGD process has shown promising results. FA treated with calcium hydroxide has been experimentally tested as a reactive adsorbent for SO_2 removal (Sun et al. 2011).

A good removal capacity for NO_x using FA, which could be related to the textural properties of the ash, has been proven by several researchers (Tang et al. 2006). It has been observed that some FAs may capture Hg, which would otherwise be emitted into the atmosphere (Sun et al. 2011). The adsorption of Hg on the unburned carbon in the FA can be explained by the physical and chemical interactions that occur between the carbon surface and Hg (Tang et al. 2006; Sun et al. 2011).

3.3.2. Use for the removal of heavy metals

Heavy metals are regarded as serious pollutants and occur in concentrations that may result in serious public health issues. Heavy metal and metalloid removal from aqueous solutions is commonly carried out by several processes such as chemical precipitation, solvent extraction, ion exchange, reverse osmosis, or adsorption. Among these processes, the adsorption process may be a simple and effective technique for the removal of heavy metals from waste water (Ahmaruzzaman 2010). FA has been widely used as a low-cost adsorbent for the removal of heavy metals, and several researchers have investigated the utilization of FA for the removal of various heavy metals from

waste water (Song et al. 2006; Li 2012; among others). The factors influencing the removal efficiency of toxic heavy metals, that is, Cu^{2+}, Pb^{2+}, and Cd^{2+}, from aqueous solution using FA have also been investigated (Lin and Chang 2001; Zhu 2006; Li 2012). The adsorption sequence is basically in accordance with the order of insolubility (at the pH values reached by adding FA) of the corresponding metal hydroxides (Li 2012).

Apart from the adsorption process, the alkalinity of the FA in water may also remove heavy metals from solution by chemical precipitation. Thus, Li (2012) investigated and compared the adsorption efficiency of FAs and synthesized zeolites for heavy metal removal from synthetic solutions. It was found that, with zeolite, the removal efficiency of FA was lower and it was necessary to use large amounts of FA to reduce the heavy metals to non-hazardous levels (Figure 6). Furthermore, the coarse fraction of the FA was more efficient in removing heavy metals than the fine fraction (Figure 6), which was ascribed to the higher proportion of unburned carbon and the higher pH of the coarse FA fraction (Li 2012).

The removal efficiency for heavy metals using FA was found to be dependent on the physical and chemical characteristics of the FA dosage, the adsorbate, and the experimental system (e.g. contact time, temperature, pH, and the initial concentration of heavy metals), among others (Lin and Chang 2001). It is apparent that by increasing the adsorbent dose, the adsorption efficiency increases, but the adsorption density, the amount adsorbed per unit mass, decreases. Moreover, the adsorption capacity of FA depends on the surface activities, such as the specific surface area available for the solute surface interaction, which is accessible to the solute. The crystalline CaO content of the FA seemed to be a significant factor influencing the adsorption of Cr^{4+} and Cd^{2+} (Ahmaruzzaman 2010).

3.3.3. Use for the removal of other pollutants

Apart from heavy metals in waste water, some other inorganic contaminants, such as P, F, and B, may also exist in waters and be of environmental concern (Zhu 2006). Phosphorous from concentrated agricultural activities, including soil fertilization, feed lots, and diaries, may accumulate in the surface and groundwater. Chen et al. (2007) has discussed phosphate immobilization from aqueous solution using Chinese FAs. Acid-modified FA was effective in the removal of phosphate from contaminated antibiotic waste water. Adsorption and chemical precipitation were the main mechanisms for the removal of phosphate with modified FA. The addition of FA to water produces insoluble or low-solubility salts when combined with phosphate. Solid-

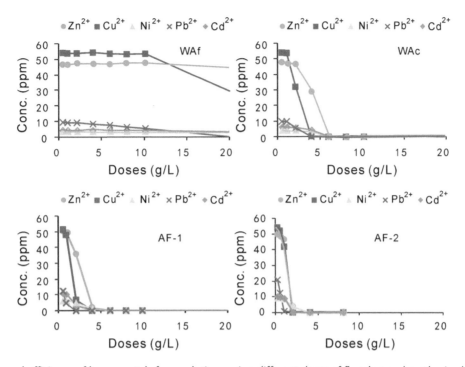

Figure 6. The removal efficiency of heavy metals from solutions using different doses of fly ashes and synthesized zeolites. WAc and Waf, coarse and fine fraction of fly ash; AF, zeolite synthesized from fly ash by alkaline fusion.

phase phosphate compounds are separated from water by sedimentation or filtration. Jia et al. (2015) analysed the mechanisms of thermal, acid, alkali, and surfactant modifications of FA. It was found that the vitreous fracture of FA was deeply altered and that the FA activities could be increased due to new products formed from the alteration of the glass matrix, whereas the adsorption ability was increased through a combination of surfactant and FA in surfactant modification.

In addition, some organic pollutants have been reported to be removed by FA. Kao et al. (2000) utilized FA for the removal of 2-chlorophenol (2-CP) and 2,4-dichlorophenol (2,4-DCP) from an aqueous solution prepared by dilution of the stock solution. Higher adsorption was reported for FAs with a higher carbon content and a larger specific surface area. Dyes and pigments may also be discharged into waste waters from the dye-manufacturing and textile finishing industries. There are various methods available for the treatment of dyes from waste water, among which the adsorption process provides an attractive alternative. FA has been reported to be an adsorbent that is cheap and does not require an additional pretreatment before application. Various researchers have examined the potential of FA for the removal of azo, thiazine, xanthene, arylmethane, and other types of dyes from waste water (Wu et al. 2005; Lin et al. 2008; Sun et al. 2010). To increase the adsorption efficiency, heat treatment and chemical treatment have also been applied to the as-received FA. Heat

treatment reduces the adsorption capacity, whereas acid treatment with HNO_3 results in an increase in the adsorption capacity of the FA. Chemical treatment using HCl will also increase the adsorption capacity. Microwave treatment is a fast and efficient method while producing ash with the highest adsorption capacity (Lin et al. 2008).

On the whole, the applicability of FA as a low-cost adsorbent for waste-water treatment depends strongly on its origin. The effectiveness of the adsorption process depends on the properties of the adsorbent and the adsorbate. It also depends on the environmental conditions and variables used for the adsorption process, such as pH, ionic strength, temperature, existence of competing organic or inorganic compounds in solution, initial adsorbent concentration, contact time, and speed of rotation, among other aspects (Ahmaruzzaman 2010).

3.4. Applications of FA in the ceramic industry

Ceramics are produced from clay, or a mixture containing clay, through the process of moulding, drying, and firing. As FA is made up of oxides such as SiO_2, Al_2O_3, CaO, and Fe_2O_3, it has received attention as a low-cost material for the manufacture of ceramics, glass-ceramics, and glass materials (Erol et al. 2008a). In addition, the fine powder form of FA makes it suitable to be directly incorporated into ceramic pastes with almost no pretreatment (Erol *et al.* 2008b). Furthermore, due to the consumption of large quantities of natural raw materials in the ceramic industry, using FA as the raw material, or as a partial replacement for clay, may reduce the consumption of natural resources (Queralt *et al.* 1997).

The main principle of manufacturing ceramic materials using FA is the temperature activation of the raw FA with vitrification in temperatures and co-reagents directing the final form of the glass or ceramic. As mentioned above, the clay minerals in coal are converted into molten spheres during combustion, and then solidified into glassy solids, typically with a mullite ($3Al_2O_3.2SiO_2$) composition. It is just this conversion of raw clays into glassy-phase solids that makes FA such a potentially attractive ceramic raw material (Cox *et al.* 2008). Nevertheless, the thermal processes to produce ceramics, glass, and glass-ceramics from FAs as well as the properties of the three ceramic materials may be different (Erol *et al.* 2008a; Blissett and Rowson 2012). It has been reported that ceramics were formed using a conventional powder processing technique based on powder compaction and sintering of FA (Erol *et al.* 2008a). FA-based glasses were prepared by melting and subsequent annealing of the ash (Blissett and Rowson 2012). Glass-ceramics were produced from the annealed glass by a one-stage heat treatment (nucleation) process (Erol *et al.* 2008a; Blissett and Rowson 2012).

In China, the use of FA in ceramic manufacture has been well investigated (Shao *et al.* 2004; Yao *et al.* 2015). Recent research has focused on the production of glass (Sheng *et al.* 2003), glass-ceramic (He *et al.* 2005; Peng *et al.* 2005), and sintered ceramic materials (Jing *et al.* 2012; Ji *et al.* 2016) prepared from coal FA.

3.4.1. Use for conventional ceramic production

Ceramic tiles are popular materials for interior and exterior decorations. Ji *et al.* (2016) have proposed an alumina-rich ceramic system (the $SiO_2-Al_2O_3-CaO-K_2O$ system) to fabricate ceramic tiles by using FA as the main raw material. The performance is far better than the standard requirements for porcelain tiles (ISO-13006, 1998). The rupture modulus of the produced ceramic tiles can reach 51.28 MPa, 47% higher than the standard requirement for porcelain tiles. Moreover, the surface and the overall texture are also fully acceptable according to this regulation standard. Jing *et al.* (2012) prepared ceramic granules using FA, clay, and diatomite, and showed that the prepared ceramic granules with high porosity and large pore area had been used as packing materials to develop a double-layer biofilter. This biofilter made of FA ceramic granules

proved efficient in the treatment of highly polluted river water with chemical oxygen demand (COD) and nitrogen (Jing *et al.* 2012).

The performance of the FA-based ceramics is highly dependent on the proportion of FA added, the sintering temperature, and on the properties of the FA, such as LOI and chemical composition (Li 2012; Ji *et al.* 2016). It is obvious that higher sintering temperature facilitates the improvement of sintering (Ji *et al.* 2016). Ji *et al.* (2016) proposed that the optimum sintering temperature range is 1150–1200°C with the FA proportion being 50–70 wt.%. Li (2012) pointed out that the higher Fe_2O_3 content in the FA may cause the occurrence of black cores and bumps during the sintering process.

3.4.2. Use for glass-ceramics production

The chemical composition of FA makes it particularly suitable for the manufacture of structural glass-ceramics (Peng *et al.* 2005; Zhao *et al.* 2010). The common glass-ceramic material, based on the $Li_2O-Al_2O_3-SiO_2$ ternary system, is usually prepared from high grade-reagents, which make it very expensive. Recently, $Li_2Al_2Si_3O_{10}$ was prepared using coal FA as a precursor, and the resulting product was well crystallized with a small grain size (Yao *et al.* 2011). Glass-ceramics based on the $CaO-Al_2O_3-SiO_2$ and $MgO-Al_2O_3-SiO_2$ systems have also been synthesized from FA (Wang *et al.* 2014). FA has been considered as a replacement for kaolinite in the manufacture of the cordierite ceramic (Shao *et al.* 2004; He *et al.* 2005). Shao *et al.* (2004) investigated the characterization of cordierite-based glass-ceramics produced from FA. Due to its low coefficient of thermal expansion (CTE), low dielectric constant, and high refractoriness, cordierite ceramic is considered as a candidate for structural materials and finds applications as catalyst carriers for exhaust gas purification, heat exchangers for gas turbine engines, electronic packing materials, and refractory metal coatings, among other applications (Yao *et al.* 2015).

The synthesis of glass-ceramics from FA not only offers lower cost, but the product also has many advantageous properties such as being more prone to crystallization (lower crystallization temperature and shorter crystallization time) and the formation of nanocrystals (fine microstructure), where the formation of nano-crystals is beneficial to mechanical (and other functional) properties (Wang *et al.* 2014). The synthesized glass-ceramics can be used as novel building decoration materials, or as infrared heating and drying materials (Cheng *et al.* 2007). Furthermore, the glass-ceramics produced from coal FA are non-hazardous materials with several promising potential applications, since many of the toxic trace elements were volatilized or fixed into the glass matrix during the process of high-temperature vitrification (Erol *et al.* 2007, 2008a).

3.4.3. Use for glass production

Because the coal FA contains a large amount of SiO_2 and Al_2O_3, which are the main glass network formers, it is feasible to use coal FA as a raw material to develop glass matrices for the fabrication of composite materials (Ellison *et al.* 1994; Erol *et al.* 2007). Sheng *et al.* (2003) have successfully synthesized glasses from coal FA by vitrification at 1200°C with the addition of Na_2O as a melting aid. Compared with commercial glasses, the prepared glasses displayed a good durability (mass loss ≤10 g/m^2) and suitable viscosity (<100 Pa s) when about 10%wt Na_2O was added (Sheng *et al.* 2003). Moreover, the heavy metals were successfully immobilized into the glass structure.

3.5. Applications of FA for catalysts

The application of FA as a material to be used in heterogeneous catalysis has received a great deal of attention (Wang 2008). Heterogeneous catalysis is appealing because it is often easier to recover the catalysts after the reaction compared with homogeneous catalysts. Metal oxides such as Al_2O_3, SiO_2, TiO_2, and MgO are the most widely used catalyst supports. For FA, most of the chemical compounds are Al_2O_3 and SiO_2, which have been treated at much higher combustion temperatures, exhibiting strong thermal stability (Wang 2008). This provides FA good quality as a catalyst support for various reactions (Wang 2008). Zhang *et al.* (2012) used nitric-acid-activated FA as a heterogeneous Fenton-like catalyst for the removal of p-nitrophenol from water, with a removal rate of up to 98%. Xuan *et al.* (2003) found that, after pretreatment and loading with transition metal elements (Fe, Cu, V, or Ni), FA can be used as a selective catalytic reduction (SCR) catalyst support to remove NO from flue gas, with a NO conversion rate of 95% for temperatures less than 350°C. It is also reported that FA can be used as an effective catalyst for reactions in gas, liquid, and solid phases, such as the gas-phase oxidation of volatile organic compounds (VOCs), aqueous-phase oxidation of organics, solid plastic pyrolysis, and solvent-free organic synthesis (Wang 2008). FA-supported catalysts have displayed good catalytic activities in $deSO_x$, $deNO_x$, and H_2 production, hydrocracking, and hydrocarbon oxidation (Yao *et al.* 2015).

3.6. Recovery for fractionated FA products

Some products may also be fractionated from FA by a variety of separation processes, such as the recovery of cenospheres, recovery of magnetic spheres, and carbon recovery from FA (Blissett and Rowson 2012; Yao *et al.* 2015).

3.6.1. Recovery of cenospheres

As mentioned above, FA consists of fine, spherical powdery particles, either solid or hollow, and mostly amorphous. The hollow spheres are called cenospheres, and are considered to be one of the most important added-value components of FA. They are similar in composition to the FA although they tend to have a larger particle size (Hirajima *et al.* 2010). Cenospheres are formed during the molten state of FA and attain spherical shape to produce minimum surface tension (Yao *et al.* 2015). Their wall thickness is relatively thin, <10% of the particle diameter. The density is typically less than 1 g/cm^3.

Cenospheres typically float on water, and are traditionally recovered from the surfaces of FA lagoons by recovery of the floating fraction (Blissett and Rowson 2012). However, some materials in the FA, such as toxic elements, are prone to dissolution upon contact with water (Borm 1997). Thus, their leaching and accumulation in water are disadvantages of the wet-separation processes for cenosphere recovery. Air classification was also proposed as an alternative technique to isolate rich-cenosphere materials from FA (Hirajima *et al.* 2010). Owing to the simple process of separation and the large amount of FA available in China, several research studies have been conducted on the separation of cenospheres from FA, and also on cenosphere applications. Yang *et al.* (2002) discussed the mechanism of formation and the physical-chemical properties of cenospheres in FA, and discussed existing problems and the developing prospects for the use of cenospheres recovered from FA. The lightweight nature of the cenospheres renders them suitable for use in lightweight composite materials. By mixing with concrete, metal alloys, and polymers, cenospheres can be used to produce different lightweight composite materials with higher strengths. Li *et al.* (2007a) discussed the development and properties of cenosphere-based insulation mortar. A new kind of exterior wall external thermal insulating mortar had been successfully produced from FA-based cenosphere mixed with expanded pearlite, using Portland cement as the inorganic binder and fibre and polymer as the additive. It was found that the properties of cenosphere-based insulating mortar, such as thermal insulation, strength, crack resistance, and climate resistance, were superior to those of expanded pearlite insulating mortar. Meanwhile, other properties, such as bulk stability, ageing resistance, climate resistance, crack resistance, fireproofing, and construction workability, were much better than those of expanded polystyrene sheet (EPS) insulating mortar.

The cenosphere properties have also been altered by coating them with various metals. A variety of different methods of plating have been used, such as electrodeless plating (Hu *et al.* 2010; Pang *et al.* 2011), magnetron sputtering (Cai *et al.* 2007; Yu *et al.* 2007a, 2007b), and heterogeneous precipitation (Meng *et al.* 2010).

The lower density of cenospheres relative to water may make it possible to prepare new photocatalysts (Nair *et al.* 1993). Cenospheres can be used as a buoyant carrier to enhance catalytic activity as they increase the exposure of the particle to light sources (Huo *et al.* 2009; Wang *et al.* 2011a). The floating-on-water feature of cenospheres makes them suitable for use in water purification applications, such as the removal of excess F^- from drinking water, which can be achieved by producing magnesia-loaded cenospheres using a relatively simple wet impregnation method with $MgCl_2$ (Xu *et al.* 2011b).

3.6.2. Magnetic recovery

The proportion of magnetic material in different FAs varies depending on the coal source and the operating conditions of the coal boiler. As stated above, the Fe_2O_3 content of Chinese FA can vary from 1.5% to as high as 19.3%. Since the magnetic separation of materials is a mature and well-established technology, the removal of magnetic concentrates from FA could also provide a significant opportunity to create added value. The basic principle of magnetic separation relies on the fact that materials with different magnetic moments experience varying forces in the presence of magnetic field gradients. In this way, an externally applied magnetic field can lift out those materials with similar magnetic properties (Yavuz *et al.* 2009). Magnetic separation is an effective approach for iron removal and it may be conducted in either a dry or a wet environment.

There are only a few studies that provide specific examples of magnetic recovery from FA in China (Jiang *et al.* 2009; Gong *et al.* 2011). Jiang *et al.* (2009) adopted the dry magnetic separation method, in which a cyclone is used as the feeding device for ferromagnetic material separation from FA. Uniform, continuous separation and extraction of ferromagnetic material from FA were achieved. A combination of wet and dry magnetic separation appears to be another viable option offering the benefits of both technologies. Gong *et al.* (2011) investigated magnetic Fe

removal from FA by using a magnetic selection instrument and obtained an Fe recovery ratio of up to 92.11%. Magnetic microspheres separated from FA have great resource utilization potential for their ferromagnetism and special microsphere structures, such as dense medium separation, waste-water separation, and magnetic composite adsorption materials (Wu *et al.* 2015).

3.6.3. Recovery of carbon

The recovered unburned carbon from coal FA could serve as a fuel or a source of carbon-derived materials. One of the most promising methods to recover unburned carbon from FA is froth flotation (Aplan 1999), which is a highly versatile method for separating particles based on the ability of air bubbles to selectively adhere to the surface of a particular mineral surface. The particles attached to the air bubbles become positively buoyant and rise through the slurry, whereas those remaining wet sink to the bottom of the vessel. In China, Yang (2000) used froth flotation to separate the carbonaceous portion from low-carbon FA, with the carbon being separated as single grains from the FA. Li *et al.* (2007b) carried out mini-floatation experiments to recover carbon from FA and found the recovery rate of carbon to be as high as 85%. Electric separation is another commonly used method for carbon separation from FA (Yang and Chen 2012). Zhai *et al.* (2014) investigated the electric decarbonization of FA from circulating fluidized-bed and pulverized-coal-fired boilers, and found that the electric separation method is not suitable for carbon separation from circulating fluidized-bed FA, due to the incomplete combustion and low proportion of well-formed spherical FA particles.

The carbon-enriched product recovered from FA has been reported to be used for a variety of applications, such as coke in the metallurgical industry, and a precursor for activated carbon preparation (Cheng 2000). FA-derived activated carbon has been experimentally used for a variety of applications, such as for the removal of SO_2 and NO_x from flue gases, and for the treatment of liquid waste (Cheng 2000). It was found that activated carbons obtained from pretreatment with an NH_4^+-salt solution prior to activation show better adsorption properties than those from conventional steam activation (Li *et al.* 2006; Wu *et al.* 2010).

The unburned carbon from FA has been used as a precursor for experimentally obtaining graphite, with the graphite products showing good physical properties and performance in Li-ion batteries (Cheng 2000).

3.7. Other advanced applications for FA

Owing to the fact that the markets for FA in the conventional construction, agriculture, and ceramic industries are nearly close to saturation and of relatively low added value, alternative high-value-added uses for FA need to be developed and commercialized. In China, innovative FA recycling technology has been developed and the scale of utilization has grown significantly. For instance, under a series of stimulation and preferential policies issued by the state and local governments, the recovery of alumina is being aggressively undertaken in Inner Mongolia and Shanxi province (Cao *et al.* 2008). Second, efforts have been directed to the developments of other commercially viable or novel products, which may yield high economic value. The utilization of FA has also been experimentally and commercially extended to several high-value-added applications, such as the synthesis of zeolites (Liu *et al.* 2007b; Hui and Chao 2006; Cui *et al.* 2009; Zhao *et al.* 2009; Li *et al.* 2014a; among others), valuable metal recovery (Ge, Ga, V, U, Al, Ti, and other elements) (Zeng 2007; Yao 2013; Yao *et al.* 2014; Sun *et al.* 2016b), production of foam glass and fire-resistant materials (Li 2012; Li *et al.* 2015), and synthesis of geopolymers (Xu *et al.* 2006, 2010), among others.

3.7.1. Synthesis of zeolites

Zeolites may be found in natural deposits, generally associated with the alkaline activation of glassy volcanic rocks, or synthesized from a wide variety of high-Si and -Al starting materials. The compositional similarity of FA to certain volcanic materials, the precursors of natural zeolites, is the main reason for synthesizing zeolites from FA. In addition, the presence of abundant reactive phases, such as aluminosilicate glass in FA, and the high specific surface area of FA make it a suitable starting material for zeolite synthesis (Höller and Wirsching 1985). Considerable research on the synthesis of zeolites from FA has been performed since the initial study by Höller and Wirsching (1985). The main types of zeolites synthesized from coal FA are shown in Table 4. Semi-quantitative estimates of synthesized zeolite percentages can be produced from their experimental cation exchange capacity (CEC) value compared with pure zeolite.

At present, there are mainly three methods for synthesizing zeolites from coal FA (Querol *et al.* 2002): (a) direct conversion, (b) alkaline fusion, and (c) two-stage synthesis. Direct conversion method is based on the mixing of FA with different doses of alkaline solutions (commonly NaOH/KOH) and the activation at different temperatures with various

Table 4. Molecular compositions of some natural and synthetic zeolites.

	Zeolite	Formula
Natural	Chabazite	$Ca_2Al_4\,Si_8O_{24} \cdot 13H_2O$
	Clinoptilolite	$Na_6Al_6Si_{30}O_{72} \cdot 24H_2O$
	Erionite	$(Ca, Mg, Na_2, K_2)_{4.5}\,Al_9Si_{27}O_{72} \cdot 27H_2O$
	Faujasite	$Na_{12}Ca_{12}Mg_{11}\,Al_{59}Si_{133}\,O_{384} \cdot 235H_2O$
	Mordenite	$Na_8\,Al_8Si_{40}O_{96} \cdot 24H_2O$
Synthetic	High CEC, with large pore size	
	Zeolite A	$Na_{86}Al_{86}Si_{106}O_{384}.264H_2O$
	Zeolite X	$Na_2Al_2Si_{2.5}O_{9.6}.2H_2O$
	Zeolite Y	$Na_{56}Al_{56}\,Si_{136}O_{384} \cdot 250H_2O$
	Zeolite P (NaP1)	$Na_6Al_6Si_{10}O_{32}.12H_2O$
	Phillipsite-K	$(K_{2.5}Na)Al_{4.7}Si_{11.3}O_{32} \cdot 13H_2O$
	Phillipsite	$K_2Ca_2Al_6Si_{10}O_{32} \cdot 12H_2O$
	Zeolite W (Merlinoite)	$K_2Al_2Si_3O_{10} \cdot 3H_2O$
	Low CEC, with small pore size	
	Analcime	$Na_{16}Al_{16}Si_{32}O_{96}.16H_2O$
	Hydroxy-sodalite	$Na_{1.08}Al_2Si_{1.68}O_{7.44}.1.8H_2O$
	Tobermorite	$Ca_5(OH)_2Si_6O_{16}.4H_2O$
	Perlialite	$K_9NaCaAl_{12}Si_{24}O_{72}.15H_2O$

reaction times (Querol *et al.* 1997a, 2002); the process commonly results in zeolite products such as NaP1 and analcime (Figure 7). Querol *et al.* (1997b) proposed a microwave-assisted method as a basis for direct conversion to greatly reduce the activation time (from hours to a few minutes). In the alkaline fusion method, mixtures of FA with alkaline solid (mainly NaOH/KOH) are subjected to high-temperature alkaline fusion (550–600°C) and long-time ageing (6–16 h) prior to the direct conversion procedure (Shigemoto *et al.* 1993; Rayalu *et al.* 2000). Zeolites A and X with high porosity, large pore size, and a high CEC are the common zeolites synthesized from this procedure (Figure 7). The two-stage synthesis procedure, developed by Hollman *et al.* (1999), enables the synthesis of >99% pure zeolite products from high-Si solutions obtained from a light alkaline attack on FA. Moreover, the solid residue from this attack can be converted into classic zeolitic products by using the direct conversion method. This method produces pure zeolitic material with a high pore size (> 95% of zeolites X and A). Moreno *et al.* (2002) optimized this method, synthesizing zeolites with high purity from SiO_2 extracts in the same process, and another zeolitic product, equivalent to zeolites obtained by the direct conversion method.

In China, several researchers have investigated the synthesis of zeolites from FA by different methods. Wang *et al.* (2005) obtained NaP1 zeolitic materials from the hydrothermal reactions of FA in NaOH solutions, with the CEC of the synthetic NaP1 being 100 times higher than that of raw FA. Yao *et al.* (2013) investigated the hydrothermal reactions of FA in $LiOH \cdot H_2O$ solutions, which acted as a strong activator and promoted the dissolution of inert phases. The contents of zeolites (ABW phases) increased at the expense

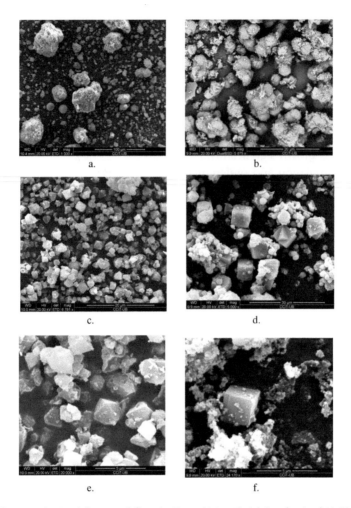

Figure 7. SEM images of zeolites synthesized from coal fly ash (From Li 2012). (a) Synthesized NaPl; (b) synthesized analcime; (c) synthesized zeolite X; (b) synthesized zeolite A; (e) synthesized zeolite X; (d) synthesized zeolite A.

of quartz and mullite, with an increase in alkaline solution concentrations. Guo et al. (2003) reported that the FA can be heated up and directly crystallized into laumontite, chabazite, and NaP1 zeolites, under conditions of hydrothermal alkaline activation by using microwaves. The conversion of FA to zeolites can be achieved in about 30 min. Several researchers have reported the synthesis of zeolite A by the alkaline fusion method using common NaOH as the activation agent (Cui et al. 2009; Li 2012; etc.). Wang et al. (2008a) synthesized single-phase zeolites NaA and NaX from FA using the alkaline fusion method, without the presence of other phases. Yao et al. (2009) proposed the synthesis of zeolite LiABW by alkaline fusion followed by hydrothermal treatment in a $LiOH \cdot H_2O$ medium, obtaining up to 97.8% of a high-crystallinity zeolite under optimal conditions. Wang et al. (2008b) prepared pure-form zeolite A using the two-step process. A single-phase zeolite NaA was synthesized from FA using the two-stage method, and the structure and performance of the products as well as the adsorption properties of

NaA zeolite to Cu^{2+}, Cr^{4+}, and Zn^{2+} in aqueous solution were discussed by Wang et al. (2009a, 2009b)).

Most studies have shown that NaOH solutions have higher conversion efficiency than KOH solutions under the same temperature conditions (Murayama et al. 2002). It is reported that the zeolite synthesis efficiency not only is largely determined by the FA characteristics, especially the particle size, glass content, and the SiO_2/Al_2O_3 ratio in the glass matrix, but also is influenced by the activation temperature, activation time, concentrations of the activation solution, and the solution/FA ratios (Wu et al. 2005; Li 2012). The activation time required for synthesis yield is inversely proportional to the Al-Si glass content of the FA (Wu et al. 2005). Thus, for ash with a higher Al-Si glass content, shorter activation periods and lower solution/FA ratios are needed to reach high yields of zeolite (Querol et al. 2002; Li 2012). Li (2012) has found that the zeolite synthesis efficiency can be improved by increasing the activation temperature and the concentration of the activation agent, but zeolites with low CEC values, such as cancrinite and sodalite, are obtained at higher temperatures (≥200°C) and concentrations (≥3 M). Conversely, low temperatures (125–150°C) and concentrations (2–3 M/L) allow the synthesis of high-CEC zeolites, such as NaP1, A, or chabazite.

As a consequence of the peculiar structural properties of zeolites, the synthesized zeolites can be used in various engineering and agricultural applications, for water purification, gas cleaning, and soil amendment (Liu et al. 2007b; Wang et al. 2009a, 2009b). Li (2012) reported the CEC of several synthesized zeolite materials (NaP1, X, and A) and estimated their effect on heavy metal removal from solutions. Wang et al. (2009a) prepared zeolites A and X from FA and evaluated the adsorption capacity of zeolites using Cu and Zn as the target heavy metal ions, which indicates that the adsorption capacity of zeolite A showed much higher value than that of zeolite X. Chen et al. (2006) reported the removal of phosphate from aqueous solution by zeolite synthesized from FA. Wu et al. (2006) and Zhang et al. (2007) investigated the simultaneous removal of ammonium and phosphate by zeolite synthesized from coal FA as influenced by salt and acid treatment, respectively. Zhou et al. (2014) synthesized zeolite NaP1 and used it as adsorbents for VOCs. Moreover, the use of these zeolites as molecular sieves for the gas cleaning technology has also been investigated by a few studies. Liu et al. (2011) prepared Zeolite A and A + X mixtures and simulated the performance of the synthesized materials for CO_2 capture. Although most of the previously referenced studies demonstrated that the NaOH solutions have higher conversion efficiency than the KOH solutions under the same temperature (Ahmaruzzaman 2010), Li et al. (2014a) prepared K-merlinoite and used it as an efficient slow-release K-fertilizer for plant growth.

Despite these advances, there are some limitations on zeolite synthesis from FA. The main limitations of the processes for synthesizing zeolites from FA are the slow speed of the reaction, the relatively high temperatures (125–200°C) needed, and the alkaline liquid waste generated.

3.7.2. Recovery of valuable metals

Some coal FA in China contains high concentrations of economically valuable metals, such as Ge, Ga, U, V, Li, and Ti, which may be extractable if an acceptable process can be developed (Dai et al. 2010a, 2014a, 2014b; Sun et al. 2016b). The potential for the recovery of valuable metals from FA is primarily based on the concentration and enrichment processes in FAs during coal combustion and gasification (Meij 1994; Querol et al. 1995). Likewise, due to the Al-Si-rich composition, FA

may also represent a potential source of these less-valuable elements (Ji et al. 2007; Wang et al. 2012; Yao et al. 2014).

3.7.2.1. Germanium recovery.
Germanium is a valuable element used in the manufacture of light emitting diodes (LEDs), photovoltaic (PV) cells, infrared devices, fibre optics, and as a catalyst for polyethylene terephthalate (PET) production. The main source of the present-world Ge production is the high-Ge lignite and subbituminous coal containing a few hundreds of µg/g to more than 1000 µg/g of Ge (Höll et al. 2007). In FAs, the Ge can be further concentrated to values up to 10 times higher than in the original coals (Font et al. 2005). Hence, the Ge content in coal and coal FAs may be regarded as a primary source once accessible natural Ge resources have been depleted. Currently, two Chinese (Lincang and Wulantuga) and one Russian (Spetzugli, Pavlovsk Coalfield) Ge-bearing coal deposits are being mined where Ge is recovered from the FA, with the Ge production from these deposits accounting for more than 50% of the total industrial Ge production in the world (Seredin and Dai 2014; Dai et al. 2014a).

Several studies have focused on Ge recovery from FA. Recovery methods have been pioneered which make use of the occurrence of Ge as water soluble species (GeS_2, GeS, and hexagonal-GeO_2) in gasification FA (Arroyo et al. 2009). The recovery process involves leaching of FA and a subsequent Ge separating process.

In China, Li et al. (2000) reported the extraction of Ge from low acidic leaching liquor of coal ash using dihydroxamic acid in sulphonated kerosene, indicating that the final germanium yield can reach up to 99% with a purity higher than 99.8%. The optimum experimental conditions and process parameters for preparing GeO_2 and Al_2O_3 by extraction and compound salt pyrolysis after leaching coal ash with sulphuric acid were reported by Liu et al. (1998). This method is a simple process, with a high recovery rate for Ge and Al. Ma and Xie (2003) described the laboratory extraction of U- and Ge-bearing coal ash by intensified leaching with reagents consisting of sulphuric acid and another acid (HA) or the salt compound of another acid. It was found that the method is not only economically reasonable but also technically feasible, with a Ge extraction yield of 85–90% (Ma and Xie 2003). Huang et al. (1999) proposed a new three-step process for Ge and Al extraction, which consists of extracting Ge by dilute acid immersion, extracting Al by concentrated acid immersion, and double salt decomposition under high temperature. In addition, the siliceous residues are used to manufacture cement. Shi and Zhu (2007) investigated a process for extracting germanium from coal ash by chlorination with ammonium chloride and extraction with dihydroxamic acid. NH_4Cl was used to chorinate the Ge from coal ash and the Cl was vaporized through hydrochloric acid and enriched, the hydrolysate of which was GeO_2. The overall Ge recovery rate from coal ash by this method was greater than 95% (Shi and Zhu 2007).

3.7.2.2. Gallium and aluminium recovery.

Gallium is widely used in applications such as optoelectronics, telecommunications, PV, aerospace, and many commercial and household items, such as alloys, computers, and DVDs. A super-large Ga ore deposit associated with coal was discovered in the Jungar Coalfield, Inner Mongolia, by Dai et al. (2006a, 2006b)). It has been found that the No. 6 Coal, which is the main seam of the Jungar deposit (average thickness about 30 m), is abnormally enriched in Al_2O_3 (up to 15%) and in gallium, which usually accompanies alumina, with a concentration of up to 76 mg/kg (Dai et al. 2006b). The Ga content can be 6–10 times further enriched in the FA after coal combustion.

There have been several approaches to the extraction of Ga metals from FAs. Zeng (2007) investigated the extraction of Ga using HCl leaching and adsorption of Ga by polyurethane foam from FAs collected from a power plant in Inner Mongolia, China. It was shown that the extraction efficiency by this method was higher than 80%. Zhao (2010) described an acid leaching process for extracting both Ga and Al from FAs collected from the Xian and Weihe power plants in Shaanxi Province, China; Ga was absorbed using polyurethane foam under acid conditions, which shows good prospects for industrial applications and may bring about very high social and economic benefits. In the case of the Jungar high-Ga-Al coals, a number of technologies, for example, acid and alkali methods or a salt activation method, have been developed to extract Ga and Al from the FAs (Guo et al. 2013; Feng et al. 2014; Li et al. 2014b, 2016). Al_2O_3, silica gel, and Ga have been successfully extracted from the FA at the lab scale and Ga recovery in the laboratory experiment was greater than 90% (Zhang 2008a). A pilot plant with an annual processing capacity of 800,000 t Al_2O_3 and approximately 150 t Ga was built at the beginning of 2011, which was operated for the extraction from coal ash on an industrial scale, not only for one rare metal (Ga), but also, for the first time, for a base metal (Al), and also silicon (Seredin 2012).

3.7.2.3. Uranium recovery.

Because of the radioactivity and the varying environmental paths of the daughter radionuclides, the concentrations of U, Th, and other radionuclides in coal and CCPs have been the subject of numerous studies (Lauer et al. 2015). On the other hand, when U is concentrated in coal to a level comparable to conventional U ore deposits, the coal has potential economic significance as a uranium source (Seredin and Finkelman 2008; Hower et al. 2016a). As mentioned above, there are some U-rich coals in Yili, Yunnan, and other places in China (Dai et al. 2015a, 2015b). Since it is not a volatile element, U is mostly concentrated in FA and BA after combustion. Uranium concentrations in CCPs are found to be more or less in proportion to the U concentration in the feed coal (Hower et al. 2016a). However, the natural radioactivity level of coal FA is 4–10 times higher than that of feed coals (Bhangare et al. 2014). As a consequence, the radioactivity of this CCP and its impact on the surrounding environment (including air, soil, and underwater) and human health may be significant (Qu et al. 2016). Hence, it is essential to treat FA with comparatively higher levels of radioactivity before reutilization and minimize its influence on ambient environment. Therefore, recovery of U from FA could be an eco-friendly way to resolve this problem.

The main method of extracting U from FA is acid leaching. However, the methodology for the extraction of U from coal ashes varies greatly for different coals and between regions, and there is no unified extraction method that has been applied to all U-rich FAs. The different U extraction methods depend on the combustion conditions (e.g. combustion temperatures, categories of raw coal, furnace types) and the modes of U occurrence in the raw coals. Ma and Xie (2003) concluded that U in a Chinese FA mainly occurs in the form of $Ca(UO_2)O_2$, and proposed an intensified leaching with reagents mixed with sulphuric acid and another acid (HA) or the salt compound of another acid, to extract both U and Ge from this U-Ge-rich FA by means of a two-stage countercurrent method. The extraction rate for U from FA may be as high as 98% under optimal conditions. Sun et al. (2016b) reported the occurrence of U into aluminosilicates (mainly glass phase) and Fe-oxides of FA. The extraction of U was achieved by Tessier sequential extraction, acid leaching, magnetic separation, and mechanical activation combined with alkali activation, as well as calcination with $CaCl_2$ followed by HNO_3 leaching. Using this method, the recovery of U reached up to 95.8%.

3.7.2.4. Aluminium recovery.

Some coal FAs are rich in Al, making them a potential source of Al_2O_3. Due to diminishing bauxite resources, as well as the increased demand for Al_2O_3, the profitable industrial utilization of coal FA in Al recovery has attracted extensive attention (Rayzman et al. 1997).

Grzymek (1976) pioneered the recovery of Al_2O_3 from coal FA, and from this study different methods have been proposed. These methods can be grouped into three types: sintering process, acid leaching process, and high-temperature chlorination process (Yao et al. 2014, 2015). The acid leaching process includes direct acid leaching, improved leaching, and a sinter–acid leaching combination (Yao et al. 2015). In China, many possible approaches have been developed for the recovery of Al from coal FA, for example, a sintering-alkalization method, an NH_4 F-solubilization-acidification method with CaO, polyacrylamide dispersant to produce $Al(OH)_3$, and solubilization by microwave (Liu and Li 2006; Fu et al. 2014; Zhang et al. 2014a). Bai et al. (2010a) described an FA with an extremely high Al concentration of 40–45% and its potential to produce aluminium commercially. An SiO_2 extraction step was carried out first using NaOH solution to increase the Al_2O_3/SiO_2 ratio of the ash, after which the desilicated ash was mixed with CaO and Na_2CO_3, and sintered before Al_2O_3 extraction with a Na_2CO_3 and NaOH solution. An Al extraction efficiency of 90% was achieved using this lime-soda sinter alkali leach method. Ji et al. (2007) calcined FA with soda at 900°C to yield soluble aluminates. The sinter was then leached with H_2SO_4 to produce a solution containing Al. The extraction efficiency was greater than 98%. Wang et al. (2012) reported the extraction of $Al(OH)_3$ from coal FA by pre-desilication and calcination methods.

3.7.2.5. Recovery of rare earth elements and yttrium (REY).
The demand for rare earth elements and yttrium (REY) has grown due to their wide applications as metal catalysts, phosphors, LEDs, permanent magnets, various components for renewable green energy equipment, and batteries (Pecht et al. 2012; Hower et al. 2016b). It has been reported that 86% of the world's REE production took place in China in 2013, and China controls 48% of the global REE supply (Taggart 2015). However, because of the supply crisis of 2010 and the price spike of 2011 (Massari and Ruberti 2013), searching for alternative REY source other than the conventional REY ore deposits that are being exhausted is of high significance. Elevated REY levels have been reported in some coals (Finkelman 1981; Seredin 1996, 1998), and REY in coals have attracted attention in recent years worldwide (Arbuzov et al. 2000; Nifantov 2003; Seredin et al. 2006, 2013; Seredin and Dai 2012; Hower et al. 2013, 2014, 2015, 2016b, 2016c; Mayfield and Lewis 2013). REY have been reported to be enriched in some Chinese coals (Daqingshan and Jungar coalfields,

Inner Mongolia; Songzo coalfield, southwestern China; eastern Yunnan) and/or CCPs (Zhuang et al. 2000; Dai et al. 2007, 2010b, 2011; Zhuang et al. 2012; Seredin et al. 2013; Zou et al. 2014; Dai et al. 2015c, 2016a, 2016b, 2017; among others). When REY concentrations in some coals and/or coal ashes are comparable to or even higher than those in conventional REY ore deposits (e.g. carbonatites, alkaline granites, and weathering crusts; Seredin and Dai 2012; Seredin et al. 2013), the REY-rich coal deposits may become important alternative sources for REY (Seredin and Dai 2012; Hower et al. 2015, 2016b).

The recovery of REY from coal or coal ash is a complicated multi-step process, which generally includes initial acid leaching of the ash material, followed by the removal or precipitation of undesired minerals, and purification using solvent extraction (Gupta and Krishnamurthy 2004; Mayfield and Lewis 2013). In China, mature procedures for industrial REY recovery from coal FA are virtually non-existent, although lab-scale REY recovery from FA has been only sporadically reported. Dai et al. (2014b) revealed that the concentrations of the critical REEs were greater in the finer-sized fractions (454 LREE mg/kg in the <25 μm fraction) of FA than in the coarser fractions (156 mg/kg in >125 μm fraction) of FA from the Jungar power plant, Inner Mongolia. However, the feasibility of recovering REEs from FA on an industrial scale needs to be explored.

3.7.2.6. Recovery of other valuable elements.
As stated above, co-extraction of Al with other valuable elements, such as Ge and Ga, from Chinese FAs has been investigated at the lab and pilot scale by several researchers (Liu et al. 1998; Huang et al. 1999; Zhang 2008a; Zhao 2010; Seredin 2012). In addition, V has been reported to be recovered from coal FAs (Zhang et al. 2008). Hou et al. (2015) proposed a leaching process to recover Li from FA using Na_2CO_3 and $CaCO_3$, from which Li_2CO_3 was recovered by calcination activation using carbonate, and leaching from the Na_2CO_3 solution. The recovery of Li can reach 70% using this method.

3.7.3. Geopolymer
Geopolymer, an inorganic polymeric material, is an end-product arising from the alkali activation of aluminosilicate materials (Davidovits 1991). Geopolymerization involves a chemical reaction between aluminosilicate and alkali metal silicate solutions under strongly alkaline conditions and atmospheric pressure at temperatures below 100°C, which yields an amorphous to semi-crystalline three-dimensional aluminosilicate polymeric structure with an empirical formula of $M_n[(SiO_2)_zAlO_2]_n$.

wH_2O, where z is the Si/Al molar ratio, M is an alkali cation, such as Na^+ or K^+, n is the polymerization degree, and w is the water content (Palomo et al. 1999). Geopolymers have similar binding performance to ordinary Portland cement (OPC), but higher strength and lower permeability than cement and concrete, which allows them to be used in the building industry and to encapsulate trace pollutants. Conceptually similar to zeolite synthesis, the alkaline activation of an aluminosilicate material to synthesize geopolymer can be described as the reaction of a liquid with a high alkaline concentration and a solid with a high proportion of reactive silicate and aluminate. The occurrence of a dominant and reactive amorphous silico-aluminate matrix in FA makes it a suitable raw material for the synthesis of geopolymer. The basic and simplified principle of the formation of an FA-based geopolymer is the alkali-facilitated decomposition of aluminosilicate in the FA and then polycondensation (Izquierdo et al. 2010; Zhuang et al. 2016). The alkali activation is usually conducted by adding NaOH, KOH, Na_2SiO_3, or K_2SiO_3 solution into FA together or individually, and the synthesis can proceed at room temperature or slightly elevated temperatures (usually<100°C), giving rise to minimal CO_2 emission (Davidovits 1991; Verdolotti et al. 2008). The simple alkaline activation of FA to produce geopolymer completely bypasses the high-temperature calcination process in OPC production. Moreover, the toxic trace metal elements in FA can be trapped and immobilized in the geopolymer structure (Li et al. 2013), so production is considered to be energy and source efficient.

In the last few decades, intensive research has been carried out on the synthesis of geopolymers from Chinese FAs. Jia (2009) used FA, NaOH, and sodium silicate as the main raw materials to prepare geopolymer materials. The effects of synthesis conditions, the ratio of the materials, setting time, and Ca additions on the mechanical properties of FA-based geopolymers were investigated. Chou (2015) prepared FA-based geopolymer through an alkali-activation method, and investigated the structure stability and performance of the products under high temperatures and different chemical environments. Zhuang et al. (2016) have summarized the scientific advances in the preparation, properties, and applications of FA-based geopolymers. Many modified methods have also been investigated to improve the reactivity of FA, as well as the production and the performance of the FA-based geopolymer. Some other waste materials, such as slag, fibre, rice husk-bark ash (RHBA), and red mud, have been added into FA to synthesize geopolymers (Zhuang et al. 2016). Use of these waste materials can also help reduce the consumption of mineral resources, such as limestone and clays, for geopolymer synthesis, and reduce the emission of GHGs. It was found that the addition of slag, which is the waste from Fe extraction processes, can enhance the reactivity of FA during geopolymerization (Li and Liu 2007).

There are numerous factors that affect the desired properties of geopolymers. The Si/Al ratios, the type and the amount of the alkali solution, the temperature, the curing conditions, and the additives are critical factors in a geopolymerization process, which may further influence the properties of geopolymers (Zhu 2005; Izquierdo et al. 2010; Chou 2015; Zhuang et al. 2016). The mechanical properties of FA-based geopolymers are influenced by the synthesis conditions, the ratio of the materials, setting time, and Ca additions (Jia 2009). Variations in the chemical composition of the coal feedstocks and details of the combustion processes applied in the different facilities may result in significant differences in the chemical and physical characteristics of the resulting FAs. The microstructure and mechanical properties of the FA-based geopolymer are also strongly affected by the concentrations and types of alkaline solution. The type and concentration of the alkaline solution influence the release of Si^{4+} and Al^{3+} from the ash during geopolymerization, and the leaching rate of Si^{4+} and Al^{3+} further determines the real available Si/Al ratio in a series of reactions to form the geopolymer and subsequently plays a pivotal role in the structure of the FA-based geopolymer (Zhuang et al. 2016). Zhu (2005) analysed the influence of different factors on the strength of the geopolymer and found that the alkali additives are the main factors having an impact on the strength of the FA geopolymer. Finally, the curing conditions exert a significant effect on several properties of the bodies formed. Open curing conditions enable water evaporation and, as a consequence, produce solid geopolymer bodies characterized by high porosity, low compressive strength, and exacerbated leaching of certain oxyanionic metalloids. By contrast, closed or protected curing promotes binder development, giving rise to higher strength and less-porous systems (Izquierdo et al. 2010). Furthermore, the geopolymerization of FA is influenced by the curing temperature. The activation of the amorphous glass in FA is not completed at room temperature (Xu et al. 2010; Chen et al. 2011). It is found that the reactivity of FA becomes higher when the curing temperature increases from 30 to 50°C, and the geopolymerization is almost complete when curing temperature lies between 60 and 90°C (Hardjito et al. 2004).

Geopolymers exhibit good physical, chemical, and mechanical properties, including low density, micro- and nano-porosity, low shrinkage, high mechanical strength, good thermal stability, durability, surface hardness, fire, and chemical resistance (Zhuang et al. 2016). Given these desirable properties, they are seen as potential alternative materials for industrial applications such as cement-based construction, transport, aerospace, mining, metallurgy, and adsorption and immobilization of toxic metals material (Jia 2009; Zhuang et al. 2016). The major focus is on the fact that they can replace the use of OPC as a binder in concrete applications (Yin et al. 2014; Zhuang et al. 2016). Second, heavy metals and hazardous elements, such as Ba, Cd, Co, Cr, Cu, Nb, Ni, Pb, Sn, and U, can be tightly and effectively fixed in the three-dimensional structure of the FA-based geopolymer matrix, based on the zeolite-like and porous structure of the geopolymer (Chou 2015). The FA-based geopolymer shows higher immobilization of metal ions than OPC and FA (Chou 2015). The mechanisms of metal immobilization in FA-based geopolymer are mainly physical encapsulation rather than chemical stabilization (Izquierdo et al. 2009, 2010). Li et al. (2013) compared the immobilization of $^{133}Cs^+$ in FA-based geopolymer and OPC. It was shown that the FA-based geopolymer showed different metal immobilization behaviours in acid and sulphate solutions. The difference in leaching of $^{133}Cs^+$ between the geopolymer and OPC in H_2SO_4 solution was small, whereas the leaching of $^{133}Cs^+$ from the geopolymer in deionized water and 5 wt.% $MgSO_4$ solution was much lower than that in OPC. In addition, FA-based geopolymer materials have been used as cement, low-temperature ceramics, chemical reactor materials for corrosion resistance, lightweight thermal insulation material, and handicrafts material (Jia 2009).

3.7.4. Foam glass

Foam glass is a porous heat-insulating and soundproof material, with true porosity up to 90 vol.%. The desirable properties of foam glass are high strength, low density, high surface area, high permeability (intercommunicating pores), low specific heat, low thermal conductivity, high thermal and acoustic insulation as well as high chemical resistance (Hurley and Consultant 2003). The principle of the foam glass manufacturing process is that, between 700 and 900°C, the glass powder changes into a viscous liquid and then the foaming agent decomposes to form a gas, which in turn forms bubbles. The glass needs to have sufficient viscosity not to allow the gas bubbles to rise through the mass of the body but remains in position during the foaming heat cycle (Hurley and Consultant 2003).

The production of foam glass may follow two distinct processes. The first, dating back to the 1930s, involves the direct introduction of gases ('blowing') into molten glass (Lytle 1940). The second is the powder method (Demidovich 1972). The essence of the powder method involves sintering a mixture of glass powders and special powder additives (foaming agents), facilitating the formation of a gaseous phase during firing. When the temperature of the mixture exceeds the softening temperature, the glass particles start sintering and form a continuous sintered body. Particles of the pore-forming agent become insulated by softening glass. After a certain temperature is reached, they start emitting gases frothing the glass melt. Due to gas emission, pores emerge in all parts of the sintered body where the particles of the pore-forming agent were blocked. The shape of pores and the properties of foam glass obtained largely depend on the concentration and type of the foaming agent used (Spiridonov and Orlova 2003). Owing to the significant energy savings (sintering of the glass occurs at a much lower temperature than that required for blowing) as well as the environmental advantages (reuse of the crushed recycled glass), currently all glass foams are produced by the powder method.

Over the past few years there has been an increasing interest in the production of foam glass from FA in China (Fang et al. 2003; Chen 2012; Li 2012; Zhang 2013; Song 2014; among others). Fang et al. (2003) introduced the principles for producing foam glass by using FA, and investigated the factors that influence the quality of the foam glass. Li (2012) investigated the softening temperature of a Chinese FA used for producing foam glass and found that due to the relatively high softening temperature of the FA, a dose of Na_2CO_3 was necessary to lower the melting temperature, and thus the production costs. In addition, SiC was selected as a foaming agent due to the oxidation reaction during the firing process (Li 2012). Zhang et al. (2014b) obtained foam glass with the best quality when foamed for 60 min at 850°C, and sintered for 60 min at 950°C, with 37% FA, 58% waste glass, and 5% clay as the main raw materials, 7.5% Na_2CO_3 and 2.5% carbon powder as a foaming agent, 1.5% Na_3PO_4 as a stabilizing agent, and 1.5% H_3BO_3 as a fluxing agent. Clay could increase the strength of the foam glass (Zhang et al. 2014b). The final properties of the foam glass are usually highly influenced by the types and doses of the foaming agent used, the foam temperature and time, and the sintering temperature and time, as well as the doses and the fineness of the FA used (Fang et al. 2003; Chen 2012; Liu et al. 2012; Zhang 2013). Temperature and heating rate are vital factors in the foaming process (Li 2012). It was found that the key to the technique is to efficiently control the foaming temperature to ensure matching of both the softening temperature of the body and the gas temperature from the foaming agent, so as to form enclosed pores of even distribution and give the structure high compressive strength (Li 2012). If the temperature is too high, the bubbles will rise and the body will collapse and not form foam (Hurley and Consultant 2003). Control of the heating rate is one of the most important factors in optimizing the foam glass product. Rapid heating can cause the foam glass feedstock to crack, whereas slow heating will lead to early release of the gas from the foaming agent before the viscosity of the glass is low enough to allow the glass to expand (Li 2012). Owing to its excellent structural properties, foam glass is suitable for use as insulation in roofs, walls, and high-traffic areas such as floors, where other insulation products may be compressed, resulting in an uneven surface and the loss of insulating properties (Fernandes *et al.* 2009). It is also used as industrial insulation for a number of purposes such as sandwich panels, or is used as a thermal insulation material in extreme environmental conditions (Gu 2007). As the product is made of glass, it is naturally inert in most environments with respect to biological, thermal, chemical, and environmental degradation (Li 2012). Furthermore, compared with the polymeric foams currently widely employed, glass foams possess significant advantages for thermal and acoustic insulation applications (Chen 2012).

4. Applications of BS

The particle size and the durability of BS are considered as the main additional values of this material. These advantages make slag suitable for various beneficial applications, such as fine aggregate in asphalt paving and structural fill, among others (NCASI 2003). Slag has also been reported to be used as a granular base, for soil stabilization/waste solidification, as a component of blasting grit and roofing granules, a fill material for embankments, as a raw material in concrete products, and as a snow and ice traction control material (NCASI 2003).

Owing to the similarity of the chemical composition of slag to that of FA, slag can also be used in roadway pavement and filling, and used for slag brick production and other low-value-added construction applications (Zhou 2009). In China, slag has been reported to be used for cement and concrete production, for roadway pavement, and for the production of briquette coal and building materials (Wu 1993). Apart from these low-value-added applications, a number of studies have been carried out to investigate the possible high-value-added utilization of slag or mixtures of slag and other waste materials. The utilization of BS generated from Chinese power plants is reported in detail below.

4.1. Application for reclamation of a burnable product

Due to incomplete coal combustion, the slag generated from power plants may contain a certain proportion of unburned carbon, which could represent a waste of otherwise useful resources and result in environmental pollution while in ash repositories (Chen 1992). If slag or coal ash has a significant amount of unburned carbon, it cannot be utilized directly in applications such as concrete and concrete products. According to ASTM C-618, the LOI of ash must not exceed 6% for concrete applications; and an upper LOI limit of 3% is more realistic. Ash with higher LOI cannot be used because of colour problems and concerns with the use of admixtures, especially for durability under freezing and thawing conditions (Tharaniyil 2013). Therefore, slag with high unburned carbon content should be treated to reduce the carbon content before being used for concrete and concrete production.

Reclamation of a burnable product is an alternative for slag utilization. Since slag from power plants may include both completely burnt coal residues and partly burnt coal, Chen (1992) proposed an ore dressing technology to separate the slag into two (burnable and unburnable) products. The reclamation of a burnable product and the utilization of the unburnable material are of social and economic interest. According to Wu (1993), the unburned carbon content in some slag may be over 30%, and have a relatively high calorific value. Hence, these kinds of slag could be used as raw materials for the production of honeycomb briquettes. Honeycomb briquettes with comparable thermal efficiency to the ordinary commercial ones were produced using mixtures of slag, coal, additive (CaO, Fe_2O_3, among others), and binder (mostly loess and clay). Almost 40% slag can be added to the mixtures, giving rise to briquettes with an acceptable thermal efficiency (Wu 1993).

4.2. Application for cement production

Slag can be used as an additive for the production of common Portland cement and composite Portland cement (Guo and Zhu 2002). The slag is ground to a suitable fineness, and then mixed with various raw materials, such as gypsum and lime, or gypsum and cement clinker, to produce different types of cement

(Wu 1993). Guo and Zhu (2002) prepared common Portland cement and composite Portland cement using mixtures of coal slag, cement clinker, local ore slag, and zeolitic materials. It was found that the mechanical strength of the common Portland cement made from coal slag and cement clinker was higher than that of the composite Portland cement made from coal slag, cement clinker, local ore slag, as well as zeolitic materials. Furthermore, incorporation of 10–15% of slag in the raw materials produced a Portland cement with good mechanical strength. The increase of slag dosage may give rise to an increase of normal consistency and setting time for cement (Guo and Zhu 2002).

The unburned carbon in slag may have an adverse influence on its activity. Hence, when used for cement production, a lower unburned carbon content is needed for slag, usually below 10%, and the lower, the better (Wu 1993).

4.3. Application for concrete production

Slag has been used as an aggregate for the production of concrete (Wu 1993; Zhang *et al.* 2006; Pan 2011). Prior to preparation, the slag is ground to 3–20 mm and 0.5–3 mm. Appropriate proportions of ground slag are then incorporated into a mortar made from cement and highly absorbent resin, and mixed to produce concrete. The proportion of slag in the raw material aggregates can range from 55 wt.% to 72 wt. %, and the produced concrete is characterized by low density, and high insulating, acoustic, water-proofing properties (Wu 1993). Yue *et al.* (2011) reported the preparation of an ecological-type cement from coal FA, FGD gypsum, cement clinker, and ore slag. Then they investigated the production of insulating masonry mortars using slag as the aggregate (55–75 wt.%) and the prepared ecological-type cement as binder. With the addition of certain additives, masonry mortars with high mechanical and insulating properties were produced (Yue *et al.* 2011). Zhang *et al.* (2006) ground slag to the fineness of first class ash, and then investigated the preparation of concrete using this ground ash as a main raw material. It was found that the ground slag can replace part (up to 42 wt.%) of the common Portland cement used for the industrial production of concrete with various mechanical strengths, which is of high economic and environmental significance. Pan (2011) found that for high-performance concrete, compared with other admixtures with the same dosage at 30%, coal slag powder concrete had higher compressive strength.

4.4. Application for building materials

Slag can be used for the production of slag bricks, hollow block bricks, and slag roofing tiles (Wu 1993). Slag has been used as an additive and mixed with clay and cement to produce slag bricks, the comprehensive and flexural strength of which is higher than the standard limit for common clay bricks (Wu 1993; Zhou 2009). In addition, slag can be used for the production of hollow block bricks because during the sintering process, unburned C in the slag is burned or decomposed to gases, giving rise to a number of pores inside the bricks (Wu 2009). It has been reported that slag could replace volcanic ash materials as aggregate, and mixed with cement to prepare hollow block bricks. The slag-based hollow block bricks are characterized by good mechanical, insulating, and acoustic properties (Wu 1993, 2009).

4.5. Application for roadway pavement

BS has occasionally been used as an unbound aggregate or granular base material for pavement construction. Slag is considered as a fine aggregate in this application. Screening or grinding of oversized particles will typically be required to use slag in paving applications (Tharaniyil 2013). Moreover, to meet the required specifications, the slag may need to be blended with other natural aggregates prior to its use as a base or sub-base material.

In China, slag has been investigated to replace sand as a raw material for the production of base coarse materials for road pavements (Wu 1993; Zhou 2009). Base coarse materials with high strength, thermal insulating, and antifreezing properties were successfully obtained using mixtures of slag with clay and FA (Wu 1993); these can also be used as roof insulating materials. In addition, slag has also been mixed with other materials, such as limestone, river sand, clay, or gravel, to produce different kinds of road pavements, including common factory roadways, footpath pavings, and stadium runway pavements (Wu 1993). Furthermore, road pavement bricks can be produced using the mixtures of slag and cement (Wu 1993; Zhou 2009).

It was worth noting that pyrite (FeS_2) is volumetrically unstable, expansive, and produces acidity and a reddish stain when exposed to water over an extended time period. Therefore, pyrite that may be present in the slag should be removed (with electromagnets) prior to use (Tharaniyil 2013).

4.6. Application for waste-water treatment

Due to the physical properties of slag, it has been tested for use in various waste-water treatments (Zhou

2009). Slag has a large surface area, strong adsorption ability, and low price, resulting in prospects for wide application in waste-water treatment (Zhang 2008b). Slag has been experimentally used as filter material to treat the waste water from paper mills (Wu 1993). It has also been used to purify biological waste water, and to decolourize and remove phosphate anions from waste water, with a decolourization rate of 70–75% (Wu 1993). Zhang (2008b) reviewed the application of slag to the treatment of dyeing and printing waste waters as well as its decolourization mechanism.

If the slag is ground to a fineness of 1.2–1.5 mm, it could be used as a decolourizing agent to treat paint waste water (Wu 1993). In addition, slag with high ash yield has been used as an efficient adsorbent to remove heavy metals in waste water (Zhou 2009).

4.7. Applications as road acoustic barriers

Since noise is becoming an increasingly significant concern because of its negative impact on human health, it is necessary and of prominent importance to develop some traffic noise barriers. One of the most common sound-absorbing materials used for noise barriers is a combination of lightweight concrete with a hard backing consisting of standard concrete (Kim and Lee 2010). Lightweight concrete is often prepared by gap grading the coarse aggregates and either eliminating or minimizing the sand volume in the matrix to develop a network of interconnected pores within the material. Consequently, the pores inside the material absorb sound energy through internal friction (Nelson and Phillips 1994).

In China, a few researchers have discussed the manufacture of acoustic barriers using slag (Jia 1994; He *et al.* 2010, 2012; Li 2012). Jia (1994) prepared a sand-free macroporous slag concrete to use in noise barriers, which highly reduced the noise from a primary school. He *et al.* (2010) obtained a porous material made of granulated furnace slag through the pressing moulding method, using silica fume or cement as a caking agent, and the results showed that the material had good sound-absorption properties with average sound-absorption coefficient over 0.70 and compressive strength over 3.0 MPa. Li (2012) prepared sound acoustic barriers from mixtures of 20% OPC and 80% slag with different grain sizes (coarse, >5 mm; medium, 1.25–5 mm; and fine, <1.25 mm). Homogeneous mixtures were mixed with appropriate amounts of water to obtain homogeneous pastes. The acoustic barriers were produced by moulding, demoulding (after 24 h), and curing at ambient temperature for 28 days. It was found that the three acoustic barrier products had similar density and compressive strength to those made of conventional aggregate materials. The obtained noise absorption coefficients were also similar to those obtained from commercial coarse porous cement. Furthermore, the mechanical and acoustic behaviours of the slag-based acoustic barriers are highly influenced by the grain size of the slag. The noise absorption coefficient increases with an increase of grain size. In contrast, the density and compressive strength of acoustic barriers decrease with an increase in slag grain size (Li 2012).

As stated above, the production of road acoustic barriers can consume large amounts of slag, and the production process is completed at ambient temperature, which drastically reduces the manufacturing cost (Jia 1994; Li 2012). Furthermore, there is no new waste generated from the production process. Hence, the slag-based road acoustic barriers can be widely used as pavement or high-traffic road bases, with low economic cost and low environmental threat.

4.8. Other applications

Apart from the above-mentioned applications, slag can also be used for soil amendment and backfill, as well as used as anti-skid materials. Slag has been used as a soil-amending agent to increase the workability and porosity of soil (Wu 1993; Zhou 2009). It has been shown that land application of slag had no negative effect on crops or soil during a five-year period of study. Furthermore, crop yield from the slag-treated soils was generally higher than that from the non-treated soils (Tharaniyil 2013).

The suitability of slag as a backfill material can be understood from its close resemblance to commonly used natural granular backfill materials. In most cases, the most critical factor is the gradation or particle size distribution of backfill material (Zhou 2009).

5. Applications of FGD gypsum

As mentioned above, FGD gypsum is similar to natural gypsum (a hydrated form of calcium sulphate, $CaSO_4.2H_2O$) as far as its overall composition is concerned, and hence can be used in many of the same manufactured products as natural gypsum. The primary value of FGD material is its chemical composition ($CaSO_4$), which renders it suitable for a variety of beneficial use applications, especially in the construction and agricultural industries (NCASI 2003). However, there are differences between the FGD and natural gypsum, which can either restrict or enhance the use of FGD gypsum in place of natural gypsum. For instance, FGD gypsum has a higher moisture content, which, combined with its fine grain size, can affect handling and processing in manufacturing facilities designed for natural gypsum. Moreover, FGD gypsum requires less grinding than natural gypsum due to its finer grain size.

Most new plants for producing wallboard are designed to accommodate FGD gypsum, either solely or in combination with natural gypsum. On the other hand, the Cl^-, F^-, Fe, Mn, and Si-Al FA contents of FGD gypsum can cause issues such as surface crystallization that can affect paper adherence and colour variation, which makes it undesirable for some products and applications (Ladwig 2006). Furthermore, although the FGD systems (typically based on the wet limestone FGD process) also act as effective systems for the removal of HF from the flue gas (Meij 1994; Álvarez-Ayuso et al. 2006), it has been found that the leachable contents of F^- and SO_4^{2-} may be very close to the prescribed limit values, which could risk the classification of FGD gypsum as acceptable at landfills for non-hazardous wastes (Álvarez-Ayuso and Querol 2008; Álvarez-Ayuso et al. 2008). This fact, in addition to the corresponding environmental concerns, represents an important issue for coal-combustion power plants because of the high cost of disposal of this CCP if it is only acceptable at landfills for hazardous wastes (Álvarez-Ayuso et al. 2008).

Currently, FGD gypsum generated in China has primarily been used as a cement retarder and for gypsum plaster board, which accounts for 96% of the total utilization of FGD gypsum (Tian et al. 2006). By contrast, approximately 88% of the natural gypsum produced was used in the cement industries (Wang and Wu 2004). Another important application of FGD gypsum is in agriculture. In addition, there are applications such as fill material for structural applications and embankments, raw material in concrete products, and solidification. Yang et al. (2011) analysed the physico-chemical properties of FGD gypsum in Xinjiang, China, and pointed out the application development of FGD gypsum, such as cement retarder and building plaster. Zhang et al. (2015) reviewed the properties of FGD gypsum and comprehensive utilization methods, and provided suggestions for the use of FGD gypsum in China. The detailed applications of FGD gypsum in China are reported as follows.

5.1. Application as a cement retarder

Portland cement is utilized throughout the construction industry in a variety of applications, one of which is the production of concrete. Portland cement consists of five major compounds: tricalcium silicate (50%), dicalcium silicate (25%), tricalcium aluminate (10%), tetracalcium aluminoferrite (10%), and gypsum (5%). When water is added to cement, each of these compounds undergoes hydration, resulting in the final hardened product.

With proper treatment, the quality of FGD gypsum can be superior to natural gypsum, which makes it more feasible to replace natural gypsum as a cement-retarding agent (Tharaniyil 2013). In China, most of the gypsum has been used as a cement retarder, accounting for 80% of the total gypsum utilization (Mao 2005). Feng et al. (2009) used FGD gypsum as a CaO resource substituting for limestone to prepare cement clinker. It was shown that clinkers could be prepared using FGD gypsum as a CaO resource at 1350°C, which is lower than the temperature when using limestone as a CaO source. Lu (2013) first prepared cement clinker by burning FGD gypsum with anthracite, SiO_2, Al_2O_3, and Fe_2O_3, under optimal temperature (1450°C) and insulation time (60 min), and then mixed the clinker prepared from the best calcination conditions with FGD gypsum to produce Portland cement. It was found that the best amount of FGD gypsum as a retarder is 5%, and the cement prepared with 5% FGD gypsum met the requirements of Chinese national standards. Zhou (2011) reported that, at the same content of SO_3, FGD gypsum had more obvious retarding activities compared with natural gypsum, and its mechanical enhancement effect on cement was also superior to the latter.

Despite the advantages of using FGD gypsum as an efficient cement retarder, there are also some limitations. First, due to the commonly used wet desulphurization technology, the FGD gypsum is characterized by a high moisture content, and moisture may adhere to the production equipment, resulting in blocking. In addition, it has been shown that an excessive water content of the FGD gypsum was deleterious for cement properties, and that a drying process was required before its application (Zhou 2011). In addition, during the power-generation process, impurities such as FA, $CaCO_3$, and $CaSO_4.0.5H_2O$ may be incorporated into FGD gypsum, which may influence the retarding effects. For instance, a higher content of $CaSO_4.0.5H_2O$ will increase the setting time of the cement (Zhang et al. 2015).

5.2. Applications in agriculture

According to statistical data (Agriculture Department of China 2015), there are around 340,000 km² of saline-alkali soil in China, amongst which around 124,000 km² can be potentially used for agricultural production after amelioration. The amelioration of saline-alkali soils over such an enormous area is one of the greatest challenges facing Chinese agriculture. For many centuries, gypsum has been used in agriculture as a soil amendment, conditioner, and fertilizer. The amelioration of saline-alkali soil using FGD gypsum would make use of tens of millions of tons of FGD gypsum, thereby boosting the application of FGD technology and the development of the pollution-control industry. In addition, the huge area of barren saline-alkali soils ameliorated with the FGD gypsum would then be suitable for growing agricultural crops. Hence, it would be of significant benefit to both agricultural development and the improvement of local ecosystems (Wang *et al.* 2016c).

5.2.1. Use as fertilizer

Calcium is an essential component of plant cell wall structures, providing strength in the plant. It plays a role counteracting the effects of alkali salts and organic acids within the plant. Owing to its chemical components, gypsum provides soluble sources of Ca and S, supplying the needed nutrients and improving plant growth. Sulphate in the gypsum is in the most favourable form for the plant roots to absorb S to enhance crop production and increase resistance to environmental stresses and pests. Hence, gypsum can be used as a S fertilizer. Furthermore, without the use of gypsum, soil compaction prevents root penetration, aeration, and water infiltration. The loss of soil permeability causes saturation of the soil with salt or other elements that can be harmful to plant growth and health.

FGD gypsum has been widely used as a fertilizer for various plants in China. Bai *et al.* (2010b) suggest that the application of 22.5 t/hm^2 FGD gypsum exerts optimal effects on increasing the rice resistance on saline-alkali land. Too low or too high a dosage of FGD gave rise to adverse effects on rice growth. Mao and Li (2016) conducted a set of pot experiments to assess the effect of application of FGD gypsum at different rates (0–50 g/kg) on the chemical properties of saline-Na soil in Shanghai tidal flats and plant growth. The results showed that soil salinity can be decreased by appropriate rates (25 g/kg) of FGD gypsum addition, promoting plant growth, whereas the excessive application of FGD gypsum (50 g/kg) is seriously harmful to plant growth (Mao and Li 2016). According to Wang *et al.* (2016c), during a test on the feasibility of FGD gypsum as a soil amendment in some alkali soils in the Ningxia Autonomous Region, the helianthus planted grew well and produced a good harvest even without additional fertilizer application. Wang *et al.* (2016b) evaluated the feasibility and effectiveness of FGD gypsum on

improving the typical takyric solonetzs in the north of the Yinchuan Plain of the Ningxia Autonomous Region, and the west of Hetao Plain and the desert steppe in N Xinjiang in China. It was found that by using FGD gypsum stimulation, the soil pH value and exchangeable Na$^+$ percentage decreased, facilitating the growth of rice, and thereby increasing the rice yield. The most suitable amount of FGD gypsum applied to reclaim takyric solonetz farmlands and improve rice quality was 31.5 t/hm^2, which had an obvious effect.

Despite the promising utilization as a fertilizer, it is worth noting that as it is a CCP generated from coal-fired power plants, some F$^-$, metalloids, and heavy metals present in FGD gypsum may be released to the soil or groundwater, or be adsorbed by plants when using it as fertilizer, causing potential environmental problems. However, previous studies have shown that such utilization of FGD would not cause secondary pollution of the soil. Wang *et al.* (2008c) found that, after using FGD gypsum as fertilizer, the concentrations of some toxic elements, Cr, Pb, Cd, As, and Hg, in the seeds of corn and lucerne were still lower than the standard limit for Chinese food security. Wang *et al.* (2016b) indicated that the proper use of FGD gypsum will be helpful to fertilize the soil, thereby increasing production and improving rice quality. Moreover, the contents of heavy metals in the soils meet the requirements of the Hygienic Standard for Grains in China (GB/T2715-2005), which implies that FGD gypsum will not deteriorate the soil environmental quality. Therefore, utilization of appropriate amounts of FGD gypsum in this way will not cause environmental pollution.

5.2.2. Use for soil amendment

In addition to nutritional benefits, FGD gypsum can provide many physical and chemical benefits to soil as well. Ca^{2+} in FGD gypsum can exchange with Na$^+$ in the soil colloid, which can change the composition of soluble salt cations in the soil, reduce the saline alkaline degree of soil, and buffer the alkalinity or acidity of the soil (Frenkel *et al.* 1989). Without adequate Ca, the biochemical uptake mechanism would fail. Soils that are Ca-deficient in humid regions have the tendency to disperse and form a stable suspension of particles in water. In other words, highly hydrated ions, such as Na$^+$ or Mg^{2+}, repel the clay particles, causing soil erosion. Adding gypsum allows amendment for sodic soil reclamation (Chen and Dick 2011).

First, FGD gypsum can improve the soil structure with flocculation effects for root growth and air and water movement. It can also prevent crusting of the soil and aid in seed emergence (Tharaniyil 2013). The Ca^{2+} that is provided by FGD gypsum exchanges with Na$^+$

and Mg^{2+}, leading to clay flocculation in the soil. Clay flocculation is the coagulation of the individual clay particles into microaggregates (Chen and Dick 2011), thus improving the soil structure for root growth and air and water movement. The flocculation also prevents crusting of the soil and aids in rapid seed emergence for no-till field crops.

Furthermore, FGD gypsum may correct for subsoil acidity and Al^{3+} toxicity (Tharaniyil 2013). Plants growing in acid soils can be chemically detrimental as they may be prone to high concentrations of soluble aluminium. Subsoil acidity prevents root exploitation of nutrients and water in the subsoil horizons (Chen and Dick 2011). Although the soil has low pH, the presence of high levels of exchangeable Al^{3+} makes it very toxic to most plant roots. As a neutral salt and not a limiting agent, gypsum does not change the soil's pH but rather enhances the root tolerance to acid subsoils. The addition of FGD gypsum can ameliorate phytotoxic conditions arising from excess soluble aluminium in acid soils by reacting with Al^{3+} to form soluble complexes ($CaSO_4$ $+Al^{3+} = Al(SO_4)^+ + Ca^{2+}$), thus removing Al^{3+} from solution in the soil and reducing its toxic effect on the plant roots (Chen and Dick 2011). This also leads to an increase in Ca^{2+} supply at lower depths for root uptake of water and nutrition from the subsoil layers. In addition, FGD gypsum can improve infiltration rates and hydraulic conductivity of soils to provide adequate drainage, and reduce erosion losses of soils and nutrients and P concentrations in surface water runoff (Tharaniyil 2013). Gypsum utilization can also improve the water-use efficiency of crops that are grown in dry areas or during times of drought (Chen and Dick 2011). Hence, with FGD gypsum application, the soluble Ca binds with the soluble reactive phosphate, forming insoluble calcium phosphate ($Ca_3(PO_4)_2$) precipitates, improving the water quality with decreased runoff.

In China, applications and effects of FGD for soil amendment have been extensively investigated at both laboratory and pilot scales. Takyric solonetz, a typical subclass of alkali soil, is widely distributed in the north of the Yinchuan Plain of the Ningxia Autonomous Region, and the west of the Hetao Plain and the desert steppe in N Xinjiang in China (Wang et al. 2016b). Due to the hard soil texture, poor permeability, and difficulty of improvement, takyric solonetz has been the main factor severely affecting the development of local agriculture, and is expected to bring a threat to the regional food security in the future. As a consequence, Wang et al. (2016b) evaluated the feasibility and effectiveness of FGD gypsum addition on improving typical takyric solonetzs. It was found that by using FGD gypsum stimulation, the soil pH value and

exchangeable Na^+ percentage decreased, facilitating the growth of rice, and thereby increasing the rice yield.

According to Mao and Li (2016), addition of appropriate rates of FGD gypsum to the saline-Na soil in the Shanghai tidal flats resulted in a decrease of the soil salinity, pH, and exchangeable Na^+ percentage to a large extent, whereas soluble Ca^{2+}, SO_4^{2-}, and available K^+ increased. Furthermore, the contents of soil organic matter and available P were reduced. Shen et al. (2016) have reviewed studies of improving saline-Na soil technology using FGD gypsum in the past few years, with the purpose of clarifying its effect by analysing the alleviating impact of constraint factors from soil physical and chemical properties, soil microbial activities, and enzymic activities in addition to salt-tolerant plant growth. It was shown that significant effects exist on soil salt content, exchangeable Na^+ percentage, soil porosity, microbial activity, soil enzyme activities, plant survival rate, root activity, photosynthesis, yield, and quality. The application amount varied from 22.5 to 37.5 tons per square hectometre, depending on the exchangeable Na^+ percentage, soil depth, and bulk density. Wang et al. (2016c) conducted research on soil amendment using FGD gypsum on around 120 km^2 of saline-alkali soil in China, and found that the productivity of the plants grown in the ameliorated soil substantially increased, and also that the physical and chemical characteristics of the soil significantly improved. Both field and laboratory studies have shown that the technology is safe considering the heavy metal concentration in the soil and plants grown on it. Furthermore, it was shown that the saline-alkali soil amelioration can fix more C, contributing to mitigation of global warming.

5.3. Applications for building plaster (β-hemihydrated gypsum)

The main component of building plaster (building gypsum products), β-hemihydrated gypsum, is obtained by calcination of natural gypsum, providing a material that could further be used for the production of gypsum block, gypsum plaster board, mould gypsum, and other wall-building materials (Zhang et al. 2015). Owing to the similarity of FGD gypsum to natural gypsum, similar calcination products are expected to be obtained (Tharaniyil 2013). Hence, instead of natural gypsum, transformation of FGD gypsum to building plaster during calcination under 150–200°C is another common utilization area for FGD gypsum (Zhang et al. 2015).

Mao (2005) summarized that the production of building plaster usually can be processed by two methods: direct processing (one-step) and indirect processing (two-step) technique. When the moisture content in FGD gypsum is low, building plaster was usually produced by direct calcination processing; when the moisture content is high, building plaster was reasonably produced by means of a pre-drying process before calcination (Mao 2005).

Yang et al. (2011) found that the dehydration from natural gypsum to β-hemihydrated gypsum occurs at around 170°C, whereas that from FGD gypsum occurred at a lower temperature, around 150°C, indicating that this might slightly reduce the production costs. Bo (2010) found that the effect of calcination technology using a two-step method is notable; first, FGD gypsum is dried at 50°C for 2 h, and then calcined at a temperature of 155°C. The effect of retarder, water-retaining/reducing agent, and water repellent on the building plaster, as well as the possibilities of using building plaster to produce wall-building materials, such as rendering material, thistle board, and gypsum plaster block, were also discussed in this study. To guarantee both good strength and retardation effect of the building plaster, citric acid, bone glue, and Na metaphosphate were added as retarders, and addition of a proper water-retaining agent was also used (Bo 2010).

It has also been reported that the building plasters produced were used to prepare FGD gypsum-based binding materials and products by mixing them with FA, cement, and slag powder (Bai et al. 2016). The influence of the admixture on the intensity of FGD gypsum-based binding materials was investigated by means of orthogonal tests. It was found that the dosage of FA and slag powder is a critical factor on the intensity. The building plaster produced from FGD gypsum has higher strength than that from natural gypsum, due to its compact columnar crystalline structure relative to the loose acicular or flaky crystalline structure of the natural gypsum-based building plaster (Tian et al. 2006). On the other hand, the relatively high impurities in the FGD gypsum, such as $CaCO_3$, F, Cl, and Na, may to some extent have an unfavourable influence on the properties of the FGD gypsum-based building plaster (Tian et al. 2006; Álvarez-Ayuso et al. 2008). Furthermore, FGD gypsum is usually yellow in colour, which makes the whiteness of the building plaster relatively low (<60), and not suitable for the production of plaster sculpture and gauge material (Tian et al. 2006).

5.4. Applications for hemihydrated gypsum whiskers (α-hemihydrated gypsum)

Apart from the applications aforementioned for FGD gypsum, it is technically and economically feasible to synthesize high-value-added $CaSO_4$ whiskers from FGD gypsum (Xu et al. 2011a; Sun et al. 2016a). $CaSO_4$ whiskers, namely α-hemihydrated gypsum, are widely used as a filler in the rubber, plastic, paint, and paper industries because of their excellent physicochemical properties (high thermal stability, chemical inertness, high mechanical strength, and good compatibility with rubber and plastics), low cost, and non-toxicity (Xu et al. 2011a; Yang et al. 2013).

Generally, $CaSO_4$ whiskers could be synthesized from FGD gypsum via hydrothermal (Yang et al. 2013) and atmospheric acidification methods (Wu et al. 2011; Sun et al. 2016a). Compared with the hydrothermal method, the atmospheric acidification method has an advantage in the mild reaction conditions (atmospheric pressure, and temperature lower than 90°C). Hence, recent research has mainly focused on the synthesis of α-hemihydrated $CaSO_4$ whiskers from FGD gypsum by the atmospheric acidification method. Cheng et al. (2016) reported the preparation of hemihydrate $CaSO_4$ whiskers by the NaCl salt solution method using FGD gypsum as a raw material and discussed the influences of NaCl concentration, reaction temperature, ratio of liquid to solid, and pH on the conversion rate. Given the typical characteristics of FGD gypsum, such as high purity, small crystal granularity, and fine powder in hygrometric state, Lin (2009) also prepared α-hemihydrate $CaSO_4$ gypsum in an aqueous salt solution using the hydrothermal method. The influences of the salt solution concentration, the reaction temperature and time, slurry concentration, and pH value on the dehydration reaction dynamics and the resulting α-hemihydrate $CaSO_4$ were studied. It was generally supposed the α-hemihydrate $CaSO_4$ takes shape through a dissolution-recrystallization process. Wu et al. (2011) investigated the direct phase transformation of FGD gypsum to α-hemihydrated $CaSO_4$ whiskers in hot salt solutions at atmospheric pressure. It was found that the transformation of $CaSO_4.2H_2O$ to α-hemihydrated $CaSO_4$ occurs because of the difference in solubility between the two solid $CaSO_4$ phases in this system, and that both the concentration of HCl and the leaching temperature exerted significant effects on the morphology of the α-hemihydrated $CaSO_4$ whiskers.

5.5. Application for the production of fire-resistant panels

Fire-resistant panels are a kind of wallboard with high thermal insulating properties. Owing to the high moisture content of FGD gypsum, it can prevent the development of high temperatures or even delay burning during fire accidents. Hence, FGD gypsum can be used for the production of fire-proof materials (Zhang *et al.* 2015).

Li *et al.* (2015) investigated the production of fire-resistant panels using mixtures of FGD gypsum and FA from a power plant in Xinjiang, NW China, following the procedure proposed by Leiva *et al.* (2010). The principle procedure for the preparation of fire-resistant panels is a pre-dehydration of FGD gypsum at an appropriate temperature (140°C) with sufficient time to promote partial loss of water, after which the material is mixed with appropriate proportions of additives (FA) and water, followed by moulding, demoulding after 24 h, and curing at ambient temperature (20°C on average) and constant humidity (an average relative humidity of 45%) for 28 days (curing time). The pre-dehydration of FGD gypsum is aimed at improving the setting time of the gypsum after mixing with water as well as the mechanical properties of the panels made from the mixtures (Leiva *et al.* 2010). It was found that up to 80% FA could be used in the raw material. Even with 80% FA (or at least 60% FA), the products can still be used as benign fire-resistant panels with no significant physico-chemical or mechanical limitations, but the pure FGD gypsum has the best properties (Li *et al.* 2015). The promising insulating and other properties of these fire-resistant products were due to the passive protection against fire of these products (Leiva *et al.* 2005, 2007; Vilches *et al.* 2005, 2007; Arenas *et al.* 2011). Passive protection against fire can be understood as a set of construction measures that are taken in order to lessen the possibility of a fire starting, and preventing it from spreading for a certain period of time before the fire is under control or has been put out (Vilches *et al.* 2005). Thus, the fire resistance is indicated by the time necessary for the non-exposed surface of the material to reach 180°C (Vilches *et al.* 2007). Among the most widely adopted measures for protection against fire are fire doors and screens using fire-resistant plates or panels based on FGD gypsum.

Based on previous research, the production process of fire-resistant panels is carried out at ambient temperature, with the addition of only appropriate amounts of water, which reduces the economic cost (Li *et al.* 2015). In addition, there are no wastes generated from the production process. These indicate that the manufacturing of fire-resistant panels is a very promising area of utilization for both FGD gypsum and FA, with low environmental implications and economic costs. It will consume large amounts of both FGD gypsum and/or FAs as well as give rise to high added value due to the thermal insulation properties of the products.

Nevertheless, there are several characteristics of FGD gypsum and ashes that can significantly affect the properties of the insulating materials made with these products, such as the physico-chemical properties, pozzolanic activity, grain-size distribution, morphology, and mineralogy (Leiva *et al.* 2005; Vilches *et al.* 2005). For instance, the compressive and flexural strength, as well as the insulating capacity, generally increases with increasing proportion of FGD in the mixture (Li *et al.* 2015). Furthermore, the detrimental impurities, such as water-soluble organic or inorganic substances, including K, Na, F, Fe, and Mg contained in the FGD gypsum, tend to have negative impacts on the utilization in gypsum board production (Wang and Wu 2004).

6. Future utilization prospects

As stated above, CCPs have been used for various applications in China, which can help reduce disposal cost, save natural resources, release useful land, or generate high economic benefits. Nevertheless, there are a number of barriers to the use of large quantities of these CCPs, such as technical, economic, regulatory issues, as well as the public perception barrier.

The technical barriers include issues related to CCP production, specifications, and standards, product commercialization, and user-related factors. The chemical and physical properties of FA, slag, and FGD gypsum vary considerably, depending upon the sources of the coal, particulate control equipment, furnace type, ash collection method, as well as the type of ash discharge system (Fu 2010). Therefore, there is no universal process formulation for applications that can match all types of these CCPs. Economic barriers to increased CCPs utilization are widely accepted as the most important elements among all other factors that affect CCPs utilization. The high cost of transportation for low unit-value CCPs and competition from locally available natural materials pose two of the most important economic barriers (Ahmaruzzaman 2010; Fu 2010). In addition, spatial variation in the supply-demand of high-grade FA and the high transportation cost for low unit-value CCPs are likely to impede the expansion of the CCP market (Lu *et al.* 2011). In the northern area of China, large amounts of high-quality CCPs are accumulated without any profitable applications, whereas the short supply of CCPs in southern China has left a broad market gap in this area. A regulatory barrier exists in the lack of State guidelines, poor execution of regulations, and weakness in market supervision. The imperfect regulatory system for CCP utilization management has provided illegal marketers a chance to hoard high-quality CCPs for speculation and monopolization of the market, which consequently causes extraordinary increase of the local market price of CCPs (Fu 2010; Lu *et al.* 2011). Other barriers to CCP use arise from the definition of CCPs as industry wastes, and the general lack of public familiarity and negative perception of CCP utilization. On the whole, high transportation costs of low unit-value CCPs, competition from available natural materials, and spatial variation in supply-demand of CCPs are recognized as the most critical challenges to the increased use of CCPs in China.

With respect to FA, the market for cement and concrete applications is limited in many cases by the current maximum blended rate (normally about 30%, Wang and Wu 2004) and tends to be saturated based on current replacement rates. If the replacement rate could be increased to a technically viable level of 50%, larger amounts of FA can be utilized beneficially. Consequently, developing high-volume FA concrete technology and increasing the amount of CCPs used in cement and concrete applications are likely to be major methods for the increased use of CCPs.

Apart from the traditional applications of FA, a number of investigations on high-value-added applications, such as valuable metal recovery, waste-water treatment, glass ceramics, zeolite synthesis, geopolymer synthesis, and foam glass, have also been carried out in China. However, most of these studies are still at a laboratory or pilot scale, or still in the early stages of commercialization, although some of the technologies have been commercialized in other developed countries for many years. For instance, Al recycling from FA is limited to laboratory-scale studies due to uneconomic energy consumption and technical restrictions. FA has great potential in environmental applications and is an interesting alternative to replace activated carbon or zeolites for adsorption in air or for water pollution treatment. However, to the best of the author's knowledge, to date, industrial-scale applications for these high-value-added products have not been completed in China. Further studies are needed to turn these high-value-added research studies into commercial reality. Economic barriers have to be overcome in terms of high-value and high-volume utilization for the industrial applications of FA.

Raw FA generally has a low adsorption capacity, and modification of FA could enhance this capacity. FA contains aluminosilicates and is a potential source for the synthesis of zeolites. Zeolites have a variety of applications as adsorbents and ion exchangers and exhibit much higher adsorption capacity than the raw FA. Further research will focus on technology to convert the bulk FA into pure zeolites, as well on the commercialization of synthetic zeolites from FA.

In the last few decades, FA-based geopolymers have been emerging as a novel binder for applications in concrete technology by the construction industries. The presence of silico-aluminate phases in FA makes it a suitable raw material for the synthesis of geopolymers. To reach the full potential of the material, further research needs to be continued on geopolymer binders, especially on curing and durability aspects.

In addition, activated carbon made from the unburned carbon of FA has a significant potential for cost advantage over other types of activated carbon. Therefore, separation of unburned carbon from FA will be beneficial to FA application, either for carbon recycling or for FA applications in cement production and zeolite synthesis. However, few investigations have been conducted on the commercial utilization of unburned FA carbon for the production of activated carbon. More effort should be directed to this area.

Furthermore, there should be a greater emphasis on the development of new technologies for more efficient and comprehensive applications. For instance, combinations of processes should be developed to simultaneously recover more valuable materials, such as unburned carbon, cenospheres, alumina, silica, gallium, and titanium. Additional work is needed to optimize the processing steps, particularly in extraction and purification of the valuable elements.

With respect to slag and FGD gypsum, future research will also focus on high-volume, high-added-value, and industrial/commercial applications, in order to maximize their consumption and economic benefits as well as to minimize the environmental impacts. In addition, traditional applications, such as cement and concrete, and road pavement for slag, as well as cement retarder, building plaster, and soil amendment for FGD gypsum, should still continue to be the most important applications for these CCPs in the future. It must be borne in mind that the leaching of elements of environmental concern, such as F^-, Cl^-, and heavy metals from FGD gypsum, must be controlled prior to major use as a landfill material in mine reclamation or transmission to landfills for disposal.

Finally, the government should sponsor high-volume and high-added-value CCPs utilization for commercial-scale demonstration projects and evaluate the viability and environmental acceptability of these applications.

7. Conclusions

This review article has introduced the characteristics of CCPs generated from coal-fired power plants in

China, and summarized the current utilization of FA, slag, and FGD gypsum. FA has been used as an important raw material for various applications. In the construction field, it can be used for the production of cement and concrete, for the production of various building bricks, and for roadway pavements. In the agriculture field, it can be used for mine backfill and soil amelioration, or used as a fertilizer. FA has been used as an adsorbent for the removal of heavy metals, organic compounds, or dyes from waste water and of gaseous pollutants from flue gases. Production of ceramics, glass-ceramics, and glass is another important area of utilization for FA. In addition, FA has also been used as a raw material for a variety of high-value-added applications, such as zeolite synthesis, geopolymer synthesis, recovery of valuable metals, recovery of unburned carbon, cenospheres, and foam glass production, which can help a great deal in the reduction of environmental pollution or make good economic sense.

BS has generally been used in common construction applications, such as concrete and cement, building bricks, roadway pavements, and backfill. It also can be used for waste-water treatment and for the production of acoustic barrier materials. Slag with a high unburned carbon content has also been used for reclamation of the burnable product, and for the production of honeycomb briquettes.

FGD gypsum has been widely used to replace natural gypsum as a cement retarder, for the production of building plaster (β-hemihydrated gypsum) and high-added-value $CaSO_4$ whiskers (α-hemihydrated gypsum), and for the production of fire-resistant panels. It also can be used as a fertilizer and soil amendment agent.

The various applications of CCPs are highly dependent on the physical, chemical, and environmental properties of the materials. The aluminosilicate phases in FA make it a suitable raw material for the synthesis of zeolites or geopolymers, or for the synthesis of foam glass. The fineness and LOI of FA are defined as the two key indicators by the GB standard (GB1596-79) for the direct utilization of FA in the cement industries. FA that is coarse or has a high carbon content, and is outside the GB standard, however, is unable to be used in the cement and concrete industries. The enrichment of valuable metals in FA is a prerequisite for their potential recovery from the ash. With respect to FGD gypsum, the high moisture content in it is favourable for the production of fire-resistant panels, but is deleterious to cement properties without a drying process.

Therefore, investigation of CCP characteristics prior to utilization is vital for the optimal utilization of CCPs. In the future, CCP utilization programmes must be extensively taken up covering various aspects at different levels to maximize CCP consumption and the associated economic benefits and to minimize the adverse environmental impacts. To achieve these ends, traditional applications, such as cement and concrete, bricks and tiles, land reclamation, and use in road and embankments, should still increase and continue to be the most important applications for CCPs in the future. However, developing high-volume FA concrete technology and increasing the amount of CCPs used in these applications need to be achieved as well. In addition, some underutilized applications, including agriculture and waste treatment, and several high-value-added applications, such as the production of ceramics and foam glass, synthesis of zeolites and geopolymers, and recovery of valuable metals, should be enhanced and transferred to the commercial stage.

Transforming such laboratory- or pilot-scale technologies into industrial productivity is of the highest priority for the increased use of CCPs. For maximum benefit, greater emphasis also should be put on the development of new technology for the efficient utilization of CCPs.

Acknowledgements

The authors would like to thank Professor Shifeng Dai and Professor Bob Finkleman for helpful suggestion on how to improve this review paper. We also appreciate the anonymous reviewers and editors for their careful reviews and constructive comments, which greatly improved the article's quality.

Disclosure statement

No potential conflict of interest was reported by the authors.

Funding

This work was supported by the National Natural Science Foundation of China [grant number 41402141], [grant number 41372166]; and the Fundamental Research Funds for the Central Universities, China University of Geosciences (Wuhan) [grant number CUGL160406].

References

ACAA, 2016, Sustainable Construction With Coal Combustion Products: https://www.acaa-usa.org/Portals/9/Files/PDFs/Sustainability_Construction_w_CCPs (Consolidated).pdf: The American Coal Ash Association.

ACAA Glossary of Terms, 2003, http://www.wwccpn.net/index_htm_files/Glossary_ of_Terms_Concerning_the_Management_and_Use_of_CCPs_2003.pdf: The American Coal Ash Association.

ADAA, 2016, http://www.adaa.asn.au/about-ccps: The Ash Development Association of Australia.

Agriculture Department of China, 2015, http://www. soil. csdb. cn/page/index vpage.

Ahmaruzzaman, M., 2010, A review on the utilization of fly ash: Progress in Energy and Combustion Science, v. 36, p. 327–363. doi:10.1016/j.pecs.2009.11.003

Álvarez-Ayuso, E., and Querol, X., 2008, Study of the use of coal fly ash as an additive to minimise fluoride leaching from FGD gypsum for its disposal: Chemosphere, v. 71, p. 140–146. doi:10.1016/j.chemosphere.2007.10.048

Álvarez-Ayuso, E., Querol, X., Ballesteros, J.C., and Giménez, A., 2008, Risk minimisation of FGD gypsum leachates by incorporation of aluminium sulphate: Science Of The Total Environment, v. 406, p. 69–75. doi:10.1016/j. scitotenv.2008.08.010

Álvarez-Ayuso, E., Querol, X., and Tomás, A., 2006, Environmental impact of a coal combustion-desulphurisation plant: abatement capacity of desulphurisation process and environmental characterisation of combustion by-products: Chemosphere, v. 65, p. 2009–2017. doi:10.1016/j. chemosphere.2006.06.070

Aplan, F.F., 1999, The historical development of coal flotation in the United States, *in* Parekh, B.K., and Miller, J.D., eds., Advances in flotation technology: Colorado, Society for Mining Metallurgy & Exploration, Inc., Littleton, CO, p. 269–287.

Arbuzov, S.I., Ershov, V.V., Potseluev, A.A., and Rikhvanov, L.P., 2000, Rare Elements in Coals of the Kuznetsk Basin: Kemerovo, Publisher House KPK, p. 248 (in Russian).

Arenas, C., Marrero, M., Leiva, C., Solís-Guzmán, J., and Arenas, L., 2011, High fire resistance in blocks containing coal combustion fly ashes and bottom ash: Waste Management, v. 31, p. 1783–1789. doi:10.1016/j.wasman.2011.03.017

Arroyo, F., Font, O., Fernández-Pereira, C., Querol, X., Juan, R., Ruiz, C., and Coca, P., 2009, Germanium recovery from gasification fly ash: evaluation of end-products obtained by precipitation methods: Journal of Hazardous Materials, v. 167, p. 582–588. doi:10.1016/j.jhazmat.2009.01.021

ASTM C 618-05, 2005, Standard specification for coal fly ash and raw or calcinated natural pozzolan for use in concrete: Annual book of ASTM Standards, 04.02: Pennsylvania, American Society for Testing Materials.

Bai, G., Teng, W., Wang, X., Qin, J., Xu, P., and Li, P., 2010a, Alkali desilicated coal fly ash as substitute of bauxite in lime-soda sintering process for aluminum production: Transactions of Nonferrous Metals Society of China, v. 20, p. 169–175. doi:10.1016/S1003-6326(10)60034-9

Bai, H., Mao, G., Li, X., Zheng, G., Yang, J., and Xu, X., 2010b, Effects of desulfurized slag on active oxygen metabolism of rice seedlings on saline -alkali land: Acta Botanica Boreali-Occidentalia Sinica: Vol. 30, p. 2225–2231 (in Chinese with English abstract).

Bai, X., Li, Z., Ma, H., Chen, J., Zhang, P., and Zhang, T., 2016, Research on preparation of gypsum-based binding material blocks by desulphurization gypsum: Non-Metallic Mines, V. 39, p. 61–64 (in Chinese with English abstract).

Bhangare, R., Tiwari, M., Ajmal, P., Sahu, S., and Pandit, G., 2014, Distribution of natural radioactivity in coal and combustion residues of thermal power plants: Journal of Radioanalytical and Nuclear Chemistry, v. 300, p. 17–22. doi:10.1007/s10967-014-2942-3

Blissett, R., and Rowson, N., 2012, A review of the multi-component utilisation of coal fly ash: Fuel, v. 97, p. 1–23. doi:10.1016/j.fuel.2012.03.024

Bo, Y., 2010, Research on the calcination technology of desulfurization gypsum and its application in wall-building materials [M.E. thesis]: Guangzhou, Jinan University (in Chinese with English abstract).

Borm, P.J.A., 1997, Toxicity and occupational health hazards of coal fly ash (CFA). A review of data and comparison to coal mine dust: The Annals of Occupational Hygiene, v. 41, p. 659–676. doi:10.1016/S0003-4878(97)00026-4

BP Statistical Review of World Energy, 2016, http://www.bp. com/content/dam/bp/pdf -/energy-economics/statistical-review-2016/bp-statistical-review-of-world-energy-2016-full-report.pdf.

Cai, C., Yu, X., Shen, Z., and Xing, Y., 2007, A comparison of two methods for metallizing fly ash cenosphere particles: electroless plating and magnetron sputtering: Journal of Physics D: Applied Physics, v. 40, p. 6026. doi:10.1088/0022-3727/40/19/038

Cao, D., Selic, E., and Herbell, J., 2008, Utilization of fly ash from coal-fired power plants in China: Journal of Zhejiang University Science A, v. 9, p. 681–687. doi:10.1631/jzus. A072163

Chen, B., 2012, Study on preparation and properties of fly ash foam glass [M.E. thesis]: Changsha, Central South University (in Chinese with English abstract).

Chen, C., Gong, W., Lutze, W., Pegg, I.L., and Zhai, J., 2011, Kinetics of fly ash leaching in strongly alkaline solutions: Journal of Materials Science, v. 46, p. 590–597. doi:10.1007/s10853-010-4997-z

Chen, J., Kong, H., Wu, D., Chen, X., Zhang, D., and Sun, Z., 2007, Phosphate immobilization from aqueous solution by fly ashes in relation to their composition: Journal of Hazardous Materials, v. 139, p. 293–300. doi:10.1016/j. jhazmat.2006.06.034

Chen, J., Kong, H., Wu, D., Hu, Z., Wang, Z., and Wang, Y., 2006, Removal of phosphate from aqueous solution by zeolite synthesized from fly ash: Journal of Colloid and Interface Science, v. 300, p. 491–497. doi:10.1016/j.jcis.2006.04.010

Chen, L., and Dick, W.A., 2011, Gypsum as an agricultural amendment: general use guidelines: Columbus, Ohio, The Ohio State University.

Chen, X., 1992, Reclamation and utilization of the slag from power plant: Journal of China University of Mining & Technology, v. 21, p. 65–68 (in Chinese with English abstract).

Cheng, T., and Chen, Y., 2004, Characterisation of glass ceramics made from incinerator fly ash: Ceramics International, v. 30, p. 343–349. doi:10.1016/S0272-8842(03)00106-8

Cheng, T., Huang, M., Tzeng, C., Cheng, K., and Ueng, T., 2007, Production of coloured glass–ceramics from incinerator ash using thermal plasma technology: Chemosphere, v. 68, p. 1937–1945. doi:10.1016/j.chemosphere.2007.02.046

Cheng, Y., Lin, M., and Zhao, M., 2016, Research on preparation of hemihydrate gypsum whiskers from desulfurized gypsum: Inorganic Chemicals Industry, v. 48, p. 63–67 (in Chinese with English abstract).

Cheng, Z., 2000, Discussion on carbon recovery from powdered coal ash of thermal electric plant: Large Scale Nitrogenous Fertilizer Industry, v. 23, p. 316–318 (in Chinese with English abstract).

Chinchón, J., Querol, X., Fernández-Turiel, J., and López-Soler, A., 1991, Environmental impact of mineral transformations undergone during coal combustion: Environmental Geology and Water Sciences, v. 18, p. 11–15. doi:10.1007/BF01704573

Chou, X., 2015, Research on immobilization of heavy metal and in situ translation into zeolite through using fly ash based geopolymer [Doctoral thesis]: Wuhan, China University of Geosciences (in Chinese with English abstract).

CIRCA, 2010, Traditional and Non-Traditional Use of Coal Combustion Products: http://www.circainfo.ca/CIRCA/WebPages/Introduction.htm: The Canadian Industries Recycling Coal Ash.

Cox, M., Nugteren, H., and Janssen-Jurkovičová, M., 2008, Combustion residues-current, novel and renewable applications: England, Wiley.

Creelman, R.A., Ward, C.R., Schumacher, G., and Juniper, L., 2013, Relation between coal mineral matter and deposit mineralogy in pulverized fuel furnaces: Energy and Fuels, v. 27, p. 5714–5724. doi:10.1021/ef400636q

Cui, X., Chen, S., Yin, H., Tan, J., and Li, R., 2009, Synthesis of Na-X zeolite from coal fly ash: Coal Conversion, v. 32, p. 85–88 (in Chinese).

Dai, S., Graham, I.T., and Ward, C.R., 2016a, A review of anomalous rare earth elements and yttrium in coal: International Journal of Coal Geology, v. 159, p. 82–95. doi:10.1016/j.coal.2016.04.005

Dai, S., Liu, J., Ward, C.R., Hower, J.C., French, D., Jia, S., Hood, M.M., and Garrison, T.M., 2015c, Mineralogical and geochemical compositions of Late Permian coals and host rocks from the Guxu Coalfield, Sichuan Province, China, with emphasis on enrichment of rare metals: International Journal of Coal Geology. doi:10.1016/j.coal.2015.12.004

Dai, S., Ren, D., Chou, C.-L., Li, S., and Jiang, Y., 2006b, Mineralogy and geochemistry of the No. 6 Coal (Pennsylvanian) in the Junger Coalfield, Ordos Basin, China: International Journal of Coal Geology, v. 66, p. 253–270. doi:10.1016/j.coal.2005.08.003

Dai, S., Ren, D., and Li, S., 2006a, Discovery of the superlarge gallium ore deposit in Jungar, Inner Mongolia, North China: Chinese Science Bulletin, v. 51, p. 2243–2252. doi:10.1007/s11434-006-2113-1

Dai, S., Ren, D., Zhou, Y., Chou, C., Wang, X., Zhao, L., and Zhu, X., 2008, Mineralogy and geochemistry of a superhigh-organic-sulfur coal, Yanshan Coalfield, Yunnan, China: Evidence for a volcanic ash component and influence by submarine exhalation: Chemical Geology, v. 255, p. 182–194. doi:10.1016/j.chemgeo.2008.06.030

Dai, S., Seredin, V.V., Ward, C.R., Hower, J.C., Xing, Y., Zhang, W., Song, W., and Wang, P., 2015a, Enrichment of U–Se–Mo–Re–V in coals preserved within marine carbonate successions: geochemical and mineralogical data from the Late Permian Guiding Coalfield, Guizhou, China: Mineralium Deposita, v. 50, p. 159–186. doi:10.1007/s00126-014-0528-1

Dai, S., Seredin, V.V., Ward, C.R., Jiang, J., Hower, J.C., Song, X., Jiang, Y., Wang, X., Gornostaeva, T., Li, X., Liu, H., Zhao, L., and Zhao, C., 2014a, Composition and modes of occurrence of minerals and elements in coal combustion products derived from high-Ge coals: International Journal of Coal Geology, v. 121, p. 79–97. doi:10.1016/j.coal.2013.11.004

Dai, S., Wang, X., Zhou, Y., Hower, J.C., Li, D., Chen, W., and Zhu, X., 2011, Chemical and mineralogical compositions of silicic, mafic, and alkali tonsteins in the late Permian coals from the Songzao Coalfield, Chongqing, Southwest China: Chemical Geology, v. 282, p. 29–44. doi:10.1016/j.chemgeo.2011.01.006

Dai, S., Xie, P., Jia, S., Ward, C.R., Hower, J.C., Yan, X., and French, D., 2017, Enrichment of U-Re-V-Cr-Se and rare earth elements in the Late Permian coals of the Moxinpo Coalfield, Chongqing, China: Genetic implications from geochemical and mineralogical data: Ore Geology Reviews, v. 80, p. 1–17. doi:10.1016/j.oregeorev.2016.06.015

Dai, S., Yan, X., Ward, C.R., Hower, J.C., Zhao, L., Wang, X., Zhao, L., Ren, D., and Finkelman, R.B., 2016b, Valuable elements in Chinese coals: A review: International Geology Review. p. 1–31. doi:10.1080/00206814.2016.1197802

Dai, S., Yang, J., Ward, C.R., Hower, J.C., Liu, H., Garrison, T.M., French, D., and O'Keefe, J.M.K., 2015b, Geochemical and mineralogical evidence for a coal-hosted uranium deposit in the Yili Basin, Xinjiang, northwestern China: Ore Geology Reviews, v. 70, p. 1–30. doi:10.1016/j.oregeorev.2015.03.010

Dai, S., Zhao, L., Hower, J.C., Johnston, M.N., Song, W., Wang, P., and Zhang, S., 2014b, Petrology, mineralogy, and chemistry of size-fractioned fly ash from the Jungar Power Plant, Inner Mongolia, China, with emphasis on the distribution of rare earth elements: Energy & Fuels, v. 28, p. 1502–1514. doi:10.1021/ef402184t

Dai, S., Zhao, L., Peng, S., Chou, C.L., Wang, X., Zhang, Y., Li, D., and Sun, Y., 2010a, Abundances and distribution of minerals and elements in high-alumina coal fly ash from the Jungar Power Plant, Inner Mongolia, China: International Journal of Coal Geology, v. 81, p. 320–332. doi:10.1016/j.coal.2009.03.005

Dai, S., Zhou, Y., Ren, D., Wang, X., Li, D., and Zhao, L., 2007, Geochemistry and mineralogy of the Late Permian coals from the Songzo Coalfield, Chongqing, southwestern China: Science in China Series D: Earth Sciences, v. 50, p. 678–688. doi:10.1007/s11430-007-0001-4

Dai, S., Zhou, Y., Zhang, M., Wang, X., Wang, J., Song, X., Jiang, Y., Luo, Y., Song, Z., Yang, Z., and Ren, D., 2010b, A new type of Nb (Ta)-Zr(Hf)-REE-Ga polymetallic deposit in the late Permian coal-bearing strata, eastern Yunnan, southwestern China: possible economic significance and genetic implications: International Journal of Coal Geology, v. 83, p. 55–63. doi:10.1016/j.coal.2010.04.002

Davidovits, J., 1991, Geopolymers inorganic polymeric new materials: Journal of Thermal Analysis and Calorimetry, v. 37, p. 1633–1656. doi:10.1007/BF01912193

Demidovich, B.K., 1972, Production and Application of Glass Foam: Minsk, Nauka i Tekhnika (in Russian).

Dilmore, R.M., and Neufeld, R.D., 2001, Autoclaved aerated concrete produced with low-NOx burner/selective catalytic reduction fly ash: Journal of Energy Engineering, v. 127, p. 37–50. doi:10.1061/(ASCE)0733-9402(2001)127:2(37)

ECOBA, 2016, http://www.ecoba.com/ecobaccpspec.html: the European Coal Combustion Products Association.

Ellison, A.J.G., Mazer, J.J., and Ebert, W.L., 1994, Effect of glass composition on waste form durability: a critical review: DuPage County, Illinois, Argonne National Laboratory ANL-94/28.

Erol, M., Küçübayrak, S., and Ersoy-Meriçboyu, A., 2007, Characterization of coal fly ash for possible utilization in glass production: Fuel, v. 86, p. 706–714. doi:10.1016/j. fuel.2006.09.009

Erol, M., Küçükbayrak, S., and Ersoy-Meriçboyu, A., 2008a, Comparison of the properties of glass, glass–ceramic and ceramic materials produced from coal fly ash: Journal of Hazardous Materials, v. 153, p. 418–425. doi:10.1016/j. jhazmat.2007.08.071

Erol, M., Küçükbayrak, S., and Ersoy-Meriçboyu, A., 2008b, Characterization of sintered coal fly ashes: Fuel, v. 87, p. 1334–1340. doi:10.1016/j.fuel.2007.07.002

European Council Decision 2003/33/EC, 2003, Council Decision of 19 December 2002 establishing criteria and procedures for the acceptance of waste at landfills pursuant to Article 16 of and Annex II to Directive 1999/31/ EC: Official Journal of the European Communities, v. 16, p. 27–49.

Fang, R., Zhou, Y., Liu, M., and Chen, D., 2003, Study of factors influencing fly ash foam glass quality: Coal Ash China, v. 1, p. 8–11 (in Chinese).

Feng, C., Li, Y., and Li, D., 2009, Preparation of clinkers using flue gas desulphurization gypsum as CaO resource substituting limestone at low sintering temperature: Bulletin of the Chinese Ceramic Society, v. 28, p. 10–15 (in Chinese with English abstract).

Feng, C., Yao, Y., Li, Y., Liu, X., and Sun, H., 2014, Thermal activation on calcium silicate slag from high-alumina fly ash: a technical report: Clean Technologies and Environmental Policy, v. 16, p. 667–672. doi:10.1007/ s10098-013-0648-9

Fernandes, H.R., Tulyaganov, D.U., and Ferreira, J.M.F., 2009, Preparation and characterization of foams from sheet glass and fly ash using carbonates as foaming agents: Ceramics International, v. 35, p. 229–235. doi:10.1016/j. ceramint.2007.10.019

Finkelman, R.B., 1981, Coal geochemistry: practical applications: Washington, D. C., US Geol. Surv. Open-File Rep, Vols. 81-99, p. 1–12.

Fisher, G., Prentice, B., Silberman, D., Ondov, J., Bierman, A., Ragaini, R., and McFarland, A., 1978, Physical and morphological studies of size-classified coal fly ash: Environmental Science & Technology, v. 12, p. 447–451. doi:10.1021/es60140a008

Font, O., Querol, X., Lopez-Soler, A., Chimenos, J.M., Fernandez, A.I., Burgos, S., and Peña, F.G., 2005, Ge extraction from gasification fly ash: Fuel, v. 84, p. 1384–1392. doi:10.1016/j.fuel.2004.06.041

Freeman, E., Gao, Y.M., Hurt, R., and Suuberg, E., 1997, Interactions of carbon containing fly ash with commercial air-entraining admixtures for concrete: Fuel, v. 76, p. 761–765. doi:10.1016/S0016-2361(96)00193-7

Frenkel, H., Gerstl, Z., and Alperoviteh, N., 1989, Exchange-induced dissolution of gypsum and the reclamation of sodic soils: Journal of Soil Science, v. 40, p. 599–611. doi:10.1111/j.1365-2389.1989.tb01301.x

Fu, J., 2010, Challenges to increased use of coal combustion products in China [M.E. thesis]: Linköping, Linköping University.

Fu, X., Xu, B., and Huang, C., 2014, Alkali dissolution of silicon and aluminum from fly ash by microwave heating: Hydrometallurgy of China, v. 33, p. 196–198.

Gao, P., Lu, X., Lin, H., Li, X., and Hou, J., 2007, Effects of fly ash on the properties of environmentally friendly dam concrete: Fuel, v. 861, p. 208–211.

GB 18598-2001, 2001, Standard for pollution control on security landfill site for hazardous wastes: Beijing, National Standard (GB), Ministry of Environmental Protection of the People's Republic of China.

GB1596-1991, 1991, National Standard of the People's Republic of China Fly Ash used for cement and concrete: Beijing, National Standard (GB), Ministry of Environmental Protection of the People's Republic of China.

Gong, Z., Wang, M., Wang, F., and Zhai, Y., 2011, Research on removing magnetic Fe from fly ash by magnetic selection: Journal of Materials and Metallurgy, v. 10, p. 257–259 (in Chinese with English abstract).

González, A., Navia, R., and Moreno, N., 2009, Fly ashes from coal and petroleum coke combustion: current and innovative potential applications: Waste Management & Research, v. 27, p. 976–987. doi:10.1177/ 0734242X09103190

Grzymek, J., 1976, Prof. Grzymek's self-disintegration method for the complex manufacture of aluminum oxide and Portland cement, in Proceedings of Sessions 105th AIME Annual Meeting, February 22–26, 1976: Las Vegas, N. V.

Gu, D., and Hu, J., 2002, Applied study present condition of coal ash: Mining Technology, v. 2, p. 1–4 (in Chinese).

Gu, Y., 2007, Manufacturing and application of the foam glass of low unit weight: China Ceramics, v. 43, p. 51–54 (in Chinese with English abstract).

Guo, C., Zou, J., Wei, C., and Jiang, Y., 2013, Comparative study on extracting alumina from circulating fluidized-bed and pulverized-coal fly ashes through salt activation: Energy & Fuels, v. 27, p. 7868–7875. doi:10.1021/ef401659e

Guo, D., and Zhu, X., 2002, Research on utilization of slag as cement mixture: China Resources Comprehensive Utilization, v. 10, p. 29–30 (in Chinese).

Guo, Y., Wang, Y., Cai, H., and Yang, Z., 2003, Synthesis of zeolites from fly ash by microwave assisted hydrothermal alkaline activation: Earth Science— Journal of China University of Geosciences, v. 28, p. 517–521 (in Chinese with English abstract).

Gupta, C.K., and Krishnamurthy, N., 2004, Extractive metallurgy of rare earths: Boca Raton, Florida, CRC Press, 484 p.

Hardjito, D., Wallah, S., Sumajouw, D., and Rangan, B., 2004, On the development of fly ash-based geopolymer concrete: ACI Materials Journal, v. 101, p. 467–472.

He, D., Guo, Z., Liao, H., Yue, C., Sain, B., and Yu, G., 2012, Research progress and development trend of porous absorption materials: Materials Review, v. 26, p. 303–307 (in Chinese with English abstract).

He, R., Guo, Z., and Li, Z., 2010, Study on making of sound-absorbing porous material using water-granulated slag of blast furnace: Chinese Journal of Environmental Engineering, v. 4, p. 2870–2874 (in Chinese with English abstract).

He, Y., Cheng, W., and Cai, H., 2005, Characterization of α-cordierite glass-ceramics from fly ash: Journal of Hazardous Materials, v. 120, p. 265–269. doi:10.1016/j. jhazmat.2004.10.028

Heidrich, C., Feuerborn, H., and Weir, A., 2013, Coal combustion products-A global perspective: VGB Power Tech. http:// www.circainfo.ca/documents/VGBPower-Tech2013-12pp46-52HEIDRICHAutorenexemplar.pdf.

Hirajima, T., Petrus, H., Oosako, Y., Nonaka, M., Sasaki, K., and Ando, T., 2010, Recovery of cenospheres from coal fly ash using a dry separation process: separation estimation and potential application: International Journal of Mineral Processing, v. 95, p. 18–24. doi:10.1016/j.minpro.2010.03.004

Höll, R., Kling, M., and Schroll, E., 2007, Metallogenesis of germanium-a review: Ore Geology Reviews, v. 30, p. 145–180. doi:10.1016/j.oregeorev.2005.07.034

Höller, H., and Wirsching, U., 1985, Zeolites formation from fly ash: Fortschr Mineral, v. 63, p. 21–43.

Hollman, G.G., Steenbruggen, G., and Janssen-Jurkovicova, M., 1999, A two-step process for the synthesis of zeolites from coal fly ash: Fuel, v. 78, p. 1225–1230. doi:10.1016/S0016-2361(99)00030-7

Hou, X., Li, Y., Dai, H., and Hou, Y., 2015, Leaching of lithium from fly ash using carbonate: Journal of Hebei University of Engineering (Natural Science Edition), v. 32, p. 58–61.

Hower, J.C., Dai, S., and Eskenazy, G., 2016a, Distribution of uranium and other radionuclides in coal and coal combustion products, with discussion of occurrences of combustion products in Kentucky power plants: Coal Combustion and Gasification Products, v. 8, p. 44–53.

Hower, J.C., Dai, S., Seredin, V.V., Zhao, L., Kostova, I.J., Silva, L.F.O., Mardon, S.M., and Gurdal, G., 2013, A note on the occurrence of yttrium and rare earth elements in coal combustion products: Coal Combustion and Gasification Products, v. 5, p. 39–47.

Hower, J.C., Eble, C.F., Dai, S., and Belkin, H.E., 2016c, Distribution of rare earth elements in eastern Kentucky coals: Indicators of multiple modes of enrichment?: International Journal of Coal Geology, v. 160-161, p. 73–81. doi:10.1016/j.coal.2016.04.009

Hower, J.C., Granite, E.J., Mayfield, D.B., Lewis, A.S., and Finkelman, R.B., 2016b, Notes on contributions to the science of rare earth element enrichment in coal and coal combustion by-products: Minerals, v. 6, p. 32. doi:10.3390/min6020032

Hower, J.C., Groppo, J.G., Henke, K.R., Hood, M.M., Eble, C.F., Honaker, R.Q., Zhang, W., and Qian, D., 2015, Notes on the potential for the concentration of rare earth elements and yttrium in coal combustion fly ash: Minerals, v. 5, p. 356–366. doi:10.3390/min5020356

Hower, J.C., Groppo, J.G., Joshi, P., Dai, S., Moecher, D.P., and Johnston, M.N., 2014, Location of cerium in coal-combustion fly ashes: implications for recovery of lanthanides: Coal Combustion and Gasification Products, v. 5, p. 73–78. doi:10.4177/CCGP-D13-00007.1

Hsu, T., Yu, C., and Yeh, C., 2008, Adsorption of Cu^{2+} from water using raw and modified coal fly ashes: Fuel Processing Technology, v. 87, p. 1355–1359.

Hu, Y., Zhang, H., Li, F., Cheng, X., and Chen, T., 2010, Investigation into electrical conductivity and electromagnetic interference shielding effectiveness of silicone rubber filled with Ag-coated cenosphere particles: Polymer Testing, v. 29, p. 609–612. doi:10.1016/j.polymertesting.2010.03.009

Huang, S., Li, F., Liu, B., Li, Y., Liu, G., and Liu, G., 1999, Study on extracting germanium and aluminium from fly-ash by acid immesion method and utilization of material: Journal of Nanchang University (Engineering & Technology), v. 21, p. 85–90 (in Chinese with English abstract).

Hui, K., and Chao, C., 2006, Effects of step-change of synthesis temperature on synthesis of zeolite 4A from coal fly ash: Microporous and Mesoporous Materials, v. 88, p. 145–151. doi:10.1016/j.micromeso.2005.09.005

Hui, K., Chao, C., and Kot, S., 2005, Removal of mixed heavy metal ions in wastewater by zeolite 4A and residual products from recycled coal fly ash: Journal of Hazardous Materials, v. 127, p. 89–101. doi:10.1016/j.jhazmat.2005.06.027

Hung, M.L., Wu, S.Y., Chen, Y.C., Shih, H.C., Yu, Y.H., and Ma, H., 2009, The health risk assessment of Pb and Cr leachated from fly ash monolith landfill: Journal of Hazardous Materials, v. 172, p. 316–323. doi:10.1016/j.jhazmat.2009.07.013

Huo, P., Yan, Y., Li, S., Li, H., and Huang, W., 2009, Preparation and characterization of cobalt sulfophthalocyanine/TiO_2/fly-ash cenospheres photocatalyst and study on degradation activity under visible light: Applied Surface Science, v. 255, p. 6914–6917. doi:10.1016/j.apsusc.2009.03.014

Hurley, J., and Consultant, S., 2003, A UK market survey for foam glass: Research and Development Final Report, The Waste and Resources Action Programme, Banbury, Oxon, (WRAP) Publications.

Iyer, R., and Scott, J., 2001, Power station fly ash– a review of value-added utilization outside of the construction industry: Resources, Conservation and Recycling, v. 31, p. 217–228. doi:10.1016/S0921-3449(00)00084-7

Izquierdo, M., and Querol, X., 2012, Leaching behaviour of elements from coal combustion fly ash: an overview: International Journal of Coal Geology, v. 94, p. 54–66. doi:10.1016/j.coal.2011.10.006

Izquierdo, M., Querol, X., Davidovits, J., Antenucci, D., Nugteren, H., and Fernández-Pereira, C., 2009, Coal fly ash-slag-based geopolymers: Microstructure and metal leaching: Journal of Hazardous Materials, v. 166, p. 561–566. doi:10.1016/j.jhazmat.2008.11.063

Izquierdo, M., Querol, X., Phillipart, C., Antenucci, D., and Towler, M., 2010, The role of open and closed curing conditions on the leaching properties of fly ash-slag-based geopolymers: Journal of Hazardous Materials, v. 176, p. 623–628. doi:10.1016/j.jhazmat.2009.11.075

Jankowski, J., Ward, C.R., French, D., and Groves, S., 2006, Mobility of trace elements from selected Australian fly ashes and its potential impact on aquatic ecosystems: Fuel, v. 85, p. 243–256. doi:10.1016/j.fuel.2005.05.028

Jha, V., Matsuda, M., and Miyake, M., 2008, Resource recovery from coal fly ash waste: an overview study: Journal of the Ceramic Society of Japan, v. 116, p. 167–175. doi:10.2109/jcersj2.116.167

Ji, H., Lu, H., Hao, X., and Wu, P., 2007, High purity aluminia powders extracted from fly ash by the calcing-leaching process: Journal of the Chinese Ceramic Society, v. 35, p. 1657–1660.

Ji, R., Zhang, Z., Yan, C., Zhu, M., and Li, Z., 2016, Preparation of novel ceramic tiles with high Al_2O_3 content derived from coal fly ash: Construction and Building Materials, v. 114, p. 888–895. doi:10.1016/j.conbuildmat.2016.04.014

Jia, Y., 2009, Synthesis and characterization of fly-ash-based Na-geopolymer [Doctoral thesis]: Beijing China University of Mining and Technology (in Chinese with English abstract).

Jia, Y., Jiang, X., Jiang, G., Zhang, H., and Zhang, L., 2015, Preparation of modified fly ash and its application progress in treating phosphorus wastewater: Bulletin of the Chinese

Ceramic Society, v. 34, p. 1921–1925 (in Chinese with English abstract).

Jia, Z., 1994, Application of sand-free macroporous slag concrete in noise barrier engineering: New Building Materials, v. 6, p. 28–30 (in Chinese).

Jiang, Q., Xi, F., and Huang, K., 2009, Research on method and equipment of recovering ferromagnetic material in fly ash: Journal of Hefei University of Technology, v. 32, p. 959–961 (in Chinese with English abstract).

Jing, Z.Q., Li, Y.Y., Cao, S.W., and Liu, Y.Y., 2012, Performance of double-layer biofilter packed with coal fly ash ceramic granules in treating highly polluted river water: Bioresource Technology, v. 120, p. 212–217. doi:10.1016/j.biortech.2012.06.069

Kalyoncu, R.S., 2001, Coal Combustion Products: http://minerals.usgs.gov/minerals/pubs/commodity/coal/874400.pdf.

Kao, P., Tzeng, J., and Huang, T., 2000, Removal of chlorophenols from aqueous solution by fly ash: Journal of Hazardous Materials, v. 76, p. 237–249. doi:10.1016/S0304-3894(00)00201-6

Ke, G., Yang, X., Peng, H., and Jia, F., 2005, Environmental characteristics of coal fly ash: Journal of China Coal Society, v. 30, p. 102–106 (in Chinese with English abstract).

Kentucky Ash Education Site, 2016, Production from CCBs: http://www.caer.uky.edu/kyasheducation/whyimportant3.shtml.

Kikuchi, R., 1999, Application of coal ash to environmental improvement: Transformation into zeolite, potassium fertilizer, and FGD absorbent: Resources, Conservation and Recycling, v. 27, p. 333–346. doi:10.1016/S0921-3449(99)00030-0

Kim, H.K., and Lee, H.K., 2010, Acoustic absorption modeling of porous concrete considering the gradation and shape of aggregates and void ratio: Journal of Sound and Vibration, v. 329, p. 866–879. doi:10.1016/j.jsv.2009.10.013

Kolbe, J.L., Lee, L.S., Jafvert, C.T., and Murarka, I.P., 2011, Use of alkaline coal ash for reclamation of a former strip mine, in the World of Coal Ash (WOCA) Conference, Denver, Colombia, 9-12 May 2011: Ash library, University of Kentucky, Lexington, Kentucky.

Koukouzas, N., Zeng, R., Perdikatsis, V., Xu, W., and Kakaras, E.K., 2006, Mineralogy and geochemistry of Greek and Chinese coal fly ash: Fuel Processing Technology, v. 85, p. 2301–2309.

Ladwig, K., 2006, A Review of manufacturing uses for gypsum produced by flue gas desulfurization systems: Palo Alto, California, Electric Power Research Institute (EPRI)Interim Report.

Lauer, N., Hower, J.C., Hsu-Kim, H., Taggart, R.K., and Vengosh, A., 2015, Naturally occurring radioactive materials in coals and coal combustion residuals in the United States: Environmental Science & Technology, v. 49, p. 11227–11233. doi:10.1021/acs.est.5b01978

Leiva, C., Arenas, C., Vilches, L.F., Vale, J., Gimenez, A., Ballesteros, J.C., and Fernández-Pereira, C., 2010, Use of FGD gypsum in fire resistant panels: Waste Management, v. 30, p. 1123–1129. doi:10.1016/j.wasman.2010.01.028

Leiva, C., Gomez-Barea, A., Vilches, L.F., ollero, P., vale, J., and Fernández-Pereira, C., 2007, Use of biomass gasification fly ash in lightweight plasterboard: Energy & Fuels, v. 21, p. 361–367. doi:10.1021/ef060260n

Leiva, C., Vilches, L.F., Fernández-Pereira, C., and Vale, J., 2005, Influence of the type of ash on the fire resistance characteristics of ash-enriched mortars: Fuel, v. 84, p. 1433–1439. doi:10.1016/j.fuel.2004.08.031

Leiva, C., Vilches, L.F., Vale, J., Olivares, J., and Fernández-Pereira, C., 2008, Effect of carbonaceous matter contents on the fire resistance and mechanical properties of coal fly ash enriched mortars: Fuel, v. 87, p. 2977–2982. doi:10.1016/j.fuel.2008.04.020

Li, H., Hui, J., Wang, C., Bao, W., and Sun, Z., 2014b, Extraction of alumina from coal fly ash by mixed-alkaline hydrothermal method: Hydrometallurgy, v. 147-148, p. 183–187. doi:10.1016/j.hydromet.2014.05.012

Li, J., 2012, Coal quality and coal combustion byproducts from the Junggar coal basin and power plants in Xinjiang province, northwest China: geochemistry, mineralogy and potential utilization of combustion byproducts [Doctoral thesis]: Wuhan, China university of Geosciences.

Li, J., Cui, J., Zhao, N., Shi, C., and Du, X., 2006, The properties of granular activated carbons prepared from fly ash using different methods: Carbon, v. 44, p. 1346–1348. doi:10.1016/j.carbon.2005.11.031

Li, J., Zhu, S., Chen, G., and Zhu, M., 2007b, Study of flotation experiments on carbon recycled from fly ash: China Resources Comprehensive Utilization, v. 25, p. 7–9 (in Chinese with English abstract).

Li, J., Zhuang, X., Font, O., Moreno, N., Vallejo, V.R., Querol, X., and Tobias, A., 2014a, Synthesis of merlinoite from Chinese coal fly ashes and its potential utilization as slow release K-fertilizer: Journal of Hazardous Materials, v. 265, p. 242–252. doi:10.1016/j.jhazmat.2013.11.063

Li, J., Zhuang, X., Leiva, C., Cornejo, A., Font, O., Querol, X., moeno, N., Arenas, C., and Fernández-Pereira, C., 2015, Potential utilization of FGD gypsum and fly ash from a Chinese power plant for manufacturing fire-resistant panels: Construction and Building Materials, v. 95, p. 910–921. doi:10.1016/j.conbuildmat.2015.07.183

Li, J., Zhuang, X., Querol, X., Font, O., Moreno, N., and Zhou, J., 2012, Environmental geochemistry of the feed coals and their combustion by-products from two coal-fired power plants in Xinjiang Province, Northwest China: Fuel, v. 95, p. 446–456. doi:10.1016/j.fuel.2011.10.025

Li, Q., Sun, Z., Tao, D., Xu, Y., Li, P., Cui, H., and Zhai, J., 2013, Immobilization of simulated radionuclide $^{133}Cs^+$ by fly ash-based geopolymer: Journal of Hazardous Materials, v. 262, p. 325–331. doi:10.1016/j.jhazmat.2013.08.049

Li, R., Yu, R., Chen, J., and Cheng, X., 2007a, Experimental study on fly ash floating bead mortar for external thermal insulation: Building Science, v. 23, p. 83–86 (in Chinese with English abstract).

Li, S., Wu, W., Li, H., and Hou, X., 2016, The direct adsorption of low concentration gallium from fly ash: Separation Science and Technology, v. 51, p. 395–402. doi:10.1080/01496395.2015.1102282

Li, Y., Liu, B., Li, F., Zhou, N., and Liu, G., 2000, Extraction of germanium from coal ash with dihydroxamic acid: Modern Chemical Industry, v. 20, p. 34–36, 39 (in Chinese with English abstract).

Li, Z., and Liu, S., 2007, Influence of slag as additive on compressive strength of fly ash based geopolymer: Journal of Materials in Civil Engineering, v. 19, p. 470–474. doi:10.1061/(ASCE)0899-1561(2007)19:6(470)

Li, Z., Ward, C.R., and Gurba, L.W., 2010, Occurrence of non-mineral inorganic elements in macerals of low-rank coals:

International Journal of Coal Geology, v. 81, p. 242–250. doi:10.1016/j.coal.2009.02.004

Lieberman, R.N., Teutsch, N., and Cohen, H., 2014, Chemical and surface transformations of bituminous coal fly ash used in israel following treatments with acidic and neutral aqueous solutions: Energy & Fuels, v. 28, p. 4657–4665. doi:10.1021/ef500564k

Lin, C., and Chang, J., 2001, Effect of fly ash characteristics on the removal of Cu(II) from aqueous solution: Chemosphere, v. 44, p. 1185–1192. doi:10.1016/S0045-6535(00)00494-X

Lin, J., Zhan, S., Fang, M., Qian, X., and Yang, H., 2008, Adsorption of basic dye from aqueous solution onto fly ash: Journal of Environmental Management, v. 87, p. 193–200. doi:10.1016/j.jenvman.2007.01.001

Lin, K., 2006, Feasibility study of using brick made from municipal solid waste incinerator fly ash slag: Journal of Hazardous Materials, v. 137, p. 1810–1816. doi:10.1016/j.jhazmat.2006.05.027

Lin, M., 2009, Research on preparation and trans-crystal technology of α- hemihydrate flue gas desulfurization gypsum with the hydrothermal method [M.E. thesis]: Chongqing, Chongqing University (in Chinese with English abstract).

Xu, L., Guo, W., Wang, T., and Yang, N., 2005, Study on fired bricks with replacing clay by fly ash in high volume ratio: Construction and Building Materials, v. 19, p. 243–247. doi:10.1016/j.conbuildmat.2004.05.017

Liu, B., Li, Y., and Liu, G., 1998, Study on the preparation of GeO_2 and Al_2O_3 from coal ash by acid leaching method: Inorganic Chemicals Industry, v. 30, p. 3–4, 7 (in Chinese with English abstract).

Liu, G., Vassilev, S.V., Gao, L., Zheng, L., and Peng, Z., 2005, Mineral and chemical composition and some trace element contents in coals and coal ashes from Huaibei coal field, China: Energy Conversion and Management, v. 46, p. 2001–2009. doi:10.1016/j.enconman.2004.11.002

Liu, G., Zhang, H., Gao, L., Zheng, L., and Peng, Z., 2004, Petrological and mineralogical characterizations and chemical composition of coal ashes from power plants in Yanzhou mining district, China: Fuel Processing Technology, v. 85, p. 1635–1646. doi:10.1016/j.fuproc.2003.10.028

Liu, L., Singh, R., Xiao, P., Webley, P.A., and Zhai, Y., 2011, Zeolite synthesis from waste fly ash and its application in CO_2 capture from flue gas streams: Adsorption, v. 17, p. 795–800. doi:10.1007/s10450-011-9332-8

Liu, X., Zhang, H., and Luo, M., 2007a, Study on production of artificial fish reef using coal fly ash and sludge: Journal of Building Materials, v. 10, p. 622–626 (in Chinese).

Liu, Y., and Li, L., 2006, Progress in research of alumina recycling from fly ash: Light Metals, v. 5, p. 20–23 (in Chinese).

Liu, Y., Ma, Y., Li, Y., Xing, C., and Lv, H., 2007b, Zeolite synthesis from coal fly ash: Coal Conversion, v. 30, p. 91–95 (in Chinese with English abstract).

Liu, Y., Xu, F., and Zhu, X., 2012, Experimental study on fly ash foam glass: China Ceramics, v. 48, p. 53–55 (in Chinese with English abstract).

Liu, Z., 2009, The preparation and application of fly ash based adsorbent: Beijing, Chemical Industry Press (in Chinese with English abstract).

Lu, G., Xue, F., and Zhao, J., 2011, Some advice to the fly ash of China: China Mining Magazine, v. 20, p. 193–200 (in Chinese with English abstract).

Lu, H., 2013, Research on preparation of Portland cement from desulphurization gypsum [M.E. thesis]: Nanjing, Najing University of Science and Technology (in Chinese with English abstract).

Lu, S., Chen, Y., Shan, H., and Bai, S., 2009, Mineralogy and heavy metal leachability of magnetic fractions separated from some Chinese coal fly ashes: Journal of Hazardous Materials, v. 169, p. 246–255. doi:10.1016/j.jhazmat.2009.03.078

Lytle, W.O., 1940, Pittsburgh Plate Glass: USA, US Patent 2, p. 215–223.

Ma, B., Qi, M., Peng, J., and Li, Z., 1999, The compositions, surface texture, absorption, and binding properties of fly ash in China: Environment International, v. 25, p. 423–432. doi:10.1016/S0160-4120(99)00010-0

Ma, M., and Xie, F., 2003, Study on intensified leaching of germanium and uranium in coal ash: Uranium Mining and Metallurgy, v. 22, p. 40–44 (in Chinese with English abstract).

Mao, S., 2005, Analysis on comprehensive utilization of flue gas desulfurization gypsum [M.E. thesis]: Hangzhou, Zhejiang University (in Chinese with English abstract).

Mao, Y., and Li, X., 2016, Amelioration of flue gas desulfurization gypsum on saline-sodic soil of tidal flats and its effects on plant growth: China Environmental Science, v. 36, p. 225–231 (in Chinese with English abstract).

Massari, S., and Ruberti, M., 2013, Rare earth elements as critical raw materials: focus on international markets and future strategies: Resources Policy, v. 38, p. 36–43. doi:10.1016/j.resourpol.2012.07.001

Mayfield, D.B., and Lewis, A.S., 2013, Environmental review of coal ash as a resource for rare earth and strategic elements, in The World of Coal Ash (WOCA) Conference, Lexington, Kentucky, 22–25 April 2013: Ash library, University of Kentucky, Lexington, Kentucky.

Meij, R., 1994, Trace element behavior in coal-fired power plants: Fuel Processing Technology, v. 39, p. 199–217. doi:10.1016/0378-3820(94)90180-5

Meng, X., Li, D., Shen, X., and Liu, W., 2010, Preparation and magnetic properties of nano-Ni coated cenosphere composites: Applied Surface Science, v. 256, p. 3753–3756. doi:10.1016/j.apsusc.2010.01.019

Mittra, B.N., Karmakar, S., Swain, D.K., and ghosh, B.C., 2005, Fly ash - a potential source of soil amendment and a component of integrated plant nutrient supply system: Fuel, v. 84, p. 1447–1451. doi:10.1016/j.fuel.2004.10.019

Moreno, N., Querol, X., Andres, J., Stanton, K., Towler, M., Nugteren, H., Janssen-Jurkovicova, M., and Jones, R., 2005, Physico-chemical characteristics of European pulverized coal combustion fly ashes: Fuel, v. 84, p. 1351–1363. doi:10.1016/j.fuel.2004.06.038

Moreno, N., Querol, X., Plana, F., Andres, J.M., Janssen, M., and Nugteren, H., 2002, Pure zeolite synthesis from silica extracted from coal fly ashes: Journal of Chemical Technology & Biotechnology, v. 77, p. 274–279. doi:10.1002/(ISSN)1097-4660

Murayama, N., Yamamoto, H., and Shibata, J., 2002, Mechanism of zeolite synthesis from coal fly ash by alkali hydrothermal reaction: International Journal of Mineral Processing, v. 64, p. 1–17. doi:10.1016/S0301-7516(01)00046-1

Nair, M., Luo, Z., and Heller, A., 1993, Rates of photocatalytic oxidation of crude oil on salt water on buoyant, ceno-sphere-attached titanium dioxide: Industrial & Engineering Chemistry Research, v. 32, p. 2318–2323. doi:10.1021/ie00022a015

National Research Council, 2006, Managing coal combustion residues in mines: Committee on Mine Placement of Coal Combustion Wastes, Board on Earth Sciences and Resources: Washington, D.C., National Academies Press, 238 p.

NCASI, 2003, Beneficial use of industrial by-products: http://www.industrialmaterials-summit.com/midwest/summit/rmt_rpt.pdf.

Nelson, P.M., and Phillips, S.M., 1994, Designing porous road surfaces to reduce traffic noise: TRL Annual Review: Wokingham, Berkshire, Transportation Research Laboratories, UK.

Nifantov, B.F., 2003, Valuable and toxic elements in coals: Coal Resources of Russia, volume II, Geoinformmark, Moscow, p. 77–91 (in Russian).

P&U Statistics, 2004-2006, Canada, Production and Use of Coal Combustion Products: http://www.circainfo.ca/documents/ProductionandUseStatistics.pdf.

P&U Statistics, 2011-2013, Canada, Production and Use of Coal Combustion Products:http://www.circainfo.ca/documents/2011_2013CCPSurvey-NRCan.pdf

Palomo, A., Blanco-Varela, M.T., Granizo, M.L., Puertas, F., Vazquez, T., and Grutzeck, M.W., 1999, Chemical stability of cementitious materials based on metakaolin: Cement and Concrete Research, v. 29, p. 997–1004. doi:10.1016/S0008-8846(99)00074-5

Pan, P., 2011, Research of coal cinder and using it as concrete admixtures [M.E. thesis]: Guangzhou, Jinan University (in Chinese with English abstract).

Pang, J., Li, Q., Wang, W., Xu, X., and Zhai, J., 2011, Preparation and characterization of electroless Ni–Co–P ternary alloy on fly ash cenospheres: Surface and Coatings Technology, v. 205, p. 4237–4242. doi:10.1016/j.surfcoat.2011.03.020

Pecht, M.G., Kaczmarek, R.E., Song, X., Hazelwood, D.A., Kavetsky, R.A., and Anand, D.K., 2012, Rare earth materials: insights and concerns: College Park, M. D., CALCE EPSC Press, 194 p.

Peng, F., Liang, K., and Hu, A., 2005, Nano-crystal glass-ceramics obtained from high alumina coal fly ash: Fuel, v. 84, p. 341–346. doi:10.1016/j.fuel.2004.09.004

Qi, L., and Yuan, Y., 2011, Characteristics and the behavior in electrostatic precipitators of high-alumina coal fly ash from the Jungar power plant, Inner Mongolia, China: Journal of Hazardous Materials, v. 192, p. 222–225.

Qu, Q., Liu, G., Sun, R., and Kang, Y., 2016, Geochemistry of tin (Sn) in Chinese coals: Environmental Geochemistry and Health, v. 38, p. 1–23. doi:10.1007/s10653-015-9686-z

Queralt, I., Querol, X., LopezSoler, A., and Plana, F., 1997, Use of coal fly ash for ceramics: A case study for a large Spanish power station: Fuel, v. 76, p. 787–791. doi:10.1016/S0016-2361(97)00024-0

Querol, X., Alastuey, A., LopezSoler, A., Plana, F., Andres, J.M., Juan, R., Ferrer, P., and Ruiz, C.R., 1997b, A fast method for recycling fly ash: Microwave-assisted zeolite synthesis: Environmental Science & Technology, v. 31, p. 2527–2533. doi:10.1021/es960937t

Querol, X., Fernandezturiel, J., and Lopezsoler, A., 1995, Trace elements in coal and their behaviour during combustion in a large power station: Fuel, v. 74, p. 331–343. doi:10.1016/0016-2361(95)93464-O

Querol, X., Moreno, N., Umana, J.C., Alastuey, A., Hernandez, E., Lopez-Soler, A., and Plana, F., 2002, Synthesis of zeolites from coal fly ash: an overview: International Journal of Coal Geology, v. 50, p. 413–423. doi:10.1016/S0166-5162(02)00124-6

Querol, X., Plana, F., Alastuey, A., and LopezSoler, A., 1997a, Synthesis of Na-zeolites from fly ash: Fuel, v. 76, p. 793–799. doi:10.1016/S0016-2361(96)00188-3

Ram, L.C., Srivastava, N.K., Tripathi, R.C., Thakur, S.K., Sinha, A.K., Jha, S.K., Masto, R.E., and Mitra, S., 2007, Leaching behavior of lignite fly ash with shake and column tests: Environmental Geology, v. 51, p. 1119–1132. doi:10.1007/s00254-006-0403-1

Rayalu, S., Meshram, S.U., and Hasan, M.Z., 2000, Highly crystalline faujasitic zeolites from fly ash: Journal of Hazardous Materials, v. 77, p. 123–131. doi:10.1016/S0304-3894(00)00212-0

Rayzman, V.L., Shcherban, S.A., and Dworkin, R.S., 1997, Technology for chemical–metallurgical coal ash utilization: Energy & Fuels, v. 11, p. 761–773. doi:10.1021/ef960190s

Saikia, N., Kato, S., and Kojima, T., 2006, Compositions and leaching behaviours of combustion residues: Fuel, v. 85, p. 264–271. doi:10.1016/j.fuel.2005.03.035

Sarode, D.B., Jadhav, R.N., Khatik, V.A., Ingle, S.T., and Attarde, S.B., 2010, Extraction and leaching of heavy metals from thermal power plant fly ash and its admixtures: Polish Journal of Environmental Studies, v. 19, p. 1325–1330.

Seredin, V., 1996, Rare earth element-bearing coals from the Russian Far East deposits: International Journal of Coal Geology, v. 30, p. 101–129. doi:10.1016/0166-5162(95)00039-9

Seredin, V., 1998, Rare earth mineralization in Late Cenozoic explosion structures (Khankai massif, Primorskii Krai, Russia): Geology Ore Deposits, v. 40, p. 357–371.

Seredin, V., Arbuzov, S., and Alekseev, V., 2006, Sc-bearing coals from Yakhlinsk deposit, Western Siberia: Doklady Earth Sciences, v. 409, p. 967–972. doi:10.1134/S1028334X06060304

Seredin, V.V., 2012, From coal science to metal production and environmental protection: a new story of success: International Journal of Coal Geology, v. 90-91, p. 1–3. doi:10.1016/j.coal.2011.11.006

Seredin, V.V., and Dai, S., 2012, Coal deposits as potential alternative sources for lanthanides and yttrium: International Journal of Coal Geology, v. 94, p. 67–93. doi:10.1016/j.coal.2011.11.001

Seredin, V.V., and Dai, S., 2014, The occurrence of gold in fly ash derived from high-Ge coal: Mineralium Deposita, v. 49, p. 1–6. doi:10.1007/s00126-013-0497-9

Seredin, V.V., Dai, S., Sun, Y., and Chekryzhov, I.Y., 2013, Coal deposits as promising sources of rare metals for alternative power and energy-efficient technologies: Applied Geochemistry, v. 31, p. 1–11. doi:10.1016/j.apgeochem.2013.01.009

Seredin, V.V., and Finkelman, R.B., 2008, Metalliferous coals: a review of the main genetic and geochemical types: International Journal of Coal Geology, v. 76, p. 253–289. doi:10.1016/j.coal.2008.07.016

Shao, H., Liang, K., Zhou, F., Wang, G., and Peng, F., 2004, Characterization of cordierite-based glass–ceramics produced from fly ash: Journal of Non-Crystalline Solids, v. 337, p. 157–160. doi:10.1016/j.jnoncrysol.2004.04.003

Shen, J., Wang, B., and Xu, X., 2016, Review on research of using desulfurized gypsum to ameliorate saline-sodic soil: Journal of Agricultural Sciences, v. 37, p. 65–69 (in Chinese with English abstract).

Sheng, J., Huang, B., Zhang, J., Zhang, H., Sheng, J., Yu, S., and Zhang, M., 2003, Production of glass from coal fly ash: Fuel, v. 82, p. 181–185. doi:10.1016/S0016-2361(02)00238-7

Shi, C., and Qian, J., 2003, Increasing coal fly ash use in cement and concrete through chemical activation of reactivity of fly ash: Energy Sources, v. 25, p. 617–628. doi:10.1080/00908310390195688

Shi, W., and Zhu, G., 2007, Study on extraction Germanium from coal ash by chlorination with ammonium chloride and extraction with dihydroxamic: Journal of Henan University (Natural Science), v. 37, p. 147–151.

Shigemoto, N., Hayashi, H., and Miyaura, K., 1993, Selective formation of Na-X, zeolite from coal fly ash by fusion with sodium hydroxide prior to hydrothermal reaction: Journal of Materials Science, v. 28, p. 4781–4786. doi:10.1007/BF00414272

Song, H., Zhai, F., and Zhang, L., 2006, Comprehensive Utilization of Coal Ash in China: Journal of Kunming University of Science and Technology (Science and Technology), v. 31, p. 71–77.

Song, Q., 2014, The preparation and performance optimization of fly foam glass [M.E. thesis]: Shihezi, Shihezi University (in Chinese with English abstract).

Spiridonov, Y.A., and Orlova, L.A., 2003, Problems of Foam Glass Production: Glass and Ceramics, v. 60, p. 313–314. doi:10.1023/B:GLAC.0000008234.79970.2c

Sun, D., Zhang, X., Wu, Y., and Liu, X., 2010, Adsorption of anionic dyes from aqueous solution on fly ash: Journal of Hazardous Materials, v. 181, p. 335–342. doi:10.1016/j.jhazmat.2010.05.015

Sun, H., Tan, D., Peng, T., and Liang, Y., 2016a, Preparation of Calcium Sulfate Whisker by Atmospheric Acidification Method from Flue Gas Desulfurization Gypsum: Procedia Environmental Sciences, v. 31, p. 621–626. doi:10.1016/j.proenv.2016.02.112

Sun, J., Liu, H., and Min, F., 2011, Research on multiple utilization of fly ash: Cleaning Coal Technology, v. 17, p. 101–104 (in Chinese with English abstract).

Sun, Y., Qi, G., Lei, X., Xu, H., and Wang, Y., 2016b, Extraction of uranium in bottom ash derived from high-germanium coals: Procedia Environmental Sciences, v. 31, p. 589–597. doi:10.1016/j.proenv.2016.02.096

Taggart, R., 2015, Recovering rare earth metals from coal fly ash, in The World of Coal Ash (WOCA) Conference, Nashville, Tennessee, 5-7 May 2015: Lexington, Kentucky, Ash library, University of Kentucky.

Tang, B., Li, J., and Wang, Y., 2006, Exploration of dentrifying adsorbent made from fly ash: Environmental Engineering, v. 24, p. 45–46 (in Chinese with English abstract).

TFHRC, 2010, Coal fly ash material description: TFHRC: http://www.tfhrc.gov/hnr20/recycle/waste/

Tharaniyil, R., 2013, Coal combustion products utilization handbook (third edition): Milwaukee, Wisconsin, We Energies.

Tian, H., Hao, J., Zhao, Z., Kong, X., Yang, C., Lu, G., Liu, H., and Xu, F., 2006, Analysis on current and potential utilization of

FGD gypsum from coal-fired power plants: Electric Power, v. 39, p. 64–69 (in Chinese with English abstract).

Van Dyk, J., Benson, S., Laumb, M., and Waanders, B., 2009, Coal and coal ash characteristics to understand mineral transformations and slag formation: Fuel, v. 88, p. 1057–1063. doi:10.1016/j.fuel.2008.11.034

Vassilev, S.V., and Vassileva, C.G., 2007, A new approach for the classification of coal fly ashes based on their origin, composition, properties, and behaviour: Fuel, v. 86, p. 1490–1512. doi:10.1016/j.fuel.2006.11.020

Verdolotti, L., Iannace, S., Lavorgna, M., and Lamanna, R., 2008, Geopolymerization reaction to consolidate incoherent pozzolanic soil: Journal of Materials Science, v. 43, p. 865–873. doi:10.1007/s10853-007-2201-x

Vilches, L.F., Leiva, C., Vale, J., and Fernández-Pereira, C., 2005, Insulating capacity of fly ash pastes used for passive protection against fire: Cement & Concrete Composites, v. 27, p. 776–781. doi:10.1016/j.cemconcomp.2005.03.001

Vilches, L.F., Leiva, C., Vale, J., Olivares, J., and Fernández-Pereira, C., 2007, Fire resistance characteristics of plates containing a high biomass-ash proportion: Industrial & Engineering Chemistry Research, v. 46, p. 4824–4829. doi:10.1021/ie061194f

Wang, B., Li, Q., Wang, W., Li, Y., and Zhai, J., 2011a, Preparation and characterization of Fe^{3+}-doped TiO_2 on fly ash cenospheres for photocatalytic application: Applied Surface Science, v. 257, p. 3473–3479. doi:10.1016/j.apsusc.2010.11.050

Wang, C., Li, J., Han, W., Sun, X., and Wang, L., 2008a, Property characterization of zeolites NaA and NaX from fly ash using alkaline fusion method: Chinese Journal of Environmental Engineering, v. 2, p. 814–819 (in Chinese with English abstract).

Wang, C., Li, J., Sun, X., Wang, L., and Sun, X., 2009a, Evaluation of zeolites synthesized from fly ash as potential adsorbents for wastewater containing heavy metals: Journal of Environmental Sciences, v. 21, p. 127–136. doi:10.1016/S1001-0742(09)60022-X

Wang, C., Li, J., Wang, L., and Sun, X., 2008b, Influence of NaOH concentrations on synthesis of pure-form zeolite A from fly ash using two-stage method: Journal of Hazardous Materials, v. 155, p. 58–64. doi:10.1016/j.jhazmat.2007.11.028

Wang, C., Li, J., Wang, L., Sun, X., and Han, W., 2009b, Adsorption kinetics of heavy metal ions on Na A zeolite synthesized from fly ash: China Environmental Science, v. 29, p. 36–41.

Wang, F., Wu, D., He, S., Kong, H., Hu, Z., and Ye, C., 2005, Property characterization of NaP1 zeolite from coal fly ash by hydrothermal synthesis, v. 8, p. 47–50 (in Chinese with English abstract). http://www.paper.edu.cn.

Wang, F., and Wu, Z., 2004, A handbook for fly ash utilization (second edition): Beijing, China Power Press (in Chinese with English abstract).

Wang, J., Liu, G., Liu, Y., Zhou, C., Wu, Y., and Zhang, Q., 2016a, Mobilization of substance around stackable fly ash and the environmental characteristics of groundwater: With particular reference to five elements: B, Ba, Pb, Sb and Zn: Fuel, v. 174, p. 126–132. doi:10.1016/j.fuel.2016.01.092

Wang, J., Xu, X., Xiao, C., and Wang, J., 2016b, Effect of typical takyr solonetzs reclamation with flue gas desulphurization gypsum and its security assessment: Transactions of the

Chinese Society of Agricultural Engineering, v. 32, p. 141–147 (in Chinese with English abstract).

Wang, M., Yang, J., Ma, H., Shen, J., Li, J., and Guo, F., 2012, Extraction of aluminum hydroxide from coal fly ash by pre-desilication and calcination methods: Advanced Materials Research, v. 396-398, p. 706–710. doi:10.4028/www.scientific.net/AMR.396-398.706

Wang, S., 2008, Application of solid ash based catalysts in heterogeneous catalysis: Environmental Science & Technology, v. 42, p. 7055–7063. doi:10.1021/es801312m

Wang, S., Chen, C., Xu, X., and Li, Y., 2008c, Amelioration of alkali soil using flue gas desulfurization byproducts: productivity and environmental quality: Environmental Pollution, v. 151, p. 200–204. doi:10.1016/j.envpol.2007.02.014

Wang, S., Chen, Q., Li, Y., Zhuo, Y., and Xu, L., 2016c, Research on saline-alkali soil amelioration with FGD gypsum: Resources, Conservation and Recycling. https://doi.org/10.1016/j.resconrec.2016.04.005.

Wang, S., and Wu, H., 2006, Environmental-benign utilisation of fly ash as low-cost adsorbents: Journal of Hazardous Materials, v. 136, p. 482–501. doi:10.1016/j.jhazmat.2006.01.067

Wang, S., Zhang, C., and Chen, J., 2014, Utilization of coal fly ash for the production of glass-ceramics with unique performances: A brief review: Journal of Materials Science & Technology, v. 30, p. 1208–1212. doi:10.1016/j.jmst.2014.10.005

Wang, W., Qin, Y., Song, D., and Wang, K., 2008d, Column leaching of coal and its combustion residues, Shizuishan, China: International Journal of Coal Geology, v. 75, p. 81–87. doi:10.1016/j.coal.2008.02.004

Wang, Y., Ren, D., and Zhao, F., 1999, Comparative leaching experiments for trace elements in raw coal, laboratory ash, fly ash and bottom ash: International Journal of Coal Geology, v. 40, p. 103–108. doi:10.1016/S0166-5162(98)00062-7

Ward, C.R., 2002, Analysis and significance of mineral matter in coal seams: International Journal of Coal Geology, v. 50, p. 135–168. doi:10.1016/S0166-5162(02)00117-9

Ward, C.R., and French, D., 2006, Determination of glass content and estimation of glass composition in fly ash using quantitative X-ray diffractometry: Fuel, v. 85, p. 2268–2277. doi:10.1016/j.fuel.2005.12.026

Ward, C.R., French, D., Jankowski, J., Dubikova, M., Li, Z., and Riley, K., 2009, Element mobility from fresh and long-stored acidic fly ashes associated with an Australian power station: International Journal of Coal Geology, v. 80, p. 224–236. doi:10.1016/j.coal.2009.09.001

Ward, C.R., French, D., Stephenson, L.G., Riley, K., and Li, Z., 2010, Testing of interactions between coal ash leachates and rock materials for mine backfill evaluations: Coal Combustion and Gasification Products, v. 2, p. 15–27. doi:10.4177/suffix

Wei, X., Chu, K., Zhai, Y., Li, F., and Zhang, D., 2004, Experimental study of the removal of sulfur dioxide with modified-character fly ash: Journal of Hunan University (Natural Science), v. 31, p. 78–80 (in Chinese with English abstract).

Wu, D., Kong, H., Zhao, T., Wang, C., and Ye, C., 2005, Effects of synthesis conditions on the formation and quality of zeolite during the hydrothermal zeolitization processes of fly ash: Journal of Inorganic Materials, v. 20, p. 1153–1158 (in Chinese with English abstract).

Wu, D., Sui, Y., Chen, X., He, S., Wang, X., and Kong, H., 2008, Changes of mineralogical–chemical composition, cation exchange capacity, and phosphate immobilization capacity during the hydrothermal conversion process of coal fly ash into zeolite: Fuel, v. 87, p. 2194–2200. doi:10.1016/j.fuel.2007.10.028

Wu, D., Zhang, B., Li, C., Zhang, Z., and Kong, H., 2006, Simultaneous removal of ammonium and phosphate by zeolite synthesized from fly ash as influenced by salt treatment: Journal of Colloid and Interface Science, v. 304, p. 300–306. doi:10.1016/j.jcis.2006.09.011

Wu, F., 2009, The process requirement of manufacture of brick with furnace slag: Coal Ash China, v. 4, p. 39–40 (in Chinese).

Wu, F., Wu, P., Tseng, R., and Juang, R., 2010, Preparation of activated carbons from unburnt coal in bottom ash with KOH activation for liquid-phase adsorption: Journal of Environmental Management, v. 91, p. 1097–1102. doi:10.1016/j.jenvman.2009.12.011

Wu, J., 1993, Comprehensive utilization of coal fire slag: Coal Processing & Comprehensive Utilization, v. 2, p. 6–9 (in Chinese).

Wu, X., Li, J., Zhu, J., Qiao, S., and Zhang, M., 2015, Advances in the resource utilization of fly ash magnetic micro-spheres: Materials Review, v. 29, p. 103–107 (in Chinese with English abstract).

Wu, X., Tong, S., Guan, B., and Wu, Z., 2011, Transformation of flue-gas-desulfurization gypsum to α-hemihydrated gypsum in salt solution at atmospheric pressure: Chinese Journal of Chemical Engineering, v. 19, p. 349–355. doi:10.1016/S1004-9541(11)60175-4

Xu, A., Li, H., Luo, K., and Xian, L., 2011a, Formation of calcium sulfate whiskers from $CaCO_3$-bearing desulfurization gypsum: Research on Chemical Intermediates, v. 37, p. 449–455. doi:10.1007/s11164-011-0283-1

Xu, H., 1999, Analysis of impact of the harmful element contentson soil and crop in farmland incurred by fly ash: Bulletin of mineralogy, Petrology and Geochemistry, v. 18, p. 29–32 (in Chinese with English abstract).

Xu, H., Li, Q., Shen, L., Zhang, M., and Zhai, J., 2010, Low-reactive circulating fluidized bed combustion (CFBC) fly ashes as source material for geopolymer synthesis: Waste Management, v. 30, p. 57–62. doi:10.1016/j.wasman.2009.09.014

Xu, J., Zhou, Y., Chang, Q., and Qu, H., 2006, Study on the factors of affecting the immobilization of heavy metals in fly ash-based geopolymers: Materials Letters, v. 60, p. 820–822. doi:10.1016/j.matlet.2005.10.019

Xu, X., Li, Q., Cui, H., Pang, J., Sun, L., An, H., and Zhai, J., 2011b, Adsorption of fluoride from aqueous solution on magnesia-loaded fly ash cenospheres: Desalination, v. 272, p. 233–239. doi:10.1016/j.desal.2011.01.028

Xuan, X., Yue, C., Li, S., and Yao, Q., 2003, Selective catalytic reduction of NO by ammonia with fly ash catalyst: Fuel, v. 82, p. 575–579. doi:10.1016/S0016-2361(02)00321-6

Zhang, Y., Sun, W., and Shang, L., 1997, MECHANICAL PROPERTIES OF HIGH PERFORMANCE CONCRETE MADE WITH HIGH CALCIUM HIGH SULFATE FLY ASH: Cement and Concrete Research, v. 27, p. 1093–1098. doi:10.1016/S0008-8846(97)00087-2

Yan, L., Wang, Y., Ma, H., Han, Z., Zhang, Q., and Chen, Y., 2012, Feasibility of fly ash-based composite coagulant for coal washing wastewater treatment: Journal of Hazardous Materials, v. 203-204, p. 221–228. doi:10.1016/j.jhazmat.2011.12.004

Yang, C., 2000, A Study on the Resources Recovery of Coal Fly Ash [M.E. dissertation]: Taiwan, Cheng Kung University.

Yang, D., and Chen, X., 2012, Carbon recovery by flotation from high-carbon fly ash: Journal of Wuhan University of Science and Technology, v. 35, p. 250–253 (in Chinese with English abstract).

Yang, J., and Qin, W., 1999, Influence of coal ash on high performance concrete strength: Journal of Building material, v. 2, p. 218–222 (in Chinese with English abstract).

Yang, L., Wang, X., Zhu, X., and Du, L., 2013, Preparation of Calcium Sulfate Whisker by Hydrothermal Method from Flue Gas Desulfurization (FGD) Gypsum: Applied Mechanics and Materials, v. 268-270, p. 823–826. doi:10.4028/www.scientific.net/AMM.268-270

Yang, X., Shi, Q., and Sun, F., 2011, Quality analysis and application development of xinjiang FGD gypsum: Journal of Xinjiang Normal University (Natural Sciences Edition), v. 30, p. 38–43 (in Chinese with English abstract).

Yang, Z., Liu, Y., Yang, Z., and Liu, H., 2002, Physical-chemical properties and utilization of floating beads in fly -ash from power plant: Research & Application of Building Materials, v. 6, p. 13–16 (in Chinese with English abstract).

Yao, Z., Ji, X., Sarker, P., Tang, J., Ge, L., Xia, M., and Xi, Y., 2015, A comprehensive review on the applications of coal fly ash: Earth-Science Reviews, v. 141, p. 105–121. doi:10.1016/j.earscirev.2014.11.016

Yao, Z., 2013, Generation, characterization and extracting of silicon and aluminium from coal fly ash, in Sarker, P., ed., Fly Ash: Sources, Applications and Potential Environmental Impacts: New York, Nova Science Publishers, p. 3–58.

Yao, Z., Xia, M., Sarker, P., and Chen, T., 2014, A review of the alumina recovery from coal fly ash, with a focus in China: Fuel, v. 120, p. 74–85. doi:10.1016/j.fuel.2013.12.003

Yao, Z., Xia, M., and Ye, Y., 2011, Dilithium dialuminium trisilicate crystalline phase prepared from coal fly ash: Journal of Materials Engineering and Performance, v. 21, p. 877–881.

Yao, Z., Xia, M., Ye, Y., and Zhang, L., 2009, Synthesis of zeolite Li-ABW from fly ash by fusion method: Journal of Hazardous Materials, v. 170, p. 639–644. doi:10.1016/j.jhazmat.2009.05.018

Yao, Z., Ye, Y., and Xia, M., 2013, Synthesis and characterization of lithium zeolites with ABW type from coal fly ash: Environmental Progress & Sustainable Energy, v. 32, p. 790–796. doi:10.1002/ep.11689

Yavuz, C.T., Prakash, A., Mayo, J.T., and Colvin, V.L., 2009, Magnetic separations: from steel plants to biotechnology: Chemical Engineering Science, v. 64, p. 2510–2521. doi:10.1016/j.ces.2008.11.018

Yin, M., Bai, H., and Zhou, L., 2014, Strength characteristic of fly ash based geopolymer concrete: Bulletin of the Chinese Ceramic Society, v. 23, p. 2723–2727 (in Chinese with English abstract).

Yu, X., Shen, Z., and Xu, Z., 2007a, Preparation and characterization of Ag-coated cenospheres by magnetron sputtering method: Nuclear Instruments and Methods in Physics Research Section B: Beam Interactions with Materials and Atoms, v. 265, p. 637–640. doi:10.1016/j.nimb.2007.10.013

Yu, X., Xu, Z., and Shen, Z., 2007b, Metal copper films deposited on cenosphere particles by magnetron sputtering method: Journal of Physics D: Applied Physics, v. 40, p. 2894. doi:10.1088/0022-3727/40/9/034

Yue, Y., Cao, S., Fu, Y., and Mu, Y., 2011, Study of the insulating masonry mortars from waste residues of power plant: Fly Ash Comprehensive Utilization, v. 1, p. 18–21 (in Chinese).

Zeng, Q., 2007, Experiment research on Ga extraction from coal fly ash [M.E. Thesis]: Beijing, Geosciences (in Chinese with English abstract).

Zhai, B., Zhang, H., Ma, C., and Liang, H., 2014, Experimental study of circulating fluidized bed and electric separation decarbonization of fly ash from pulverized coal fired boiler: Coal Ash China, v. 6, p. 9–12 (in Chinese).

Zhang, A., Wang, N., Zhou, J., Jiang, P., and Liu, G., 2012, Heterogeneous Fenton-like catalytic removal of p-nitrophenol in water using acid-activated fly ash: Journal of Hazardous Materials, v. 201-202, p. 68–73. doi:10.1016/j.jhazmat.2011.11.033

Zhang, B., Wu, D., Wang, C., He, S., Zhang, Z., and Kong, H., 2007, Simultaneous removal of ammonium and phosphate by zeolite synthesized from coal fly ash as influenced by acid treatment: Journal of Environmental Sciences, v. 19, p. 540–545. doi:10.1016/S1001-0742(07)60090-4

Zhang, J., Liu, Y., and Wang, L., 2008, Current utilization of coal fly ash in China: West-China Exploration Engineering, v. 9, p. 215–216 (in Chinese).

Zhang, J., Song, L., and Yan, C., 2006, The Study and Practice of Comprehensive Utilization of Baotou No. 3 Power Plant's Slags: Multipurpose Utilization of Mineral Resources, v. 4, p. 40–43 (in Chinese with English abstract).

Zhang, J., Zhao, Y., Wei, C., Yao, B., and Zheng, C., 2010, Mineralogy and microstructure of ash deposits from the Zhuzhou coal-fired power plant in China: International Journal of Coal Geology, v. 81, p. 309–319. doi:10.1016/j.coal.2009.12.004

Zhang, L., 2008b, Research on decolorization technology of dyeing and printing wastewater [M.E. thesis]: Suzhou, Suzhou University (in Chinese with English abstract).

Zhang, Y., 2008a, Extraction method for gallium from high-aluminum fly ash [M.E. thesis]: Beijing, China University of Mining and Technology (in Chinese with English abstract).

Zhang, Y., 2013, Study on preparation of foam glass using desulfurization fly ash and waste glass as raw material [M.E. thesis]: Dalian, Dalian Polytechnic University (in Chinese with English abstract).

Zhang, Y., Ren, Y., Liu, B., and Zhang, Z., 2015, Comprehensive utilization status of sintering Flue Gas Desulfurization gypsum: Bulletin of the Chinese Ceramic Society, v. 34, p. 3563–3566, 3567 (in Chinese with English abstract).

Zhang, Y., Wang, Z., and Liu, B., 2014a, Preparation of foam glass using fly ash and waste glass as the main raw material: Journal of Dalian Polytechnic University, v. 33, p. 127–130 (in Chinese with English abstract).

Zhang, Y., Wang, Z., Xu, X., Chen, Y., and Qi, T., 1999, Recovery of heavy metals from electroplating sludge and stainless steel pickle waste liquid by ammonia leaching method: Journal of Environmental Sciences-China, v. 11, p. 381–384.

Zhang, Z., Qiao, X., and Yu, J., 2014b, Aluminum extraction from coal fly ash ultra-fast activated by microwave: China Science Paper, v. 9, p. 981–984.

Zhao, H., 2010, Study on comprehensive technique of recycling gallium and alumina from coal ash [M.E. thesis]: Xi'an, Chang'an university (in Chinese with English abstract).

Zhao, Y., Ye, J., Lu, X., Liu, M., Lin, Y., Gong, W., and Ning, G., 2010, Preparation of sintered foam materials by alkali-activated coal fly ash: Journal of Hazardous Materials, v. 174, p. 108–112. doi:10.1016/j.jhazmat.2009.09.023

Zhao, Y., Zhao, X., and Jiang, Y., 2009, Research progress in the synthesis of zeolite from fly ash: Bulletin of the Chinese Ceramic Society, v. 28, p. 1008–1012.

Zhou, G., 2011, Research on the influence of the flue gas desulfurization gypsum on the properties of Portland cement [M.E. thesis]: Wuhan, Wuhan University of Technology (in Chinese with English abstract).

Zhou, L., Chen, Y.L., Zhang, X.H., Tian, F.M., and Zu, Z.N., 2014, Zeolites developed from mixed alkali modified coal fly ash for adsorption of volatile organic compounds: Materials Letters, v. 119, p. 140–142. doi:10.1016/j.matlet.2013.12.097

Zhou, W., 2009, Method research of waste material complex utilization of Huaibei power plant and economic benefits analysis [M.E. thesis]: Hefei, Hefei University of Technology (in Chinese with English abstract).

Zhu, W., 2006, Wastewater treatment with coal fly ash: Coal Engineering, v. 4, p. 73–75 (in Chinese).

Zhu, X., 2005, Study on the fly ash geopolymer materials [M.E. thesis]: Nanchang, Nanchang University (in Chinese with English abstract).

Zhuang, X., Chen, L., Komarneni, S., Zhou, C., Tong, D., Yang, H., Yu, W., and Wang, H., 2016, Fly ash-based geopolymer: clean production, properties and applications: Journal of Cleaner Production, v. 125, p. 253–267. doi:10.1016/j.jclepro.2016.03.019

Zhuang, X., Querol, X., Zeng, R., Xu, W., Alastuey, A., Lopez-Soler, A., and Plana, F., 2000, Mineralogy and geochemistry of coal from the Liupanshui mining district, Guizhou, South China: International Journal of Coal Geology, v. 45, p. 21–37. doi:10.1016/S0166-5162(00)00019-7

Zhuang, X., Su, S., Xiao, M., Li, J., Alastuey, A., and Querol, X., 2012, Mineralogy and geochemistry of the Late Permian coals in the Huayingshan coal-bearing area, Sichuan Province, China: International Journal of Coal Geology, v. 94, p. 271–282. doi:10.1016/j.coal.2012.01.002

Zou, J.-H., Liu, D., Tian, H.-M., Li, T., Liu, F., and Tan, L., 2014, Anomaly and geochemistry of rare earth elements and yttrium in the late Permian coal from the Moxinpo mine, Chongqing, southwestern China: International Journal of Coal Science & Technology, v. 1, p. 23–30. doi:10.1007/s40789-014-0008-3

Mineralogy and geochemistry of ash and slag from coal gasification in China: a review

Shuqin Liu ⓘ, Chuan Qi ⓘ, Zhe Jiang ⓘ, Yanjun Zhang ⓘ, Maofei Niu ⓘ, Yuanyuan Li ⓘ, Shifeng Dai ⓘ and Robert B. Finkelman ⓘ

ABSTRACT

China is the largest coal gasification market in the world. The gasifiers have been commercialized for producing syngas for coal-to-liquids, coal-to-olefin, and coal-to-substitute natural gas. During coal gasification at high temperature and pressure, the differences in reactivity of organic matter in coal are almost negligible; however, the transformation behaviour of minerals is important to the stability of the process. Detailed knowledge of the mineralogical properties of ash and slag is essential for optimizing the operational parameters. It is also significant for determining the influence of the solid wastes on the environment: for example, trace elements carried in bottom slag, fly ash, condensate water, and gaseous products migrate to natural water and soil and into the atmosphere. With this in mind, the mineralogy of ash and slag from typical coal gasification processes in China are reviewed, including the transformation of minerals typically found in coal, and the effects of operating conditions on mineral transformation and melting/slagging behaviour of minerals during gasification process. In addition, trace element migration behaviour in commercialized gasifiers is discussed.

1. Introduction

China is the largest producer and consumer of coal in the world (Dai *et al.* 2016), with coal production reaching 3740 million tonnes in 2015, accounting for 47.4% of global coal production (BP 2016). Coal will be the principal source of energy in China until 2050 and for this reason clean and efficient utilization of coal is essential for diversification and low-carbon energy development. One method of using coal in a more environmentally benign way is coal conversion, especially coal gasification, which is not only a technology for the clean utilization of coal but also a necessary approach for developing a modern coal chemical industry (Li and Bai 2013).

The coal chemical industry in China, with coal gasification as the predominant technology, has developed rapidly in recent years. Annual coal consumption for coal gasification exceeds 200 Mt (Wang 2016). China is the largest coal gasification marketplace in the world: the market share of various gasifiers is given in Table 1.

especially for coal-to-liquids, coal-to-olefins, and coal-to-substitute natural gas (Wang 2016).

Commercialized gasification is generally restricted to the entrained flow, fluidized-bed and fixed-bed gasification processes. Entrained-flow gasification processes are subdivided by feed type into dry pulverized coal pressurized gasification and wet coal water slurry (CWS) pressurized gasification. Typical dry pulverized coal pressurized gasification technologies include the Shell gasifier, Gaskombimat Schwarze Pumpe (GSP) gasifier, and independently developed gasifiers in China such as Hang Tian gasifier, Wu Huan gasifier, two-stage gasifier, and the Sinopec/ECUST (SE)-Oriental gasifier. Typical wet CWS pressurized gasification technologies include the GE (formerly Texaco) gasifier, opposed multi-burner CWS gasifier, multi-component slurry gasifier, Tsinghua gasifier, and the E-gas (formerly Destec) gasifier. Typical fluidized-bed gasifiers include the U-gas®gasifier, Synthesis Energy Systems lignite gasifier, and the Coal Ash-Agglomerated Gasification™ gasifier. Typical fixed-

Table 1. The market share of various gasifiers in China's coal chemical industry (Wang 2016).

Gasifier	GE	OMB[a]	MCS[b]	Tsinghua	Shell	GSP	HT-LZ	TS[c]	Lurgi	Total
User	55	37	23	23	22	7	31	10	11	219
User share (%)	25.12	16.90	10.50	10.50	10.05	3.20	14.16	4.57	5.00	100
Gasified coal (Mt day^{-1})	13.32	13.45	4.22	3.65	7.00	7.60	8.65	1.61	7.28	66.78
Gasified share (%)	19.95	20.14	6.32	5.47	10.48	11.38	12.95	2.41	10.90	100
Standby-coal (Mt day^{-1})	6.22	7.11	2.48	3.20	0.00	1.20	0.80	0.00	1.17	22.18
Gasifier number	159	107	57	43	28	48	71	10	130	655

Parts of gasifiers are under design or construction, and the gasifiers with a few users are not listed.
[a]Coal-water slurry gasifier with opposed multi-burners.
[b]Multi-component slurry gasifier.
[c]Two-stage gasifier.

lignite gasifier and British Gas/Lurgi (BGL) gasifier (Wang 2016).

Entrained-flow gasifiers inject the gas into pulverized coal or coal slurry in the furnace by special nozzles. Oxygen/feed mixtures are flash-ignited and burn rapidly, generating considerable radiated heat. The high temperature (average >1300°C) and strong turbulence give this type of gasifier a greater transformation capacity than is possible in either fluidized-bed or fixed-bed gasifiers. GE CWS gasifier, employing typical entrained-flow gasification technology, uses coal slurry feed, with temperature 1300–1500°C and pressure 4.0–6.5 MPa. The ash fusion temperature of the coal is less than 1350°C (Liang 2005; Gao 2010; Jiao 2015). In the GSP and Shell gasifiers, the feed incorporates dry pulverized coal with the oxygen/steam. Operating pressure is 2.5–4.0 MPa and gasification temperature is 1350–1750°C, higher than for wet CWS pressurized gasification (Wu and Zhang 2013).

In fluidized-bed gasifiers, the coal particles are fluidized by controlling the velocity of the introduced gas, such that the contact, chemical reaction and heat transfer between the coal and gas are greatly enhanced by the violent agitation and back mixing. Once fluidization becomes inhomogeneous, local high-temperature zones occur in the gasifier, causing partial slagging and unstable operation. To avoid slagging, the fluidized-bed gasifier is usually operated at temperatures below 950°C (Li and Bai 2013).

Fixed-bed gasifiers, with larger lump coal as feedstock, operate at lower temperatures than entrained-flow or fluidized-bed gasifiers. The coal is fed at the top of the gasifier, and then moves slowly to the bottom counter current to the updraft gas and gaseous products. The coal undergoes physical and chemical changes including drying, pyrolysis, gasification, and combustion. Dust, gas and pyrolysis products are discharged as exhaust from the top of the gasifier at 300–450°C. Solid ash and slag are discharged from the bottom of the gasifier at 350–450°C (Schobert 1992; Van Dyk et al. 2009; Guo and Hu 2012). Pressurized Lurgi gasifiers are widely used, usually operated at temperatures of 950–1150°C and pressures of 2.45–2.94 MPa (Wu and Zhang 2013). The feedstock comprises bulk coal in the 5–50 mm range mixed with an oxygen/steam.

The trend of large-scale gasification is to improve single-furnace production by using higher operating temperatures and pressures. In these conditions, from the coal chemistry point of view, the differences in reactivity of organic matters in coal are almost negligible, whereas the transformation behaviour of the inorganic minerals in the coal encompasses a series of complex physical and chemical changes: volatilization, melting, crystallization, deposition and so on, all of which are closely related to stability of the gasifier operation (e.g. for gasifiers designed for slag discharge, the fluidity of slag is fundamental to the design and operation). Hence, a comprehensive understanding of mineral transformation is essential for successful operation and expanded commercialization of coal-based technologies (Raask 1985; Zhang et al. 2002; Shannon et al. 2009). Detailed information about the mineralogical properties of ash and slag (e.g. classification, transformation and physical and/or chemical properties of the various minerals) is essential for optimization of operational parameters, and is also significant for improving the efficiency of coal utilization as well as for assessing the environmental impact of generated solid waste products.

Coal is a complex and combustible organic rock formed from organic materials, predominantly land plants, deposited during various geological periods (O'Keefe et al. 2013). Of the 90 naturally occurring elements, 66 have been detected in coal by current available analytical techniques (Finkelman 1993; Dai et al. 2012a), and 86 have been detected in coal combustion products and coalbed gas (Zhang et al. 2006a). These include dozens of harmful and potentially harmful alkaline elements, alkaline earths, metallic elements, non-metals, and transition elements (Finkelman 1993; Dai et al., 2012a). During coal utilization, the harmful elements and their compounds migrate to bottom ash and slag, or are concentrated in the fly ash, condensate water, and gaseous products through a series of physical and chemical changes (Li and Bai 2013). The trace element enriched in ash and slag may be water-soluble and leached by rain, surface or ground water, and transported to the surrounding environment, leading to ecological deterioration (Liu 2009; Dai *et al.*, 2012a). Trace elements in fly ash and in coal gas will be partly transported in condensate water and then discharged as wastewater (Liu 2009). That portion that remains in the coal gas is discharged into the atmosphere. Such harmful trace elements can affect human and ecological health through the food chain (Li *et al.* 2007; Zhao *et al.* 2007). An investigation of the distribution, mode of occurrence and migration behaviour of trace elements associated with coal gasification is, therefore, of significance for trace-element risk assessment with the aim of minimizing the effect of pollution by trace elements on the environment and humans (Kong *et al.* 2002).

In China, many studies have been published on the mineralogical and geochemical processes taking place during coal combustion (Yao *et al.* 2002; Liu *et al.* 2003; Li *et al.* 2009; Dai *et al.*, 2012b; Dai *et al.* 2014; Wang *et al.* 2015), but there are few studies of the mineral and trace element associated with coal gasification (Bai 2003; Huang *et al.* 2004, 2005, 2006, 2008; Bai *et al.* 2008, 2009, 2010; Li and Bai 2013; Liu *et al.* 2016). The present study reviews and discusses the mineralogical properties of ash and slag from typical coal gasification plants in China (including the effects of gasification conditions on mineral transformation and the melting and slagging of minerals, etc.) and summarizes the migration behaviour of trace elements during commercial gasification processes.

2. Mineral transformation during coal gasification in China

2.1. Classification of mineral matter in coal

Ash-related problems in coal utility systems depend exclusively upon the quantity and associations of mineral matter in coal (Benson *et al.* 1993). The term 'mineral matter' includes all the minerals and other inorganic substances in coal (Ward 2002). A number of classifications have been used in the literature for mineral matter in coal (Ward 2002, 2016). Mineral matter is broadly classified into three groups: (1) dissolved salts and other inorganic substances in the pore water of the coal, (2) inorganic elements within the organic structure of the coal macerals, and (3) discrete inorganic particles (Ward 2002). Categories (1) and (2), often termed non-mineral inorganics, are usually abundant in low-rank coals and their contents decrease with

increase in coal rank (Given and Spackman 1978; Miller and Given 1986; Ward 2002). Intimately dispersed inorganic minerals within the coal matrix are referred to as 'included minerals', whereas 'extraneous minerals', generally minerals in ash-rich partings, cleat, or fractures, have little or no association with the coal itself (Yan *et al.* 2002). In this review, minerals that are liberated from the coal during grinding are also considered to be 'extraneous minerals'.

Unlike non-mineral inorganics, discrete mineral particles may occur in coal of any rank and are usually the predominant components of mineral matter in high-rank coals (Ward 2002). The differentiation between the different kinds of mineral content is necessary since the ash or slag formation conditions are different for included minerals, non-mineral inorganics and extraneous minerals (Raask 1984, 1985).

2.2. Transformation of typical minerals in coal

Ash is the product of thermal conversion of minerals at high temperatures below the minimum melting point. The products of mineral decomposition and the reactions between minerals and organics, and between minerals and gaseous reactants, are collectively called 'ash', as well as unreacted minerals (Li and Bai 2013). The composition of ash formed during coal utilization depends upon the mineral species in the coal, as well as the temperature, atmosphere and pressure (Li and Bai 2013). Of these, temperature is the most crucial external condition (Li and Bai 2013). The reaction atmosphere only partly transforms minerals. The effect of pressure is related to simultaneous changes of temperature and atmosphere.

Slag is a thermal conversion product of minerals at temperatures above the minimum melting point, that is, molten ash; minerals in coal first form ash, then higher temperatures melt the ash to form slag along with other reactions and transformations of minerals (Li and Bai 2013). The transformation process from minerals to ash or slag is illustrated in Figure 1. The liquid-phase formation of slag means that it is almost free of non-metal elements except oxygen, and contains fewer alkali metal elements. Furthermore, the slag is substantially more amorphous than ash.

Depending on the reactants involved, the reactions of the minerals contained in coal ash or slag may be divided into three types: (1) mineral decomposition; (2) inter-mineral reactions; and (3) mineral–atmosphere reactions. A particular mineral may take part in one, two or all three types of reaction simultaneously during thermal conversion, depending on its nature and associations (Li and Bai 2013). Different coal ranks with ash

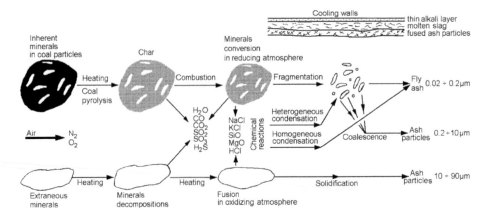

Figure 1. Mechanism of mineral transformation (Tomeczek and Palugniok 2002).

of different chemical composition result indifferent mineral reactions. Li and Bai (2013) divided coal in China into four groups based on their chemical composition: (1) high silica–alumina (HSA) coal ($SiO_2 + Al_2O_3 > 80\%$); (2) high silica–alumina ratio (HSAR) coal ($SiO_2/Al_2O_3 > 2$); (3) high calcium (HC) coal ($CaO > 15\%$); and (4) high iron (HI) coal ($Fe_2O_3 > 15\%$). When the classification criteria overlap, the priority is HC $>$ HI $>$ HSAR $>$ HSA. The thermal conversion of mineral mainly includes transformation of clay minerals, carbonate minerals, pyrite, and quartz at high temperature (Srinivasachar et al. 1990; Slade and Davies 1991; Hu et al. 2006; Ranjan et al. 2010). In this review, mineral conversion in coal during gasification is investigated by comparing the transformation of kaolinite, illite, calcite, gypsum, and quartz, among which kaolinite and illite are clay minerals, quartz is a silicate mineral, gypsum belongs sulphate mineral, and calcite is a carbonate mineral.

2.2.1. Kaolinite

Kaolinite [$Al_2Si_2O_5(OH)_4$] starts to lose water of crystallization to form metakaolin ($SiO_2 \cdot 2Al_2O_3$) at about 327°C, and further decomposes into SiO_2 and Al_2O_3 at around 827°C (Li and Bai 2013). In a reducing atmosphere, the SiO_2 and Al_2O_3 from the decomposition of metakaolin react to form other minerals at higher temperatures. A study of the mineral transformation in HSA coal (typical of Huainan coal (Anhui, China)) (Bai et al. 2009) found that the SiO_2 and Al_2O_3 from the decomposition of metakaolin predominantly transform to mullite ($3SiO_2 \cdot 2Al_2O_3$) and sillimanite ($SiO_2 \cdot Al_2O_3$) as follows:

$$Al_2O_3 + SiO_2 \rightarrow Al_2O_3 \cdot SiO_2,$$

$$Al_2O_3 + SiO_2 \rightarrow 3Al_2O_3 \cdot 2SiO_2.$$

Mullite is thermodynamically stable; sillimanite is regarded as its precursor. At temperatures higher than

1000°C, sillimanite content decreases and mullite content increases due to its greater thermodynamic stability:

$$Al_2O_3 \cdot SiO_2 + 2Al_2O_3 + SiO_2 \rightarrow 3Al_2O_3 \cdot 2SiO_2.$$

Ma et al. (2012) studied HSAR coal (typified by Fugu coal (Shannxi, China)) and found that metakaolin reacted with CaO formed by the decomposition of calcium-bearing minerals in the coal, transforming into anorthite at around 1100°C. The anorthite content peaked at about 1200°C then began to melt at higher temperatures, and its crystalline content gradually decreased and disappeared by 1400°C:

$$Al_2O_3 \cdot 2SiO_2 + CaO \rightarrow CaO \cdot Al_2O_3 \cdot SiO_2.$$

For HC coal (typical of Xiaolongtan coal (Yunnan, China) (Ma et al. 2012) and Shenfu coal (Shaanxi, China) (Bai et al. 2010)), the CaO from the decomposition of calcium-bearing minerals in the coal reacts with metakaolin, or SiO_2 and Al_2O_3 from the decomposition of metakaolin, mainly to form gehlenite in considerable quantities in the temperature range 1200–1400°C, then starts to decrease above 1400°C, but still exists at 1500°C:

$$Al_2O_3 \cdot 2SiO_2 + CaO \rightarrow Ca_2Al_2SiO_7 + SiO_2,$$

$$CaO + Al_2O_3 + SiO_2 \rightarrow Ca_2Al_2SiO_7.$$

Among HI coals, the metakaolin in Yanzhoucoal (Shandong, China) principally transforms to anorthite and gehlenite, while that in Guizhoucoal (Guizhou, China) converts into sekaninaite ($Fe_2Al_4Si_5O_{18}$), which forms from the reaction between Fe_2O_3, SiO_2, and Al_2O_3. The SiO_2 and Al_2O_3 are derived from metakaolin decomposition at 1200°C and exists in large quantities above 1450°C. The reaction is

$$Fe_2O_3 + Al_2O_3 + SiO_2 \rightarrow Fe_2Al_4Si_5O_{18}.$$

2.2.2. Illite

Illite (K,H$_2$O)Al$_2$[OH]$_2$[AlSi$_3$O$_{10}$] loses adsorbed water at around 110°C, and begins to lose water of crystallization at 650°C, forming a dehydroxylated illite-like phase. Above 1000°C, the hydroxyls have been completely removed, and interlayer cations generate a new coordination resulting in reduced commutativity, while the structural framework remains unchanged. At temperatures above 1200°C, the illite-like phase is transformed into mullite. Above 1300°C, it converts into an amorphous substance (Li and Bai 2013).

2.2.3. Calcite

The pyrogenic decomposition of calcite (CaCO$_3$·MgCO$_3$) is characterized by two independent reactions (Li and Bai 2013):

$$\text{(Calcium carbonate)} CaCO_3 \rightarrow CaO + CO_2,$$

$$\text{(Magnesium carbonate)} MgCO_3 \rightarrow MgO + CO_2.$$

CaCO$_3$ starts to decompose at about 650°C, then rapidly decomposes between 812°C and 928°C. The decomposition temperature is higher during gasification due to the inhibiting presence of a CO$_2$ atmosphere. Figure 2 shows that the initial decomposition temperature rises to 900°C with increasing CO$_2$ content (Chi et al. 2006). The initial decomposition temperature of MgCO$_3$ is around 590°C, which rises in a CO$_2$ atmosphere in a similar manner to CaCO$_3$ (Li and Bai 2013).

CaO reacts with H$_2$O in air to form Ca(OH)$_2$, or reacts with sulphur-bearing compounds which result from mineral transformation to form gypsum or anhydrite, or reacts with aluminosilicate at high temperatures to form anorthite (CaAl$_2$Si$_2$O$_8$) – and so on (Li and Bai 2013). Above 1250°C, the structure of the anorthite loses some calcium, becoming de-calcium

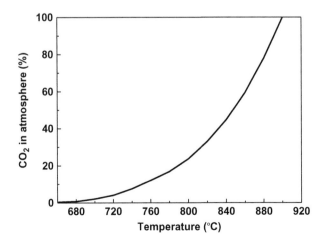

Figure 2. The initial decomposition temperature of calcium carbonate variation as function of CO$_2$ content (Chi et al. 2006).

(Ca$_{0.88}$Al$_{1.77}$Si$_{2.23}$O$_8$) with a structure similar to that of anorthite (Bai et al. 2010). MgO reacts with CaO and quartz to form akermanite (Bai et al. 2010).

2.2.4. Gypsum

A study by Li and Bai (2013) showed that gypsum (CaSO$_4$·2H$_2$O) firstly loses water of crystallization under heating, then the CaSO$_4$ reacts in a reducing atmosphere to form CaS above 900°C. The temperature at which this reaction occurs is affected by the CO$_2$ content. On the other hand, the CaSO$_4$ decomposes to form CaO when the temperature reaches 1000°C:

$$CaSO_4 \rightarrow CaO + SO_3.$$

When a gypsum particle is wrapped in coal, the dominant reaction is

$$CaSO_4 + 4CO \rightarrow CaS + 4CO_2.$$

When the gypsum particle is exposed to the atmosphere during coal fragmentation, the following reaction takes place simultaneously:

$$CaSO_4 \cdot 2H_2O \rightarrow CaSO_4 + 2H_2O.$$

As discussed in Section 2.2.3, at higher temperatures CaO may react in several ways to form anorthite, etc. (Ma et al. 2012; Li and Bai 2013).

2.2.5. Pyrite

Generally, pyrite (FeS$_2$) is present in coal both as an inherent mineral and as an extraneous mineral (Yao et al. 2003). Inherent minerals follow different thermal transformation paths than extraneous minerals (Figure 3), and their transformation temperatures are usually lower (Raask 1985). The two mineral species are subject to different physical and chemical changes (Li 2004). Inherent pyrite is affected by the char and by other materials in the char. Unlike extraneous pyrite, inherent pyrite does not begin to oxidize until it is exposed by the receding char surface (Bool et al. 1995).

Inherent and extraneous pyrite both decomposes to form pyrrhotite (Fe$_{1-x}$S) in the temperature range 327–627°C (Srinivasachar and Boni 1989). Some of the Fe^{2+} is replaced by Fe^{3+}, leaving a vacant Fe^{2+} position to maintain valency balance. In the general formula Fe$_{1-x}$S of pyrrhotite, x represents the missing Fe atom; the value of x ranges between 0 and 0.223. With increasing reaction time, all the pyrite eventually turns into pyrrhotite (Fe$_{0.887}$S), which further reacts with oxidizing gas during gasification to form magnetite (Fe$_3$O$_4$) at higher temperatures (Srinivasachar and Boni 1989).

An investigation (Bai et al. 2008) of the mineral change in HI coal (Yanzhou) at 1100–1500°C found that the predominant iron oxide component at 1100°C

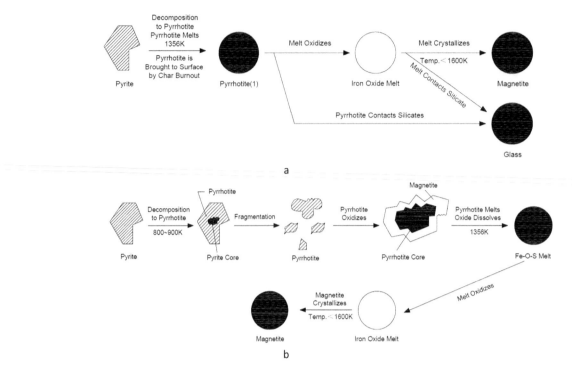

Figure 3. Transformation of inherent and extraneous pyrite. (a) Reaction pathway for inherent mineral (Srinivasachar and Boni 1989). (b) Reaction pathway for extraneous inherent mineral (Bool *et al.* 1995).

was haematite (Fe_2O_3); above 1250°C, some of the Fe_2O_3 had transformed into Fe_3O_4; above 1350°C, only Fe_3O_4 was found. Almost all of the $CaSO_4$ in the Yanzhou coal decomposes, then reacts with SiO_2 to form $CaSi_2O_5$, which further reacts with Fe_2O_3 to form the more stable augite ($CaFeSi_2O_6$). A small amount of sekaninaite ($Fe_2Al_4Si_5O_{18}$) is also generated.

In HI coal (Guizhou) (Ma *et al.* 2012), Fe_2O_3 in low-temperature (815°C) ash gradually transforms to Fe_3O_4. The iron oxide content decreases with increasing temperature and in most cases becomes ferrosilicates and iron aluminosilicates above 1450°C. At 1200°C, $Fe_2Al_4Si_5O_{18}$ and $CaFeSi_2O_6$ are generated by the following reactions:

$$Fe_2O_3 + Al_2O_3 + SiO_2 \rightarrow Fe_2Al_4Si_5O_{18},$$

$$Fe_2O_3 + CaO + SiO_2 \rightarrow CaFeSi_2O_6.$$

$Fe_2Al_4Si_5O_{18}$ is the principal component of the minerals in Guizhou coal, increasing in content with rising temperature. $CaFeSi_2O_6$ content gradually decreases with rising temperature, and is not found at 1350°C. In addition, some hercynite ($FeAl_2O_4$) is generated at 1250°C, but its content decreases with increasing temperature:

$$Fe_2O_3 + Al_2O_3 \rightarrow FeAl_2O_4.$$

Studies of ash and slag from underground coal gasification (UCG) operations (McCarthy *et al.* 1988; Liu *et al.*

2016) found that sekaninaite is the product of SiO_2 and hercynite at high temperatures.

2.2.6. Quartz

Quartz (SiO_2) has a high melting temperature (~1800°C) and is relatively stable; therefore it is found in coals heated to 1100°C. From a phase-state perspective, quartz reacts with char to form $SiO(g)$, and then transforms to $SiO_2(g)$. When the partial pressure of $SiO_2(g)$ in the gaseous phase is greater than that of $SiO_2(l)$, $SiO_2(g)$ occurs as core aggregation and forms $SiO_2(l)$. When liquid drops of $SiO_2(l)$ enter a reducing atmosphere, it is transformed into $SiO(g)$, as follows:

$$SiO_2(g) \rightarrow SiO_2(l),$$

$$SiO_2(l) + CO(g) \rightarrow SiO(g) + CO_2(g).$$

From a mineral-form perspective, SiO_2 undergoes the following changes at high temperature:

$$Quartz \overset{870}{\rightarrow} Tridymite \overset{1470}{\rightarrow} Cristobalite.$$

Quartz, tridymite, and cristobalite have identical elemental composition (SiO_2) but different mineral forms (Li 2008). Tridymite and cristobalite are generated by slow solid-phase reaction at high temperature; therefore, quartz, tridymite, and cristobalite may be found in slag contemporaneously. Above 1300°C, quartz reacts with other compounds to form silicates (anorthite, gehlenite, millite, sekaninaite, augite, etc.) (Bai *et al.* 2008, 2009, 2010; Ma *et al.* 2012). A study of the HSAcoal (Huainan) (Bai *et al.* 2009) found that, above 1450°C, the crystalline content of SiO_2 decreases and the amorphous content accordingly increases. Furthermore, the conversion from crystalline to amorphous SiO_2 depends on the amounts of alkali metal and alkaline-earth metal in the system (Li and Bai 2013).

2.3. Effects of gasification conditions on mineral transformation

2.3.1. Temperature

The reaction temperature is the foremost external factor in determining the mineral transformation in coal during gasification. The primary effect of temperature is that the minerals undergo different physical and chemical changes at different temperatures, forming different mineral types and crystalline states. This is discussed at length in Section 2.2.

2.3.2. Atmosphere

The effect of the atmosphere on mineral reactions falls into two categories: (1) affecting chemical reaction equilibrium and (2) reacting with minerals (Li and Bai 2013). For example, CO_2 retards the decomposition of $CaCO_3$ during gasification, thereby raising its decomposition temperature (Klein *et al.* 1983). This reflects the fact that, in this case, the atmosphere affects the chemical reaction equilibrium. CO reacts with $CaSO_4$ to generate CaS, with the CO_2 concentration effect depending on the reaction temperature (Ma *et al.* 2011). This example shows that the atmosphere can react with the minerals, and can also affect the reaction equilibrium. The composition of the atmosphere has its most marked influence on changes in the iron-bearing minerals. The reaction degree is characterized by the valence of the iron.

An investigation (Klein *et al.* 1983) of $Fe^{3+}/\Sigma Fe$ of iron-bearing minerals in different atmospheres reported that (1) in an oxidizing atmosphere (air), $Fe^{3+}/\Sigma Fe < 0.25$; (2) in a weakly reducing atmosphere (95% N_2/5% H_2), $Fe^{3+}/\Sigma Fe > 0.27$; and (3) in a fully reducing atmosphere (carbon dust), $Fe^{3+}/\Sigma Fe > 0.50$.

A study of mineral transformation in the Jincheng coal (Shanxi, China) when in a CO_2 atmosphere, an H_2O atmosphere, an oxidizing atmosphere, and during pressurized fluidized-bed gasification (Jing 2013) found that the composition of the atmosphere had some influence on mineral transformation, and different mineral types were formed in the different atmospheres. In a reducing atmosphere, some Fe^{3+} was reduced to Fe^{2+} whose compounds facilitated eutectic melting in the presence of alkaline metallic oxides in ash (CaO, Na_2O, etc.), forming solid solutions of fayalite and hercynite, and resulting in a lower sintering temperature than was the case in an oxidizing atmosphere. In addition, the high-temperature mineral content increased at higher temperatures, both in combustion and gasification. In the gasification process, more molten minerals and feldspars such as muscovite, anhydrite and K-feldspars were formed, lowering the ash fusion temperature (Jing *et al.* 2013).

2.3.3. Pressure

A study of the effects of pressure on the mineral properties of the Jincheng coal (Jing *et al.* 2013) indicated that the effect of pressure was more complex and also depended on changes in temperature and atmosphere. At 900°C, the decomposition of low-temperature minerals such as muscovite, anhydrite and oldhamite were suppressed with increasing pressure, resulting in lower fusion temperatures. On the other hand, at 1000°C, increased pressure transformed low-temperature minerals into high-temperature minerals such as mullite and sanidine. With increasing pressure under gasification conditions, oldhamite was transformed into anhydrite, and quartz reacted with other oxides to form sanidine andmullite, and the decomposition of fluxing minerals (anhydrite) was retarded. Also, increasing pressure under gasification conditions transformed muscovite into mullite, and microcline into albite (Jing 2013). Thus it may be concluded that the increasing pressure facilitated the formation of high-temperature minerals and suppress the decomposition of fluxing minerals.

2.4. Melting and slagging of minerals in gasification process

As gasification technology has developed, gasification temperatures have increased significantly; the temperatures in fluidized-bed and entrained-flow gasifiers even exceed the melting point of ash. Hence, the transformation behaviour of the mineral content of coal during gasification is a significant factor when selecting a gasification technology: the mineral behaviour determines the design of the ash discharge, and at the same time it is significant when guaranteeing the normal operation of gasifier (Li and Bai 2013). A clear understanding of the interaction of minerals and gasification is essential when selecting a suitable gasification technology for a specific coal, or when expanding the coal variety for a specific gasifier.

As in the combustion process, mineral matter transformation during gasification generally includes char/ash fragmentation, ash coalescence and volatilization

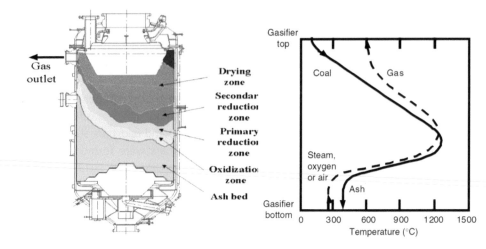

Figure 4. Sasol-Lurgi gasification process and the temperature distribution of gasifier (Van Dyk *et al.* 2009).

and condensation of inorganics. In the process, minerals in coal first form ash which is then transformed into slag. Ash and slag have similar chemical compositions, whereas the compositions and morphologies of the minerals vary depending on temperature, pressure and retention time (Li and Bai 2013). The process of forming ash and slag clearly differs in the various types of gasifiers. This section focuses on the commercial dry-ash gasifiers and slagging gasifiers currently used in China, and discusses their relationship with the properties of the mineral content of the coal.

2.4.1. Dry-ash gasifier

The category of dry-ash gasifier includes predominantly fixed-bed gasifiers and fluidized-bed gasifiers; however, some low-melting-point minerals still melt during gasification, resulting in the agglomeration and slagging of ash.

Partially molten minerals are agglomerated during gasification and enclose particles of unburned carbon. This carbon loss increases with higher ash content. The probability of aggregation (i.e. in terms of the proportion of solids to liquids) decreases in the lower liquid phase. Therefore, minerals tend to melt at lower ash fusion point, resulting in greater loss of unburned carbon. To summarize: in fixed-bed and fluidized-bed gasifiers, a lower ash content and higher ash fusion point of the raw coal are beneficial for stable operation (Li and Bai 2013).

Partially melted minerals in the ash also result in slagging in dry-ash gasifiers. This occurs when the ash contains a large content of alkali metal and alkaline-earth metal. Therefore, for the gasification of coal with relatively high alkali metal and alkaline-earth metal content, the optimal operating temperature must be carefully ascertained to avoid slagging. In addition, some

coal treatments, such as washing, reduce the alkali metal content in the coal (Li and Bai 2013).

The Lurgi fixed-bed gasifier is the most commonly used commercial reactor of this type (see Figure 4). The minerals in the coal decompose by dehydration and decomposition: for example, kaolinite undergoes dehydration, dolomite releases sodium, and illite releases potassium. The decomposition and volatilization of low-melting-point minerals promote the formation of fly ash. Char is consumed gradually and fragmented after entering the oxidation zone, and fine particles of minerals and carbon residue are carried by the airflow to form gasification dust. The mineral matter starts to sinter when the temperature reaches its deformation temperature. If it is below the melting point, the ash and slag will be partially sintered, which enhances the permeability for the gas and stabilizes the pressure inside the furnace (Li and Bai 2013).

An investigation of the Lurgi gasifier (Fu 2015) has classified the ash and slag products into four types based on particle size, and examined the variation in their mineral content. Table 2 shows that with increasing particle size, the proportion of amorphous substance increased gradually, and the amount of crystalline mineral material decreased. A possible reason is that the large particles remained in the gasifier for a longer period of time, adequate for the reaction to be completed during gasification.

Table 2. Mineral composition of ash and slag from commercial Lurgi gasifier (Klein *et al.* 1983).

Particle size (mm)	Amorphous	Augite	Anorthite	Haematite	Gehlenite
<3	30.1	19.7	20.9	11.8	17.5
3–6	32.9	14.2	25.7	14.8	12.4
6–9	35.2	17.7	23.2	12.3	11.6
>9	49.5	9.7	21.9	7.9	11.0

A study of the effect of the temperature on the fusibility of coal ash in fluidized bed gasification using a high pressure thermogravimetric analyser apparatus (Jing *et al.* 2013) indicated that at pressures below 0.5 MPa, samples of the ash prepared at 800°C and at 900°C both contained quartz, anhydrite, muscovite and sanidine. When the temperature reached 1000°C, no muscovite and less sanidine were found, and haematite and mullite appeared. At 1090°C, no sanidine or haematite was found, and anorthite and hercynite appeared. This indicated that the sanidine had been transformed into anorthite, and haematite had been transformed into hercynite at high temperatures. Anorthite and hercynite are high-temperature minerals which react with other minerals to generate low-temperature copolymers in a reducing atmosphere, and thus decrease the melting point of ash and slag.

2.4.2. Slagging gasifier

Entrained-flow gasification frequently utilizes slagging gasifiers with ash discharge, designed for slagging conditions. Their basic operation is described in Section 1.

The temperature of the flame centre reaches 2000°C, sufficient to melt the mineral matter. During the char gasification process, the airflow causes the droplets of molten mineral to be deposited on the furnace wall, forming a liquid slag layer, which is discharged along with some undeposited particles from the bottom of the gasifier to form the gasification slag. Some particles (gasification ash) are discharged from the top of the gasifier with the airflow (Li and Bai 2013).

The slag flows down the furnace wall and discharged from the slag hole. The operating temperature must be 100–150° above the flow temperature of the slag to guarantee its smooth discharge. The flow temperature of coal ash cannot exceed 1500°C, due to the limits of the operating temperature of the gasifier. The flow temperature should not be too low in gasifiers with refractory lining material on the inner wall, since lower temperatures imply the presence of alkali metal, alkaline-earth metal or Fe_2O_3 in relatively large quantities, which may react with, and cause loss of refractory material. In addition, slag with high CaO and Fe_2O_3 content is commonly characterized by low viscosity, good fluidity and wetting ability on the surface of the refractory materials, resulting in obvious scouring action during slag flow (Li and Bai 2013). Viscosity lower than 2 Pa s causes abrasion and erosion of the refractory material at rates of 4–20 m/h (Strobel and Hurley 1995). Therefore, it is recommended that the CaO + Fe_2O_3 content of coal ash should not exceed 35%.

In general, the heat loss caused by melting slag decreases with decreasing coal ash. Gasifiers with water-cooled walls require the coal ash content to be high enough to guarantee the dynamic balance of the slag layer and maintain a moderate layer, and prevent it from burning through the wall. For example, the ash content is adjusted to 8–35% in Shell gasifiers, several of which are currently operating in China. This implies that 8% is the minimum ash content necessary to maintain the slag layer (Li and Bai 2013).

The temperature of molten slag for the discharge process in slagging fix-bed gasifiers (e.g. BGL gasifier) must be higher than the flow temperature of the coal ash. At the same time, the viscosity of molten slag should be less than 5 Pa s to maintain a moderate discharge rate (Patterson and Hurst 2000).

CWS also is affected by the mineral composition in the coal during preparation. It has been concluded (Zhu 1996; Zhang *et al.* 2006b) that the mineral content of the coal reduces its volume fraction, and thus partly affects the concentration of the coal slurry. The ions released by the dissolution of soluble minerals retard the dispersion of coal particles, reducing the slurrying ability of coal. It is generally recognized that high ash content, which lowers the reactivity, uniformity and dispersivity of the CWS, is detrimental to the preparation of CWS. A regression analysis (Li and Bai 2013) has found that clay minerals and iron-bearing minerals in coal ash significantly resist the dispersion of the coal.

Data from a Texaco gasifier operation (Liang 2005) showed that the mineral content of the slag ranged from 11% to 48%. The predominant minerals were mullite, quartz, haematite, magnetite, calcium aluminate and melilite, with mullite accounting for the largest proportion (6–15%).

In another study (Wu *et al.* 2015), slag from a Texaco gasifier of Shanghai Coking Co. Ltd. was divided into coarse and fine particles, and the physical and chemical properties were determined by morphological and mineralogical analysis. Figure 5 shows that fully molten and partly molten inorganic components in the coarse slag (CS) mainly tended to co-exist in the overlapping bulk of lumps and agglomerates, whereas in the fine slag (FS) they almost always tended to appear separately, either as spherical particles or as agglomerates, respectively.

The inorganic constituents in gasification slags predominantly contained crystalline components (silicates, aluminosilicates, and Ca–Fe and Fe oxides) and vitreous components (Ca–Fe–aluminosilicate glass), indicating a distinct catalytic action in the carbon gasification, mainly due to the predominant presence of these catalytic components (Ca–Fe and Fe oxides). The greater gasification reactivity of the CS was mainly ascribed to the presence of more catalytic components than was found in the FS.

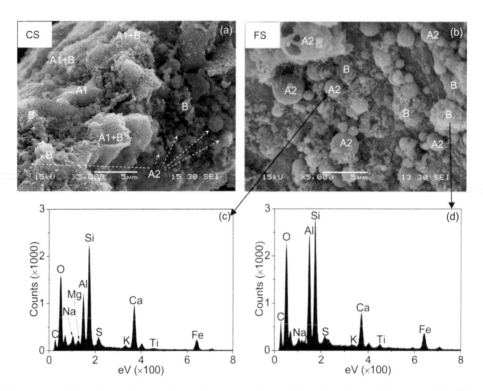

Figure 5. SEM/EDS patterns of the inorganic constituents in coarse slag and fine slag from a Texaco gasifier (Li and Bai 2013).

In addition, many studies have investigated methods of adding fluxes to meet the requirements of slagging gasifiers. For instance, the high ash, high sulphur and high ash melting point properties of Huainan coal (Anhui, Province) has prompted studies of the fluxing performance of calcium-based, iron-based, and magnesium-based fluxes for Huainan coal gasification (Wei and Li 2008; Li et al. 2010; Lu et al. 2011). The fluxes decreased the melting temperature of the coal ash by altering its mineral composition at high temperature. It was found that when the ash content of typical Shanxi Province anthracite was about 20%, the addition of 2–3% of the limestone or sillimanite in coal met the requirement of slag tapping co-current with the flow of gas and slag, while 5–7% was sufficient for counter-current flow (Bai et al. 2013). The viscosity–temperature curve obtained for added sillimanite was shown to be similar to that of vitreous slag; therefore, sillimanite was suitable for slag tapping. In addition, 3–5% of CaO and Fe_2O_3 also reduced the melting point and viscosity.

2.5. Mineralogy of gasification residue from UCG

UCG, which is a new strategic industry in China, is the process of *in situ* conversion of coal directly into combustible gaseous products. Since UCG is always performed in coal seams several hundreds of metres below ground surface, only the injection and production parameters

are obtainable. It is particularly difficult to determine the actual reaction conditions, especially the temperature field distribution and thermal equilibrium of an underground gasifier. Therefore, it is necessary to investigate the relationship between the gasification technology and the mineralogical characteristics of ash and slag by means of UCG simulation experiments (Liu et al. 2016).

Artificial UCG slags (Liu et al. 2016) were prepared following simulations of oxygen-enriched gasification in atmospheres containing 40/60% O_2/N_2, 60/40% O_2/N_2, and 80/20% O_2/N_2. The minerals in the different slags were then identified using X-ray diffraction (XRD) and Siroquant™ software. It was found that the typical minerals in the 40/60% O_2/N_2 laginclude anorthite, mullite, sekaninaite, and a large quantity of amorphous material. The predominant minerals in the 60/40% and 80/20% O_2/N_2 slags were crystalline anorthite, pyroxene and gehlenite, with very little amorphous material (Table 3). The higher oxygen concentration was associated with the re-formation of the mineral crystals after ash fusion, and the crystal structure also tended to be more orderly.

Mullite evidently reacted with CaO to generate anorthite at oxygen concentrations above 40%, accounting for the disappearance of mullite and the marked increase in anorthite in the 60/40% O_2/N_2 slag. As the scanning electron microscope (SEM) image in Figure 6 shows that, overall, 80/20% O_2/N_2 slag is mainly composed of two types of material, phases 'a'

Table 3. Mineral composition of UCG slag by XRD analysis and Siroquant (Liu *et al.* 2016).

Mineral composition	Raw coal	40%-O_2 slag	60%-O_2 slag	80%-O_2 slag
Quartz	29.6	1.5	9.3	1.8
Illite	43.3	–	–	–
Kaolinite	21.5	–	–	–
Chlorite	1.4	–	–	–
Pyrite	4	–	–	–
Anorthite	–	13.7	59.7	45.4
Pyroxene	–	–	25.7	26.3
Gehlenite	–	–	1.3	26.4
Sekaninaite	–	11.5	–	–
Mullite	–	22.3	–	–
Amorphous	–	49.0	3.6	–

'–' Indicates less than 1%.

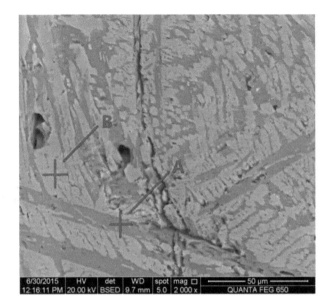

Figure 6. SEM-EDS analysis of minerals in 80%/20% O_2/N_2 slag. (a) Anorthite crystals. (b) Solid solution of gehlenite and augite (Liu *et al.* 2016).

and 'b'. Energy dispersive X-ray spectroscopy (EDS) quantitative analysis indicated that phase 'a' material contained anorthite crystals, and phase 'b' was a solid solution of gehlenite and augite. The XRD analysis showed less anorthite but large quantities of gehlenite in the 80/20% O_2/N_2 slag. Therefore, it was inferred that large quantities of anorthite formed at 1200°C and melted at higher temperatures. In addition, anorthite may have been a source of Ca for the formation of gehlenite, accounting for the loss of anorthite in the 80/20% O_2/N_2 slag.

Of the iron-bearing minerals, sekaninaite was found in the low-oxygen slag, but mainly pyroxene was present in the higher-oxygen slags. From this observation, the reaction sequence of iron-bearing minerals at high temperature was inferred as follows: the iron-bearing mineral is oxidized to magnetite (Fe_3O_4), then converts to Fe^{2+} as hercynite in a reducing atmosphere. Hercynite

then reacts with SiO_2 to form sekaninaite. Finally, with increasing oxygen concentration, sekaninaite is further oxidized to form pyroxene.

3. Geochemistry of coal during gasification

The migration of trace elements in coal may be divided into three categories (Figure 7):

(1) Elements that are not volatilized during gasification, and are well distributed throughout the fly ash and slag.
(2) Elements that are partially volatilized and undergo complex physical and chemical changes, becoming enriched in the heat-transfer units by the air flow; or are enriched in the fly ash, with only small amounts present in the slag.
(3) Elements with relatively low melting point (e.g. mercury, halogens) are completely volatilized and are enriched in the gas.

The escalating decline in the volatility of elements in the three classes is shown in Figure 7. The volatility of any given element is not necessarily related to its atomic mass at high temperature; rather, it is mainly related to its chemical forms (elementary substance, and oxide) (Clarke 1993). Ash and slag produced by different gasification technologies show significantly different chemical and physical properties. As a result, the transformation and enrichment behaviours of trace elements in ash and slag also differ. In this section, the enrichment characteristics and leaching behaviour of trace elements in several typical gasifiers are reviewed and discussed.

3.1. Trace element migration during fixed-bed gasification

The residence time of coal in a fixed-bed gasifier is up to one hour, ensuring that the reactions are complete, and the efficiency of carbon conversion and gasification is correspondingly high.

A study of the migration and enrichment of the trace elements in ash and slag from a Lurgi fixed-bed pressurized gasifier in Inner Mongolia (Fu 2015) indicated that the migration rates of trace elements to ash and slag can be divided into three categories: (1) highly volatile elements (e.g. Hg, Pb, Cd, Cu, and Ba) migration rate >50%; (2) moderately volatile elements (e.g. Li, Zn, Cr, and Ni) migration rates <50%; and (3) relatively stable elements (e.g. As, Be, and U) migration rates very low.

Elements	Boiling point/ ^{0}C
F	-188.1
Cl	-34.1
SeO$_2$	317
Hg	357
As$_2$O$_3$	465
As	613
Se	684.4
Zn	907
MoO$_3$	1155
Sb$_2$O$_3$	1425
B$_2$O$_3$	1860
Mn	2061
Cu	2570
Sc	2836
Co	2870
Ni	2913
CoO	3800
Cr$_2$O$_3$	3000~4000
Mo	4639

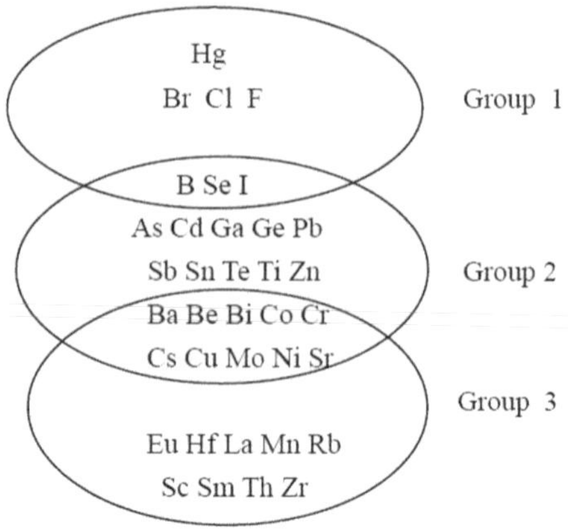

Group 1: Hg, Br Cl F

Group 2: B Se I, As Cd Ga Ge Pb, Sb Sn Te Ti Zn, Ba Be Bi Co Cr, Cs Cu Mo Ni Sr

Group 3: Eu Hf La Mn Rb, Sc Sm Th Zr

Figure 7. Volatile of trace elements in coal (Clarke 1993).

The relative enrichment (R_{RE}) of elements with high migration rate was less than 1 generally, where R_{RE} <1 implies that the trace elements in the bottom slag or fly ash tended to be expelled with the exhaust gas. R_{RE} <0.5 for Hg, Pb, Cd, Cu, and Ba. R_{RE} >1 for As, Be, and U, indicating that these tended to be enriched in the bottom slag or fly ash.

A study of the leaching characteristics, morphological distribution and environmental stability of heavy metals in slag from the Lurgi fixed-bed gasifier (He *et al.* 2014) using sulphuric–nitric acid, acetate buffer solution, and horizontal-oscillation leaching solutions found that the greatest number of species and largest quantities of heavy metals were leached by the acetate buffer solution, indicating that weak acid was the most effective for leaching heavy metals. The study also included an investigation of the acid extractable state, the reduced state, the oxidation state, and the residue state of Cr, Zn, Cu, Pb, Ni, As, and Cd in the slag, using modified Community Bureau of Reference sequential extraction procedure (Rauret *et al.* 1999) which was used to obtain the information about the non-residual speciation of metals. Of these, the acid extractable state is the most harmful to the environment, because it is easily leached to the surroundings. The residue state posed the least threat to the environment. It was indicated that Cd and Cr present the highest potential risk to the environment, followed by Cu. Trace elements of As, Ni, Pb, and Zn mainly occurred in residual slag form, and were the least harmful to the environment.

A study was also carried out on the effect of the extraction agent pH on leaching trace elements in the ash and slag (Fu 2015). Extraction agents with pH values

of 3.2, 5.2, and 7 were prepared using a 2:1 mass ratio of concentrated sulphuric acid and nitric acid, respectively. The slag leaching solutions using the three extraction agents were weakly alkaline (pH 8–10), which exceeded the buffer capacity of the extraction agent. This phenomenon was possibly due to the relatively high alkali metal content in the ash and slag. However, the initial pH of the extraction agent very effectively leached the trace elements. As the initial pH decreased, the leaching rates for As, Pb, Cd, Ni, Ba, etc., increased significantly, whereas the leaching rate for Cr decreased. In addition, the leached quantity of trace elements increased with decreasing particle size, especially for As, Cd, Be, and Cr, possibly because these tended to be enriched in the fine slag.

3.2. Trace element migration during fluidized-bed gasification

The trace elements in coal migrate into bottom slag, fly ash and gas during gasification after undergoing complex physical and chemical changes. The redistribution process is associated with such factors as temperature, pressure, atmosphere, air/coal ratio, etc., during gasification (Huang *et al.* 2004, 2005, 2006, 2008; Shen *et al.* 2016).

An investigation of the relationship between trace element migration and gasification temperature using a laboratory scale fluidized-bed gasifier at atmospheric pressure (Huang *et al.* 2006) found that the trace elements in both high- and low-temperature char showed identical tendencies with variation of temperature.

When the temperature was increased, R_{RE} of As, Cd, Co, Cr, Cu, Mg, Ni, Pb, and Se in high- and low-temperature char increased. With increasing gasification temperature, many molecular structures of the substances in coal are destroyed and, in particular, adsorbed and organic trace elements, and those bound by electrostatic attraction, migrate. In addition, more ash particles were formed during coal gasification, shortening the distances between the gasified trace elements and melting mineral and coal char particle surfaces, thus weakening the diffusion resistance of trace elements in the ash, and intensifying volatilization. Since more ash content was volatilized at higher temperatures, more trace elements were released. In addition, low-melting-point salts formed from different minerals facilitated volatilization of trace elements. For these reasons, the concentration of As, Cd, Co, Cr, Cu, Mg, Ni, Pb, and Se increased in the outlet gas, increasing the R_{RE} of the trace elements absorbed in the char. Higher gasification temperatures accelerated the volatilization and mass transfer rate of trace elements in the mineral, reducing the trace element content of bottom slag.

Investigation of the effects of gasification pressure on trace element migration during gasification (Huang et al. 2008) showed that the modified geochemical enriched factors (MGEF) of Cr, Cd, Zn, and Pb in the char in a cyclone separator decreased with increasing gasification pressure. Four reasons for this behaviour include:

(1) With increasing pressure, the partial pressure of gaseous trace elements increases, reducing the rate of element volatilization from mineral.

(2) From molecular thermal motion, with increasing ambient pressure, the diffusion fluxes of the trace element steam leaving the mineral surface decreases, and the collision rate of the volatilized gaseous trace elements accelerates, increasing the rate of condensation of trace elements returning to the volatilizing surface.

(3) Higher ambient pressure increases the size of particles condensed and prolongs gas–solid contact time, favouring the adsorption of trace elements by the fly ash (captured in the cyclone dust extractor, in this case).

(4) The free energy of the reaction between active Cl with Zn or Pb in gas is much greater than between Cl and H atoms. Therefore, with increasing pressure, the H_2 content in the gas increases, reducing the likelihood that low-boiling-point $ZnCl_2$ and $PbCl_2$ are generated, which in turn reduces the probability that Zn and Pb will volatilize.

Other volatile elements with a high volatilization rate (e.g. Se and Hg) were only slightly affected by changing gasification pressure (Huang et al. 2008).

A study of the temporal release and kinetics of Se during coal gasification and coal combustion (Shen et al. 2016) showed that Se was more volatile during coal gasification than during coal combustion, and determined the second-order kinetic law for Se release. During coal gasification, a high proportion of $H_2Se(g)$ was present at gas temperatures from 750°C to 1550°C. The transformation of $H_2Se(g)$ to $Se(g)$ increased above 1150°C. When the gas temperature dropped below 750°C, Se was mainly present in its atomic form. Unlike in coal combustion, however, almost no calcium selenate ($CaSeO_4$) was formed during gasification, indicating that the gaseous Se species were not likely react with CaO in fly ash during coal gasification; CaO was more likely to react with CO_2 to form $CaCO_3$. Selenium species in flue gas and the reactions of gaseous Se species with fly ash accounted for the different kinetics of Se release in coal combustion and coal gasification.

Variation of the air/coal ratio affects trace element migration in many ways. When the air/coal ratio increases, the share of coal participating in combustion also increases at first and the temperature of coal particles and gasifier bed also increase. The CO_2 content in the gas increases, resulting in higher airflow rate. The H_2 and CO content in the gas decreases, producing a weaker reducing atmosphere. Higher temperatures accelerate volatilization of the trace elements from the coal particles. Increased airflow rate shortens the gas residence time and the contact time of trace elements with fly ash at the cyclone outlet. In addition, the type of atmosphere affects the morphology of the trace elements (Huang et al. 2004).

Pollution control experiments were conducted by adding solid materials to the raw coal feed to capture trace elements generated during gasification (Huang et al. 2005). Comparisons of the effectiveness of limestone, dolomite and sodium carbonate indicated that each of the additives had a distinctly different inhibitory effect on the volatilization of particular trace elements. When limestone and dolomite was added to the raw coal, the H_2 and CO content of the gas was increased and the reducing atmosphere was strengthened, and almost all of the modified geochemical enriched factors of the trace elements in the ash were increased. The CaO decomposition product of limestone and dolomite physically and chemically adsorbed As, Co, Cr, Fe, and Zn, and porous CaO physically adsorbed Cd, Cu, Hg, Pb, V and Sr. Dolomite demonstrated a greater inhibitory effect for most trace elements, for the reason that dolomite bursts easily, producing a powder with greater specific area than limestone. Sodium carbonate captured trace elements mainly by chemical adsorption (Huang *et al.* 2005).

3.3. Trace element migration during entrained-flow gasification

As stated in Section 2.4.2, the temperature at the flame centre reached 2000°C, sufficient to melt the minerals. During gasification, a liquid slag layer and partly undeposited ash particles formed the gasification slag which was discharged at the bottom of the gasifier. The particles discharged from the top of the gasifier with the airflow consisted of fly ash (Li and Bai 2013).

A study of inorganic components enrichment in fine and coarse slags was conducted for the Texaco gasifier of Shanghai Coking Co. Ltd. (Wu *et al.* 2015) found that highly volatile trace elements such as Pb, As, and Zn were readily enriched in fine slag, which generally comprised fly ash discharged with the gas. Therefore, during gasification, highly volatile trace elements tended to volatilize in the upper gasifier and became enriched in the fly ash.

An investigation and analysis of a Texaco CWS gasifier and a GSP pressurized gasifier (Wang *et al.* 2016) studied the trace element content in fine and coarse slag and filter cake, and the enrichment of trace elements in coal and residue. Results indicated that, although V, Cr, Cu, Zn, Ba, Se, Pb, and As tended to become enriched during gasification in both operations, the enrichment mechanism was different in the two gasification technologies. The course slag (CS) from Texaco gasifier was enriched in Cr, Ba, and Pb, and the fine slag (FS) was enriched only in Ba. The filter cake was enriched in Ba, V, Cr, Pb, Cu, and Zn. Both the FS and filter cake from GSP gasifier were enriched in Cr, Cu, and Ba. Residue enrichment was 2–10 times more than that of the coal. The Cr, Cu, Zn, As, and Pb residue contents of both gasifiers conformed with Chinese Environmental Quality Standard for Soils (GB15618-1995 1995), while V, Se, and Ba require further investigation and treatment.

Research on the speciation and volatility of As and Cr during pyrolysis and gasification at high temperature and pressure has been conducted (Li 2011) by the high temperature and pressure thermobalance method, using a Cahn TherMax 500 thermogravimetric analyser to simulate an entrained-flow gasifier. The results showed that temperature had the predominant effect on release of As, which was released in increasing amounts with rising temperature. Cr was only weakly volatile due to its high boiling- and melting points. During pyrolysis, the As began to volatilize at 800°C in low-ash coal, and volatilized gradually above 800°C in

high-ash coal. Longer residence time was beneficial to the further release and oxidation both of As and Cr. During gasification, the volatilization rate of As and Cr increased to different extents at higher temperatures: for example, the volatilization rate of As increased gradually with rise in temperature, peaking at 1200 °C. Overall, the volatility of trace elements during gasification was greater than during combustion, mainly in reducing atmospheres which encouraged the generation of volatile chlorides, sulphides, and hydroxides; however, oxides and sulphates were present to a lesser extent in oxidizing atmospheres. By comparison, higher pressures produced only slight increases in As and Cr volatility.

Another study of the speciation and volatility of Hg (Li *et al.* 2008; Li 2011) used temperature programming pyrolysis apparatus (Model KSY-6D-16 electric furnace temperature controller) with the Ontario-Hydro method (1999) to capture gaseous mercury. It was found that Hg emission depended upon the temperature. The volatilization rate of Hg increased during gasification, especially in gasification processes using steam. During pyrolysis and gasification, gaseous mercury mainly included Hg^0. With increasing temperature and residence time, the Hg^0 content decreased as divalent mercury (Hg^{2+}) content increased. Moderate steam promoted further oxidation of the Hg at high temperature. Halogens and acidic metal compound components in coal facilitated the release of Hg. Volatilization and transformation of Hg was affected by the maceral content, microstructure and mineral composition of coal, coal rank, and the presence of elements and unburned carbon in the char.

A study of the relationship between the volatility and mode of occurrence of trace elements during coal pyrolysis and gasification (Bai 2003) showed that Hg had relatively higher mobility during pyrolysis regardless of its form in coal, whereas the mobility of other elements was markedly affected by their mode of occurrence. During CO_2 gasification of chars, some sulphophile elements, such as Hg, As, Pb, Zn, Co, Mo, Ni, Se, Cu, etc., and certain elements with higher organic affinity, including V and Ge, were more volatile. Most of these were more mobile in the CO_2 gasification of chars than in coal pyrolysis. The key factors lay in the occurrence of CO_2 and the further decomposition of the structure of the char and minerals at higher temperatures. Elements dispersed in clay minerals in coal, including Be, Sc, P, Ti, Mn, Ga, Th, Sr, Cs, W, Ba, etc., did not readily migrate during either coal pyrolysis or CO_2 gasification.

Bai *et al.* (2010) and Ma *et al.* (2012) found that, at above 1500°C all minerals in gasification slag converted into amorphous vitrification matter. In addition,

Moustakas et al. (2012) investigated the leaching property of vitrification slag, and found that the property of vitrified slag was significantly stable, and the leached concentration of harmful trace elements was so low that it didn't have the potential for leaching pollution.

3.4. Trace element migration during UCG

Hazardous trace elements in coal are enriched in gas, fly ash, and slag during UCG, then discharged into the atmosphere, or transported into groundwater by long-term leaching (Chu et al. 2001; Liu et al. 2007; Liu 2009; Wang et al. 2012), resulting in pollution of air and groundwater. In addition, during gas utilization, the presence of certain trace elements pollutes and corrodes the equipment and pipeline, etc. (Liu 2009).

An investigation of migration and enrichment of trace elements during UCG using a UCG laboratory model gasifier (Liu 2009) considered the cases for lignite, bituminous coal and anthracite. The sulphophile elements Hg, As, and Se were mainly present as sulphides in the three types of coal, and Pb, Sb, Cr, and Cd existed mainly as residual speciation; other elements had various modes of occurrence. Sulphides and organic combinations decomposed extremely easily during gasification; carbonates and Fe–Mn oxides decomposed at high temperature. All of these materials therefore showed a significant tendency to migrate at high temperatures. Depending on their volatilization and enrichment properties, 11 trace elements were divided into the following three groups. Group 1 included Hg, As, and Se, which volatized readily and were mostly released from the gas phase. Group 2 included Pb, Sb and Co, which possessed moderate volatility and enrichment in underground residue. Group 3 included Mn, Ni, Cr, Cd, and Be with low volatility and R_{RE} >10, and were generally enriched in the residue during high-temperature gasification. Trace elements migrated during UCG following mineral transformation during coal pyrolysis and fragmentation. Trace elements present as either mineral or non-mineral inclusions were volatilized, condensed, adsorbed and redistributed in various forms during gasification, and eventually distributed in gas, residue char, ash, and slag.

Due to the volatility and toxicity of mercury, an investigation of its leaching and reaction mechanism in the gas produced by UCG was conducted (Liu 2009). This showed that the Hg carried in the gas was mainly present as $Hg^0(g)$ together with a small amount of $Hg^{2+}(g)$. The quantities of $Hg^0(g)$ and $Hg^{2+}(g)$ vary depending on gasification time. The steam resists oxidation of $Hg^0(g)$ and, because the temperature drops as the gas flows through the gasification channel, some $Hg^0(g)$ is adsorbed onto the surface of residual particles of carbon, ash or slag, remaining there in particulate form. Unlike above-ground gasification or combustion operations, it was found that only slight amounts of Hg were entrained in UCG gas.

In addition, the speciation and migration of trace elements during UCG has been predicted using MTDATA thermodynamic equilibrium model software (Liu 2009). It was found that high volatility elements Hg, As, and Se would be present mainly in gaseous form, even in a condensation system. In UCG gas, the dominant speciation of mercury is $Hg^0(g)$; of selenium, it is $H_2Se(g)$; and of arsenic, it is $As_4(g)$. Pb, Cd, and Sb are volatile elements that volatilize and enter the gas at relatively low temperatures, and are present in the gas in the forms $PbS(g)$, $Cd(g)$, and $SbCl(g)$, respectively, during high-temperature gasification. The gaseous speciation, flowing through the air-flow channel with the coal gas, were cooled gradually and re-condensed into the solid phase. In contrast, it was predicted that Be, Ni, Cr, Co, and Mn, with low volatility, would be present as gases during high-temperature gasification only, and would mainly be enriched in ash and slag at temperatures below 100000B0°C. A study of the effect of other factors on the volatility of trace elements found that, when potassium was added into the gasification system, it tended to react with arsenic to form K_3AsO_4, which is a stable solid compound, thus decreasing the amount of volatilized arsenic. The presence of sulphur greatly facilitates the volatilization of trace elements during air gasification, because sulphides are a dominant species of volatile elements. Increased gasification pressure enhances reducing atmosphere and increases the proportion of hydride, raising the freezing point of elements.

Long-term leaching of hazardous residual trace elements is a potential source of groundwater pollution. An investigation of the leaching tendency of trace elements in gasification residue (Liu 2009) was carried out using a column leaching test on residue from UCG, including raw lignite, char, ash, and slag. When the pH of the leaching agent was increased, the concentration of trace elements in the leachate decreased. According to the leaching index (Wang et al. 1994), the leaching behaviour of various elements is divided into different levels. Among the elements, Na, Ca, and Cd with an above-medium leaching level, need further attention. Aside from the alkali metal and alkaline-earth metal, other trace elements in the gasification residue have a low leaching level, implying a weak leaching tendency and that the long-term leaching effects needs to be monitored. In addition, the pH of the leaching agent affects the leaching level of the elements, especially those with medium leaching levels. The factors influencing the leaching behaviour of trace elements from UCG residue are summarized as follows.

- For elements enriched on the surface of solids, the leaching behaviour is related to its concentration and chemical species (initial phase (e.g. formed during gasification or combustion) and secondary phase (e.g. formed by the reaction of water with ash and slag).)
- For elements enriched in matrix, the leaching behaviour is related to the dissolution rate of the matrix, the diffusion rate of the lattice ions, and the formation of a secondary phase.

The Eh of the leaching agent and the mode of occurrence of the element were also found to influence its leaching behaviour. When the leaching agent was replaced by groundwater, the concentration of trace elements leached from coal, char, ash, and slag decreased markedly, implying that the leaching behaviour in natural conditions is less than when an acid leaching agent is used.

4. Conclusions

China has the largest coal gasification industry in the world, and annual coal consumption for coal gasification exceeds 200 Mt. Commercialized gasification is generally restricted to the entrained flow, fluidized-bed and fixed-bed gasification processes with quite different reaction conditions.

Mineral transformation during coal gasification and the composition of ash depend upon the mineral species in the coal, as well as the temperature, atmosphere and pressure. Of these, temperature is the most crucial external condition. The thermal conversion of the minerals mainly occurs in a series of transformations of clay minerals, carbonate minerals, pyrite, and quartz with increasing temperature. The effect of the atmosphere on mineral transformation falls into two categories: affecting chemical reaction equilibrium, and reacting with minerals. The effect of pressure was more complex and also depended on changes in temperature and atmosphere. It is concluded that the decomposition of low-temperature minerals such as muscovite, anhydrite, and oldhamite were suppressed with increasing pressure, resulting in lower fusion temperatures.

Slag is a thermal conversion product of minerals at temperatures above the minimum melting point. Minerals in coal first form ash, then higher temperatures melt the ash to form slag along with other reactions and transformations of minerals. Slag flow is tightly related to reaction temperature, coal ash content, ash melting point, and mineral composition of coal.

The volatility of trace element during gasification is not necessarily related to its atomic mass; rather, it is mainly related to its morphology at high temperatures. The transformation and enrichment behaviours of trace elements in ash and slag differ distinctly in typical gasifiers. Increasing temperature always promotes the volatility of trace elements, while increasing pressure will decrease the enriched factor of volatile trace elements in the fly ash. In the reducing atmosphere during gasification, it is easy to form gaseous chloride, sulphide, and hydroxide of trace element, and in turn promote their migration. In weak acidity conditions, leaching rate and leached quantity of trace elements in ash from Lurgi gasifier and underground gasification increase. The leached concentrations of harmful trace elements from vitrified slag produced by entrained-flow gasifiers are so low that they don't present potential environmental pollution.

Disclosure statement

No potential conflict of interest was reported by the authors.

Funding

This work was supported by the National Key Basic Research Program of China [grant number 2014CB238905]; National Natural Science Foundation of China [grant number 51476185]; and Fundamental Research Funds for the Central Universities [grant number 2009QH13].

ORCID

Shuqin Liu ⓘ http://orcid.org/0000-0002-2831-6600
Chuan Qi ⓘ http://orcid.org/0000-0002-0694-2846
Zhe Jiang ⓘ http://orcid.org/0000-0003-3629-2482
Yanjun Zhang ⓘ http://orcid.org/0000-0003-4878-4847
Maofei Niu ⓘ http://orcid.org/0000-0001-6003-7671
Yuanyuan Li ⓘ http://orcid.org/0000-0003-1249-6421
Shifeng Dai ⓘ http://orcid.org/0000-0002-9770-1369
Robert B. Finkelman ⓘ http://orcid.org/0000-0002-7295-5952

References

Bai, J., Kong, L.X., Li, H.Z., Guo, Z.X., Bai, Z.Q., Yu, C.W., and Li, W., 2013, Adjustment in high temperature flow property of ash from Shanxi typical anthracite: Journal of Fuel Chemistry and Technology, v. 41, p. 805–813. [in Chinese with English abstract]
Bai, J., Li, W., Bai, Z.Q., Li, B.Q., and Li, C.Z., 2008, Transformation of mineral matter in Yanzhou coal ash at high temperature: Journal of China University of Mining and Technology, v. 37, p. 369–372. [in Chinese with English abstract]

Bai, J., Li, W., Li, C.Z., Bai, Z.Q., and Li, B.Q., 2009, Influence of coal blending on mineral transformation at high temperatures: Mining Science & Technology, v. 19, p. 300–305.

Bai, J., Li, W., Li, C.Z., Bai, Z.Q., and Li, B.Q., 2010, Influences of Minerals Transformation on the Reactivity of High Temperature Char Gasification: Fuel Processing Technology, v. 91, p. 404–409.

Bai, X.F., 2003, The distributions, modes of occurrence and volatility of trace elements in coals of China, China Coal Research Institute. [in Chinese with English abstract]

Benson, S.A., Hurley, J.P., Zygarlicke, C.J., Steadman, E.N., and Erickson, T.A., 1993, Predicting ash behavior in utility boilers: Energy & Fuels, v. 7, p. 746–754.

Bool, L.E.B., Peterson, T.W., and Wendt, J.O.L., 1995, The partitioning of iron during the combustion of pulverized coal : Combustion & Flame, v. 100, p. 262–270.

BP, 2016, Statistical review of world energy: bp.com/statisticalreview.

Chi, B.H., Zheng, Y., Wang, B.W., and Zheng, C.G., 2006, Decomposition of CaCO3 in CO2 atmosphere, *in* China national symposium on combustion 2006, Wuhan, China: Chinese Society of Engineering Thermophysics, p. 832–835. [in Chinese].

Chu, M., Li, H.M., Yu, L., and Liang, J., 2001, Underground coal gasification: An effective way of recovering left-over resouces in abandoned coal mines: China Coal, v. 27, p. 22–23. [in Chinese with English abstract]

Clarke, L.B., 1993, The fate of trace elements during coal combustion and gasification: An overview: Fuel, v. 72, p. 731–736.

Dai, S.F., Jiang, Y.F., Ward, C.R., Gu, L.D., Seredin, V.V., Liu, H.D., Zhou, D., Wang, X.B., Sun, Y.Z., and Zou, J.H., 2012b, Mineralogical and geochemical compositions of the coal in the Guanbanwusu Mine, Inner Mongolia, China: Further evidence for the existence of an Al (Ga and REE) ore deposit in the Jungar Coalfield: International Journal of Coal Geology, v. 98, p. 10–40.

Dai, S.F., Ren, D.Y., Chou, C.L., Finkelman, R.B., Seredin, V.V., and Zhou, Y.P., 2012a, Geochemistry of trace elements in Chinese coals: A review of abundances, genetic types, impacts on human health, and industrial utilization: International Journal of Coal Geology, v. 94, p. 3–21.

Dai, S.F., Seredin, V.V., Ward, C.R., Jiang, J.H., Hower, J.C., Song, X.L., Jiang, Y.F., Wang, X.B., Gornostaeva, T., and Li, X., 2014, Composition and modes of occurrence of minerals and elements in coal combustion products derived from high-Ge coals: International Journal of Coal Geology, v. 121, p. 79–97.

Dai, S.F., Yan, X.Y., Ward, C.R., Hower, J.C., Zhao, L., Wang, X.B., Zhao, L.X., Ren, D.Y., and Finkelman, R.B., 2016, Valuable elements in Chinese coals: A review: International Geology Review, p. 1–31. 10.1080/00206814.2016.1197802

Finkelman, R.B., 1993, Trace and minor elements in coal: NewYork, Plenum, Organic Geochemistry, p. 593–607.

Fu, Y.G., 2015, Study on the mineralogy properties and leaching toxicity experiment of Lurgi gasification slag: Beijing, China University of Mining Science and Technology. [in Chinese with English abstract]

Gao, L., 2010, Application of Texaco coal water slurry presurized gasification: Coal Technology, v. 29, p. 161–162. [in Chinese with English abstract]

GB15618-1995 (National Standard of P.R. China), 1995. Environmental quality standard for soils. [in Chinese]

Given, P.H., and Spackman, W., 1978, Reporting of analyses of low-rank coals on the dry, mineral-matter-free basis: Fuel, v. 57, p. 319–319.

Guo, S.C., and Hu, H.Q., 2012, Chemical Technology of coal: Beijing, Chemical Industry Press, p. 367. [in Chinese]

He, X.W., Cui, W., Wang, C.R., Shi, Y.T., and Zhang, J., 2014, Analysis on leaching characteristics and chemical speciation of heavy metals in gasification slag: Environmental Protection of Chemical Industry, v. 34, p. 499–502. [in Chinese with English abstract]

Hu, G.L., Dam-Johansen, K., Wedel, S., and Hansen, J.P., 2006, Decomposition and oxidation of pyrite: Progress in Energy & Combustion Science, v. 32, p. 295–314.

Huang, Y.J., Jin, B.S., Zhong, Z.P., and Xiao, R., 2004, The influence of air-coal ratio on the migration mechanism of trace elements during a gasification process: Journal of Engineering for Thermal Energy and Power, v. 19, p. 402–407. [in Chinese with English abstract]

Huang, Y.J., Jin, B.S., Zhong, Z.P., and Xiao, R., 2008, Effect of gasification pressure on the occurrence of trace elements: Journal of Southeast University (Natural Science Edition), v. 38, p. 92–96. [in Chinese with English abstract]

Huang, Y.J., Jin, B.S., Zhong, Z.P., Xiao, R., and Zhou, H.C., 2005, Effects of solid additives on the control of trace elements during coal gasification: ACTA Scientiae Circumstantiae, v. 25, p. 507–511. [in Chinese with English abstract]

Huang, Y.J., Jin, B.S., Zhong, Z.P., Xiao, R., and Zhou, H.C., 2006, The relationship between occurrence of trace elements and gasification temperature: Proceedings of the CSEE, v. 26, p. 10–15. [in Chinese with English abstract]

Jiao, H.L., 2015, Brief analysis of large-scale coal gasification technologies: China New Technologies and Products, p. 57–57. [in Chinese with English abstract]

Jing, N.J., 2013, Ash fusion behaviours of coal and biomass during pressurized fluidized bed gasification process, Zhe Jiang University. [in Chinese with English abstract]

Jing, N.J., Wang, Q.H., Cheng, L.M., Luo, Z.Y., Cen, K.F., and Zhang, D.K., 2013, Effect of temperature and pressure on the mineralogical and fusion characteristics of Jincheng coal ash in simulated combustion and gasification environments: Fuel, v. 104, p. 647–655.

Klein, L.C., Fasano, B.V., and Wu, J.M., 1983, Viscous flow behavior of four iron-containing silicates with alumina, effects of composition and oxidation condition: Journal of Geophysical Research Solid Earth, v. 88, p. A880–A886.

Kong, H.L., Zeng, R.S., and Zhuang, X.G., 2002, The new advance in researches of trace elements in coals: Bulletin of Mineralogy: Petrology and Geochemistry, v. 21, p. 121–126.

Li, C.Z., 2004, Advances in the science of victorian brown coal: Advances in the science of victorian brown coal, p. viii.

Li, J.B., Shen, B.X., Zhao, J.G., and Wang, J.M., 2010, Effect of flux on the melting characteristics of coal ash for the Liuqiao No. 2 Coal Mine: Journal of China Coal Society, v. 35, p. 140–144. [in Chinese with English abstract]

Li, M., Lv, S., and Jiao, X.W., 2009, Experimental study on influences of included minerals on the char combustion characteristics: Coal Conversion, v. 32, p. 33–36. [in Chinese with English abstract]

Li, S.R., 2008, Crystallography and Mineralogy: Beijing, Geological Publishing House, p. 346. [in Chinese]

Li, W., and Bai, J., 2013, Chemistry of Ash from Coal: Beijing, Science Press, p. 337. [in Chinese]

Li, Y., 2011, Experimental study on emission and high temperature control of trace elements during coal gasification, Huazhong University of Science and Technology. [in Chinese with English abstract]

Li, Y., Zhang, J.Y., He, B.H., Zhao, Y.C., Yu, C., Wang, Z.H., and Zheng, C.G., 2008, Mercury speciation and volatility during coal pyrolysis and gasification: Journal of Engineering Thermophysics, v. 29, p. 1775–1779. [in Chinese with English abstract]

Li, Y., Zhang, J.Y., Zhao, Y.C., Wu, Y.Q., Gao, J.S., and Zheng, C.G., 2007, Influence of pyrolysis conditions on volatility of trace elements in coals: Journal of Engineering Thermophysics, v. 28, p. 189–192. [in Chinese with English abstract]

Liang, G.Z., 2005, Mineral composition of coal ash: Mining Science & Technology, p. 27–29. [in Chinese with English abstract]

Liu, G.J., Wang, J.X., Yang, P.Y., and Peng, Z.C., 2003, Minerals in coal and their changes during combustion: Journal of Fuel Chemistry and Technology, v. 31, p. 215–219. [in Chinese with English abstract]

Liu, S.Q., 2009, Migration and enrichment of harmful trace elements during underground coal gasification, China Coal Industry Publishing House. [in Chinese]

Liu, S.Q., Li, J.G., Mei, M., and Dong, D.L., 2007, Groundwater Pollution from Underground Coal Gasification: Journal of China University of Mining & Technology, v. 17, p. 467–472.

Liu, S.Q., Qi, C., Zhang, S.J., and Deng, Y.P., 2016, Minerals in the Ash and Slag from Oxygen-Enriched Underground Coal Gasification: Minerals, v. 6, p. 27.

Lu, H.Q., Li, H.X., Ma, F., Meng, Y., and Jia, C.L., 2011, Study on fly ash fusibility affected by calcium base flux and fusion mechanism: Coal Science and Technology, v. 39, p. 111–114. [in Chinese with English abstract]

Ma, Z.B., Bai, Z.Q., Bai, J., Li, W., and Guo, Z.X., 2012, Evolution of coal ash with high Si/Al ratio under reducing atmosphere at high temperature: Journal of Fuel Chemistry and Technology, v. 40, p. 279–285. [in Chinese with English abstract]

Ma, Z.B., Bai, Z.Q., and Li, W., 2011, Transformation behaviors of mineral matters in coal ash with different Si/Al ratio at high temperature under reducing atmosphere, in The 11th China-Japan Symposium on Coal and C1 Chemistry, Yinchuan, China, Chinese Academy of Sciences, p. 45–46.

McCarthy, G.J., Stevenso, R.J., and Oliver, R.L., 1988, Mineralogical characterization of the residues from the TONO I UCG experiment, in The Fourteenth Annual Underground Coal Gasification Symposium, Chicago, Illinois, the Morgantown Energy Technology Center, p. 41-50.

Miller, R.N., and Given, P.H., 1986, The association of major, minor and trace inorganic elements with lignites. I. Experimental approach and study of a North Dakota lignite: Geochimica Et Cosmochimica Acta, v. 50, p. 2033–2043.

Moustakas, K., Mavropoulos, A., Katsou, E., Haralambous, K.J., and Loizidou, M., 2012, Leaching properties of slag generated by a gasification/vitrification unit: The role of pH, particle size, contact time and cooling method used: Journal of Hazardous Materials, v. 207-208, p. 44–50.

O'Keefe, J.M.K., Bechtel, A., Christanis, K., Dai, S.F., DiMichele, W.A., Eble, C.F., Esterle, J.S., Mastalerz, M., Raymond, A.L.,

Valentim, B.V., Wagner, N.J., Ward, C.R., and Hower, J.C., 2013, On the fundamental difference between coal rank and coal type: International Journal of Coal Geology, v. 118, p. 58–87.

Patterson, J.H., and Hurst, H.J., 2000, Ash and slag qualities of Australian bituminous coals for use in slagging gasifiers: Fuel, v. 79, p. 1671–1678.

Raask, E., 1984, Creation, capture and coalescence of mineral species in coal flames: Journal of the Institute of Energy, v. 57, p. 231–239.

Raask, E., 1985, Mineral impurities in coal combustion, behavior, problems, and remedial measures: Washington, Hemisphere Publishing Corporation.

Ranjan, S., Sridhar, S., and Fruehan, R.J., 2010, Reaction of FeS with simulated slag and atmosphere: Energy Fuels, v. 24, p. 5002–5007.

Rauret, G., López-Sánchez, J.F., Sahuquillo, A., Rubio, R., Davidson, C., Ure, A., and Quevauviller, P., 1999, Improvement of the BCR three step sequential extraction procedure prior to the certification of new sediment and soil reference materials: Journal of Environmental Monitoring Jem, v. 1, p. 57–61.

Schobert, H.H., 1992, Clean Utilization of Coal. NATO ASI Series 370: Dordrecht, Kluwer Academic Publishers, 337 p. [Chinese]

Shannon, G.N., Matsuura, H., Rozelle, P., Fruehan, R.J., Pisupati, S., and Sridhar, S., 2009, Effect of size and density on the thermodynamic predictions of coal particle phase formation during coal gasification: Fuel Processing Technology, v. 90, p. 1114–1121.

Shen, F.H., Liu, J., Zhang, Z., and Yang, Y.J., 2016, Temporal measurements and kinetics of selenium release during coal combustion and gasification in a fluidized bed: Journal of Hazardous Materials, v. 310, p. 40–47.

Slade, R.C.T., and Davies, T.W., 1991, Evolution of structural changes during flash calcination of kaolinite. A 29Si and 27Al nuclear magnetic resonance spectroscopy study: Journal of Materials Chemistry, v. 1, p. 361–364.

Srinivasachar, S., and Boni, A.A., 1989, A kinetic model for pyrite transformations in a combustion environment: Fuel, v. 68, p. 829–836.

Srinivasachar, S., Helble, J.J., and Boni, A.A., 1990, Mineral behavior during coal combustion 1. Pyrite transformations: Progress in Energy & Combustion Science, v. 16, p. 281–292.

Strobel, T.M., and Hurley, J.P., 1995, Coal-ash corrosion of monolithic silicon carbide-based refractories: Fuel Processing Technology, v. 44, p. 201–211.

Tomeczek, J., and Palugniok, H., 2002, Kinetics of mineral matter transformation during coal combustion: Fuel, v. 81, p. 1251–1258.

US EPA, 1999. Standard test method for mercury from coal-fired stationar (Ontario-Hydro Method).

Van Dyk, J.C., Benson, S.A., Laumb, M.L., and Waanders, B., 2009, Coal and coal ash characteristics to understand mineral transformations and slag formation: Fuel, v. 88, p. 1057–1063.

Wang, G.M., Luo, Z.X., Zhang, J.Y., and Zhao, Y.C., 2015, Modes of Occurrence of Fluorine by Extraction and SEM Method in a Coal-Fired Power Plant from Inner Mongolia, China: Minerals, v. 5, p. 863–869.

Wang, L., Reddy, K.J., and Munn, L.C., 1994, Geochemical modeling for predicting potential solid phases controlling the dissolved molybdenum in coal overburden, Powder River Basin, WY, U.S.A: Applied Geochemistry, v. 9, p. 37–43.

Wang, S.J., 2016, Development and application of modern coal gasification technology: Chemical Industry and Engineering Progress, v. 35, p. 653–664. [in Chinese with English abstract]

Wang, X.S., Tang, Y.G., Heng, B.B., and Xu, J.J., 2016, Some trace elements enrichment features during coal gasification process in Ningdong coal gasification plant: Coal Geology of China, v. 28, p. 14–18. [in Chinese with English abstract]

Wang, Y.Y., Liu, H.T., Li, J.G., Pan, X., Yao, K., Zhao, J., and Jin, Z.W., 2012, Experimental study on leaching behavior of trace elements in the residues "three zones" from underground coal gasification: Journal of China Coal Society, v. 37, p. 2092–2096. [in Chinese with English abstract]

Ward, C.R., 2002, Analysis and significance of mineral matter in coal seams: International Journal of Coal Geology, v. 50, p. 135–168.

Ward, C.R., 2016, Analysis, origin and significance of mineral matter in coal: An updated review: International Journal of Coal Geology, v. 165, p. 1–27.

Wei, Y.J., and Li, H.X., 2008, Melting behavior of flux MgO in high temperature and less reductive atmosphere: Journal of Anhui University of Science and Technology (Natural Science), p. 74–77. [in Chinese with English abstract]

Wu, G.G., and Zhang, R.G., 2013, Coal Gasification Technology: Xuzhou, China University of Mining and Technology Press, p. 265. [in Chinese]

Wu, S.Y., Huang, S., Wu, Y.Q., and Gao, J.S., 2015, Characteristics and catalytic actions of inorganic constituents from entrained-flow coal gasification slag: Journal of the Energy Institute, v. 88, p. 93–103.

Yan, L., Gupta, R.P., and Wall, T.F., 2002, A mathematical model of ash formation during pulverized coal combustion: Fuel, v. 81, p. 337–344.

Yao, D.X., Zhi, X.C., and Zheng, B.S., 2002, Evolutionary feature of minerals in the process of coal combustion: Mining Safety & Environmental Protection, v. 15, p. 10–11. [in Chinese with English abstract]

Yao, D.X., Zhi, X.C., and Zheng, B.S., 2003, Evolutionary features of minerals in coal during coal combustion: Coal Geology of China, v. 15, p. 10–11. [in Chinese with English abstract]

Zhang, D.K., Jackson, P.J., and Vuthaluru, H.B., 2002, Low-rank coal and advanced technologies for power generation: Impact of Mineral Impurities in Solid Fuel Combustion: New York, Springer US, 45–64 p.

Zhang, J.Y., Zhao, Y.C., Huang, W.C., Li, Y., Song, D.Y., Dai, S.F., Zhao, F.H., and Zheng, C.G., 2006a, Arsenic emission of high-arsenic coal combustion from southwestern Guizhou, China: Chinese Journal of Geochemistry, v. 25, p. 49–50.

Zhang, L.H., Wang, W., Wang, W., Zhou, T.X., Song, D.G., and Chen, Z., 2006b, Effect of minerals in coal on the properties of coal water slurry: Coal Processing & Comprehensive Utilization, p. 26–28. [in Chinese with English abstract]

Zhao, Y.C., Zhang, J.Y., Li, Y., Gao, Q., and Zheng, C.G., 2007, A model for pyrite partition during coal combustion: Journal of Engineering Thermophysics, v. 28, p. 1050–1052. [in Chinese with English abstract]

Zhu, S.Q., 1996, Review on the influence of coal properties on its formation: Coal Processing & Comprehensive Utilization, p. 5–8. [in Chinese with English abstract]

REVIEW ARTICLE

Stone coal in China: a review

Shifeng Dai ⓘ, Xue Zheng, Xibo Wang ⓘ, Robert B. Finkelman ⓘ, Yaofa Jiang, Deyi Ren, Xiaoyun Yan and Yiping Zhou

ABSTRACT

Stone coal is defined as a combustible, low-heat value, high-rank black shale of early Paleozoic (in a few cases, Permian) age, widely distributed in southern China. Attention has been focused on stone coals because (1) they can be used as fuel energy (for power plants and daily use in some villages) mainly in southern China; (2) they are enriched in critical elements and are currently industrially (economic extraction of V) and agriculturally (such as Se) utilized or have such a great potential (e.g. Au, platinum group elements, Mo, and Ni); (3) they are the sources for some toxic elements that have caused environmental pollution (e.g. SO_2 emission during their combustion) and endemic diseases such as selenoisis and fluorosis; and (4) they can provide useful information for geological events and regional geological setting (e.g. hydrothermal activities). This article reviews stone coal's definition; occurrence and distribution; petrologic properties, mineralogy, and geochemistry; adverse impacts on environment and human health; and by-products of critical elements as well as major challenges remaining from point of view of determining element enrichment mechanisms, utilization of critical elements, and control of toxic elements released during stone coal utilization.

1. What is stone coal?

Based on Chinese Standard (GB/T15663.1-2008 GB/T. 2008), stone coal (also referred to as stone-like coal) is a combustible, low-heat value, high-rank sedimentary rock mainly derived from early Paleozoic bacteria and algae after saprofication and coalification in a marine-influenced environment (e.g. epicontinental sea, lagoon, or bay). Stone coals usually formed before the Middle Devonian and have an ash yield generally higher than 50%. The heat values mostly range from 3.5 to 10.5 MJ/kg and typically have an anthracite rank. However, the carbonaceous siliceous rocks and carbonaceous shales in the Permian in Enshi, Hubei Province, are also known locally as 'stone coal' (Figures 1(a–c), 2(a–c)). These are unique in that they are highly enriched in Se (Song 1989; Zhang et al. 2011a, 2011b; Peng 2015; Dai et al. 2016b; Dai and Finkelman 2017). The term 'stone coal' is almost exclusively used in China (e.g. Coveney et al. 1994; Křibek et al. 2007; GB/T15663.1-2008 GB/T. 2008; Dai et al. 2012); in a very few cases, it has been used outside China (e.g. Tovarovsky et al. 2011; Vidanovic et al. 2011; Łapcik

et al. 2016) but its meaning is technically and/or scientifically different from the definition used in China. For example, Łapcik et al. (2016) refers stone coal to a clast component of pebbly mudstone. Essentially, stone coals are black shales but the term specifically refers to those organic-rich black shales deposited in the early Paleozoic in a marine-influenced, anoxic, and reducing environment, with a high rank and characterized by combustible properties (with the content of total organic carbon generally >10%).

2. Samples and analytical methods

In addition to a comprehensive review on previously published literature and scientific reports, particularly a report by Geology Research Institute of Geology Prospection Branch, Coal Science Academy (GRI-CSA 1982), a total of 30 stone coal samples (sample nos. L1–L30) were collected from Enshi, Hubei Province (Figure 1(a-b)), for analyses of rare earth elements and mineralogical compositions. Additionally, 10 stone coal

ⓑ Supplemental data for this article can be accessed here.

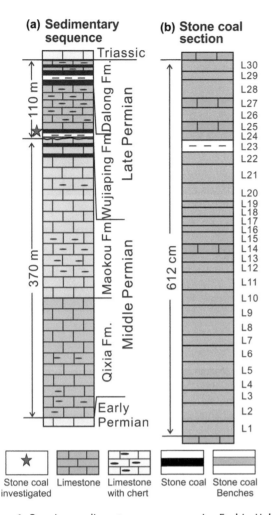

(a) Sedimentary sequence **(b) Stone coal section**

Figure 1. Permian sedimentary sequences in Enshi, Hubei Province (a) and stone coal seam section where samples were collected in this study (b).

stone coal samples were subjected to quantitative mineralogical analysis using Siroquant™, commercial interpretation software developed by Taylor (1991) based on the principles for diffractogram profiling set out by Rietveld (1969). Further details indicating the use of this technique for coal-related materials are given by Ward et al. (1999, 2001) and Ruan and Ward (2002).

Samples were crushed and ground to pass 75 μm for analysis of rare earth elements and Y (REY, or REE if Y is not included) using inductively coupled plasma mass spectrometry (ICP-MS, Thermo Scientific, X series II) in a pulse counting mode (three points per peak). Prior to ICP-MS analysis, stone coal samples were digested using an UltraClave High Pressure Reactor (Milestone). The basic load for the digestion tank consists of distilled H_2O (330 ml), H_2O_2 (v/v 30%, 30 ml), and H_2SO_4 (v/v 98%, 2 ml). The initial nitrogen pressure for the microwave reactor was set at 50 bars and the temperature was set at 60°C, through 125°C, 160°C to 240°C. The 240°C digestion condition with microwave power (MP) 1200 W lasted for 15 min and then was retained for 60 min under 1000-W MP and 240°C. The reagents for digestion of ~50-mg stone coal sample were 5 ml 40% HF and 2 ml 65% HNO_3. Multi-element standards (Inorganic Ventures CCS-1) were used for calibration of trace element concentrations. Analysis and sample microwave digestion programs for coal and coal-related materials are outlined by Dai et al. (2011). Fluorine concentration in the samples was determined by pyrohydrolysis in conjunction with an ion-selective electrode according to the method of ASTM D5987-96 (2015).

samples as shown in Figure 2(c) (sample nos. ES-1 to ES-10; core drill no. Xiandi 1#) for fluorine analysis.

A field emission-scanning electron microscope (FE-SEM, FEI Quanta™ 650 FEG), in conjunction with an EDAX energy-dispersive X-ray spectrometer (Genesis Apex 4), was used to study the morphology of the minerals and to determine traces of minerals that may be under detection limit of X-ray powder diffraction (XRD) technique. Samples were carbon-coated using a Quorum Q150T ES sputtering coater. The working distance of the FE-SEM-EDS was 10 mm, beam voltage 20.0 kV, aperture 6, and micron spot size 5.0. The images were captured via a retractable solid state backscatter electron detector. The mineralogy was also supplemented by optical microscopic observation and XRD. XRD analysis of the stone coal samples was performed on a powder diffractometer (D/max-2500/PC XRD) with Ni-filtered Cu-Kα radiation and a scintillation detector. The XRD pattern was recorded over a 2θ interval of 2.6–70°, with a step size of 0.01°. X-ray diffractograms of the

3. Physical appearance and occurrence in China

Stone coals commonly are rock-like hard, combustible like coal, and are mined as an energy source. They are black or grey black (Figure 2(a–f)), with dull lustre; uniform texture and conchoidal (Figure 2(d)), flat-shaped (Figure 2(e)), or ladder-shaped fractures. Stone coals usually have a density of 2.2–2.3 g/cm^3 (Jiang et al. 1994) and macroscopically look like sapropelic coal (Figure 2(d,e); Han et al. 1996). The thickness of stone coal seams generally varies from 5 to 30 m (Jiang et al. 1994), but in some cases, they occur as lenses and thin beds interbedded within limestones (Figure 2(a–c); O'Keefe et al. 2013). The roof and floor limestone strata contain calcite veins (Figure 2(c)) in the Enshi, Hubei Province.

Stone coals are mainly distributed in southern, and to less extent, middle China (Figure 3), including Hunan, Zhejiang, Hubei, Guangxi, Anhui, Jiangxi, Henan, Guangdong, Guangxi, Sichuan, Guizhou, and Shaanxi

Figure 2. Stone coals in southern China. (a), (b), and (c) Permian Se-rich stone coals in Enshi, Hubei Province; (b) is the enlargement of the red rectangle in (a). (d) Conchoidal fracture of Cambrian stone coals in Pingli, Shaanxi Province. (e) Flat-shaped fracture in Cambrian stone coals in Pingli, Shaanxi Province; (f) nodular textures shown by the Ni-Mo sulphides in Cambrian stone coals from Zunyi (Guizhou Province) and Dayong (Hunan Province). (d) and (e) are from Chen *et al.* (1996); (f) is from Coveney and Nansheng (1991).

Provinces. The total reserves of the stone coal are 61,876.7 Mt (Table 1), with Hunan Province having the most abundant reserves (18,716.37 Mt), covering an area of about 30,000 km² (Liu *et al.* 2016). The age of the stone coal spans from the Proterozoic to Middle Devonian (Table 2, Figure 4); the early Cambrian and Ediacaran (Sinian) stone coals are most widely distributed and the former has the most abundant reserves, amounting to 99% of the total reserves (Geology Research Institute of Geology Prospection Branch, Coal Science Academy (GRI-CSA) 1982).

4. Chemical, mineralogical, and petrological properties of stone coal

4.1. Chemistry and mineralogy

In addition to high ash yields and low heat values as mentioned earlier, the stone coals of early Paleozoic age are characterized by higher sulphur (usually 2–4% and up to 13.5%) and lower organic carbon contents (TOC, 15–25% on average) relative to common coals (GRI-CSA 1982; Han *et al.* 1996). However, low-ash (20–40%) and higher heat-value (16.74–25.12 MJ/kg) stone coals locally occur in western Hunan, southern

Anhui, and northern Shaanxi Provinces (Han *et al.* 1996). Wang *et al.* (2017) performed a comprehensive investigation on the chemical and mineralogical composition of stone coals of Permian age from Enshi, Hubei Province. They showed that TOC varies from 3.88% to 46.99% (17.83% on average).

Minerals identified in the stone coals include quartz (35–75%) and clay minerals, and to a lesser extent, carbonate minerals, sericite, pyrite, feldspar, and collophane (Wu *et al.* 1999a, 1999b). Studies by GRI-CSA (1982) show that minerals in the stone coals in western Zhejiang Province mainly consist of quartz, sericite, carbonate minerals, clay minerals, pyrite, and collophane. A numerous other minerals have also been found in stone coals, including elemental sulphur, sulphide minerals (e.g. chalcopyrite, sphalerite), sulphate minerals (e.g. barite, celestobarite, and halotrichite), carbonate minerals (e.g. carbonate and dolomite), phosphate minerals (e.g. carbonate-apatite, vashegyite, and barrandite), and aluminosilicate minerals (feldspar, muscovite, kaolinite, hydromica, smectite, allophane, zeolite, chlorite, tremolite, epidote, and garnet).

Minerals in the Enshi stone coals mainly consist of authigenic quartz (abundance greatly varies from 1% to 89.2%, mostly 40–88%) (Figures 5 and 6(a)), followed by mixed-

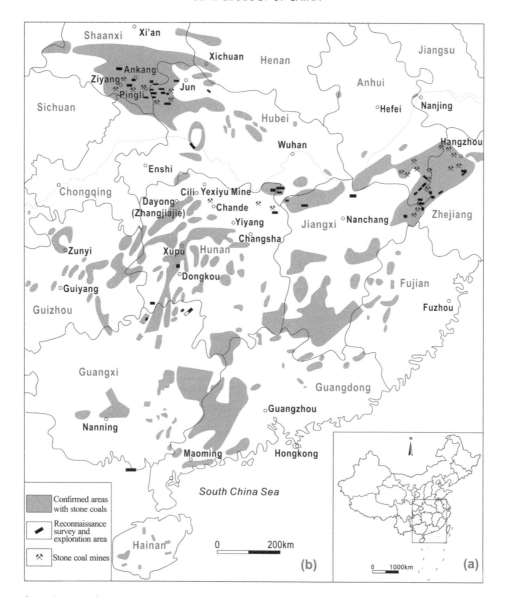

Figure 3. The confirmed areas of stone coal occurrence in southern China. (a) China map. (b) Enlargement of the red rectangle in (a). Modified from GRI-CSA (1982).

Table 1. Reserves of stone coals in southern China (Mt).

Province	Total reserves	Ensured reserves	Inferred reserves*	Inferred reserves* in different periods			
				Ediacaran	Early Cambrian	Silurian	Others
Zhejiang	10,635.3	123,707	9398.23		9398.23		
Anhui	7459.1		7459.1		7459.1		
Jiangxi	6834.03		6834.03		6834.03		
Henan	438.12		438.12	337	101.12		
Hubei	2564.15	46,421	2099.94	35.42	1701.96	239.79	122.77
Hunan	18,716.37	159,272	17,123.65	563.9	16,559.75		
Guangdong	5.09		5.09		5.09		
Guangxi	12,879.62	58,389	12,295.73		12,295.73		
Guizhou	829.28		829.28	94.23	735.05		
Shaanxi	1515.64	1941	1496.23		217.31	1039.61	239.31
Total	61,876.7	389,730	57,979.4	1030.55	55,302.28	1279.4	367.17

*Inferred reserves are referred to Comprehensive Investigation Reserves in GRI-CSA (1982).

layer smectite/illite (mostly 8–33%), detrital quartz of terrigenous origin (Figure 6(a)), and traces of gypsum, calcite, feldspar, pyrite, dolomite, albite, marcasite, and jarosite (Wang *et al.* 2017), as well as traces of chalcopyrite (Figure 6(b)), autunite [Ca(UO$_2$)$_2$(PO$_4$)$_2$ · 11H$_2$O], sphalerite (Figure 6(c)), pyrophyllite (Figure 6(e)), and roscoelite

Table 2. Stone coal-bearing strata in the provinces of southern China.

Period	Zhejiang	Anhui	Jiangxi	Henan	Hubei	Hunan	Guangdong	Guangxi	Sichuan	Guizhou	Shaanxi
Middle Devonian							Donggangling F	Donggangling F			
Middle–Late Silurian					Zhuxi F						Middle-Up Silurian
Early Silurian					Meiziya F Daguiping F					Longmaxi F	Meiziya F, Daguiping F
Late Ordovician					Low Ordovician	Baishuixi F		Shengping F, Huang'ai F		Wufeng F	
Early Ordovician						Tianjiaping F					
Late Cambrian								Shuimen Group			Up Cambrian
Middle Cambrian					Maobaguan F	Middle Cambrian					Baguamiao F Maobaguan F Niunaichong F
Early Cambrian	Hetang F	Hetang F, Huangboling F Huanglishu F, Houjiashan F	Wangyinpu F, Hetang F, Niujiaohe F	Shuigoukou F,	Xiaoquangou F	Shuijingtuo F Shuigoukou F Dongkeng F Yangloudong F Dayanjiao F	Niutitang F Xiaoyanxi F Shuijingtuo F				
Bacun Group	Qingxi F,	Shuikou Group the second F	Shuijingtuo F	Niutitang F Qiongzhusi F Zhalagou F	Shuijingtuo F Lujiaping F, Shuigoukou F Donghe Group						
Ediacaran	Xifengsi F				Dengying F, Qingshanzhai F	Dengying F, Liuchapo F, Doushantuo F	Ediacaran	Doushantuo F	Doushantuo F	Doushantuo F	Doushantuo F
Cryogenian				Meiyaogou F, Nannihu F, Baishugou F	Liantuo F	Liantuo F					Liantuo F
Mesoproterozoic				Dagou F	Qingshangang F Shitanghe F Hezi F Qijiaoshan F	Madiyi F					

F, Formation.

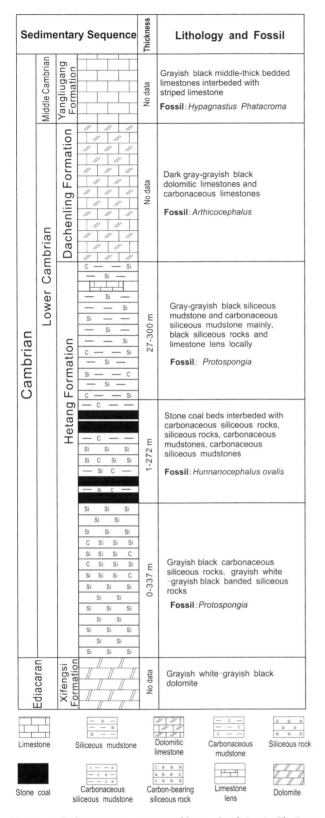

Sedimentary Sequence			Thickness	Lithology and Fossil
Cambrian	Lower Cambrian	Middle Cambrian / Yangliugang Formation	No data	Grayish black middle-thick bedded limestones interbeded with striped limestone **Fossil**: *Hypagnastus Phatacroma*
		Dachenling Formation	No data	Dark gray-grayish black dolomitic limestones and carbonaceous limestones **Fossil**: *Arthicocephalus*
		Hetang Formation	27-300 m	Gray-grayish black siliceous mudstone and carbonaceous siliceous mudstone mainly, black siliceous rocks and limestone lens locally **Fossil**: *Protospongia*
			1-272 m	Stone coal beds interbeded with carbonaceous siliceous rocks, siliceous rocks, carbonaceous mudstones, carbonaceous siliceous mudstones **Fossil**: *Hunnanocephalus ovalis*
			0-337 m	Grayish black carbonaceous siliceous rocks, grayish white–grayish black banded siliceous rocks **Fossil**: *Protospongia*
	Ediacaran	Xifengsi Formation	No data	Grayish white–grayish black dolomite

Limestone — Siliceous mudstone — Dolomitic limestone — Carbonaceous mudstone — Siliceous rock

Stone coal — Carbonaceous siliceous mudstone — Carbon-bearing siliceous rock — Limestone lens — Dolomite

Figure 4. Sedimentary sequences of lower Cambrian in Zhejiang Province (modified from Geology Research Institute of Geology Prospection Branch, Coal Science Academy (GRI-CSA) 1982).

apatite, U-bearing xenotime, xenotime, U-bearing rhabdophane, and rhabdophane (Figure 6(b–f)).

4.2 Petrography

The Cambrian, Ordovician, and Silurian stone coals in southern China are of anthracite-rank with $R_{o,max}$ values ranging from 3.75% to 8.40% in the homogeneous sapropelic groundmass (Zhu 1983). Petrological compositions and textures of stone coals in southern China are presented in Figure 7(a–k). These stone coals are also referred to as boghead coals with an anthracite-rank (O'Keefe *et al.* 2013). In addition to normal anthracite-rank stone coals, natural coke, semi-graphite, and graphite derived from stone coals occur in the metamorphic sedimentary sequences. Evidence of thermally alteration for rank advance, such as fine and coarse mosaic textures (Figure 7(d,e)) and leafy and spherical anisotropic carbon (Figure 7(h)), can be observed under the microscope. The $R_{o,max}$ of two stone coals with coarse and fine mosaic textures in Ziyang of Shaanxi Province can be up to 10.39% and 14.22%, respectively (China National Administration of Coal Geology 1996).

The organic matter in these anthracite-rank stone coals is dominated by sapropelic groundmass (Figure 7(a)) derived from the degradation and alteration of algae and, to a lesser extent, cyanobacteria and acritarchs from shallow marine environments, as well as rare sponge spicules (Figure 7(b,c,f,g,i–k),) (China National Administration of Coal Geology 1996). The algae that formed organic matter in the early Cambrian stone coal were mainly cyanophytes with trace amounts of brown algae (GRI-CSA) 1982; Zhu 1983; China National Administration of Coal Geology 1996). The organic matter in the Silurian anthracite rank stone coal was derived from cyanophytes and red algae, indicating a shallower marine environment in the Silurian than in the Cambrian. Numerous microfossils have been found in the Early and Middle Cambrian and Early Ordovician anthracite-rank stone coals and their roof and floor strata in western Hunan (Lu and Zhu 1983). The fossils in the Cambrian stone coals include filamentous algae (Figure 7(d,e cyanobacteria (Figure 7(f)), and acritarchs. The Ordovician stone coals contain fossil cyanobacteria (Figure 7(g)) and acritarchs.

Wang *et al.* (2017) have identified some typical marine microfossils in the Enshi stone coals in Hubei Province, well-preserved *Nodosaria longa Lipina* (a typical *Foraminifera*) (Figure 5(a)) and *poriferan*

(Figure 6(b,f)). The occurrence of roscoelite indicates input of hydrothermal fluids (Dai *et al.* 2017). Phosphorous-bearing minerals identified in the Enshi stone coals include

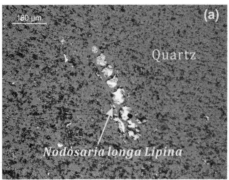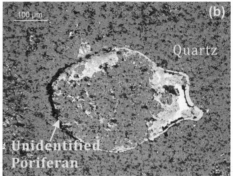

Figure 5. Quartz and well-preserved *Nodosaria longa Lipina* (a) and *poriferan* (b) in Enshi stone coals (reflected white light). From Wang *et al.* (2017).

(Figure 5(b)); algae are also observed in samples investigated.

4.3. Geochemistry

The major-element oxides in stone coals in southern China are mainly composed of SiO_2 (generally 70–80%), followed by Al_2O_3 (generally 5–15%), Fe_2O_3 (mostly 4–6%), MgO (mostly 0.8–1.2%), K_2O (~2%), and P_2O_5 (generally >0.1%) (GRI-CSA 1982). Geochemical studies on Enshi stone coals by Wang *et al.* (2017) showed that the ratio of SiO_2/Al_2O_3 ranges from 3.27 to 35.9 (12.5 on average), indicating that the percentage of SiO_2 is significantly higher than that of Al_2O_3 due to the high content of authigenic quartz.

Stone-like coals are always characterized by highly elevated V, followed by Mo, Ni, U, P, and Cd, and to a lesser extent, Au, Ag, Cu, Co, Zn, Se, Ga, Ge, Sc, Ti, rare earth elements, and Y (Geology Research Institute of Geology Prospection Branch, Coal Science Academy (GRI-CSA) 1982). Platinum group elements (PGEs) are enriched in some stone-like coals in southern China (Mao *et al.* 2002; Coveney 2003a; 2003b; Křibek *et al.* 2007; Murowchic *et al.* 1994; Pašava *et al.* 2004). Geochemical studies by Wang *et al.* (2017) showed that the stone coals in Enshi of Hubei Province are characterized by elevated concentrations of a number of elements including 2031 ppm V, 856 ppm Cr, 350 ppm Ni, 245 ppm Mo, 140 ppm Se, 526 ppm Zn, 16.5 ppm As, 77.8 ppm Cd, and 39.2 ppm U. A study by Zheng *et al.* (2006) also showed the enrichment of these elements in the Enshi stone coals.

4.3.1. Vanadium

Concentration of V_2O_5 in stone coals generally ranges from 0.13% to 1.2% (Zhang *et al.* 2011a; 2011b; Wang *et al.* 2017), the highest concentration of V_2O_5 (4.6%)

was reported from Yangjiabao, Jun County, Hubei Province (GRI-CSA 1982). For comparison, the average concentration of V for common Chinese coals and world coals is 35.1 (Dai *et al.* 2012) and 25 ppm (Ketris and Yudovich 2009), respectively. Vanadium in stone coals primarily occurs as isomorphous substitution (mainly as V^{3+}) for Al in clay minerals (such as illite, muscovite, kaolinite, smectite, sericite, and chlorite) (GRI-CSA, 1982); as vanadium minerals (Ti-V-garnet, Cr-V-garnet, germanite, and V-bearing tourmaline); as adsorbed complex anion $(VO_4)^{3-}$; and as V porphyrinin the organic matter of stone coals (GRI-CSA 1982). The contents of vanadium minerals usually are very low and are commonly detrital materials of terrigenous origin and, to a lesser extent, from epigenetic igneous hydrothermal solutions (GRI-CSA 1982).

It was reported that quartz, pyrite, and some hypergenic minerals (e.g. childrenite $Fe^{2+}Al(PO_4)(OH)_2 \cdot H_2O$, collophane, and limonite) may contain trace V. The V in quartz and pyrite is probably related to the mineralization of marine organisms, and thus, the V-bearing quartz and pyrite mainly occurs in silicified (particularly in spongy spicules) and pyritized bioclastic rocks, respectively (GRI-CSA 1982). Figure 5(a,b) showed that the quartz partially replaced *Nodosaria long*a Lipina and *poriferan*.

4.3.2. Uranium and Se

Uranium in stone coals in southern China is generally below 100 ppm. However, in some cases, it is higher than 100 ppm and up to 0.1% in some stone coals (Zhou 1981). Uranium mainly occurs as uranyl ion $(UO_2)^{2+}$ as adsorbed forms in the stone coals (GRI-CSA 1982). In addition to association with organic matter (Wang *et al.* 2017), U in the Enshi stone coals occurs in U-bearing xenotime and U-bearing rhabdophane. The concentration of U in Chinese stone coals is much higher than those in common Chinese coals and world

Figure 6. SEM backscattered electron images of minerals in the Enshi stone coals. (a) Authigenic and terrigenous detrital quartz; (b) roscoelite, xenotime, rhabdophane (Rha), U-bearing rhabdophane (U-Rha), chalcopyrite (Chal), and apatite; (c) autunite, sphalerite, U-bearing rhabdophane, and apatite; (d) apatite; (e) U-bearing xenotime (U-Xen), apatite, and pyrophyllite; (f) rhabdophane, apatite, and roscoelite.

coals, 2.43 and 2.4 ppm, as reported by Dai *et al.* (2012) and Ketris and Yudovich (2009), respectively.

Selenium in the Lower Cambrian stone coals in Shaanxi and Hunan Provinces generally varies from 16 to 40 ppm, and can be up to 370 ppm (Luo *et al.* 1995). The Se concentration in Enshi stone coals in Hubei Province greatly varies from >10 ppm to several thousands of ppm (Yang *et al.* 1983; Song 1989; Mao *et al.* 1990; Zhu *et al.* 2004a). The Se concentration in two stone coal samples collected by Yang *et al.* (1981, 1983) and Song (1989) was 8290 and 84,123 ppm, respectively. The average concentration of Se for common Chinese coals and world coals is 2.47 (Dai *et al.* 2012) and 1.3 ppm (Ketris and Yudovich 2009), respectively. The Se in the stone coals in this area occurs in pyrite, eskebornite, and mandarinoite (Zheng *et al.* 1992, 1993; Belkin *et al.* 2010); in a Cu-Fe selenide, krutaite, klockmannite, and naumannite (Zhu *et al.* 2004a, 2004b); also

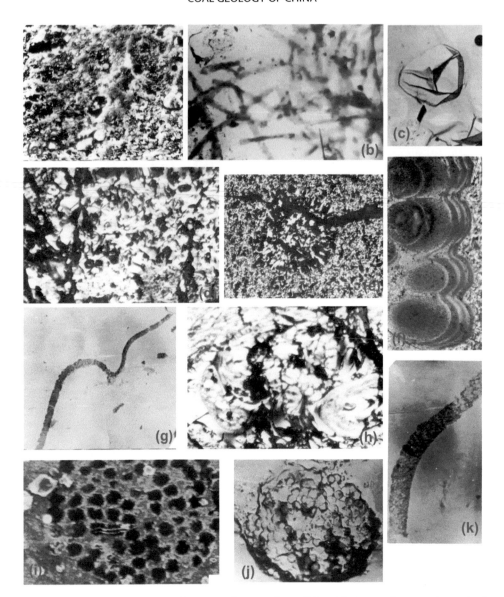

Figure 7. Petrological compositions and textures of stone coals in southern China. (a) Sapropelic groundmass in the early Paleozoic stone coal (anthracite boghead) from Xupu, Hunan Province (SEM, field width 85 μm); from Zhang *et al.* (2003). (b) *Nocardia* ? sp. in the middle Cambrian stone coal from Pingli, Shaanxi (TEM, field width 15 μm); from China National Administration of Coal Geology (1996). (c) The acritarch *Leiosphaeridia* in the Cambrian stone coal from Pingli, Shaanxi (TEM; field width 4 μm); Chen *et al.* (1996). (d) Coarse mosaic texture of the middle Cambrian stone coal from Ziyang, Shaanxi (reflected light, oil immersion; field width 160 μm); from China National Administration of Coal Geology (1996). (e) Fine mosaic texture of the early Silurian stone coal from Ankang, Shaanxi (reflected light, oil immersion; field width 380 μm); from China National Administration of Coal Geology (1996). (f) Conical columnar stromatelite (algae) in the early Silurian stone coal from Ankang, Shaanxi (reflected light, crossed polars, oil immersion; field width 375 μm); fromChen *et al.* (1996). (g) The filamentous alga *Oscillatoriopsis* tenuis in the early Cambrian stone coal from Xupu, Hunan Province (TEM; field width 8.3 μm); Chen *et al.* (1996). (h) Leafy and spherical anisotropic carbon in the early Silurian stone coal from Ziyang, Shaanxi (reflected light, oil immersion; field width 160 μm); from China National Administration of Coal Geology (1996). (i) The cyanobacterium *Sphaerocongrgus wufengensis* in the middle Cambrian stone coal from Xupu, Hunan Province (SEM; field width 11 μm); Chen *et al.* (1996). (j) The cyanobacterium *Sphaerocongrgus yinpingensis* in the Middle Ordovician stone coal from Dongkou, Hunan Province (SEM; field width 4 μm); from China National Administration of Coal Geology (1996). (k) The filamentous alga *Oscillatoriopsis minuta* in the early Cambrian from Xupu, Hunan Province (TEM; field width 11.4 μm); Chen *et al.* (1996).

as adsorbed form in the organic matter (Song 1989); and as native Se (Zhu and Zheng 2001; Zhu *et al.* 2004a).

4.3.3. Rare earth elements and Y

The concentration of rare earth elements in stone coals from western Zhejiang Province, Yexiyu of northwestern Hunan, and Chongqing is 58 ppm (on average; Gao and Ye 1999), 146–275 ppm (Jiang 2010), and 268 ppm (on average; Li *et al.* 2005), respectively. The concentrations of Y_2O_3 in stone coals from Hunan, Jiangxi, and Hubei Provinces are 400–600, 100–400, and 170–340 ppm, respectively, and Y mainly occurs in xenotime (GRI-CSA 1982). The Enshi stone coals have REE and Y concentrations ranging from 31.4 to 201 ppm, with an average of 80.4 ppm (Wang *et al.* 2017). The average concentration of rare earth elements and Y in common Chinese coals and world coals is 138 (Dai *et al.* 2012) and 68 ppm (Ketris and Yudovich 2009), respectively.

Based on La_N/Ce_N (1.82 on average; normalized to Post-Archean Australian Shale, Taylor and McLennan (1985)) and $Al_2O_3/(Al_2O_3+Fe_2O_3)$ (0.57–0.82), Wang *et al.* (2017) suggested that the Enshi stone coals were deposited in a pelagic environment with admixture of detrital input from continental margin, rather than a spreading ridge-proximal environment that has La_N/Ce_N values not less than 3.5 and low $Al_2O_3/(Al_2O_3+Fe_2O_3)$ values (Murray 1994). Indeed, such an environment is also supported by Y/Ho values in the stone coals, which varies from 31.77 to 44.87, with an average of 39.52. This indicates a mixed sediment of detrital siliciclastics and marine sediments (Figure 8; the stone coal section is presented in Figure 1(b)). The Y/Ho ratio values for detrital siliciclastic and seawater inputs are ~25–30 and ~60–70, respectively (Webb and Kamber 2000; Chen *et al.* 2015).

The distribution patterns of rare earth elements and Y in the complete section of Enshi stone coals (Figure 9; 6.12 m in total thickness, 30 bench samples; the stone coal section presented in Figure 1(b); concentrations of rare earth elements and Y are listed in the Supplementary Electron Table) are characterized by positive Y and Gd, weakly negative Ce and Eu anomalies, and heavy enrichment type ($La_N/Lu_N<1$; Seredin and Dai 2012), further indicating the influence of seawater and hydrothermal fluids (Seredin and Dai 2012; Dai *et al.* 2016a). Both seawater and hydrothermal fluids could lead to positive Y anomalies in coal and sedimentary rocks (Zhang *et al.* 1994; Bau *et al.* 1995; Dai *et al.* 2016a).

4.3.4. Nickel, Mo, Au, and PGEs

The concentrations of Ni in the stone coals in western Hunan Province is generally higher than 92 ppm (Wu *et al.* 1999b) and is up to ~0.2% in stone coals from Cili in Hunan Province and Ziyang in Shaanxi Province (GRI-CSA 1982). The average concentration of Ni in common Chinese coals and world coals is 13.7 (Dai *et al.* 2012) and 13 ppm (Ketris and Yudovich 2009), respectively. It

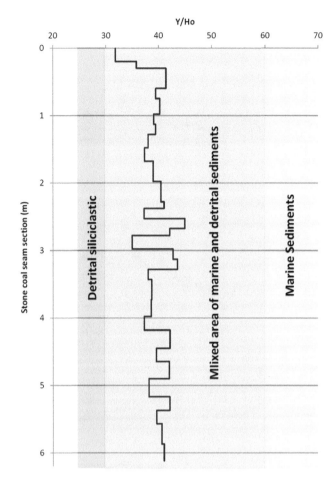

Figure 8. Variation of Y/Ho through the stone coal section in Enshi of Hubei Province.

was reported that Ni is related to organic matter (nickel porphyrin) in stone coals (GRI-CSA 1982). Belkin and Luo (2008) reported an example of Ni and Co enrichment in stone coals from Guizhou Province, China. High concentrations of 0.5–2.0% Ni and 0.1–0.5% Co were determined in the latest pyrite in the stone coals (Belkin and Luo 2008).

The concentrations of Mo in stone coals in southern China generally vary from 200 to 400 ppm (GRI-CSA 1982). Molybdenum in stone coals in western Hunan Province is 187 ppm on average and correlates positively with organic matter in stone coals (GRI-CSA 1982), suggesting an organic association. To a lesser extent, Mo also occurs in sulphide minerals (Wu *et al.* 1999b). For example, the concentration of Mo in stone coals from Tianmenshan of Dayong, Hunan Province, ranges from 0.06% to 8.8%, and mainly occurs as molybdenum sulphide (MoS) (GRI-CSA 1982; Jiang and Gao 2013). The average concentration of Mo in common Chinese coals and world coals are 3.08 (Dai *et al.* 2012) and 2.2 ppm (Ketris and Yudovich 2009), respectively.

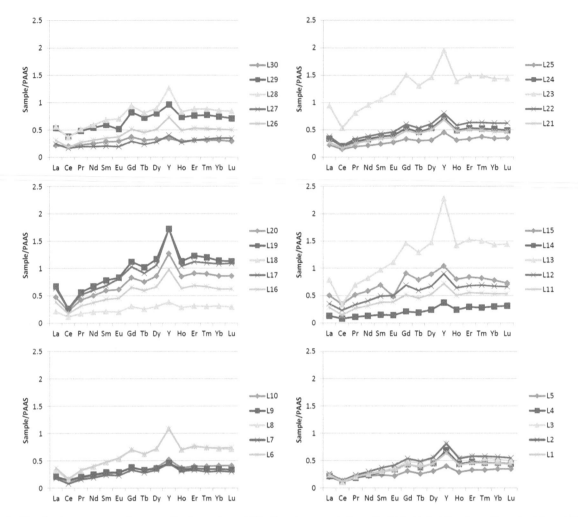

Figure 9. Distribution patterns of rare earth elements and Y in the stone coals from Enshi in Hubei Province. Samples L1 and L30 are the lowermost and upmost stone coal benches, respectively (the stone coal section is represented in Figure 1(b)). Original data are from Wang *et al.* (2017) and also listed in the Supplementary Electron Table. REE and Y are normalized to data of Post-Archean Australian Shale (PAAS) (Taylor and McLennan 1985).

Stone coals in the lowermost Cambrian Niutitang Formation have attracted much attention because a discontinuous sulphide- and phosphate-rich stratiform body was hosted by the stone coal and is highly enriched in a Mo-Ni-PGE assemblage. With the exceptions of areas of land, shallow shelf, and lagoonal facies, the stone coals in the Niutitang Formation cover almost all the Yangtze craton, approximately 1×10^6 km^2 (Li 1986; Emsbo *et al.* 2005). The thickness of the ore body ranges from <1 cm to 2 m and covers the whole paleocoastline of the Yangtze craton. The P$_2$O$_5$ content in the ores ranges from 8% to 17.2% (Pašava *et al.* 2004). The distinct characteristics of the inorganic matter in the stone coals are characterized by highly elevated native Au and PGEs, a MoS-C phase (a complex mixture of MoS$_2$ and carbonaceous matter nanocrystallites), as well as sulphide minerals including pyrite, vaesite, bravoite, chalcopyrite, covelline, sphalerite, millerite, polydimite, gersdorffite, and violarite (Coveney

et al. 1994; Murowchick *et al.* 1994; Lott *et al.* 1999; Orberger *et al.* 2007). The ore in Zunyi of Guizhou Province and Dayong of Hunan Province (Figure 2(f)) consists mainly of a mixture of nodular pyrite, nodular phosphorite, and pellets of solid organic matter, quartz, and shale (Coveney and Nansheng 1991). The highly elevated concentrations of Mo-Ni-PGE are generally attributed to hydrothermal venting of basinal sedex brines (Coveney *et al.* 1994; Murowchick *et al.* 1994; Lott *et al.* 1999; Steiner *et al.* 2001; Coveney 2003a; Emsbo *et al.* 2005). On the other hand, the influence of seawater is at least partly responsible for elevated elements such as V, Se, Mo, Ni, Au, and U (Mao *et al.* 2002). Křibek *et al.* (2007) attributed this Mo-Ni-PGE mineralized body to a remnant of phosphate- and sulphide-rich subaquatic hardground supplied with organic material derived from plankton and benthic communities as well as with algal/microbial oncolite-like bodies that originated in a wave-agitated, shallow-water, nearshore environment. The origin of these enriched metals is controversial and has been attributed either to input from hydrothermal fluids during or after sedimentation or to slow deposition from sea water over long periods (Holland 1979; Coveney 2003a).

The geochemistry of stone coals is quite similar to the Late Permian coals preserved within marine carbonate successions in South China, which are characterized by highly elevated concentrations of Se, V, Mo, Re, U, and in some cases, F, Cr, Ni, Cd (Zeng et al. 2005; Dai et al. 2015; Liu et al. 2015), and rare earth elements (Dai et al. 2015). The enrichment of these elements is predominantly attributed to hydrothermal activity (submarine exhalation) and the euxinic environment. The anomalous geochemistry of stone coals and coals preserved within marine carbonate successions in return could provide useful information for geological events and regional geological setting.

5. Adverse effects on environment and human health

The adverse effects on environment and human health caused by stone coal utilization are mainly due to its high yield of combustion residues; highly elevated concentrations of toxic elements including S, Se, Cd, F, Mo, As, Mo, Pb, and Be (GRI-CSA 1982; Luo et al. 1995; Tian and Luo 2017a, 2017b; Long and Luo 2017); and radiation (Zhou 1981). As with fly ash derived from coal (Izquierdo and Querol 2012; Jones and Ruppert 2017), some toxic elements as listed earlier contained in combustion residues of stone coals are likely to be released from the storage/disposal/application site when ash comes in contact with water. In addition to toxic trace elements above, one example of Ni enrichment in Lower Cambrian stone coal from Guizhou Province reported by Belkin and Luo (2008) showed that Ni and Co contents are up to 0.5–2.0% and to 0.1–0.5 %, respectively, in the pyrite in the stone coals.

SO_2 emission is one of environmental issues of stone-coal combustion in power plants (Bu 1988) because stone coals usually contain a high sulphur content. The stone coals in Ziyang of Shaanxi Province contain 510–2800 ppm F, which is the cause of endemic fluorosis in this area (Luo et al. 1996); 98–99% of the fluorine was released during stone coal combustion and then ingested and inhaled by villagers through food and respiration (Luo et al. 1996). In the present study, 10 stone coal samples from Enshi, Hubei Province, contain high concentrations of F, varying from 2031 to 3831 ppm, with an average of 2803 ppm. This indicates that the cautions should be taken while utilizing F- and Se-rich stone coals from this area.

Zheng et al. (1992) reported nearly 500 cases of endemic selenosis in southwest China that were attributed to the use of selenium-rich stone coal (carbonaceous shales). Symptoms of selenium poisoning include discoloration of the skin and loss of hair and nails (Finkelman 2004). Endemic selenosis in Enshi, Hubei Province, occurred from 1958 to 1987 and was attributed to the use of weathered 'stone coal' and stone-coal ashes as fertilizer, which led to the Se enrichment in the soil and then taken up by crops (Mou et al. 2007; Zhu et al. 2008; Zheng et al. 2010; Dai et al. 2012). Measures had been taken to prevent selenosis since the 1960s, these include migration of residents, forest conservation, closing five kinds of small industrial factories (small steel, cement, farming machine, fertilizer factories, and coal mines), changing farming styles and living habits, and education (Dai et al. 2012). These measures had achieved expected results and no human cases of selenium toxicity have been reported since 1987 in these areas. Detailed review of this stone-coal-related arsenosis can be found in Finkelman and Tian (2017) in this special issue and others such as Dai et al. (2012) and Chen et al. (2014).

The endemic selenosis in Ziyang of Shaanxi Province had once been attributed to high Se in stone coals (Mei 1985), but a study by Luo (1996, 2003) showed the Se source for selenosis in this area was from Se-rich black carbonaceous siliceous slates and tuffs, which served as bed rocks for soil and provided Se source for crops.

Due to high concentrations of U (e.g. Zhou 1981, Wang et al. 2017) and K (e.g. Bu 1988; Liu et al. 2016) in the stone coals, the radionuclides in residues derived from stone coal combustion should also be paid much attention and the residues should be properly disposed of.

6. Utilization and by-product potential

Stone coal mainly occurs in southern China, where coal resources are inadequate relative to northern China. Stone coal had been mined for use as household energy and stone coal ash had been used for brickmaking as early as 500 years ago (Ming Dynasty) (Han et al. 1996). Nowadays, the stone coal is still used for household energy in villages and is also used in power plant furnaces as feed fuel (e.g. Yiyang and Changde Power Plants in Hunan Province; Liu et al. 2016). As with fly ash derived from coal, the combustion residues of stone coal are mainly used as building materials, concrete pavements, soil stabilization, road base, and calcining lime. Due to the high contents of P and K, the stone coal and its combustion ash in some cases are used as fertilizers (Bu 1988; Liu et al. 2016).

Much attention has been paid to its potential as an economic source for critical elements V, Se, Mo, and PGEs (Peng 2015; Dai *et al.* 2016b). A number of techniques have been developed for extraction of V, Se, Mo, and Ni from stone coals (e.g. Duan *et al.* 2006; Huo *et al.* 2011; Peng *et al.* 2011; Zhang *et al.* 2011a). Utilization of V from stone coals represents a successful example of industrial by-products. The total reserves of V_2O_5 in stone coals are 118 Mt, accounting for 87% of total V reserves in China (GRI-CSA 1982; Zhang *et al.* 2011a).

Different technologies have been developed for V extraction from stone coals since the 1960s in China (Lu 2002; Duan *et al.* 2006; He *et al.* 2007; Zhang *et al.* 2011b; Wang *et al.* 2014a; 2014b, 2015). Because V in stone coals largely occurs as V^{3+} in the crystal lattice of minerals and is insoluble in acid or water, it is impossible to recover it from stone coal by direct leaching. However, roasting V^{3+} in stone coals can oxidize V^{3+} to acid-soluble V^{4+} and/or water-soluble V^{5+}, which could be leached by acid and water/acid, respectively (He *et al.* 2008; Li *et al.* 2009, 2010, 2013; Liu *et al.* 2010; Zhang *et al.* 2011b; Hu *et al.* 2012; Zhu *et al.* 2012; Wang *et al.* 2013, 2014a, 2014b), and thus, the traditional method for V extraction from stone coals includes the roasting under high temperature to volatize organic matter and to destroy the structure of V minerals (and thus the V^{3+} is oxidized to V^{5+}). After that, the residues are subjected to water or acid leaching (Zhang *et al.* 2011b, 2015). This extraction technology is usually used in small-capacity plants (usually less than annual 300 t; Cai 2010). A more advanced technology for V extraction was developed by Beijing Institute of Chemical Engineering and Metallurgy and includes several procedures of 'raw coal grinding, counter-flow acid leaching, solvent extraction, ammonia leaching, pyrolysis for refined V making' and, based on this extraction techniques, a plant with annual production of 660 t V_2O_5 was designed (Lu 2002). Zhang *et al.* (2011a) comprehensively reviewed different techniques of V extraction from stone coals, including the blank roasting-acid leaching, the blank roasting-alkali leaching, the calcified roasting-carbonate leaching, the direct acid leaching, and the low salt roasting-cyclic oxidation.

A successful example of Se utilization in agriculture from stone-like coal as a soil amendment is represented by the Enshi Se ore deposit hosted by stone-like coal in Hubei Province of southern China. The stone-coal-hosted ore deposit is now considered as a unique Se deposit (Dai *et al.* 2016b). In addition to Se, stone coal in Enshi is also highly enriched in V and Mo that could be potentially extracted. The Se reserves (with Se concentration > 800 ppm) in Enshi are ~45.6 t, and the ~121 t of the co-existed V_2O_5 (concentration 0.38–

0.52%) and ~12 t Mo (concentration 0.036–0.054%) (Peng 2015). The Enshi stone coals are mainly composed of early and middle Permian carbonaceous siliceous rocks and carbonaceous shales, which are different from the terms defined by Chinese Standard (GB/T15663.1-2008 GB/T. 2008). The Se concentration varies considerably from tens of ppm to thousands of ppm (Mao *et al.* 1990; Song 1989; Yang *et al.* 1983; Zhu *et al.* 2004a, 2004b; Dai *et al.* 2016b) and can be up to 84,123 ppm (Song 1989). Dai *et al.* (2016b)'s paper in this Special Issue has reviewed this Se ore deposit in great detail. Only a few studies have been developed for Se extraction from stone coals. For example, Tian *et al.* (2014) used high-efficient Ca-adsorbent as fixative for Se volatized from stone-coal burning and Na_2SO_3 was used as reducing agent for Se compound. Based on this recovering technology, the recovery rate of Se can be up to 98.5%. Extraction of Se from Enshi stone coals has been successfully developed as early as in 1991 (http://www.esdasz.org/); however, Se in the Enshi stone coals is now mainly used for Se-rich crops, such as tea, vegetables, and traditional Chinese medicinal materials.

Jiang and Gao (2013) developed technology for Mo extraction from stone coals. They first crushed the stone coals to pass 0.74-mm screen and pelletized with sodium carbonate, and then the pellets were roasted under oxidizing conditions; after that, the pellets were leached by water to recover Mo. The results showed that the leaching recovery rate of Mo can be up to 96.32% under the optimum conditions of 650°C roasting temperature, 3-h roasting time, and 5–15-mm diameter of pellets.

7. Conclusions

The stone coal in China is summarized in Figure 10. Stone coals occur in a large area covering most of southwestern China and were mostly deposited before the Middle Devonian. They are organic-rich black shales with high sulphur contents, high ash yields, low heat values, high-rank, and combustible properties.

Stone coals have distinctive textural features and geochemical compositions, and the latter is characterized by a distinctive assemblage of trace elements (elevated V, Se, Mo, Ni-Mo-PGEs). In some cases, the trace elements may represent potential hazards to the environment and human health, but in others, they may represent potential economic sources of critical metals that are increasingly in demand for modern industrial development.

Although many issues related to stone coal have been investigated in many studies, including their geographic distribution, mineralogical and

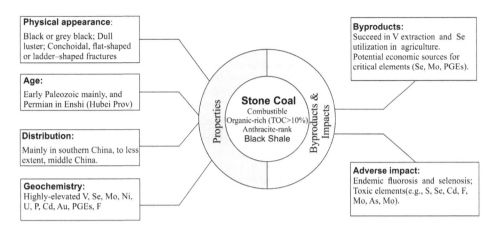

Figure 10. The summarization of stone coals in China.

geochemical assemblages, adverse effects on human health, and extract technologies for critical elements, some questions require further investigation to evaluate more fully the nature and significance of stone coals, including the sources and deposition environments of critical elements, the role of organic matter in the enrichment of trace elements, the modes of occurrence of critical elements, the controls of the tectonic framework and geodynamic processes on elemental and mineralogical compositions in stone coals, and co-recovery methods for co-production of critical elements, particularly for Se, V, Ni, Mo, and PGEs. In addition, the abundance, origin, modes of occurrence of Re, and its potential by-products should be investigated not only because it has rarely been investigated but also because it would be expected to be enriched in stone coals based on its geochemical properties.

Acknowledgments

This work was supported by the National Natural Science Foundation of China (No. 41420104001), the National Key Basic Research Program of China (No. 2014CB238902), and the "111" Project (No. B17042). We are grateful to Profs. Robert Stern and Jim Hower, and another anonymous reviewer for their careful review and useful comments, which greatly improved the quality of the manuscript.

Disclosure statement

No potential conflict of interest was reported by the authors.

Funding

This work was supported by the National Natural Science Foundation of China [No. 41420104001],National Key Basic

Research Program of China [No. 2014CB238902], and '111' Project [No. B17042].

ORCID

Shifeng Dai http://orcid.org/0000-0002-9770-1369
Xibo Wang http://orcid.org/0000-0002-8758-2845
Robert B. Finkelman http://orcid.org/0000-0002-7295-5952

References

ASTM Standard D5987-96, 2015, Standard Test Method for Total Fluorine in Coal and Coke by Pyrohydrolytic Extraction and Ion Selective Electrode or Ion Chromatograph Methods: West Conshohocken, PA, ASTM International.

Bau, M., Dulski, P., and Möller, P., 1995, Yttrium and holmium in South Pacific seawater: vertical distribution and possible fractionation mechanisms: Chemie der Erde : Beitrage zur chemischen Mineralogie, Petrographie und Geologie, v. 55, p. 1–15.

Belkin, H.E., Zheng, B., and Zhu, J., 2010, First occurrence of mandarinoite in China: Acta Geologica Sinica – English Edition, v. 77, p. 169–172. doi:10.1111/acgs.2003.77.issue-2

Belkin, H.E., and Luo, K., 2008, Late-stage sulfides and sulfarsenides in Lower Cambrian black shale (stone coal) from the Huangjiawan mine, Guizhou Province, People's Republic of China: Mineralogy and Petrology, v. 92, p. 321–340. doi:10.1007/s00710-007-0201-9

Bu, Y., 1988, Characteristics and environmental problems of stone coal: Energy Environmental Protection, no. 2, p. 23–25. [in Chinese.]

Cai, J., 2010, Latest development of process and equipment for vanadium extraction from coal stone: Rare Metals and Cemented Carbides, v. 38, no. 2, p. 67–71. [in Chinese with English abstract.]

Chen, J., Liu, G., Kang, Y., Wu, B., Sun, R., Zhou, C., and Wu, D., 2014, Coal utilization in China: Environmental impacts and human health: Environmental Geochemistry and Health, v. 36, p. 735–753. doi:10.1007/s10653-013-9592-1

Chen, J.B., Algeo, T.J., Zhao, L.S., Chen, Z.-Q., Cao, L., Zhang, L., and Li, Y., 2015, Diagenetic uptake of rare earth elements by bioapatite, with an example from lower Triassic conodonts of South China: Earth-Science Reviews, v. 149, p. 181–202. doi:10.1016/j.earscirev.2015.01.013

Chen, P., Sun, D., Ding, P., and Luo, J., 1996, Coal Petrologic Atlas of China: China Coal Ind. Publ. House, Beijing, China, 22–26 p. [in Chinese with English abstract.]

China National Administration of Coal Geology, 1996, Atlas for Coal Petrography of China: China University of Mining and Technology Press, Xuzhou, China. p. 21–23. [in Chinese with English abstract.]

Coveney, R.M., 2003a, Re–Os dating of polymetallic Ni–Mo–PGE–Au mineralization in lower Cambrian black shales of south China and its geological significance; discussion: Economics Geological, v. 98, p. 661–662.

Coveney, R.M., Grauch, R.I., and Murowchick, J.B., 1994, Metals, phosphate and stone coal in the Proterozoic and Cambrian of China: the geologic setting of precious metal-bearing Ni–Mo ore beds: Social Applications Geological (SGA) Newsletter, v. 18, p. 1–11.

Coveney, R.M., 2003b, Metalliferous Paleozoic black shales and associated strata, in Lenz, D.R., ed., Geochemistry of Sediments and Sedimentary Rocks, Geotext 4: Geological Association of Canada, p. 135–144.

Coveney, R.M., and Nansheng, C., 1991, Ni-Mo-PGE-Au-rich ores in Chinese black shales and speculations on possible analogues in the United States: Mineralium Deposita, v. 26, p. 83–88. doi:10.1007/BF00195253

Dai, S., and Finkelman, R.B., 2017, Coal as a promising source of critical elements: Progress and future prospects: International Journal of Coal Geology. doi:10.1016/j.coal.2017.06.005

Dai, S., Graham, I.T., and Ward, C.R., 2016a, A review of anomalous rare earth elements and yttrium in coal: International Journal of Coal Geology, v. 159, p. 82–95. doi:10.1016/j.coal.2016.04.005

Dai, S., Ren, D., Chou, C.-L., Finkelman, R.B., Seredin, V.V., and Zhou, Y., 2012, Geochemistry of trace elements in Chinese coals: a review of abundances, genetic types, impacts on human health, and industrial utilization: International Journal of Coal Geology, v. 94, p. 3–21. doi:10.1016/j.coal.2011.02.003

Dai, S., Seredin, V.V., Ward, C.R., Hower, J.C., Xing, Y., Zhang, W., Song, W., and Wang, P., 2015, Enrichment of U–Se–Mo–Re–V in coals preserved within marine carbonate successions: geochemical and mineralogical data from the Late Permian Guiding Coalfield, Guizhou, China: Mineralium Deposita, v. 50, p. 159–186. doi:10.1007/s00126-014-0528-1

Dai, S., Wang, X., Zhou, Y., Hower, J.C., Li, D., Chen, W., Zhu, X., and Zou, J., 2011, Chemical and mineralogical compositions of silicic, mafic, and alkali tonsteins in the late Permian coals from the Songzao Coalfield, Chongqing, Southwest China: Chemical Geology, v. 282, p. 29–44. doi:10.1016/j.chemgeo.2011.01.006

Dai, S., Xie, P., Jia, S., Ward, C.R., Hower, J.C., Yan, X., and French, D., 2017, Enrichment of U-Re-V-Cr-Se and rare earth elements in the Late Permian coals of the Moxinpo Coalfield, Chongqing, China: Genetic implications from geochemical and mineralogical data: Ore Geology Reviews, v. 80, p. 1–17. doi:10.1016/j.oregeorev.2016.06.015

Dai, S., Yan, X., Ward, C.R., Hower, J.C., Zhao, L., Wang, X., Zhao, L., Ren, D., and Finkelman, R.B., 2016b, Valuable elements in Chinese coals: a review: International Geology Review. doi:10.1080/00206814.2016.1197802

Duan, L., Tian, Q., and Guo, X., 2006, Review on production and utilization of vanadium resources in China: Hunan Nonferrous Metals, v. 22, no. 6, p. 17–20. [in Chinese with English abstract.]

Emsbo, P., Hofstra, H., Johnson, C.A., Koenig, A., Grauch, R., Zhang, X., Hu, R., Su, W., and Pi, D., 2005, Lower Cambrian metallogenesis of south China: interplay between diverse basinal hydrothermal fluids and marine chemistry, in Mao, J., and Bierlein, F.P., eds., Mineral Deposit Research: Meeting the Global Challenge. Proceedings of the Eighth Biennial SGA Meeting Bejing. Berlin, Springer, pp. 115–118.

Finkelman, R.B., 2004, Potential health impacts of burning coal beds and waste banks: International Journal of Coal Geology, v. 59, p. 19–24. doi:10.1016/j.coal.2003.11.002

Finkelman, R.B., and Tian, L., 2017, The health impacts of coal use in China: International Geology Review, p. 1–11. doi:10.1080/00206814.2017.1335624

Gao, C., and Ye, D., 1999, REE geochemical characteristics of three types of coals in the early palaeozoic in southern China: Experimental Petroleum Geology, v. 21, no. 3, p. 270–272. [in Chinese with English abstract.]

GB/T15663.1-2008 GB/T., 2008, Terms relating to coal mining-Part 1. Coal geology and prospecting, p. 4. [in Chinese.]

Geology Research Institute of Geology Prospection Branch, Coal Science Academy (GRI-CSA), 1982, Comprehensive Investigation Report of Stone Coal Resources in southern China: 254 p. [unpublished data; in Chinese.]

Han, D., Ren, D., Wang, Y., Jin, K., Mao, H., and Qin, Y., 1996, Coal Petrology of China: Xuzhou, Publishing House of China University of Mining and Technology, 599 pp. [in Chinese with English abstract.]

He, D., Feng, Q., Zhang, G., Ou, L., and Lu, Y., 2007, An environmentally-friendly technology of vanadium extraction from stone coal: Minerals Engineering, v. 20, no. 12, p. 1184–1186. doi:10.1016/j.mineng.2007.04.017

He, D.S., Feng, Q.M., Zhang, G.F., Ou, L.M., and Lu, Y.P., 2008, Study on leaching vanadium from roasting residue of stone coal: Miner Metall Proceedings, v. 25, p. 181–184.

Holland, H.D., 1979, Metals in black shales; a reassessment: Economic Geology, v. 74, p. 1676–1680. doi:10.2113/gsecongeo.74.7.1676

Hu, Y.-J., Zhang, Y.-M., Bao, S.-X., and Liu, T., 2012, Effects of the mineral phase and valence of vanadium on vanadium extraction from stone coal: International Journal Miner Metall Materials, v. 19, p. 893–898. doi:10.1007/s12613-012-0644-9

Huo, G., Miao, J., and Zhu, H., 2011, The methods for extraction and separation of Ni and Mo from Ni-Mo-bearing stone coals.Patent No. CN201110116387.3. [in Chinese.]

Izquierdo, M., and Querol, X., 2012, Leaching behaviour of elements from coal combustion fly ash: An overview: International Journal of Coal Geology, v. 94, p. 54–66. doi:10.1016/j.coal.2011.10.006

Jiang, S., and Gao, Z., 2013, Study on molybdenum extraction from molybdenum-bearing stone coal: Nonferrous Metals (Smelting Part), no. 5, p. 35–43. [in Chinese with English abstract.]

Jiang, X., 2010, Vanadium contained stone-like coal geological characteristics and REE analysis in Yexiyu, northwestern Hunan: Coal Geology of China, v. 22s, p. 4–7. [in Chinese with English abstract.]

Jiang, Y., Yue, W., and Ye, Z., 1994, Characteristics, sedimentary environment and origin of the lower Carbrian stone-like coal in southern China: Coal Geology in China, v. 6, no. 4, p. 26–31. [in Chinese with English abstract.]

Jones, K.B., and Ruppert, L.F., 2017, Leaching of trace elements from Pittsburgh coal mill rejects compared with coal combustion products from a coal-fired power plant in Ohio, USA: International Journal of Coal Geology, v. 171, p. 130–141. doi:10.1016/j.coal.2017.01.002

Ketris, M.P., and Yudovich, Y.E., 2009, Estimations of Clarkes for Carbonaceous biolithes: world averages for trace element contents in black shales and coals: International Journal of Coal Geology, v. 78, p. 135–148. doi:10.1016/j.coal.2009.01.002

Křibek, B., Sýkorová, I., Pašava, J., and Machovič, V., 2007, Organic geochemistry and petrology of barren and Mo–Ni–PGE mineralized marine black shales of the Lower Cambrian Niutitang Formation (South China): International Journal of Coal Geology, v. 72, p. 240–256. doi:10.1016/j.coal.2007.02.002

Łapcik, P., Kowal-Kasprzyk, J., and Uchman, A., 2016, Deep-sea mass-flow sediments and their exotic blocks from the Ropianka Formation (Campanian-Paleocene) in the Skole Nappe: A case study of the Wola Wola Rafałowska section (SE Poland): Geological Quarterly, v. 60, no. 2, p. 301–316.

Li, D., Tang, Y., Chen, K., Deng, T., Cheng, F., and Yang, J., 2005, Research on geochemistry of Rare Earth Elements in coals from Chongqing, China: Journal of China University of Mining & Technology, v. 34, no. 3, p. 312–317. [in Chinese with English abstract.]

Li, M.T., Wei, C., Fan, G., Li, C.X., Deng, Z.G., and Li, X.B., 2009, Extraction of vanadium from black shale using pressure acid leaching: Hydrometallurgy, v. 98, p. 308–313. doi:10.1016/j.hydromet.2009.05.005

Li, M.T., Wei, C., Fan, G., Wu, H.L., Li, C.X., and Li, X.B., 2010, Acid leaching of black shale for the extraction of vanadium: International Journal of Mineral Processing, v. 95, p. 62–67. doi:10.1016/j.minpro.2010.04.002

Li, W., Zhang, Y.M., Liu, T., Huang, J., and Wang, Y., 2013, Comparison of ion-exchange and solvent extraction in recovering vanadium from sulfuric acid leach solutions of stone coal: Hydrometallurgy, v. 131-132, p. 1–7. doi:10.1016/j.hydromet.2012.09.009

Li, Y., 1986, Proterozoic phosphorite-regional review: China, in Cook, P., and Shergord, J.H., eds., Phosphates of the World, Proterozoic and Cambrian Phosphorite: New York, Willey, p. 42–62.

Liu, J., Yang, Z., Yan, X., Ji, D., Yang, Y., and Hu, L., 2015, Modes of occurrence of highly-elevated trace elements in super-high-organic-sulfur Coals: Fuel, v. 156, p. 190–197. doi:10.1016/j.fuel.2015.04.034

Liu, Y.-H., Yang, C., Li, P.-Y., and Li, S.-Q., 2010, A new process of extracting vanadium from stone coal: International Journal Miner Metall Materials, v. 17, p. 381–388. doi:10.1007/s12613-010-0330-8

Liu, Z., Dai, H., Liu, J., and Ge, Z., 2016, Suggestions and current situation of exploration, development, and utilization of stone-coal resources in China: China Mining Magazine, v. 25, no. suppl. 1, p. 18–21. [in Chinese with English abstract.]

Long, J., and Luo, K., 2017, Trace element distribution and enrichment patterns of Ediacaran-early Cambrian, Ziyang selenosis area, Central China: Constraints for the origin of Selenium: Journal of Geochemical Exploration, v. 172, p. 211–230. doi:10.1016/j.gexplo.2016.11.010

Lott, D.A., Coveney, R.M., Murowchick, J.B., and Grauch, R.I., 1999, Sedimentary exhalative nickel–molybdenum ores in South China: Economic Geology, v. 94, p. 1051–1066. doi:10.2113/gsecongeo.94.7.1051

Lu, X., and Zhu, L., 1983, Microfossils and ultramicrofossils in the Early Paleozoic highly metamorphosed boghead coal from western Hunan province: Geological Reviews, v. 29, p. 451. [in Chinese.]

Lu, Z., 2002, Investigation and industrial practice on extraction of V_2O_5 from stone coal containing vanadium by acid process: Hydrometallurgy of China, v. 21, no. 4, p. 175–183. [in Chinese with English abstract.]

Luo, K., 2003, The age of rock distribution in the selenosis region, South Shaanxi Province: Geological Review, v. 49, no. 4, p. 383–388. [in Chinese with English abstract.]

Luo, K., 1996, Distribution and environmental and health effects of Se, As, and F in stone coals of the Early Paleozoic age in northern Dabashan area, in Ren, D., Zhao, F., Dai, S., Zhang, J., and Luo, K., eds., 1996, Geochemistry of Trace Elements in Caol: Beijing, Scientific Press, p. 486–507. [in Chinese with English abstract.]

Luo, K., Su, W., Du, M., and Lei, F., 1995, Some microelements of lower Paleozoic stone coal in South Qinling: Journal of Xi'an Mining Institute, v. 15, no. 2, p. 131–135. [in Chinese with English abstract.]

Luo, K., Yang, J., Chen, D., and Zhang, Z., 1996, Preliminary study on distribution regularity of fluorosis in Ziyang County, Shaanxi Province: Shaanxi Environment, v. 3, no. 2, p. 25–28. [in Chinese.]

Mao, D., Su, H., Yan, L., Wang, Y., and Yu, X., 1990, Investigation and analysis of endemic selenois in Exi Autonomous Prefecture of Hubei province: Chinese Journal of Endemiology, v. 9, p. 311–314. [in Chinese.]

Mao, J.W., Lehmann, B., Du, A.D., Zhang, G.D., Ma, A.D., Wang, Y.T., Zeng, M.G., and Kerrich, R., 2002, Re-Os Dating of Polymetallic Ni-Mo-PGE-Au Mineralization in Lower Cambrian Black Shales of South China and Its Geologic Significance: Economic Geology, v. 97, p. 1051–1061. doi:10.2113/gsecongeo.97.5.1051

Mei, Z., 1985, Review on two Se-rich areas in China: Chinese Journal of Endemic, no. 4, p. 379–385. [in Chinese with English abstract.]

Mou, S.H., Hu, Q.T., and Yan, L., 2007, Progress of researches on endemic selenosis in Enshi District, Hubei Province: Chinese Journal of Public Health, v. 23, no. 1, p. 95–96. [in Chinese.]

Murowchick, J.B., Coveney, R.M., Grauch, R.I., Eldridge, C.S., and Shelton, K.L., 1994, Cyclic variations of sulfur isotopes in Cambrian stratabound Ni-Mo-(PGE-Au) ores of southern China: Geochimica et Cosmochimica Acta, v. 58, p. 1813–1823. doi:10.1016/0016-7037(94)90538-X

Murray, R.W., 1994, Chemical criteria to identify the depositional environment of chert: general principles and applications: Sedimentary Geology, v. 90, no. 3–4, p. 213–232. doi:10.1016/0037-0738(94)90039-6

O'Keefe, J.M.K., Bechtel, A., Christanis, K., Dai, S., DiMichele, W. A., Eble, C.F., Esterle, J.S., Mastalerz, M., Raymond, A.L.,

Valentim, B.V., Wagner, N.J., Ward, C.R., and Hower, J.C., 2013, On the fundamental difference between coal rank and coal type: International Journal of Coal Geology, v. 118, p. 58–87. doi:10.1016/j.coal.2013.08.007

Orberger, B., Vymazalova, A., Wagner, C., Fialin, M., Gallien, J.P., Wirth, R., Pasava, J., and Montagnac, G., 2007, Biogenic origin of intergrown Mo-sulphide- and carbonaceous matter in Lower Cambrian black shales (Zunyi Formation, southern China): Chemical Geology, v. 238, p. 213–231. doi:10.1016/j.chemgeo.2006.11.010

Pašava, J., Kříbek, B., Žák, K., Vymazalova, A., Deng, H., Luo, T., Ch., L., and Zeng, M., 2004, New data on the origin of Lower Cambrian Mo-Ni-PGE black shales in Guizhou Province, south China, Proceedings 32nd IGC, Florence, Italy. pp. A189–15.

Peng, X., Liu, J., Xie, W., Wang, Y., and Su, Z., 2011, The technology for extraction and separation of Ni and Mo from stone coals using vacuum carbon thermal reduction. Patient No. CN201110275576.5. [in Chinese.]

Peng, X.-G., 2015, A preliminary analysis of the potential utilization of Se resources in Enshi, Hubei Province: Resources Environment & Engineering, v. 29, no. 4, p. 436–441. [in Chinese.]

Rietveld, H.M., 1969, A profile refinement method for nuclear and magnetic structures. Journal Of Applied Crystallography, v. 2, p. 65-71. doi:10.1107/S0021889869006558

Ruan, C.-D., and Ward, C.R., 2002, Quantitative X-ray powder diffraction analysis of clay minerals in Australian coals using Rietveld methods: Applied Clay Science, v. 21, p. 227–240. doi:10.1016/S0169-1317(01)00103-X

Seredin, V.V., 2012, From coal science to metal production and environmental protection: a new story of success: International Journal of Coal Geology, v. 90-91, p. 1–3. doi:10.1016/j.coal.2011.11.006

Song, C., 1989, A brief description of the Yutangba sedimentary type Se mineralized area in southwestern Hubei: Mineral Deposits, v. 8, p. 83–88. [in Chinese with English abstract.]

Steiner, M., Wallis, E., Erdtmann, B.-D., Zhao, Y.L., and Yang, R. D., 2001, Submarine–hydrothermal exhalative ore layers in black shales from South China and associated fossils - insights into a Lower Cambrian facies and bio-evolution: Palaeogeography, Palaeoclimatology, Palaeoecology, v. 169, p. 165–191. doi:10.1016/S0031-0182(01)00208-5

Taylor, J.C., 1991, Computer programs for standardless quantitative analysis of minerals using the full powder diffraction profile: Powder diffract, v. 6, p. 2-9. doi:10.1017/S0885715600016778

Taylor, S.R., and McLennan, S.M., 1985, The Continental Crust: Its Composition and Evolution: Oxford, Blackwell, 312 p.

Tian, H., Shuai, Q., Xu, S., Bao, Z., and Xie, S., 2014, Novel technology of preparation of crudes selenium from Se-rich stone coal: Earth Science-Journal of China University of Geosciences, v. 39, no. 7, p. 880–888. [in Chinese with English abstract.]

Tian, X., and Luo, K., 2017b, Selenium, arsenic and molybdenum variation and bio-radiation in the Ediacaran-Cambrian interval: Precambrian Research, v. 292, p. 378–385. doi:10.1016/j.precamres.2017.02.007

Tian, X.L., and Luo, K.L., 2017a, Distribution and enrichment patterns of selenium in the Ediacaran and early Cambrian strata in the Yangtze Gorges area, South China: Science China Earth Sciences, v. 60, no. 7, p. 1268–1282. doi:10.1007/s11430-016-9045-1

Tovarovsky, I.G., Bolshakov, V.I., and Lyalyuk, V.P., 2011, Alternative coke saving technologies are the prospect of blast-furnace practice development: Metallurgical and Mining Industry, v. 3, no. 2, p. 33–35.

Vidanovic, N., Tokalic, R., Ognjanovic, S., Savic, L., and Savic, L., 2011, Techno-economic assessment of cost-effectiveness of boron minerals exploitation: Technics Technologies Education Management, v. 6, no. 4, p. 1053–1057.

Wang, F., Zhang, Y.-M., Huang, J., Liu, T., Wang, Y., Yang, X., and Zhao, J., 2013, Mechanisms of aid-leaching reagent calcium fluoride in the extracting vanadium processes from stone coal: Rare Metals, v. 32, p. 57–62. doi:10.1007/s12598-013-0013-5

Wang, F., Zhang, Y.-M., Liu, T., Huang, J., Zhao, J., Zhang, G.-B., and Liu, J., 2014a, Comparison of direct acid leaching process and blank roasting acid leaching process in extracting vanadium from stone coal: International Journal of Mineral Processing, v. 128, p. 40–47. doi:10.1016/j.minpro.2013.12.010

Wang, L., Sun, W., Liu, R.-Q., and Gu, X.-C., 2014b, Flotation recovery of vanadium from low-grade stone coal: Transactions of Nonferrous Metals Society of China, v. 24, no. 4, p. 1145–1151. doi:10.1016/S1003-6326(14)63173-3

Wang, X., Tang, Y., Jiang, Y., Xie, P., Zhang, S., and Chen, Z., 2017, Mineralogy and geochemistry of an organic- and V-Cr-Mo-U-rich siliceous rock of Late Permian age, western Hubei Province, China: International Journal of Coal Geology, v. 172, p. 19–30. doi:10.1016/j.coal.2016.12.006

Ward, C.R., Spears, D.A., Booth, C.A., Staton, I., and Gurba, L.W., 1999, Mineral matter and trace elements in coals of the Gunnedah Basin, New South Wales, Australia: International Journal of Coal Geology, v. 40, p. 281–308. doi:10.1016/S0166-5162(99)00006-3

Ward, C.R., Taylor, J.C., Matulis, C.E., and Dale, L.S., 2001, Quantification of mineral matter in the Argonne Premium Coals using interactive Rietveld-based X-ray diffraction: International Journal of Coal Geology, v. 46, p. 67–82. doi:10.1016/S0166-5162(01)00014-3

Webb, G.E., and Kamber, B.S., 2000, Rare earth elements in Holocene reefal microbialites: a new shallow seawater proxy: Geochimica et Cosmochimica Acta, v. 64, p. 1557–1565. doi:10.1016/S0016-7037(99)00400-7

Wu, C., Chen, Q., and Lei, J., 1999a, The genesis factors and organic petrology of black shale series from the upper Sinian to the lower Cambrian, southwest of China: Acta Petrologica Sinica, v. 15, no. 3, p. 453–462. [in Chinese with English abstract.]

Wu, C., Yang, C., and Chen, Q., 1999b, The origin and geochemical characteristics of Upper Sinain and Lower Cambrian Black Shales in Western Hunan: Acta Petrologica Et Mineralogica, v. 18, no. 1, p. 26–39. [in Chinese with English abstract.]

Yang, G., Wang, S., Zhou, R., and Sun, S., 1983, Endemic selenium intoxication of humans in China: The American Journal of Clinical Nutrition, v. 37, p. 872–881.

Yang, G., Zhou, R., Sun, S., and Wang, S., 1981, Research on the etiology of an endemic disease characterized by loss of nails and hair in Enshi county. Journal of the Chinese Academy Medicine, v. 3, no: Suppl, v. 2, p. 1–6. [in Chinese.]

Zeng, R., Zhuang, X., Koukouzas, N., and Xu, W., 2005, Characterization of trace elements in sulphur-rich Late

Permian coals in the Heshan coal field, Guangxi, South China: International Journal of Coal Geology, v. 61, p. 87–95. doi:10.1016/j.coal.2004.06.005

Zhang, H., Li, X., Hao, Q., He, D., and Zhuang, J., 2003, Study on Coal in China by Scanning Electron Microscope: Geological Press, Beijing, China. 27 p. [in Chinese with English abstract.]

Zhang, J., Amakawa, H., and Nozaki, Y., 1994, The comparative behaviors of yttrium and lanthanides in the seawater of the north Pacific: Geophys Researcher Letters, v. 21, p. 2677–2680. doi:10.1029/94GL02404

Zhang, X.Y., Yang, K., Tian, X.D., and Qin, W.Q., 2011b, Vanadium leaching from carbonaceous shale using fluosilicic acid: International Journal of Mineral Processing, v. 100, p. 184–187. doi:10.1016/j.minpro.2011.04.013

Zhang, Y., Bao, S., Liu, T., Huang, J., and Chen, T., 2015, Research Status and Prospect of Vanadium Extraction from Stone Coal in China: Nonferrous Metals (Extractive Metallurgy), no. 2, p. 24–30. [in Chinese with English abstract.]

Zhang, Y.-M., Bao, S.-X., Liu, T., Chen, T.-J., and Huang, J., 2011a, The technology of extracting vanadium from stone coal in China: history, current status and future prospects: Hydrometallurgy, v. 109, no. 1–2, p. 116–124. doi:10.1016/j.hydromet.2011.06.002

Zheng, B., Hong, Y., Zhao, W., Zhou, H., and Xia, W., 1992, Se-rich carbonaceous-siliceous rocks of West Hubei and local Se poisoning: Chinese Science Bulletin, v. 37, p. 1027–1029.

Zheng, B., Wang, B., and Finkelman, R.B., 2010, Medical geology in China: Then and now. In medical geology: A regional synthesis, in Selinus, O., Centeno, J.A., and Finkelman, R.B., eds., Medical geology – A regional synthesis: New York, Springer, p. 303–327.

Zheng, B., Yan, L., Mao, D., and Thornton, I., 1993, The Se resource in southwestern Hubei province, China, and its exploitation strategy: Journal of Natural Resources, v. 8, p. 204–212. [in Chinese.]

Zheng, X., Qian, H.D., and Wu, X.M., 2006, Geochemical and genetic characteristics of selenium ore deposit in Shuanghe, Enshi, Huibei Province: Geological Journal China University, v. 12, no. 1, p. 83–92. [in Chinese with English abstract.]

Zhou, Z., 1981, Discussion on radiation protection in mining and utilization of uranium-bearing coal: Radiation Protection, no. 5, p. 74–78. [in Chinese.]

Zhu, J., Li, S.H., Zuo, W., Sykorova, I., Su, H.C., Zheng, B.-S., and Pesek, J., 2004a, Mode of occurrence of selenium in black Se-rich rocks of Yutangba: Geochimica, v. 33, no. 6, p. 634–640. [in Chinese with English abstract.]

Zhu, J., Wang, N., Li, S., Li, L., Su, H., and Liu, C., 2008, Distribution and transport of selenium in Yutangba, China: Impact of human activities: Science of the Total Environment, v. 392, p. 252–261. doi:10.1016/j.scitotenv.2007.12.019

Zhu, J., and Zheng, B., 2001, Distribution of selenium in a mini-landscape of Yutangba, Enshi, Hubei Province, China: Applied Geochemistry, v. 16, p. 1333–1344. doi:10.1016/S0883-2927(01)00047-6

Zhu, J., Zuo, W., Liang, X., Li, S., and Zheng, B., 2004b, Occurrence of native selenium in Yutangba and its environmental implications: Applied Geochemistry, v. 19, p. 461–467. doi:10.1016/j.apgeochem.2003.09.001

Zhu, L., 1983, Petrography of Early Paleozoic highly metamorphosed boghead coal and its geological significance: Geological Reviews, v. 29, p. p 245–261. [in Chinese with English abstract.]

Zhu, X.B., Zhang, Y.M., Huang, J., Liu, T., and Wang, Y., 2012, A kinetics study of multi-stage counter-current circulation acid leaching of vanadium from stone coal: International Journal Miner Processing, v. 114–117, p. 1–6. http://www.esdasz.org/ (accessed August 2017 21).

ARTICLE

CO$_2$ storage in coal to enhance coalbed methane recovery: a review of field experiments in China

Zhejun Pan, Jianping Ye, Fubao Zhou, Yuling Tan, Luke D. Connell and Jingjing Fan

ABSTRACT

Coal reservoirs especially deep unminable coal reservoirs, are viable geological target formations for CO$_2$ storage to mitigate greenhouse gas emissions. An advantage of this process is that a large amount of CO$_2$ can be stored at relatively low pressure, thereby reducing the cost of pumping and injection. Other advantages include the use of existing well infrastructure for CO$_2$ injection and to undertake enhanced recovery of coalbed methane (ECBM), both of which partially offset storage costs. However, ECBM faces difficulties such as low initial injectivity and further permeability loss during injection. Although expensive to perform, ECBM field experiments are essential to bridge laboratory study and large-scale implementation. China is one of the few countries that have performed ECBM field experiments, testing a variety of different geological conditions and injection technologies. These projects began more than a decade ago and have provided valuable experience and knowledge. In this article, we review past and current CO$_2$ ECBM field trials in China and compare with others performed around the world to benefit ECBM research and inform future projects. Key aspects of the ECBM field projects reviewed include the main properties of target coal seams, well technologies, injection programmes, monitoring techniques and key findings.

1. Introduction

Coal reservoirs, especially those that are deep and unmineable, are viable target formations for CO$_2$ storage to mitigate greenhouse gas emissions (e.g. White et al. 2005; Pan and Connell 2010; Pashin 2016). Compared with other target formations, coal can store a significant amount of CO$_2$ at relatively low pressure via adsorption, therefore reducing compression and injection costs. Other economic benefits include enhanced coalbed methane recovery (ECBM) and the use of existing well infrastructure. Hence, much effort has been directed towards CO$_2$ storage in coal/ECBM research as an important area of carbon capture, utilization, and storage.

The concept of CO$_2$ storage in coal to enhance coalbed methane (CBM) recovery has been considered since the early 1990s (White et al. 2005), following the idea of using N$_2$ for the same purpose (Puri and Yee 1990). Much laboratory work has been conducted to investigate the feasibility of ECBM, initially focusing more on pure gas adsorption behaviour and competitive adsorption behaviour between CO$_2$ and CH$_4$ (e.g.

Arri et al. 1992; Hall et al. 1994; Busch et al. 2003; Fitzgerald et al. 2005), followed by studies on permeability change behaviour due to difference in CO$_2$ and CH$_4$ adsorption-induced coal swelling (e.g. Robertson 2005; Mazumder et al. 2006; Pini et al. 2009; Pan et al. 2010). Core-flooding experiments were also designed to simulate CO$_2$ injection (e.g. Tsotsis et al. 2004; Mazumder and Wolf 2008; Connell et al. 2011; Sander et al. 2014). Modelling studies have also described permeability behaviour during ECBM (e.g. Pekot and Reeves 2002; Shi and Durucan 2005; Cui et al. 2007). Although laboratory experiments can provide knowledge of the ECBM process, field conditions are more complex (van Bergen et al. 2009). Therefore, many field tests have been performed in parallel with the laboratory studies to test CO$_2$ storage and ECBM behaviour under real-world conditions.

The first field ECBM test was carried out in December 1993 by Amoco in the San Juan Basin of Colorado; Meridian also conducted a CO$_2$ ECBM pilot in the San Juan Basin in 1995. However, no results for these tests were released (Gunter et al. 1997). The largest-scale CO$_2$

ECBM field trial was performed in the Allison Unit in San Juan Basin from 1995 to 2001 (Reeves et al. 2003). About 336,000 t of CO_2 were injected into four wells, with the high CO_2 injection rate facilitated by the high permeability of coal in the San Juan Basin Fairway (Reeves 2009). Since then, more than a dozen ECBM field trials have been orchestrated at much smaller scales in low-permeability coal reservoirs around the world.

As an important research topic in the CO_2 storage area, ECBM has been reviewed many times. The first comprehensive review looked at factors such as CO_2 storage capacity of the coal and coal properties relevant to ECBM (White et al. 2005). A short review of the ECBM process and key technical aspects, especially typical chemical engineering phenomena such as adsorption, swelling and permeability, was provided by Mazzotti et al. (2009). Field tests up to 2004 were briefly reviewed in the same paper, focusing on reservoir simulation modelling. These included field tests at San Juan Basin, United States (1995–2001), Fenn-Big Valley, Canada (1999), Upper Silesian Basin, Poland (2004), Ishikari Coal Field, Japan (2004), and South Qinshui Basin, China (2004). Liu et al. (2009) reviewed the same field projects as Mazzotti et al. (2009), but with a focus on flow behaviour and permeability loss. More recently, Li and Fang (2014) reviewed ECBM technology, including a brief review on the status of field projects with the focus on the injection programme. Godec et al. (2014) reviewed ECBM field projects carried out in the United States, including a few small-scale projects carried out under regional carbon capture and storage partnerships. Other more topical reviews relate to the combination of ECBM with CBM. For example, Busch and Gensterblum (2011) reviewed gas adsorption and diffusion processes related to the CBM and ECBM, while Pan and Connell (2012) reviewed coal permeability modelling work related to CBM and ECBM processes as well as ECBM field data with a focus on permeability modelling testing.

To date, no detailed reviews have been undertaken of ECBM field projects carried out in China that focus on ECBM injection programmes and coal reservoir properties. This is partly because the details of some field tests have been published only in Chinese-language literature, thus limiting access by international researchers. In this article, we provide a detailed review of ECBM field experiments in China in chronological order. Key aspects of the projects reviewed include the properties of the target coal seams, well technologies, injection programmes, monitoring techniques and key findings. We also compare the Chinese projects with others around the world, including aspects such as CO_2 injectivity and enhanced methane recovery.

Note that this article focuses on reviewing field trials of CO_2 ECBM using CBM wells. Enhanced gas drainage in coal mines using CO_2 or other gases is not reviewed. ECBM using other gases, such as the San Juan Tiffany N_2-ECBM field trial (Reeves and Oudinot 2005b), is likewise not considered. We acknowledge that other Chinese ECBM projects are underway, such as the Huaneng ECBM trial (Li and Fang 2014), but as yet, no reports are available for review.

2. CO_2 ECBM field projects in China

To date, four CO_2 ECBM field projects in China have been completed; three in the Qinshui Basin and one at the eastern margin of Ordos Basin. Although the projects are in two different basins, they are all located in Shanxi Province. The Qinshui Basin and the eastern margin of Ordos Basin are China's main CBM production basins. The detailed geology of these two coal basins, as well as other Chinese coal basins, is reviewed by Qin et al. (2017) in this Special Issue and is therefore not repeated in this article. All four ECBM field projects were performed by or in partnership with China United Coalbed Methane Co. Ltd (CUCMB); their locations are given in Figure 1. Note that in each project described below, the date given is the CO_2 injection date, not the project starting date.

2.1. Qinshui Basin ECBM (2004)

The first ECBM field trial in China was a collaborative, micro-scale, single-well project between China and Canada, operated by CUCBM and Alberta Research Council. The project was funded under the Canadian Climate Change Development Fund to develop ECBM technology in China (Wong et al. 2007). The purpose of the single-well test was mainly to quantify reservoir properties, particularly injectivity, sorption rate and permeability changes; the results would inform a decision about whether to extend to multi-well trials. The details of the project were published by Wong et al. (2007) and Ye et al. (2007). The key elements of the project are summarized in the following.

The field project was in the Southern Shizhuang CBM Block, which is located in the Southeastern Qinshui Basin (Figure 1). The target formation was the Permian Shanxi formation #3 coal seam, which is anthracite with a vitrinite reflectance of 3.4% in this area; the target coal seam at the injection well is 472.3–478.7 m deep and the coal is about 6.3 m thick (Ye et al. 2007). The gas adsorption capacity of this coal is relatively high for CH_4, with a Langmuir volume of 28.9 m^3/t and Langmuir pressure of 0.98 MPa; Langmuir volume for CO_2 adsorption is not much higher at 31.3 m^3/t, and Langmuir pressure is 0.35 MPa, all on a dry-ash-free basis (Wong et al. 2007). The CO_2/CH_4 adsorption ratio at about 1.3 MPa is approximately 1.5 (Wong et al. 2007), which is consistent with published results of high-rank coals usually having a low CO_2/CH_4 adsorption ratio. The permeability of the coal reservoir was 0.95 md in Ye et al.

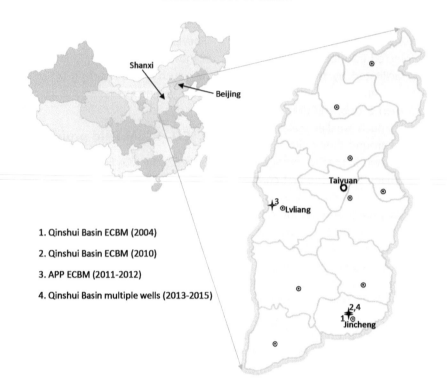

Figure 1. Location of the enhanced recovery of coalbed methane (ECBM) field trial sites in China.

1. Qinshui Basin ECBM (2004)

2. Qinshui Basin ECBM (2010)

3. APP ECBM (2011-2012)

4. Qinshui Basin multiple wells (2013-2015)

(2007) but was estimated at 12.6 md absolute and 1.8 md effective for gas in Wong *et al.* (2007).

The vertical injection well, TL-003, was hydraulically fractured for CBM production in March 1998 (Ye *et al.* 2007). The well was completed in the #15 coal seam of the Carboniferous Taiyuan formation and the #3 coal seam of the Permian Shanxi formation. Before the well was converted for CO_2 injection in 2004, it produced about 1 million m^3 of gas and about 35,000 m^3 of water, with a substantial shut-in period from April 1999 to December 2000 (Wong *et al.* 2007). As the water rate averaged 30–40 m^3/d and was believed to be produced from a sandstone formation just above the lower #15 coal seam, a bridge plug was placed just below the #3 coal seam and gas was produced from the #3 coal seam from October 2003 for the baseline production rate (Wong *et al.* 2007). The original reservoir pressure was about 3.36 MPa in the #3 coal seam, but dropped to 0.96 MPa after 4 years of CBM production (Ye *et al.* 2007). CO_2 was injected in April 2004 with 13 injection cycles, with about 13–16 t of CO_2 injected within 8–10 h in each cycle; in total, 192.8 t of CO_2 was injected (Ye *et al.* 2007). After each injection, the well was shut in until CO_2 was delivered by trucks the next day. Figure 2 shows the bottomhole pressure (BHP) during and after CO_2 injection. BHP increased from 1.19 MPa after the first day of injection to 3.27 MPa after the 13th day of injection, and the maximum BHP reached 6.7 MPa during the 13th and final day of injection. The BHP then dropped to 2.98 MPa after 1 day and to about 2.0 MPa after 2 weeks

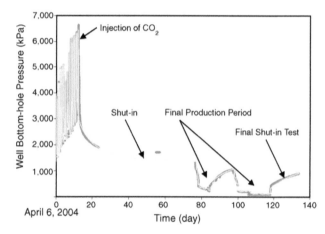

Figure 2. Well pressure behaviour during and after CO_2 injection in Qinshui Basin enhanced recovery of coalbed methane (2004) (from Wong et al. 2007).

(Ye *et al.* 2007). The BHP dropped to 1.06 MPa 63 days after injection was completed (Ye *et al.* 2007), showing the process of gas being adsorbed to the matrix of the coal.

Injection pressure and rate, pre and post-injection production rate, BHP, and gas composition were monitored (Ye *et al.* 2007). The gas and water production rates before and after CO_2 injection are shown in Figure 3. Before CO_2 injection, the gas rate was between 218 and 824 m^3/d with an average of 490 m^3/d (Ye *et al.* 2007), noting that this was produced from the #3 coal seam only. The water rate was below 1 m^3/d before CO_2 injection. After CO_2 injection, the gas rate was about 998–1466 m^3/d and averaged

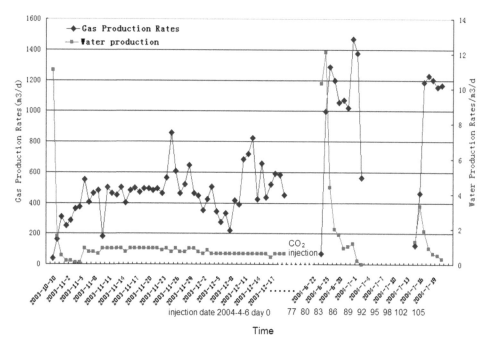

Figure 3. Gas and water production rate before and after CO_2 injection in Qinshui Basin enhanced recovery of coalbed methane (2004) (Ye and Zhang 2016) (6 April 2004 as day 0).

1186 m^3/d; the increase of the gas rate was partially because the produced gas was a mixture of CO_2 and CH_4. The composition of the produced gas is shown in Figure 4, which shows that initial CO_2 composition was about 60–70%, but dropped to about 43% after 1.5 months of production (Ye et al. 2007). The gas-production rate was about 1200 m^3/d 2 years after CO_2 injection, and it was mainly CH_4 (Ye et al. 2007). However, the reason for the increase was not provided by the authors. Produced water chemistry was also analysed; the pH was 8.8 and the mineral content was 1223 mg/L before CO_2

injection, changing to 7.0 and 3086 mg/L, respectively, after CO_2 injection (Ye et al. 2007). The increased acidity and mineral content was directly related to CO_2 injection.

The project achieved its goal of demonstrating that CO_2 injection into a coal seam was feasible at this scale (Ye et al. 2007). The rise in BHP during CO_2 injection showed permeability loss due to coal swelling was incremental when CO_2 was displacing CH_4, and the slow BHP fall-off demonstrated that CO_2 diffusion/adsorption is slow in the matrix of high-rank coal.

2.2. Qinshui Basin ECBM (2010)

This project was also operated by CUCBM; the field trial site was in Northern Shizhuang CBM block, Southern Qinshui Basin (Figure 1). This site is to the north of the ECBM field trial described in Section 2.1. This field trial was the same 'huff-and-puff' test as the 2004 project described in the previous section, with the purpose of testing ECBM technology in a deeper coal seam. This project was summarized in the Chinese literature by Ye et al. (2012), and the key elements of the project are summarized in the following.

The target CO_2 injection formation was the Permian Shanxi formation #3 coal seam. The Permian Shanxi formation #3 coal seam in this area is anthracite with vitrinite reflectance of 2.7%; the coal seam at the injection well SX-001 is about 923 m deep and the coal seam is about 6.05 m thick. Reservoir pressure in #3 seams was about 2.4–6.1 MPa. The permeability of the coal reservoir ranges from

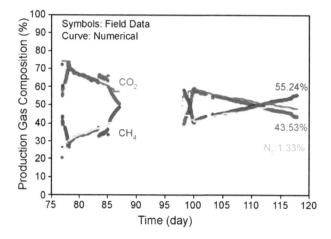

Figure 4. Post-injection produced gas composition in Qinshui Basin enhanced recovery of coalbed methane (2004) (6 April 2004 as day 0) (from Wong et al. 2007; note that the numerical result was part of, Wong et al.'s 2007 work and is not discussed in this review article).

0.8 to 0.002 md based on well-test results from two CBM wells in the area (Ye *et al.* 2012).

The SX-001 well was hydraulically fractured and put into production from January to March 2010. It was then converted for injection in April 2010, and 233.6 t of liquid CO_2 were injected during a 2-month period with 17 CO_2 injection days (Ye *et al.* 2012). The last CO_2 injection date was 13 May 2010. During the injection, the BHP increased from 4.0 to 7.0 MPa (Ye *et al.* 2012), as shown in Figure 5. The well was shut in for CO_2 to be adsorbed and was put back into production in late June 2010 (Ye *et al.* 2012). The gas-production rate before and after CO_2 injection (Figure 6) was below 170 m³/d and averaged about 80 m³/d before injection; average water production rate was 18 m³/d. It was

suspected that a hydraulic fracture may have connected the sandstone aquifer above the #3 coal seam (Ye *et al.* 2012). After CO_2 injection, the gas rate increased to 2.45 times that before injection. This was partly attributed to the production of a mixture of CO_2 and CH_4, whose composition change is shown in Figure 7. The CO_2 composition was about 70% initially and dropped to about 30% after 5 months of production, demonstrating that CO_2 had adsorbed to the coal during the shut-in period (Ye *et al.* 2012). Moreover, because the pre-injection production period was short, the production rate was still peaking.

No other monitoring was reported in this project, but reservoir simulation was carried out. The authors suggested that the hydraulic fracture connecting the coal seam and

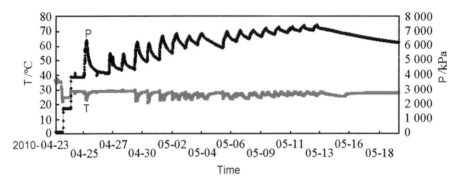

Figure 5. Well pressure behaviour during CO_2 injection in Qinshui Basin enhanced recovery of coalbed methane (2010) (from Ye *et al.* 2012).

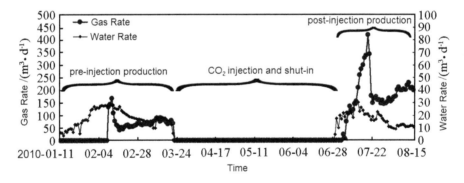

Figure 6. Gas and water production rates in Qinshui Basin enhanced recovery of coalbed methane (2010) (from Ye *et al.* 2012).

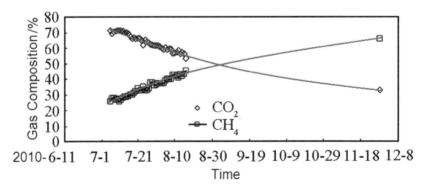

Figure 7. Post-injection produced gas composition in Qinshui Basin enhanced recovery of coalbed methane (2010) (from Ye *et al.* 2012).

the above aquifer may cause high-pressure CO_2 injection into the aquifer formation. They also concluded that single-well huff-and-puff field experiments are useful to evaluate CO_2 storage behaviour in coal, but that it is difficult to evaluate the effect of enhanced methane production. Therefore, a multi-well field test was suggested for ECBM purposes (Ye *et al.* 2012).

2.3 APP ECBM (2011–2012)

In this project, a trial was carried out at Liulin CBM block in the east margin of the Ordos Basin (Figure 1). This was a collaborative project between China and Australia under the Asia–Pacific Partnership on Clean Development and Climate (APP). The main aim of the project was to investigate CO_2 injectivity and transport behaviour in a horizontal injection well. The details of the project are provided in Connell *et al.* (2014) and Pan *et al.* (2013), and the key elements of the project are summarized in the following.

The target coal seam was the Permian #3 + 4 coal seam. The target coal seam is the same age as targeted in the previous two field trials, and the rank of the coal is bituminous with a vitrinite reflectance of 1.6%. Langmuir volumes for CH_4 and CO_2 are 30.1 and 46.3 m^3/t, respectively, and Langmuir pressures for CH_4 and CO_2 are 2.1 and 0.6 MPa, respectively, measured on dry samples at 35°C, which is higher than the reservoir temperature of 20°C. The depth at the vertical part of the multilateral horizontal injection well is 560.8–565.5 m with a net coal thickness of about 4.7 m and a permeability of about 0.6 md (Connell *et al.* 2014).

Based on experience of previous ECBM tests in China and from the globe, CO_2 injectivity often significantly reduces as CO_2 adsorption to displace CH_4 induces extra coal swelling, which reduces permeability, especially near the wellbore. Therefore, the purpose of this project was to investigate CO_2 injectivity using a multilateral horizontal well, which has a large contact area with the coal seam. The multilateral well used in this project has four branches with a total in-seam length of about 2305 m; the trajectory of the well in the seam is shown in Figure 8 (Connell *et al.* 2014). The diameter of the horizontal well is 120 mm. Therefore, the well can provide a contact area with coal of up to 870 m^2, assuming the entire well length was open and in-seam. A vertical monitoring well was drilled about 25 m away from the injection well, from which gas and water samples were collected using a U-tube system from three perforated zones including the target coal seam, #5 coal seam 2.9 m below, and a sandstone formation 4 m above the target coal seam (Figure 8). During the project, gas and water samples were collected daily from the monitoring well using the U-tube system from the three perforated zones. Pressure was also measured.

The multilateral horizontal well was in production for about 1 year, and was then converted to injection well in September 2011. The injection pressure and rate are shown in Figures 9 and 10, respectively. Before injection, the reservoir pressure was about 2.1 MPa; the maximum injection pressure at bottomhole was kept below 5.5 MPa to avoid fracturing the coal. Injection commenced on 15 September 2011 and ceased in March 2012. In total, there were 69 tanker deliveries to the site and 460 t of CO_2 was injected

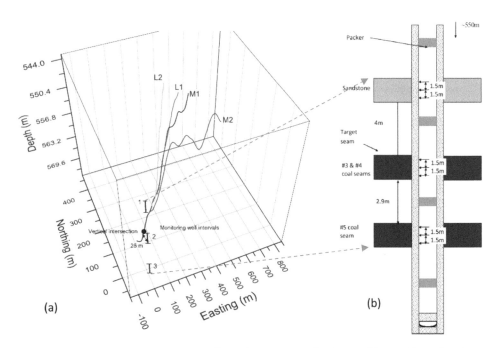

Figure 8. Multilateral injection well and monitoring well used in Asia–Pacific Partnership on Clean Development and Climate enhanced recovery of coalbed methane (2011–2012) (from Connell *et al.* 2014).

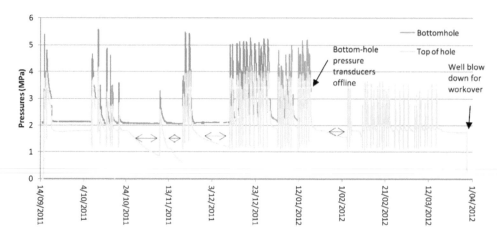

Figure 9. Injection well pressure during CO_2 injection in Asia–Pacific Partnership on Clean Development and Climate enhanced recovery of coalbed methane (2011–2012) (from Connell *et al.* 2014).

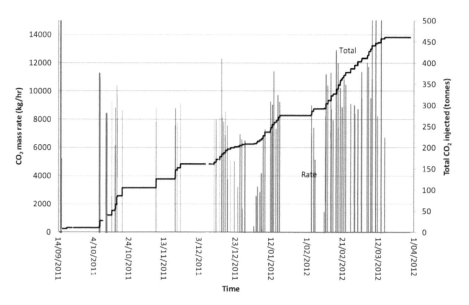

Figure 10. CO_2 injection rate and cumulative in Asia–Pacific Partnership on Clean Development and Climate enhanced recovery of coalbed methane (2011–2012) (from Connell *et al.* 2014).

during the trial. The average injection rate was 3.44 t/h, and the maximum rate was 6.9 t/h, as shown in Figure 10 (Connell *et al.* 2014). The injection rate was higher than those in the previous two field tests and was achieved at much lower injection pressure, demonstrating that a horizontal well can significantly improve injectivity. Note that the horizontal well itself had a volume of about 26.0 m³, assuming the entire well was open; therefore, part of the reason for the high injectivity could be a storage effect in the horizontal well. The well was initially filled with water, and after early injection cycles, some water may have been pushed back to the coal seam. Therefore, the well itself can provide storage space for injected high-density CO_2. However, this should be analysed further, as the storage and phase change behaviour of water and CO_2 are complex; for example, the bottomhole temperature fluctuated between 0°C and 20°C and the pressure swung between 2

and 5 MPa. This is evidenced by the analysis showing that the injectivity improved in December 2011 where injection events were closely spaced, although it could also be that there were fewer two-phase (gas–water) flow effects, with the gas from the closely spaced injection events progressively displacing water further from the injection well (Connell *et al.* 2014).

Gas compositions monitored from the monitoring well are shown in Figure 11. The main peak of CO_2 arrived in the monitoring well after about 200 t of CO_2 was injected. CO_2 composition remained high over the injection period and then dropped after injection was completed. To study the connectivity between the wells sulfur hexafluoride (SF_6) tracer gas was injected into the injection well with CO_2 in September 2011 and March 2012. Because SF_6 is not adsorbed to coal and is slightly soluble in water, the peak of SF_6 arrival time in the monitoring well was about 8.6 and

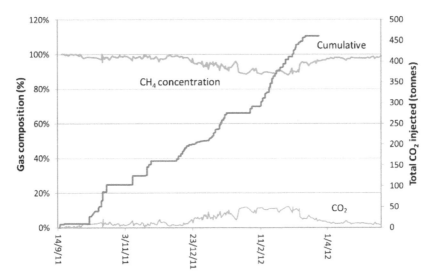

Figure 11. Gas composition from the monitoring well in Asia–Pacific Partnership on Clean Development and Climate enhanced recovery of coalbed methane (2011–2012) (from Connell *et al.* 2014).

3.3 days for the two SF_6 injections (Figure 12), and was much faster than the CO_2 arrival time. The different arrival time of the two SF_6 injections may be related to the amount of CO_2 injected after the tracer (Connell *et al.* 2014).

The multi-lateral horizontal well was put into production from September 2012. The gas composition of the produced gas was measured, but this ceased prior to the second workover on 16 February 2013. CO_2 composition in the produced gas slowly declined during the production period, averaging only about 10% (Connell *et al.* 2014), while CO_2 composition in the produced gas dropped from about 70 to 30% in the previous two tests, showing that CO_2 was adsorbed to coal more, as the injection and shut-in periods were both much longer than those in the previous tests. Up until 16 February 2013, the cumulative amount of CO_2 produced from the well was about 26 t, approximately 6% of that injected (Connell *et al.* 2014).

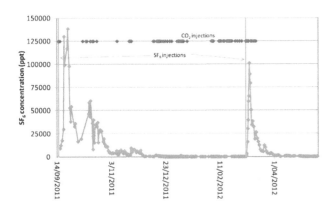

Figure 12. Monitoring well SF_6 concentration in Asia–Pacific Partnership on Clean Development and Climate enhanced recovery of coalbed methane (2011–2012) (from Connell *et al.* 2014).

2.4 Qinshui Basin multiple wells (2013–2015)

As the single-well tests achieved their goals, CUCBM carried out a multiple-well ECBM project to further investigate the potential and technology of CO_2 storage in coal and ECBM. This project aimed to better evaluate the enhanced CBM production from production wells, instead of using the huff-and-puff design. The site of this field project was the same gas field as the second Chinese ECBM field test in Northern Shizhuang CBM block (Figure 1). There have been no publications on this project, although an unpublished project report is available in Chinese (Ye and Zhang 2016).

As in the previous projects, the target formation was also the Permian Shanxi formation #3 coal seam. The coal seam in the trial area is about 5.8–6.8 m thick, with a trend of increasing thickness from the north to the south. Thickness at the injection SX006 well group is about 6.2 m and depth is about 972 m. The initial reservoir pressure was about 6.2–6.7 MPa. Langmuir volumes for CH_4 and CO_2 were 30.53 and 31.31 m^3/t, respectively, and Langmuir pressures for CH_4 and CO_2 were 1.99 and 0.7 MPa, respectively. Reservoir temperature was about 27.5°C. Gas content of the virgin coal seams was about 15.4–18.4 m^3/t; vitrinite reflectance is about 2.6% and the coal is anthracite (Ye and Zhang 2016).

The layout of the wells used in the project is shown in Figure 13. Three were used as injection wells, and eight offset CBM wells were used as the production/monitoring wells. The typical well spacing is about 300 m, and the shortest well spacing is about 200 m between one of the injection wells, SX006-1, and a production well, SX006-6. All wells in this well group were drilled from late 2010 to mid-2012 and hydraulically fractured before production. As the maximum stress direction in the area is north and south, the hydraulic fracture was expected to extend in the same

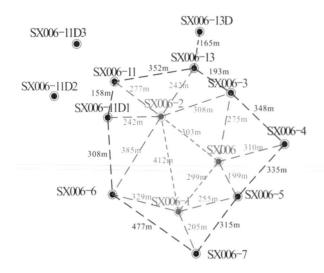

Figure 13. Well pattern of the Qinshui Basin multiple wells enhanced recovery of coalbed methane (2013–2015) (Ye and Zhang 2016); red dots are injection wells and black dots are production/monitoring wells.

direction. It was also evidenced by a large amount of sand observed from well SX006-3 while fracturing well SX006. The average gas-production rate for these wells was about 400 m³/d and water rate about 2 m³/d (Ye and Zhang 2016).

The injection programme had two phases; Phase 1 was a single-well injection using SX006-1, while Phase 2 involved injecting CO_2 into two wells, SX006 and SX006-2.

In Phase 1, well SX006-1 was used for CO_2 injection after production for about 10 months. Phase 1 had two injection stages: Stage 1 was from March 2013 to December 2013 with 103 injection days and 779 t of liquid CO_2 injected; Stage 2 was from April to December 2014 with 80 injection days and 842 t of liquid CO_2 injected. The highest injection pressure at bottomhole was about 19 MPa, which is below the hydraulic

fracturing pressure of the coal seam. The injection rate was 5000–6000 m³/d at standard condition (std); the injection rate and bottomhole pressure are shown in Figure 14.

In Phase 2, CO_2 was first injected into well SX006 from October 2014 to December 2015, with 956 t of liquid CO_2 injected in 90 injection days. The injection rate and BHP are shown in Figure 15. There were three injection stages. Stage 1 was in October 2014; the BHP was controlled between about 5 and 7 MPa as an initial injection test; 73 t CO_2 were injected. Stage 2 was from January to March 2015; the average injection rate was only a few hundred std m³/d and the injection BHP was maintained at 7 MPa. Stage 3 was from June 2015 to December 2015; the average injection rate was about 6000–7000 std m³/d and the injection pressure was also controlled at 7 MPa (Ye and Zhang 2016). The report's authors speculated that in Stage 3, the hydraulic fracture between well SX006 and SX006-3 was the reason for maintaining injection rate without raising the BHP (Ye and Zhang 2016). CO_2 was injected into SX006-2 well from March to November 2015 in 90 injection days. BHP was 5–6 MPa and the injection rate was about 7000 std m³/ d, as shown in Figure 16. Therefore, from June 2015 to November 2015, both well SX006 and SX006-2 were injecting CO_2.

An onsite CO_2 storage tank was installed to ensure more continuous injection in this project, while the previous three projects relied on trucks only. For the whole project, 4491 t of liquid CO_2 were injected. There was no CO_2 breakthrough during the course of injection. A transient electromagnetic method was used to monitor CO_2 migration in the coal, with the results showing that CO_2 migrated 80–120 m away from the injection well. Water samples were also collected from

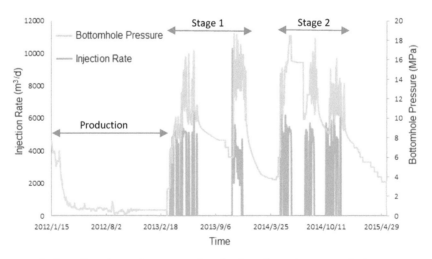

Figure 14. Qinshui Basin multiple wells enhanced recovery of coalbed methane (2013–2015) Phase 1: SX006-1 injection rate and pressure (Ye and Zhang 2016).

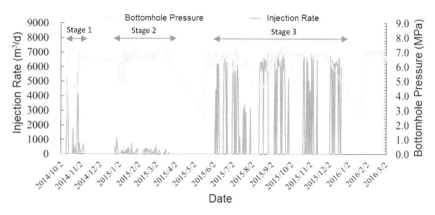

Figure 15. Qinshui Basin multiple wells enhanced recovery of coalbed methane (2013–2015) Phase 2: SX006 injection rate and pressure (Ye and Zhang 2016).

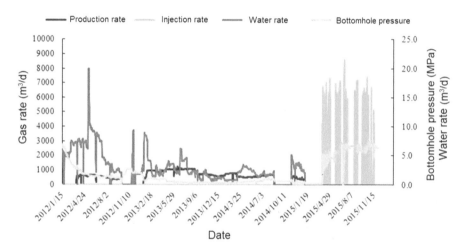

Figure 16. Qinshui Basin multiple wells enhanced recovery of coalbed methane (2013–2015) Phase 2: SX006-2 gas rates and pressure (Ye and Zhang 2016).

the production wells to analyse the changes in CH_4, CO_2 and ion concentrations in the water samples (Ye and Zhang 2016).

3. Discussion

To gain more insight from Chinese field tests and assist future ECBM projects, the coal reservoir properties and injection data are summarized in Tables 1 and 2, respectively. As knowledge from ECBM field trials around the world is also valuable, we have collected and summarized data in two tables to facilitate better comparison. Note that some coal properties were not reported and thus obtained from other sources, so they could be different from the properties at the injection location.

The international projects include eight CO_2-ECBM field projects of different scales with publically available

data in the United States. They are listed in the following and the locations of the projects are shown in Figure 17.

(1) Allison Unit ECBM, San Juan County, New Mexico, San Juan Basin (1995–2000)
(2) Pump Canyon, San Juan County, New Mexico, San Juan Basin (2008–2009), through the Southwest Regional Partnership on Carbon Sequestration
(3) Tanquary Farms, Wabash County, Illinois, Illinois Basin (2008), through the Midwest Geological Sequestration Consortium
(4) Russell County, Virginia, Central Appalachian Basin (2009), through the Southeast Regional Carbon Sequestration Partnership
(5) Burke County, North Dakota, Williston Basin (2009), through the Plains CO_2 Reduction Partnership

Table 1. Summary of coal reservoir properties in enhanced recovery of coalbed methane (ECBM) field projects.

No.	Project name	Coal age	Rank	R_o (%)	Depth (m)	Reservoir temperature (°C)	Coal seam thickness (m)	Permeability (md)	Adsorption Langmuir parameters	References
1	Qinshui Basin ECBM 2004	Shanxi formation #3 coal, Permian	Anthracite	3.35	~472	25	6.3	12.6 (Wong et al. 2007) 0.95 (Ye et al. 2007)	CH_4: 28.9 m³/t; 0.98 MPa CO_2: 31.3 m³/t; 0.35 MPa dry, ash-free basis	Wong et al. (2007); Ye et al. (2007)
2	Qinshui Basin ECBM 2010	Shanxi formation #3 coal, Permian	Anthracite	2.7	~923	24	6.05	0.002–0.8	CH_4:30.5 m³/t; 1.99 MPa CO_2: 31.3 m³/t, 0.7 MPa	Ye et al. (2012)
3	Asia-Pacific Partnership on Clean Development and Climate ECBM, Ordos Basin	Shanxi formation #3 + 4 coal, Permian	Medium volatile bituminous	1.6	~560	20	4.7	0.64	CH_4:30.1 m³/t; 2.1 MPa CO_2: 46.3 m³/t, 0.6 MPa, dry sample at 35°C	Connell et al. (2014)
4	Qinshui Basin multiple wells	Shanxi formation #3 coal, Permian	Anthracite	2.6	~972	27.5	6.22	0.002–0.8	CH_4:30.5 m³/t; 1.99 MPa CO_2: 31.3 m³/t, 0.7 MPa	Ye and Zhang (2016)
5	Allison Unit, San Juan Basin	Fruitland Formation coal seams, Upper Cretaceous	Bituminous	0.42–1.54[a]	945 Top coal depth	48.9	3 seams Upper: 6.7 Middle: 3.0 Lower: 3.4 Total: 13.1	30–150	Upper coal: CH_4: 14.0 m³/t; 3.6 MPa CO_2: 20.4 m³/t; 1.8 MPa Middle coal: CH_4: 9.5 m³/t; 3.3 MPa CO_2: 13.8 m³/t; 1.5 MPa Lower coal: CH_4: 12.3 m³/t; 4.1 MPa CO_2: 18.0 m³/t; 1.8 MPa	Reeves et al. (2003); Reeves and Oudinot (2005a); Reeves (2009); Michael et al. (1993)[a]
6	Pump Canyon, San Juan Basin	Fruitland Formation coal seams, Upper Cretaceous	Bituminous	0.42–1.54[a]	918 Top coal depth	52.2	3 seams Upper: 4.9 Middle: 4.6 Lower: 8.8 Total: 18.3	146–550	Upper coal: CH_4: 18.6–23.9 m³/t; 2.8–4.6 MPa CO_2: 38.5 m³/t; 2.2 MPa Middle coal: CH_4: 17.6–26.6 m³/t; 2.8–5.6 MPa CO_2: 38.8 m³/t; 1.8 MPa Lower coal: CH_4: 17.5–27.8 m³/t; 2.9–4.3 MPa CO_2: 39.8–47.0 m³/t; 1.7–3.4 MPa	Oudinot et al. (2009, 2011); Michael et al. (1993)[a]
7	Tanquary Farms test site, Illinois Basin	Carbondale Springfield Coal Member, Pennsylvanian Late Carboniferous	High volatile bituminous B	0.63	~274	21	~2.1	2–7.0	CH_4: 14.9–15.5 m³/t; 2.99–3.37 MPa CO_2: 41.4–43.8 m³/t, 2.09–2.26 MPa Measured at 20.6°C	Finley (2011; 2012); Morse et al. (2010)
8	Virginia Central Appalachian Basin Coal Test	Pocahontas Formation and Lee Formation, Pennsylvanian, Late Carboniferous	Bituminous	0.54[a]	427–671	23.33[c]	Each about 0.61 – 0.79 m Total 2.07 m[c]	5–20		NETL (2012c); Ripepi (2012); Eble and Greb (2016)[a]; SECARB 2013[c]
9	Lignite Field Validation Test, Williston Basin	Fort Union Group, Tertiary	Lignite	0.17–0.34	335	10–17	~3.0 to 3.6	Up to 1 core permeability	CH_4: 2.9 m³/t CO_2: 28.0 m³/t Measured at 16°C	Botnen et al. (2009)

(Continued)

Table 1. (Continued).

No.	Project name	Coal age	Rank	R_o (%)	Depth (m)	Reservoir temperature (°C)	Coal seam thickness (m)	Permeability (md)	Adsorption Langmuir parameters	References
10	Black Warrior Basin Coal Test	Black Creek, Mary Lee and Pratt coal seams, Pennsylvanian, late Carboniferous	Bituminous	0.9–2.0[a]	Black Creek: 548 Mary Lee: 437 Pratt: 287	Black Creek: 24	<0.1 – >3.0 m[c]	Black creek: 0.2 – 0.1; Pratt: 10–16	Adsorption capacity CO_2:CH_4 about 2:1. CH_4: 15–31 m³/t (on a dry, ash-free basis, 27–30°C)[b]	NETL (2012c); SECARB (2008); Pashin et al. (2015); Pashin (2007)[a]; Pashin et al. (2014)[c]; Pashin (2009)[b]
11	Marshall County project, Northern Appalachian Basin	Upper Freeport coal seam, Pennsylvanian Late Carboniferous	Bituminous	1.25[a]	400–530	20–30[c]	0.3–1.8 m	1	CH_4: 14.1 m³/t; 2.8 MPa CO_2: 27.8 m³/t; 1.7 MPa	Locke and Winschel (2015); Wilson et al. (2012); He et al. (2013); Locke et al. (2011); Ruppert et al. (1991)[a], Valero Garces et al. (1997)[b]
12	Buchanan County, Central Appalachian Basin	Pocahontas Formation and Lee Formation, Pennsylvanian, Late Carboniferous	Bituminous	0.54[a]	274–671	23.33[c]	15–20 coal seams 4.6–6.1 m net	5–27[b]		Ripepi and Karmis (2015); Eble and Greb (2016)[a], SECARB 2013[c]; Hunt and Steele (1991)[b]
13	FBV 4A Micro-Pilot Test, Western Canada Sedimentary Basin	Medicine River (Mannville) coal seams, Lower Cretaceous	High volatile C/B bituminous	0.51–0.54	1260	47.2	3.97	3.65 absolute 0.53 gas effective	CH_4: 15.1 m³/t; 4.7 MPa CO_2: 31.0 m³/t; 1.9 MPa	Mavor et al. (2004); Gentzis et al. (2008)
14	RECOPOL, upper Silesian coal basin, Poland	Carboniferous	High volatile bituminous	0.8–0.85	1012–1076	35–50[a]	Three seams, each about 1.0 – 3.5 m thick	0.4–1.5 for high-perm seams	Excess sorption capacities of moisture-equilibrated coals CH_4: 0.3–0.8 mmol/g daf CO_2: 0.8–1.2 mmol/g, daf[c]	Van Bergen et al. (2009); Pagnier et al. (2005); Malolepszy and Ostaficzuk (1999)[a]; Weniger et al. (2012)[c]
15	Yubari field test, Ishikari Coal Basin, Japan	Yubari coal seam, Tertiary	High volatile bituminous	0.74–0.81	890	30	5.66	1.0	CH_4: 28.0 m³/t; 1.78 MPa CO_2: 44.0 m³/t; 0.97 MPa	Fujioka et al. (2010)
16	CSEMP, Western Canadian Sedimentary Basin	Ardley Coal, Tertiary	Sub-bituminous	0.75–1.92[a]	~430		5–15	1–10		JPT (2005); Smith et al. (2008)[a]

[a-c]References from literature that was not part of the ECBM project.

Table 2. Summary of injection programme of enhanced recovery of coalbed methane (ECBM) field projects.

No.	Project name	Site location	Injection dates	Total amount of CO_2 injected	Injected CO_2 phase	Injection rates	Maximum injection pressure, bottomhole (MPa)	Reservoir pressure before injection (MPa)	Well type	Monitoring technique	References
1	Qinshui Basin ECBM 2004	Southern Shizhuang CBM Block, Qinshui Basin, China	April–June 2004	192 t injected in 13 days	Liquid at well head	15–18 t in 8–10 h; about 1.5–2.25 t/h	6.7	1.296	Single vertical well huff and puff	Pressure, water chemistry, gas composition	Wong et al. (2007); Ye et al. (2007)
2	Qinshui Basin ECBM 2010	Northern Shizhuang CBM Block, Qinshui Basin, China	April–May 2010	233.6 t injected in 17 days	Liquid at well head		~7	~3.8–3.9	Single vertical well huff and puff	Pressure, gas composition	Ye et al. (2012)
3	APP ECBM, Ordos Basin	Liulin CBM Block, East Margin of Ordos Basin, China	September 2011–March 2012	460 t in about 70 injection days	Liquid shipped by trucks	Maximum 8-12 t/h, typically 2 t/h	~5.5	~2.1	1 Multilateral horizontal injection well, in-seam length about 2305 m; 1 monitoring well	Vertical monitoring well ~20 m away from injection well, U-tube system Tracer: SF6	Connell et al. (2014)
4	Qinshui Basin multiple wells	Northern Shizhuang CBM Block, Qinshui Basin, China	2013–2015	4491 t of liquid CO_2 in about 460 injection days	Liquid	8–10 t in 6–8 h	19.0	6.2–6.7	3 injection 8 production or monitoring wells	Transient electromagnetic, water sampling	Ye and Zhang (2016)
5	Allison Unit, San Juan Basin	San Juan Basin, United States, San Juan County southern New Mexico, United States	April 1995 and August 2001	336,000 t	Pipeline CO_2	~28 t/d to ~17 t/d per well	10.3 (wellhead)	11.4 (original)	4 injection wells, 16 production wells and 1 pressure observation well	Gas composition	Reeves and Oudinot (2005a); Shi and Duracan (2004)
6	Pump Canyon, San Juan Basin	San Juan Basin, United States	July 2008–August 2009	16,699 t	Pipeline CO_2	1.06×10^5 m³/d to 1.4×10^4 m³/d	7.7		1 injection well, 3 production well	Atmospheric shallow Groundwater CO_2 levels	Oudinot et al. (2009); (2011)
7	Tanquary Farms test site, Illinois Basin	Wabash County, southeastern Illinois, United States	Summer 2008	92.3 t	Liquid heated to 10–38°C at surface	Average at 0.93 t/d	5.34	2.79	1 injection well, 3 monitoring wells		NETL (2012a); Finley (2011, 2012)
8	Virginia Central Appalachian Basin Coal Test	Russell County, Virginia, United States	15 January 2009–9 February 2009	~900 t	Gas at 37.8°C	36–45 t/d dropped to 20t/d	6.9		1 injection, 7 adjacent production wells	Tracers	NETL (2012c); Ripepi (2012)
9	Lignite Field Validation Test, Williston Basin	Burke County, North Dakota, United States	March 2009	90 t 16 days		3815 m³/d	<5.0	2.4	1 injection well, 4 monitoring well	Seismic imaging and downhole instruments	NETL (2012b); Botnen et al. (2009)
10	Black Warrior Basin Coal Test	Tuscaloosa County, Alabama, United States	June–August 2010	252 t to the Black Creek and Mary Lee, Pratt coal seams	Liquid at wellhead	113–136 t/d to Black Creek	Pratt:3.9 Black Creek: 7.1	Pratt: 2.7 Black Creek: 5.3 (original)	1 injection well, hydraulic fractured. 3 monitoring wells	Pressure logging, gas and water sampling	NETL (2012c); Pashin et al. (2015)

(Continued)

Table 2. (Continued).

No.	Project name	Site location	Injection dates	Total amount of CO$_2$ injected	Injected CO$_2$ phase	Injection rates	Maximum injection pressure, bottomhole (MPa)	Reservoir pressure before injection (MPa)	Well type	Monitoring technique	References
11	Marshall County project, Northern Appalachian Basin	Marshall County, West Virginia, United States	September 2009–December 2013	4500 t		8.39 t/d in well one and 7.63 t/d in the other, on average	7.7 for one well and 6.8 for the other		2 horizontal injection wells, adjacent production wells	Gas and water produced, ground and stream water monitoring, soil gas monitoring and seismic and tiltmeter observations.	Locke and Winschel (2014; 2015)
12	Buchanan County, Central Appalachian Basin	Buchanan County, Virginia, United States	July–Aug 2015	1470 t		4.5–22.5 t/d	2.9 in one well and 1.7 in the other Wellhead		3 injection wells	Gas/water composition, tracers/isotopes, formation logging, microseismic and surface deformation measurement (GPS + InSAR)	Ripepi and Karmis (2015)
13	FBV 4A Micro-Pilot Test, Western Canada Sedimentary Basin	Fenn and Big Valley, Alberta, Canada	1998	21 t pre-injection 180 t in 12 cycles 201 t in total	Liquid CO$_2$ injected, vaporized in well, vapour	~2 to 4 t/h	6.1 MPa to 2.7 MPa above reservoir pressure	7.9	1 CO$_2$ injection well		Mavor et al. (2004)
14	RECOPOL, upper Silesian coal basin, Poland	Kaniow village, about 40 km south of Katowice, Poland	August 2004–May 2005	692 t	Liquid	1–1.3 t/d before fracturing, 12–15 t/day after fracturing	14.0 MPa		1 injection well, 1 production well	Gas composition, Isotopes and water chemistry	Van Bergen et al. (2009); Pagnier et al. (2005)
15	Yubari field test, Ishikari Coal Basin, Japan	Yubari, Hokkaido, Japan	Late 2004–September 2007	~800 t	Liquid	1.7–3.8 t/d	15.5 in 2005 and 2006; 19.0 in 2007	10.2	1 injection well, 1 production well	Pressure, gas composition, etc.	Yamaguchi et al. (2004); Shi et al. 2008; Yamaguchi et al. 2006; Fujioka et al. 2010
16	CSEMP, Western Canadian Sedimentary Basin	Alder Flats, Alberta, Canada	June 2006	2 injections with unknown amount							Deng et al. (2008); ;

Figure 17. Enhanced recovery of coalbed methane field trial sites in the United States. 1 = Allison Unit, 2 = Pump Canyon, 3 = Tanquary Farms, 4 = Virginia Central Appalachian Basin, 5 = Lignite Field Validation Test, Williston Basin, 6 = Black Warrior Basin Coal Test, 7 = Marshall County project, Northern Appalachian Basin, 8 = Buchanan County, Central Appalachian Basin.

(6) Tuscaloosa County, Alabama, Black Warrior Basin (2010), through the Southeast Regional Carbon Sequestration Partnership
(7) Marshall County, West Virginia, Northern Appalachian Basin (2009–2013)
(8) Buchanan County, Virginia, Central Appalachian Basin (2015)

There are also four ECBM field trials outside China and the United States, including two in Canada, one in Poland and one in Japan.

(1) Fenn and Big Valley (FBV 4A) (1998), near the towns of Fenn and Big Valley in Alberta, Canada
(2) RECOPOL EU projects (2004–2005), about 40 km south of Katowice, upper Silesian coal basin, Poland
(3) Yubari CO_2/N_2-ECBM, Japan (2004–2007), near the town of Yubari, Hokkaido, Japan
(4) CSEMP project (2006), located near the town of Alder Flats in west-central Alberta, Canada

Of all the above projects, including the four in China, five projects have an injection amount exceeding 1000 t, including one over the 10,000 tonne mark and

one over the 100,000 tonne mark (Table 2). Both of these larger projects were in the San Juan Basin high-permeability area. For the other three projects over 1000 t, all involved using multiple injection wells: two injecting about 4500 t (including one in China) and one injecting about 1500 t. The remaining 10 projects (except for the CSEMP in Canada, where no information on injection amount is available) have injected less than 1000 t of CO_2, with four projects between 460 and 900 t and six projects between 90 and about 250 t (including two in China). Therefore, except for the two large-scale projects in the San Juan Basin, the remainder are all small scale, with some of them much smaller than CO_2 storage projects in other types of formations, such as saline aquifers.

3.1. CO_2 injectivity and permeability change

It is well understood that injecting CO_2 will significantly reduce coal permeability due to adsorption-induced coal swelling (Pekot and Reeves 2002; Shi and Durucan 2005). At the same pressure, the amount of CO_2 adsorption is higher than that of CH_4, leading to a net effect of coal swelling as CO_2 displaces CH_4 (Pan and Connell 2007). Therefore, the injectivity and permeability change behaviour are the first things to be investigated in a field trial: especially in low-permeability reservoirs, in which injectivity is an issue. CUCBM and its project partners performed four ECBM trials in Qinshui Basin and Eastern Ordos Basin, which are most important CBM producing basins in China. The first three projects mainly tested injectivity in a single well. The first project was a single-well huff-and-puff test at relatively shallow coal formation at around 472 m; the second was at a deep coal seam at around 972 m; the third tested injectivity using a multilateral horizontal well, which has a large contact area with coal. The fourth project evaluated enhanced recovery using multiple injection and production wells.

The permeability of the coal reservoirs in the four Chinese ECBM trials was mostly less than 1 md (Table 1), except for the first ECBM trial in Qinshui Basin, in which Wong et al. (2007) reported a permeability of 12.6 md and Ye et al. (2007) reported 0.95 md. All projects reported permeability loss due to swelling; however, injectivity change during injection varied in the four projects. In the first trial, the injectivity to CO_2 decreased initially but was stabilized during the injection of 13 slugs of CO_2 (Wong et al. 2007). In the second trial, Ye et al. (2012) reported a permeability increase and attributed it to the coal-fracture cleaning effect caused by the sweep of CO_2. In the third project, injectivity improved in the middle period where injection events were closely spaced; the injectivity increase could be due to fewer two-phase (gas–water) flow effects, with the gas progressively displacing water further from the injection well (Connell et al. 2014). In the fourth trial, the injectivity in well SX006 did not decrease, because a hydraulic fracture may have connected injection well SX006 with the production well SX006-03. Hydraulic fracture connection leading to injectivity maintenance/increase was also observed in the RECOPOL project (van Bergen et al. 2009).

Of all the ECBM field trials, the two in San Juan Basin had much higher permeability, and permeability decrease due to swelling clearly contributed to the decrease of injectivity. The absolute permeability at Allison Unit was 30–150 md and the gas effective permeability was about half (Reeves et al. 2003). About 336,000 t of CO_2 were injected over a 6-year period into four injection wells (Reeves 2009), but injectivity loss was significant (Reeves et al. 2003). The injection rate was allowed to vary; the initial injection was 500 Mcf/d (1.4 × 10^4 m^3/d), which dropped to 300 Mcf/d (0.84 × 10^4 m^3/d) (Reeves et al. 2003). The reduction in injection rate was presumably due to coal swelling and permeability reduction, while the rebound in injectivity during later times was believed to be due to overall reservoir pressure reduction and resulting matrix shrinkage that occurred near the injector wells (Reeves and Oudinot 2005a). The permeability at Pump Canyon was about 582 md for the basal coal and 146 md for the upper and middle coals by history matching the production data. However, the permeability had to be changed to 10 md for the upper and middle coals to match the injection results, due to CO_2-induced matrix swelling and permeability reduction (Oudinot et al. 2011). The CO_2 injection rate decreased from more than 3750 Mcf/d (1.06 × 10^5 m^3/d) at the early stages to around 500 Mcf/d (1.4 × 10^4 m^3/d) after about 8 months of CO_2 injection. The drop in rate of about 85% was speculated to be due to matrix swelling and permeability reduction as CO_2 was adsorbed onto the coal (Oudinot et al. 2009, 2011).

For ECBM cases in low-permeability coals, permeability and injectivity loss are often observed. For instance, in the Russell County trial, the CO_2 injection pressure increased up to 1000 psia (6.9 MPa) to maintain an injection rate of 40–50 tons/day (British tons, equivalent to 36–45 t/day), however, the rate dropped below 20 tons/day (18 t/day) while the injection pressure still maintained at 1000 psia (6.9 MPa) towards the end of injection when 1000 tons (900 t) of CO_2 were injected (Ripepi 2012). Injectivity increases were also reported in some projects, such as in the Fenn and Big Valley (FBV 4A) trial (Mavor et al. 2004). In this 200-tonne injection project, the pressure differential required for each injection period decreased from 6.1 to 2.7 MPa, a fall of about 55% over the 12 injection periods. Injectivity increased by 147%, from 5.5 m^3/d-kPa for the first injection period to 13.6 m^3/d-kPa for the final injection period. The injectivity for the 11th injection period was the greatest, as was the injection rate, suggesting the injectivity was also rate dependent. The greater CO_2 injectivity was likely caused by a combination of greater permeability and/or effective fracture length during injection due to ballooning (Mavor et al. 2004).

In summary, although most field projects observed injectivity loss, some projects observed injectivity increase for at least a period of time during injection. This suggests that field conditions are complex and other factors may contribute to injectivity increase that overcome the injectivity loss due to decreasing permeability. These factors may include hydraulic fractures connecting production wells, as suggested in Qinshui Basin multi-well ECBM (Ye and Zhang 2016) and RECOPOL project (van Bergen et al. 2009); increase of gas effective permeability as water is displaced with continuous CO_2 injection, as suggested in the APP ECBM project (Connell et al. 2014); or caused by possible fracture length increase during injection due to ballooning, as suggested in the FBV 4A micro-pilot test project (Mavor et al. 2004).

3.2. Enhanced methane recovery

In the three huff-and-puff tests in China, gas production resumed after CO_2 injection ceased. All three projects showed greater production than the pre-injection production rate. The cause of the production increase was not clearly provided in the literature, but was related to CO_2 injection. Although swelling reduces coal permeability, which will reduce production rate, higher pressure around the well and higher gas saturation – and therefore, higher gas effective permeability – will contribute to higher gas-production rate.

In the Qinshui Basin multiple-wells field test, the aim was to investigate enhanced recovery from production wells. Due to the large well distance of above 200 m and the amount of CO_2 injected (about 4500 t in three wells), the effect of enhanced recovery was difficult to directly evaluate. Enhanced recovery from the production wells was also not directly observed from the three larger CO_2 injection field trials: (1) the Allison Unit ECBM, with 336,000 t CO_2 injected in four injection wells over 6 years (Reeves 2009); (2) the Pump Canyon trial, with about 16,700 t CO_2 injected in one well (Oudinot et al. 2008); and (3) the Marshall County ECBM trial, West Virginia, in which 4500 t of CO_2 was injected to two horizontal wells (Locke and Winschel 2015). Enhanced recovery or incremental methane production due to CO_2 injection was verified using reservoir simulation in the Allison Unit ECBM (Reeves et al. 2003) and Pump Canyon (Schepers-Cheng 2010), and the incremental increase was marginal compared to the original gas in place (Reeves et al. 2003).

Reservoir simulation is still the most widely used method to verify enhanced recovery in most projects, as direct observation of production enhancement from production wells is difficult considering the small amount of CO_2 injected in most ECBM trials. Therefore, depleted coal reservoirs are perhaps a better candidate for ECBM trials for direct observation of gas-production enhancement.

3.3. Coal rank and adsorption capacity

Adsorption capacity is a controlling factor of CO_2 storage capacity, as well as the exchange ratio of CO_2 and CH_4 for ECBM. Adsorption capacity is strongly related to coal rank. All four projects in China were performed in Permian coals; however, the rank of coal was from anthracite to bituminous, and the vitrinite reflectance was 3.4%, 2.7%, 2.6%, and 1.6% for the four field trials, respectively (Table 1). For the three anthracite coals, the Langmuir volumes for CH_4 and CO_2 were almost 1:1, making them better candidates for ECBM with less CO_2 needed to displace each molecule of CH_4 adsorbed. For the APP ECBM on bituminous coals, the ratio of CO_2 and CH_4 Langmuir volume was about 1.5:1.

Note that these ratios are based on Langmuir volumes, while the true ratios are also related to gas pressure.

Coal reservoirs at different age were tested. Five reservoirs in the eastern states of the United States were in Carboniferous Pennsylvanian coals, two were in Cretaceous coals in the San Juan Basin and one was in Tertiary coal in North Dakota (Table 1). RECOPOL project was in Carboniferous coal in Poland, FBV 4A Micro-Pilot was in Cretaceous coal in Canada, and the Yubari project in Japan and CSEMP in Canada were in the Tertiary coals. In terms of coal ranks, three ECBM projects in China were anthracite, one in the United States was lignite and the rest were bituminous (Table 1). Adsorption capacity is clearly related to coal rank, and the CO_2 and CH_4 preferential adsorption ratio decreases as coal rank increases (Busch and Gensterblum 2011). As mentioned above, the Langmuir volume ratio between CO_2 and CH_4 for the three anthracite coals was around 1:1, while it was about 1.5:1 to 2.5:1 for bituminous coals. However, for lignite in the Williston Basin, the ratio of CO_2 and CH_4 Langmuir volume was about 10:1. A high Langmuir volume ratio between CO_2 and CH_4 means that much more CO_2 is required to displace one molecule of CH_4, making this type of coal a better candidate for CO_2 storage in terms of molecule-exchange ratio.

3.4. Injection pressure and CO_2 phases

Supercritical CO_2 has a density close to liquid and a viscosity close to gas phase. It is therefore preferred for CO_2 injection, because it can maximize injectivity. To stay in the supercritical phase, CO_2 pressure should be higher than 7.3 MPa and temperature should be above 31.0°C. In all four Chinese ECBM field projects, liquid CO_2 was transported to the site using trucks. The reservoirs were all at temperatures less than 27.5°C (Table 1), but the reservoir and injection pressures varied. For the first Qinshui Basin (2004) project, the depth was 472 m and the initial reservoir pressure was about 3.36 MPa. Before CO_2 injection, the reservoir pressure around injection well TL-003 had dropped to 0.96 MPa due to 4 years of CBM production (Ye et al. 2007). Injection pressure at the bottomhole was kept around 6.7 MPa (Table 2). For the second Qinshui Basin (2010) project, the well was in production for about 3 months before being converted to an injection well, the reservoir pressure was close to virgin pressure at about 3.8 MPa, and the injection pressure was controlled below 7 MPa. For the APP ECBM project, the reservoir pressure was 2.1 MPa before injection and the injection pressure was controlled below 5.5 MPa. For the Qinshui Basin multiple-wells project, the reservoir pressure was about 6.2–6.7 MPa and the highest injection pressure was 19 MPa. Figure 18 shows the injected CO_2 phase using the BHP and reservoir temperature. It reveals

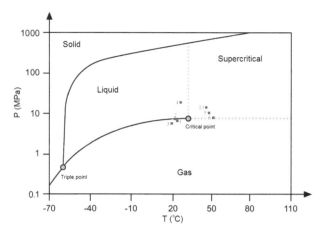

Figure 18. Phase of injected CO_2. 1 = Qinshui Basin (2004), 2 = Qinshui Basin (2010), 3. APP ECBM, 4 = Qinshui Basin Multiple wells, 5 = Allison Unit, United States, 6 = Pump Canyon, United States 13 = FBV 4A Micro-Pilot Test, Canada.

that injected CO_2 for the first three ECBM projects in China was in gas phase, while it was in liquid phase for the fourth project. However, note that CO_2 pressure in the formation was lower than the BHP, and the temperature could be also lower, as liquid CO_2 was injected. Therefore, the CO_2 phase behaviour was more complicated in these projects.

Compared with all other projects performed outside China, reservoir temperatures are mostly below 30°C (Table 1), except for the two projects in the San Juan Basin Fairway and one in Canada, in which the depth was more than 900 m and the reservoir temperature was about 50°C. In Allison Unit ECBM, CO_2 was delivered from the pipeline at a pressure of approximately 2200 psi (15.2 MPa), and it was reduced to approximately 1500 psi (10.3 MPa) at wellheads (Reeves 2009). In the Pump Canyon trial, CO_2 was also delivered using pipeline, and the injection pressure was 7.7 MPa; it was about 14 MPa in the FBV 4A micro-pilot test in Canada. As shown in Figure 18, CO_2 was able to stay in the supercritical phase. A few other projects, including the Tanquary Farms test and Russell County project, heated the CO_2 to about 38°C before injection.

Phase changes in the well bore and/or the coal formation would make analyses and modelling difficult. Non-isothermal models will be required to analyse the results; however, most of the current reservoir simulation software for ECBM are isothermal. The complex CO_2 phase behaviour, combined with the impact from water and methane, must be represented in the reservoir simulation to better understand ECBM in the field.

3.5 Well technology

Well technology is important for CO_2 injectivity, because coal seams are often thin and multi-layered.

The net coal thickness in most ECBM trials is less than 10 m, and typically around 5 m, except for the two projects in the San Juan Basin, where the net total coal thickness is 13 and 18 m (Table 1). All four ECBM trials in China injected CO_2 into a single seam. Note that in the APP ECBM field trial, the #4 coal seam was combined with the #3 seam; therefore, the #3 + 4 was a single seam. Other projects are associated with multiple seams, quite typically three seams or coal intervals, such as the Allison Unit (Reeves *et al.* 2003) and Pump Canyon projects (Oudinot *et al.* 2009, 2011), Black Warrior coal test project (Pashin *et al.* 2014) and RECOPOL project (van Bergen *et al.* 2009). The benefit of injecting into multiple coal seams is to maximize CO_2 storage capacity in a single well, although the complexity may lie in the resulting interpretation of where the CO_2 is stored. In the Black Warrior coal test project, for example, the CO_2 was injected into each single seam using packers to direct the flow.

Vertical wells were used most widely in ECBM trials. Vertical wells without fracturing were initially used in the RECOPOL project (van Bergen *et al.* 2009) and Yubari Project (Fujioka *et al.* 2010); injection rates were low in both tests. Hydraulic fracturing was then performed and injection rate improved significantly. Hydraulic fractured vertical wells or lateral wells were used in the Chinese ECBM projects, and typically 10 t/day injection rates were achieved in each project (Table 2). However, as coal seams are thin, hydraulic fractures may extend out of the target seams. Injecting those wells may cause CO_2 leakage to other formations, which may have significant safety and environmental impact. Two projects, the Marshall County test and APP ECBM in Ordos Basin, used horizontal wells to test CO_2 injection capacity, because they are usually good CBM producers.

The number of wells used in the projects also varied significantly among projects. Three projects used a single well, including the two early trials in China and the FBV 4A pilot in Canada. The remainder involved at least one injection well and one production or monitoring well (Table 2). One advantage of using multiple wells is the ability to examine the effect of ECBM; the other is monitoring gas migration.

3.6 Monitoring and verification

The monitoring techniques used in each project are summarized in Table 2. In the early two field trials in China, which were single-well tests, monitoring mainly focused on injection rates and pressure, water chemistry before and after CO_2 injection, and gas composition change in post-injection produced gas. Verification mainly relied on reservoir simulation. These were typical monitoring techniques also used in the other early projects around the world, such as the Allison Unit ECBM, RECOPOL project and Yubari project, when other monitoring techniques were not readily deployable.

In the APP ECBM field trial in 2011–2012, a monitoring well was used to obtain more monitoring information and verify CO_2 migration in the reservoir. A U-tube sampling system was installed in the target coal seam as well as in an underlying coal seam and an overlaying sandstone layer, which verified that CO_2 stayed in the target zone (Connell *et al.* 2014). SF6 tracer gas was used as another means of flow path assurance. Tracer gas was also used in other projects, such as Pump Canyon ECBM trial, in which two tracer injections were conducted on 18 September 2008 with 90% PMCH +10% o-PDMCH and 9 October 2008 with 100% PTCH (Schepers-Cheng 2010). Tracer gas was injected and detected in the adjacent CBM production wells in the Russell County project, and a significant amount of tracer was produced from seven close CBM wells, with the farthest well 700 m away from the injection well (Ripepi 2012).

Other monitoring techniques were also used. Atmospheric CO_2 levels, plant stress, gas composition at wellheads and shallow groundwater characteristics were used in Tanquary Farms project in Wabash County, Illinois to verify that there was no CO_2 leakage into groundwater or CO_2 escape at the surface (NETL 2012a). Pre and post-injection vertical seismic profiling and tilt-meters were also used in the Pump Canyon project, with the results suggesting that changes due to the CO_2 injection were minimal (Wilson *et al.* 2010; Oudinot *et al.* 2011). Seismic imaging and downhole instruments were used in the Burke County project to determine the fate of the injected CO_2. The indications were that the injected CO_2 migrated along the path of the coal and was contained within the expected injection zone during the 3-month monitoring period (NETL 2012b). Isotope analysis was used in the RECOPOL project (van Bergen et al. 2009), while monitoring techniques such as gas/water composition, tracers/isotopes, formation logging, microseismic and surface deformation measurement (GPS + InSAR) were used in the Buchanan County Project (Ripepi and Karmis 2015). These recent monitoring techniques can be considered and applied to future ECBM projects in China.

4. Conclusions

As an important CO_2 storage target formation, CO_2 injection into coal with enhanced CBM recovery has been tested in many places worldwide in the past 20 years.

This article focused on reviewing four Chinese ECBM projects and comparing them with 12 other projects across the world, aiming to provide insights that will help the design of future ECBM projects in China. The main findings include but are not limited to the following:

(1) Coal reservoirs for the ECBM projects in China are all Permian coals, with three anthracite and one bituminous coal. This makes them better candidates than low rank coals for enhanced CBM recovery, since less CO_2 is required to displace methane. Permeability loss due to coal swelling was observed in each project. However, injectivity increased on some occasions, indicting the complex nature of field conditions, where factors such as gas effective permeability and well-bore storage behaviour can contribute to injectivity change.

(2) Enhanced CBM recovery was not directly detected in the Chinese ECBM projects, due to the small amount of CO_2 injected. As for other projects around the world, enhanced recovery can be demonstrated through reservoir simulation. However, direct measurement of enhanced recovery is highly desirable. This may be achieved using depleted CBM reservoirs with more closely spaced wells.

(3) The main monitoring techniques used in the Chinese ECBM projects were injection rates and pressure, water chemistry before and after CO_2 injection, and gas composition change in post-injection produced gas. Monitoring technologies used in other ECBM projects or other CO_2 storage projects, such as air sampling, micro seismic and tilt-meters, are good candidates for future ECBM projects.

Acknowledgements

Support from CSIRO Energy and the Fundamental Research Funds for the Central Universities (Grant No. 2015XKZD03) is acknowledged. The Program for Changjiang Scholars and Innovative Research Team in University (IRT_17R103) is also acknowledged.

Disclosure statement

No potential conflict of interest was reported by the authors.

Funding

Support from CSIRO Energy and the Fundamental Research Funds for the Central Universities (Grant No. 2015XKZD03) is acknowledged. The Program for Changjiang Scholars and Innovative Research Team in University (IRT_17R103) is also acknowledged.

References

Arri, L.E., Yee, D., Morgan, W.D., and Jeansonne, M.W., 1992, Modeling coalbed methane production with binary gas sorption, Presented at the SPE Rocky Mountain Regional Meeting, Casper, WY, USA, May 1992. SPE 24363.

Botnen, L.S., Fisher, D.W., Dobroskok, A.A., Bratton, T.R., Greaves, K.H., McLendon, T.R., Steiner, G., Sorensen, J.A., Steadman, E.N., and Harju, J.A., 2009. Field test of co2 injection and storage in lignite coal seam in north dakota. Energy Procedia 1, 2013-2019.

Busch, A., and Gensterblum, Y., 2011, CBM and CO_2-ECBM related sorption processes in coal: A review: International Journal of Coal Geology, v. 87, no. 2, p. 49–71. doi:10.1016/j.coal.2011.04.011

Busch, A., Gensterblum, Y., and Krooss, B.M., 2003, Methane and CO_2 sorption and desorption measurements on dry Argonne premium coals: Pure components and mixtures: International Journal of Coal Geology, v. 55, no. 2, p. 205–224. doi:10.1016/S0166-5162(03)00113-7

Connell, L.D., Pan, Z., Camilleri, M., Meng, S., Down, D., Carras, J., Zhang, W., Fu, X., Guo, B., Briggs, C., and Lupton, N., 2014, Description of a CO_2 enhanced coal bed methane field trial using a multi-lateral horizontal well: International Journal of Greenhouse Gas Control, v. 26, p. 204–219. doi:10.1016/j.ijggc.2014.04.022

Connell, L.D., Sander, R., Pan, Z., Camilleri, M., and Heryanto, D., 2011, History matching of enhanced coal bed methane laboratory core flood tests: International Journal of Coal Geology, v. 87, no. 2, p. 128–138. doi:10.1016/j.coal.2011.06.002

Cui, X., Bustin, R.M., and Chikatamarla, L., 2007, Adsorption-induced coal swelling and stress, implications for methane production and acid gas sequestration into coal seams: Journal of Geophysical Research–Solid Earth, v. 112, p. B10202. doi:10.1029/2004JB003482

Deng, X., Mavor, M., Macdonald, D., Gunter, B., Wong, S., Faltinson, J., and Li, H., 2008, ECBM technology development at Alberta Research Council. Presented at Coal-Seq VI Forum, Houston, April: http://www.coal-seq.com/Proceedings2008/Presentations/Xiaohui%20Deng_ARC.pdf

Eble, C.F., and Greb, S.F., 2016, Palynologic, petrographic and geochemical composition of the Vancleve coal bed in its type area, Eastern Kentucky Coal Field, Central Appalachian Basin: International Journal of Coal Geology, v. 158, p. 1–12. doi:10.1016/j.coal.2016.02.004

Finley, R., 2011, Sequestration and enhanced coal bed methane: Tanquary Farms test site, Wabash County, Illinois: Final Report, 1 October 2007 – 6 January 2012.

Finley, R., 2012, An assessment of geological carbon storage options in the Illinois Basin: Validation phase: Technical Report, 1 October 2007 – 30 March 2012.

Fitzgerald, J.E., Pan, Z., Sudibandriyo, M., Robinson, R.L., Jr, and Gasem, K.A.M., 2005, Adsorption of methane, nitrogen, carbon dioxide and their mixtures on wet Tiffany coal: Fuel, v. 84, p. 2351–2363. doi:10.1016/j.fuel.2005.05.002

Fujioka, M., Yamaguchi, S., and Nako, M., 2010, CO_2-ECBM field tests in the Ishikari Coal Basin of Japan: International Journal of Coal Geology, v. 82, no. 3–4, p. 287–298. doi:10.1016/j.coal.2010.01.004

Gentzis, T., Goodarzi, F., Cheung, F.K., and Laggoun-Defarge, F., 2008, Coalbed methane producibility from the Mannville coals in Alberta, Canada: A comparison of two areas: International Journal of Coal Geology, v. 74, no. 3–4, p. 237–249. doi:10.1016/j.coal.2008.01.004

Godec, M., Koperna, G., and Gale, J., 2014, CO_2-ECBM: A review of its status and global potential: Energy Procedia, v. 63, p. 5858–5869. doi:10.1016/j.egypro.2014.11.619

Gunter, W.D., Gentzis, T., Rottenfusser, B.A., and Richardson, R.J.H., 1997, Deep coalbed methane in Alberta, Canada: A fuel resource with the potential of zero greenhouse gas emissions: Energy Conversion and Management, v. 38, p. S217–S222. doi:10.1016/S0196-8904(96)00272-5

Hall, F.E., Zhou, C., Gasem, K.A.M., Robinson, R.L., Jr, and Yee, D., 1994, Adsorption of pure methane, nitrogen, and carbon dioxide and their binary mixtures on wet Fruitland coal. SPE Eastern Regional Meeting, Charleston, West Virginia, November 1994. SPE-29194.

He, Q., Mohaghegh, S.D., and Gholami, V., 2013, A field study on simulation of CO_2 injection and ECBM production and prediction of CO_2 storage capacity in unmineable coal seam: Journal of Petroleum Engineering, v. 2013, p. 1–8. Article ID 803706. doi:10.1155/2013/803706

Hunt, A.M., and Steele, D.J., 1991, Coalbed methane development in the Appalachian Basin: Quarterly Review of Methane from Coal Seams Technology, v. 1, p. 10–19.

Journal of Petroleum Technology staff, 2005, Techbits: Coalbed-methane recovery and CO_2 sequestration raise economic, injectivity concerns: Journal of Petroleum Technology, v. 57, no. 3. doi:10.2118/0305-0026-JPT

Li, X., and Fang, Z., 2014, Current status and technical challenges of CO_2 storage in coal seams and enhanced coalbed methane recovery: An overview: International Journal of Coal Science and Technology, v. 1, no. 1, p. 93–102. doi:10.1007/s40789-014-0002-9

Liu, S., Harpalani, S., and Pillalamarry, M.R., 2009, Flow behavior in deep coals at carbon sequestration pilot sites, in Rock Mechanics: Proceedings of the Second Thailand Symposium, Fuenkajorn, K., & Phien-wej, N., (eds), pp. 121–129. Suranaree University of Technology, Thailand.

Locke, J., Winschel, R., Bajura, R., Wilson, T., Siriwardane, H., Rauch, H., and Mohaghegh, S., 2011, CO_2 sequestration in unmineable coal with enhanced coal bed methane recovery: The Marshall County Project, in Proceedings of the 2011 International Pittsburgh Coal Conference, Pittsburgh, PA, Curran Associates, Inc. September, 22 p.

Locke, J.E., and Winschel, R.A., 2014. CO_2 sequestration in unmineable coal with enhanced coal bed methane recovery (The Marshall County Project): https://www.netl.doe.gov/File%20Library/Events/2014/carbon_storage/1–James-Locke.pdf

Locke, J.E., and Winschel, R.A., 2015. CO_2 sequestration in unmineable coal with enhanced coal bed methane recovery (The Marshall County Project): https://www.netl.doe.gov/File%20Library/Events/2015/carbon%20storage/proceedings/08-20_08_CO2-Sequestration-in-Unmineable-Coal-with-ECBM-2015.pdf

Malolepszy, Z., and Ostaficzuk, S., 1999, Geothermal potential of the Upper Silesian Coal Basin, Poland: Bulletin d'Hydrogéologie, v. 17, p. 67–76.

Mavor, M.J., Gunter, W.D., and Robinson, J.R., 2004, Alberta multiwell micro-pilot testing for CBM properties, enhanced methane recovery and CO_2 storage potential. SPE Annual Technical Conference and Exhibition, Houston, September. SPE-90256.

Mazumder, S., Karnik, A.A., and Wolf, K.-H.A.A., 2006, Swelling of coal in response to CO_2 sequestration for ECBM and its effect on fracture permeability: SPE Journal, v. 11, no. 3, p. 390–398. doi:10.2118/97754-PA

Mazumder, S., and Wolf, K.H., 2008, Differential swelling and permeability change of coal in response to CO_2 injection for ECBM: International Journal of Coal Geology, v. 74, no. 2, p. 123–138. doi:10.1016/j.coal.2007.11.001

Mazzotti, M., Pini, R., and Storti, G., 2009, Enhanced coalbed methane recovery: The Journal of Supercritical Fluids, v. 47, p. 619–627. doi:10.1016/j.supflu.2008.08.013

Michael, G., Anders, D.E., and Law, B.E., 1993, Geochemical evaluation of Upper Cretaceous Fruitland formation coals, San Juan Basin, New Mexico and Colorado: Organic Geochemistry, v. 20, p. 475–498. doi:10.1016/0146-6380 (93)90094-R

Morse, D.G., Mastalerz, M., Drobniak, A., Rupp, J.A., and Harpalani, S., 2010, Variations in coal characteristics and their possible implications for CO_2 sequestration: Tanquary injection site, southeastern Illinois, USA: International Journal of Coal Geology, v. 84, p. 25–38. doi:10.1016/j.coal.2010.08.001

National Energy Technology Laboratory (NETL), 2012a, Midwest Geological Sequestration Consortium—Validation phase: https://www.netl.doe.gov/publications/factsheets/project/NT42588-P2.pdf

NETL, 2012b, Plains CO_2 reduction partnership—Validation phase: https://www.netl.doe.gov/publications/factsheets/project/NT42592-P2.pdf

NETL, 2012c, Southeast regional carbon sequestration partnership—validation phase: https://www.netl.doe.gov/publications/factsheets/project/NT42590-P2.pdf

Oudinot, A.Y., Koperna, G.J., Jr., Philip, Z.G., Liu, N., Heath, J.E., Wells, A., Young, G.B., and Wilson, T., 2009, CO_2 injection performance in the Fruitland Coal Fairway, San Juan Basin: Results of a field pilot. SPE International Conference on CO_2 Capture, Storage, and Utilization, San Diego, CA, November.

Oudinot, A.Y., Koperna, G.J., Jr., Philip, Z.G., Liu, N., Heath, J.E., Wells, A., Young, G.B., and Wilson, T., 2011, CO_2 injection performance in the Fruitland Coal Fairway, San Juan Basin: Results of a field pilot: SPE Journal, v. 16, p. 864–879. SPE-127073. doi:10.2118/127073-PA

Oudinot, A.Y., Schepers, K.C., Gonzalez, R.J., and Reeves, S., 2008, An integrated reservoir characterization, geostatistical analysis, optimized history-matching and performance forecasting study of the 9-section, 30-well Pump Canyon CO_2-ECBM/sequestration demonstration site, San Juan Basin, New Mexico. Paper No. 0804, 2008 International Coalbed and Shale Gas Symposium, Tuscaloosa, AL, May 2008.

Pagnier, H., van Bergen, F., Krzystolik, P., Skiba, J., Jura, B., Hadro, J., Wentink, P., De-Smedt, G., Kretzschmar, H.-J., Frobel, J., Muller-Syring, G., Krooss, B., Busch, A., Wolf, K.-H., Mazumder, S., Bossie-Codreanu, D., Choi, X., Grabowski, D., Hurtevent, D., Gale, J., Winthaegen, P., Van Der Meer, B., Kobiela, Z., Bruining, H., Reeves, S., and Stevens, S., 2005, Field experiment of ECBM-CO_2 in the Upper Silesian Basin of Poland (RECOPOL): Greenhouse Gas Control Technologies, v. 7, p. 1391–1397.

Pan, Z., and Connell, L.D., 2007, A theoretical model for gas adsorption-induced coal swelling: International Journal of Coal Geology, v. 69, no. 4, p. 243–252. doi:10.1016/j.coal.2006.04.006

Pan, Z., and Connell, L.D., 2010, Comparison of adsorption models in reservoir simulation of enhanced coalbed methane recovery and CO_2 sequestration in coal: International Journal of Greenhouse Gas Control, v. 3, p. 77–89. doi:10.1016/j.ijggc.2008.05.004

Pan, Z., and Connell, L.D., 2012, Modelling permeability for coal reservoirs: A review of analytical models and testing data: International Journal of Coal Geology, v. 92, p. 1–44. doi:10.1016/j.coal.2011.12.009

Pan, Z., Connell, L.D., and Camilleri, M., 2010, Laboratory characterisation of coal reservoir permeability for primary and enhanced coalbed methane recovery: International Journal of Coal Geology, v. 82, no. 3–4, p. 252–261. doi:10.1016/j.coal.2009.10.019

Pan, Z., Connell, L.D., Meng, S., Sander, R., Camilleri, M., Down, D.I., Carras, J., Lu, M., Fu, X., Zhang, W., Guo, B., Ye, J., Briggs, C., and Lupton, N., 2013, CO_2 injectivity in a multi-lateral horizontal well in a low permeability coal seam: Results from a field trial: Energy Procedia, v. 37, p. 5834–5841. doi:10.1016/j.egypro.2013.06.507

Pashin, J. C., 2016, Geologic considerations for CO2 storage in coal, in Geologic Carbon Sequestration: Understanding Reservoir Behavior, Singh, T. N., ed.,: Berlin, Springer, p. 137-159.

Pashin, J.C., 2007, Hydrodynamics of coalbed methane reservoirs in the Black Warrior Basin: Key to understanding reservoir performance and environmental issues: Applied Geochemistry, v. 22, no. 10, p. 2257–2272. doi:10.1016/j.apgeochem.2007.04.009

Pashin, J.C., 2009, Variable gas saturation in coalbed methane reservoirs of the Black Warrior Basin: Implications for exploration and production: International Journal of Coal Geology, v. 82, p. 135–146. doi:10.1016/j.coal.2009.10.017

Pashin, J.C., Clark, P.E., McIntyre-Redden, M.R., Carroll, R.E., Esposito, R.A., Oudinot, A.Y., and Koperna, G.J., Jr., 2015, SECARB CO_2 injection test in mature coalbed methane reservoirs of the Black Warrior Basin, Blue Creek Field, Alabama: International Journal of Coal Geology, v. 144–145, p. 71–87. doi:10.1016/j.coal.2015.04.003

Pashin, J.C., McIntyre-Redden, M.R., Mann, S.D., Kopaska-Merkel, D.C., Varonka, M., and Orem, W., 2014, Relationships between water and gas chemistry in mature coalbed methane reservoirs of the Black Warrior Basin: International Journal of Coal Geology, v. 126, p. 92–105. doi:10.1016/j.coal.2013.10.002

Pekot, L.J., and Reeves, S.R., 2002, Modeling coal matrix shrinkage and differential swelling with CO_2 injection for enhanced coalbed methane recovery and carbon sequestration applications: Topical Report, Contract No. DE-FC26-00NT40924: Washington, DC, United States Department of Energy.

Pini, R., Ottiger, S., Burlini, L., Storti, G., and Mazzotti, M., 2009, Role of adsorption and swelling on the dynamics of gas injection in coal: Journal of Geophysical Research, v. 114, no. B4, p. B04203. doi:10.1029/2008JB005961

Puri, R., and Yee, D., 1990, Enhanced coalbed methane recovery. SPE Annual Technical Conference and Exhibition, New Orleans, LA, September 1990. SPE 20732.

Qin, Y., Shen, J., Yang, Z., Moore, T.A., and Wang, G.G.X., 2017. The geology and resources of coalbed methane in China: A review. International Geology Review, Submitted.

Reeves, S.R., 2009, An overview of CO_2-ECBM and sequestration in coal seams, in Grobe, M., Pashin, J.C., and Dodge, R. L., eds., Carbon dioxide sequestration in geological media —State of the science: AAPG studies in geology, Volume 59, p. 17–32.

Reeves, S.R., and Oudinot, A., 2005a, The Allison Unit CO_2-ECBM pilot – A reservoir and economic analysis. Paper No. 0522, 2005 International Coalbed Methane Symposium, Tuscaloosa, AL, May 2005.

Reeves, S.R., and Oudinot, A., 2005b, The Tiffany unit N_2-ECBM pilot – A reservoir and economic analysis. Paper 0523, 2005 International Coalbed Methane Symposium, Tuscaloosa, AL, May 2005.

Reeves, S.R., Taillefert, A., Pekot, L., and Clarkson, C., 2003, The Allison Unit CO_2-ECBM pilot: A reservoir modeling study: Topical Report, DOE Contract No. DEFC26-00NT40924.

Ripepi, N., 2012, Central Appalachian Basin ECBM field projects: Results and future plans. 8th International Forum on Geologic Sequestration of CO2 in Coal Seams and Gas Shale Reservoirs "Coal-Seq VIII" and "Shale-Seq I". October 23–24. https://www.adv-res.com/Coal-Seq_Consortium/ECBM_Sequestration_Knowledge_Base/Coal-Seq%20VIII%20Forum%202012/10.NinoRipepi.pdf.

Ripepi, N., and Karmis, M., 2015, Central Appalachian Basin unconventional (coal/organic shale) reservoir small-scale CO_2 injection test: https://www.netl.doe.gov/File%20Library/Events/2015/carbon%20storage/proceedings/08-20_15_VirginiaTech.DOEReview.8.20.2015.pdf

Robertson, E.P., 2005, Measurement and modeling of sorption-induced strain and permeability changes in coal [Doctoral dissertation]: Colorado School of Mines, Arthur Lakes Library. INL/EXT-06-11832.

Ruppert, L.F., Stanton, R.W., Cecil, C.B., Eble, C.F., and Dulong, F. T., 1991, Effects of detrital influx in the Pennsylvanian Upper Freeport peat swamp: International Journal of Coal Geology, v. 17, p. 95–116. doi:10.1016/0166-5162(91)90006-5

Sander, R., Connell, L.D., Pan, Z., Camilleri, M., Heryanto, D., and Lupton, N., 2014, Core flooding experiments of CO_2 enhanced coalbed methane recovery: International Journal of Coal Geology, v. 131, p. 113–125. doi:10.1016/j.coal.2014.06.007

Schepers-Cheng, K.C., 2010, Southwest Partnership on Carbon Sequestration (SWP) Pump Canyon CO_2 enhanced coalbed methane (ECBM)/CO_2 sequestration demonstration test site. Research Experience in Carbon Sequestration, Albuquerque, NM, July 2010.

SECARB, 2008, Southeast Regional Carbon Sequestration Partnership (SECARB) black warrior basin coal seam project: https://www.netl.doe.gov/publications/proceedings/08/rcsp/factsheets/1-SECARB_Black%20Warrior%20Basin_Coal.pdf

SECARB, 2013, Central Appalachian Coal Seam Project: Summary of field test site and operations: https://www.energy.vt.edu/secarb/FT2a.pdf

Shi, J.Q., and Durucan, S., 2004. A numerical simulation study of the Allison Unit CO2-ECBM pilot: the effect of matrix shrinkage and swelling on ECBM production and CO2 injectivity. Proceedings of the 7th International Conference on Greenhouse Gas Control Technologies (GHGT 7), September 5–9, Vancouver, Canada, V.1, pp. 431–442.

Shi, J.Q., and Durucan, S., 2005, A model for changes in coalbed permeability during primary and enhanced methane recovery: SPE Reservoir Evaluation and Engineering, v. 8, no. 4, p. 291–299. doi:10.2118/87230-PA

Shi, J.Q., Durucan, S., and Fujioka, M., 2008. A reservoir simulation study of co2 injection and n2 flooding at the ishikari coalfield co2 storage pilot project, japan. International Journal Of Greenhouse Gas Control 2, 47-57.

Siriwardane, H.J., Bowes, B.D., Bromhal, G.S., Gondle, R.K., Wells, A.W., and Strazisar, B.R., 2012, Modeling of CBM production, CO_2 injection, and tracer movement at a field CO_2 sequestration site: International Journal of Coal Geology, v. 96, p. 120–136. doi:10.1016/j.coal.2012.02.009

Smith, G.G., Cameron, A.R., and Bustin, R.M., 2008, Chapter 33: Coal resources of the Western Canada Sedimentary Basin, in Coal resources. Alberta Geological Survey. http://www.cspg.org/documents/Publications/Atlas/geological/atlas_33_coal_resources.pdf

Tsotsis, T.T., Patel, H., Najafi, B.F., Racherla, D., Knackstedt, M.A., and Sahimi, M., 2004, Overview of laboratory and modelling studies of carbon dioxide sequestration in coal beds: Industrial and Engineering Chemistry Research, v. 43, p. 2887–2901. doi:10.1021/ie0306675

Valero Garces, B.L., Glerlowski-Kordesch, E., and Bragonier, W.A., 1997, Pennsylvanian continental cyclothem development: No evidence of direct climatic control in the Upper Freeport Formation (Allegheny Group) of Pennsylvania (northern Appalachian Basin): Sedimentary Geology, v. 109, p. 305–319. doi:10.1016/S0037-0738(96)00071-1

van Bergen, F., Krzystolik, P., van Wageningen, N., Pagnier, H., Jura, B., Skiba, J., Winthaegen, P., and Kobiela, Z., 2009, Production of gas from coal seams in the Upper Silesian Coal Basin in Poland in the post-injection period of an ECBM pilot site: International Journal of Coal Geology, v. 77, p. 175–187. doi:10.1016/j.coal.2008.08.011

Weniger, P., Franců, J., Hemza, P., and Krooss, B.M., 2012, Investigations on the methane and carbon dioxide sorption capacity of coals from the SW Upper Silesian Coal Basin, Czech Republic: International Journal of Coal Geology, v. 93, p. 23–39. doi:10.1016/j.coal.2012.01.009

White, C.M., Smith, D.H., Jones, K.L., Goodman, A.L., Jikich, S.A., LaCount, R.B., DuBose, S.B., Ozdemir, E., Morse, B.I., and Schroeder, K.T., 2005, Sequestration of carbon dioxide in coal with enhanced coalbed methane recovery: A review: Energy & Fuels, v. 19, no. 3, p. 659–724. doi:10.1021/ef040047w

Wilson, T., Nutt, L., Smith, R., Gulati, J., Coueslan, M., Peters, D., Wells, A., Hartline, C., Koperna, G., and Akwari, B., 2010, Pre- and post-injection vertical seismic profiling over the Southwest Regional Partnership's Phase II Fruitland Coal CO_2 pilot. Presented at AAPG Rocky Mountain Section 58th Annual Rocky Mountain Rendezvous, Durango, CO, June 2010.

Wilson, T.H., Siriwardane, H., Zhu, L., Bajura, R.A., Winschel, R. A., Locke, J.E., and Bennett, J., 2012, Fracture model of the Upper Freeport coal: Marshall County West Virginia pilot ECBMR and CO_2 sequestration site: International Journal of Coal Geology, v. 104, p. 70–82. doi:10.1016/j.coal.2012.05.005

Wong, S., Law, D., Deng, X., Robinson, J., Kadatz, B., Gunter, W. D., Ye, J., Feng, S., and Fan, Z., 2007, Enhanced coalbed methane and CO_2 storage in anthracitic coals—Micro-pilot test at South Qinshui, Shanxi, China: International Journal of Greenhouse Gas Control, v. 1, no. 2, p. 215–222. doi:10.1016/S1750-5836(06)00005-3

Yamaguchi, S., Ohga, K., Fujioka, M., and Muto, S., 2004, Prospect of CO_2 sequestration in the Ishikari coalfield, Japan, *in* Rubin, E.S., Keith, D.W., and Gilboy, C.F., eds., Proceedings of the Seventh International Conference on Greenhouse Gas Control Technologies, vol. I: Peer reviewed papers and overviews: Oxford, Elsevier Ltd, p. 423–430.

Yamaguchi, S., Ohga, K., Fujioka, M., and Nako, M., 2006, Field test and history matching of the CO_2 sequestration project in coal seams in Japan: International Journal of the Society of Materials Engineering for Resources, v. 13, no. 2, p. 64–69. doi:10.5188/ijsmer.13.64

Ye, J., Feng, S., Fan, Z., Wang, G., Gunter, W.D., Wong, S., and Robinson, J.R., 2007, Micro-pilot test for enhanced coalbed methane recovery by injecting carbon dioxide in south part of Qinshui Basin: Acta Petrolei Sinica, v. 28, no. 4, p. 77–80. [in Chinese with English abstract.]

Ye, J., and Zhang, B., 2016, Deep coal reservoir CO_2 ECBM research and equipment development: Final Report– National Science and Technology major projects, China United Coalbed Methane Corp. [in Chinese.].

Ye, J., Zhang, B., and Wong, S., 2012, Test of and evaluation on elevation of coalbed methane recovery ratio by injecting and burying CO_2 for 3# coal seam of north section of Shizhuang, Qingshui Basin, Shanxi: China Engineering Science, v. 14, no. 2, p. 38–44. [in Chinese with English abstract.]

ARTICLE

Resources and geology of coalbed methane in China: a review

Yong Qin [ID], Tim A. Moore, Jian Shen, Zhaobiao Yang, Yulin Shen and Geoff Wang

ABSTRACT

China produced 17.1 billion cubic meters (BCM) of methane sourced from coal seams in 2015, of which 4.43 BCM is from wells drilled from the surface. This level of production is a clear indicator that China has gone into early stage large-scale coalbed methane (CBM) development. CBM resources in China have been extensively investigated since the 1980s. Research has focused on the geological controls of reservoir character. There have been significant advances over the last 37 years that have aided China's CBM industry. CBM resources less than 2000 m in depth in China are estimated to be 36.81 trillion cubic meters, of which more than 84% occur in nine large-scale basins, such as the Qinshui, Ordos, Junggar, Qianxi, Erenhot, and Hailar. CBM accumulation and coal reservoir characteristics are controlled by the deposition, structure, coal rank, hydrology as well as other geological factors. Each basin has its own unique geological controls that influence the character of CBM reservoirs in both subtle and obvious ways. Coal reservoir geology in some basins or regions in China are still not well understood because of the complexity of the geological settings. At present, large-scale CBM production in China only occurs within the Qinshui and Ordos basins, mostly sourced from middle-to-high rank coal reservoirs. The CBM geology in other basins needs further investigation in order to achieve large-scale commercial production. To this end, the geological research in the paper should address issues such as how to stimulate economic gas flow from deep low permeability reservoirs and how best to efficiently produce from multiple horizons simultaneously. This paper summarizes these and other key issues that are significant scientific and technical challenges for the CBM industry within China.

1. Introduction

Commercial exploitation of coalbed methane (CBM) as a natural gas resource is an important milestone in the history of the global oil and gas industry (Flores 1998). The consumption in non-OECD countries is projected to grow an average of 1.9%/year from 2015 to 2040 in contrast to 0.9%/year in OECD countries (EIA 2017). The future of CBM will play an important role in fulfilling a significant proportion of this demand (Flores 2014).

The United States and Canada have been successful in developing their CBM resources. Since the 1970s, CBM has gone from being perceived as only a mining hazard to a viable commercially and economically sustainable commodity in basins such as the Black Warrior, San Juan, and Powder River (Repine 1990; Flores 1998, 2014; Moore 2012). But the US CBM production has been in decline since 2009. In contrast, China looks to be in the initial phases of its CBM development. China has abundant CBM resources as well as mine safety

issues. Investigations into exploitation of CBM in China began in the 1980s (Yang 1987) and then later in the 1990s production testing of wells began.

The first commercial production from CBM wells in China was in 2003 (Qin 2006). Production of methane from coal in China comes from two distinct sources: (1) methane extracted from wells drilled from the surface into a coal seam that is not expected to be mined, termed in this paper as 'CBM' for short, and (2) methane extracted during the mining process either from in-seam drilling in the mine or as wells drilled from the surface just a-head of mining, that is coal mine methane or 'CMM'. The annual CBM production in China increased from 0.01 billion cubic meters (BCM) in 2004 to 4.5 BCM in 2016, while CMM production rose from 1.93 BCM in 2004 to 12.8 BCM in 2016. The rise in CMM production is also thought to have helped, at least in part, in decreasing the number of mine deaths caused by coal mine gas, which is reported to have dropped from 2984 persons in

2004 (Liu and Zhang 2005) to 183 in 2016 (NEAC 2017). About 37.5% of the CMM produced in 2016 was utilized for power generation, conversation to LNG, or used as locally as domestic fuel.

As the CBM industry has grown, research on the CBM geology in China has also gradually increased (Qin 2009, 2012; Han *et al.* 2015). From the 1980s to about 2000, research was mainly focused on the investigation, prediction, and verification of regional CBM resource potential (Qin *et al.* 1999, 2001a; Stevens 1999; Ye *et al.* 1999; Zhang *et al.* 2002a, 2002b). During this time, small-scale well tests were carried out in some blocks such as Liulin, Jincheng, and Dacheng in North China (Stevens *et al.* 1992; Chi and Yang 2001; Hou 2002; Qin 2006; Chakhmakhchev 2007; also seen in Figure 1). Between 2000 and 2010, more specific studies relating to coal reservoir properties and how sweet spots are defined and developed were investigated (Wei *et al.* 2007, 2010; Liu *et al.* 2009b; Fu *et al.* 2009a; Qin 2008, 2012; Song *et al.* 2012; Zhou *et al.* 2012; Moore 2012). These studies contributed to the development of two CBM production areas, that is the Qinshui and Ordos basins (Yao *et al.* 2009; Jie *et al.* 2011; Yun *et al.* 2012; Zhou *et al.* 2015). Since 2011, more efforts have been made to address the geological challenges encountered during CBM exploitation, such as coal reservoir heterogeneity, fluid system complexity,

productivity simulation, and drainage dynamics (e.g. Keim *et al.* 2011; Lv *et al.* 2011; Luo *et al.* 2012; Tao *et al.* 2012; Zhang *et al.* 2013; Li *et al.* 2014; Zhou and Yao 2014; Wang *et al.* 2015; Yang *et al.* 2015). Studies have provided understanding of CBM geology has enabled development and, in particular, the optimization of single-well CBM production in China (Wong *et al.* 2010; Lv *et al.* 2012; Meng *et al.* 2014; Chen *et al.* 2015).

Geological settings of CBM resources in China are varied and complex (Ye *et al.* 1999; Qin *et al.* 1999, 2001b, 2001c, 2001d; Li *et al.* 2012a). This complexity translates into higher risks for exploration and development in many of the CBM fields in China. Three examples of the character of CBM reservoirs are given below.

In a first example, tectonically deformed coals (TDCs) occur widely in some regions of China such as southern North China and west Guizhou in South China (Gao *et al.* 2009; Liu *et al.* 2009b; Jiang *et al.* 2010). These locations account for 29% of China's CBM resources (Qin *et al.* 2012a). TDC reservoirs have characteristically high gas content, extremely low permeability, and a 'soft coal structure' (Fu *et al.* 2009b; Li *et al.* 2012b; Li *et al.* 2015), which can lead to extremely poor fracturing feasibility as well as result in reservoir damage and fines generation during gas extraction. The chemical structure in TDCs is also affected depending on whether there is ductile or tensile stress (Li *et al.* 2017).

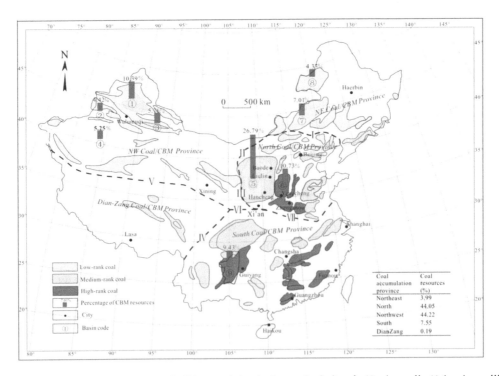

Figure 1. Coal/CBM accumulation provinces of China mainland. Orogenic belts: I: Yinshan; II: Helanshan; III: Liupanshan; IV: Longmenshan; V: Kunlunshan; VI: Qinling; VII: Dabieshan. Basin: ① Junggar basin; ② Tianshan basin group; ③ Tuha basin; ④ Santanghu basin; ⑤ Ordos basin; ⑥ Qinshui basin; ⑦ Erenhot basin group; ⑧ Hailar basin group; and ⑨ Qianxi (Western Guizhou)–Diandong basin group.

A second type of coal reservoir system occurs over multiple seams (Yang *et al.* 2011). These types of CBM reservoirs occur widely in western Guizhou and eastern Yunnan, where there are up to 30–60 coal seams in the upper Permian strata having a total net coal thickness of approximately 200 m. In this area, there are approximately 10 coal seams where gas could potentially be recovered (Gao *et al.* 2009). However, it has been noted that in this area, the vertical superposition of multiple CBM systems (Qin *et al.* 2008; Yang *et al.* 2015; Shen *et al.* 2016; Zhang *et al.* 2016) often makes it difficult to co-produce gas from multiple coal reservoirs because of the strong fluid interference between the coal seams (Qin *et al.* 2016a).

A third CBM reservoir type in China is those below 1000 m, where considerable resources occur (Qin *et al.* 2012a; Li *et al.* 2016b). Deep coal reservoirs have distinctive characteristics as a result of the higher compressibility, leading to lower permeability and lower elasticity under high temperature and pressure conditions (Qin *et al.* 2012a; Zhang *et al.* 2014; Qin and Shen 2016b). In order to successfully exploit these resources, new ideas, methods, and understanding of how these reservoirs behave will have to be first gained.

This paper introduces China's CBM resources and the associated geology. As such, there are four major goals of this paper: (1) to outline Chinese CBM in terms of the geological challenges faced and how to improve current production, (2) to thoroughly review past and ongoing research achievements, (3) to discuss exploration and development trends, and (4) using the Chinese experience in CBM research, exploration, and development, to inform and improve the understanding of how CBM systems work in other regions of the world.

2. Coal and CBM resources of China

Understanding CBM plays requires a thorough geological understanding of those resources. Over the last 30 years, the understanding of CBM in has increased hugely. This section will outline the CBM resources and their distribution within China.

2.1. Coal resources and their occurrence

The China National Administration of Coal Geology conducted a nationwide coal resource evaluation in 2013. This survey estimates that there are 5.89 trillion tons (TT) of coal resources, where 2.02 TT of explored resources and 3.87 TT of predicted resources (Cheng *et al.* 2016). Within the predicted coal resources, those buried to a depth of less than 1000 m (coal mining in China generally does not exceed this depth at present)

are estimated to be 1.44 TT or 37% of the total predicted resources, indicating that more than one-third of the predicted coal resources have the potential to be available for mining.

China has five geological coal provinces across the country based on three major tectonic belts (Figure 1). Within these areas, the resource categories can be subdivided according the palaeo-tectonics and palaeo-climate. Two are large EW-trending orogenic belts termed Yinshan and Kunlunshan–Qinling–Dabieshan. The third is the NNE-trending Helanshan–Liupanshan–Longmenshang orogenic belt (Han and Yang 1979; Cheng *et al.* 2016). About 44% of the total coal resources in China are estimated to occur in the northwestern province. Another 44% is also estimated to occur in the northern province while those in the southern province are about 8%. The coal resources in the northeastern province are calculated to have only 4% while those in the Dian-Tibet province are very small, probably less than 0.5% (Cheng *et al.* 2016). As can be noted from these estimates, the coal resources in China are mostly located in the northern and northwestern provinces (Table 1).

Distribution of coal resources through time in China is also not uniform (Table 2). From the early Palaeozoic to Cenozoic, coal accumulation in China has been influenced by five tectonic cycles (Figure 2). Coal-forming environments in China evolved according to a general rule, that is the older coals formed mainly in paralic settings but the younger coals formed in fluvial–deltaic settings. Paralic settings existed mostly in the basins of eastern China while fluvial–deltaic settings more in the west. During the early Palaeozoic, prior to the existence of land plants, organic accumulations were exclusively off shore, while during the late Palaeozoic, after the evolution of land plants, organic material, i.e. coal, formed in coastal or paralic settings. Finally, coal formed mostly in fluvial–deltaic settings during the Mesozoic and Cenozoic.

Table 1. Distribution of coal resources in China with coal rank (Cheng *et al.* 2016).

Coal accumulation province	Low-rank coal (%)	Medium-rank coal (%)				High-rank coal (%)
		Flame and gas coals	Fat and coking coals	Lean and meagre coals	Subtotal	
NE	80.67	16.03	2.61	0.46	19.10	0.23
North	0.35	26.49	16.44	41.01	85.05	15.70
South	1.34	2.97	31.44	8.23	42.64	56.02
NW	2.16	57.43	1.50	38.36	97.30	0.54
TZ	17.46	0.40	2.60	24.71	27.72	54.82
Average (%)	4.47	18.64	10.41	35.73	84.02	11.51

Notes: The classification of coals in the table follows the Chinese Coal Industry Standard 'MT/T 1158-2011' (SAWSC 2011). Compared with the classification standard proposed by the United Nations – Economic Commission for Europe (UN-ECE 1992), the sub-bituminous coal is included in the low rank coal in the table.
NE: Northeast; NW: northwest; TZ: Tian-Zang.

Table 2. Distribution of coal resources in China with geological era (Cheng *et al.* 2016).

Coal province	Late Palaeozoic (%)			Mesozoic (%)			Cenozoic (%)	
	C	C-P	P	T	J	K	E	N
NE	–	–	0.01	–	5.05	86.82	7.11	1.01
North	0.06	48.61	4.17	0.03	46.76	0.08	0.16	0.13
South	2.17	–	90.48	5.72	0.13	–	0.17	1.34
NW	0.52	0.21	–	0.02	99.20	0.04	–	–
DZ	50.82	–	19.29	11.58	–	0.27	0.08	17.96
Average (%)	0.52	21.50	8.71	0.47	64.67	3.52	0.37	0.23

C: Carboniferous; C-P: Permo-Carboniferous; P: Permian; T: Triassic; J: Jurassic; K: Cretaceous; E: Palaeogene; N: Neogene; NE: Northeast; NW: northwest; DZ: Dian-Zang.

Palaeozoic coal-bearing strata occur mainly in the north and south coal accumulation provinces while coal of Mesozoic and Cenozoic age exist mainly in the northwest, northeast, and north provinces.

In the Cambrian and Silurian strata in South China, there is a kind of combustible black shale, termed stone 'coals' (Lott *et al.* 1999). Stone 'coals', as a product of the Caledonian 'coal' accumulation, originated from marine algae in the neritic to bathyal slope environments, similar, in some aspects, to the how sapropelic coals form (Han and Yang 1979). Stone 'coals' are characterized by high ash yield, generally 40–90%, but a small amount can be as low as 20–40% (Han and Yang 1979; EBDES 2005). Stone 'coals' were once a fuel source in parts of southern China during the last century (Zhu 1979; Xu *et al.* 1990).

The most sustained coal accumulation in China, however, occurred during two periods: the Early to Middle Jurassic and the late Carboniferous to early Permian. About 65% of the total coal resources in China are found in Early to Middle Jurassic sequences, while 30% are of late Carboniferous to early Permian age (Cheng *et al.* 2016). Geographically, coal resources are abundant in the north and west of China and relatively poor in the south and east. In particular, it is in the northwest of China where coal resources are the most abundant (Figure 1 and Table 1).

The full rank suite of coals is present in China, with the highest rank found in the southeast (Figure 1). Lignite or brown coal mainly exists in the northeast, most notably in the eastern part of Inner Mongolia. Bituminous coal occurs mainly in the north and north-west provinces with some occurrence in the south province. Most anthracite occurs in the south and north provinces, with a maximum vitrinite reflectance of over 4.5% in the Fujian, southeastern China. In terms of geological periods, the older the coal-forming age, the higher the degree of coalification. Coal rank is dominantly low for the Cenozoic coals but are of medium rank for the Mesozoic coals though there are both low and high rank Mesozoic coal; finally, Palaeozoic coals tend to medium to high rank (Yang *et al.* 2005).

2.2. CBM resources and their occurrence

The first evaluation report on CBM resources of China was submitted by the Institute of Petroleum Geology, Ministry of Geology, and Mineral Resources of China in 1985. The report estimated nationwide CBM resources

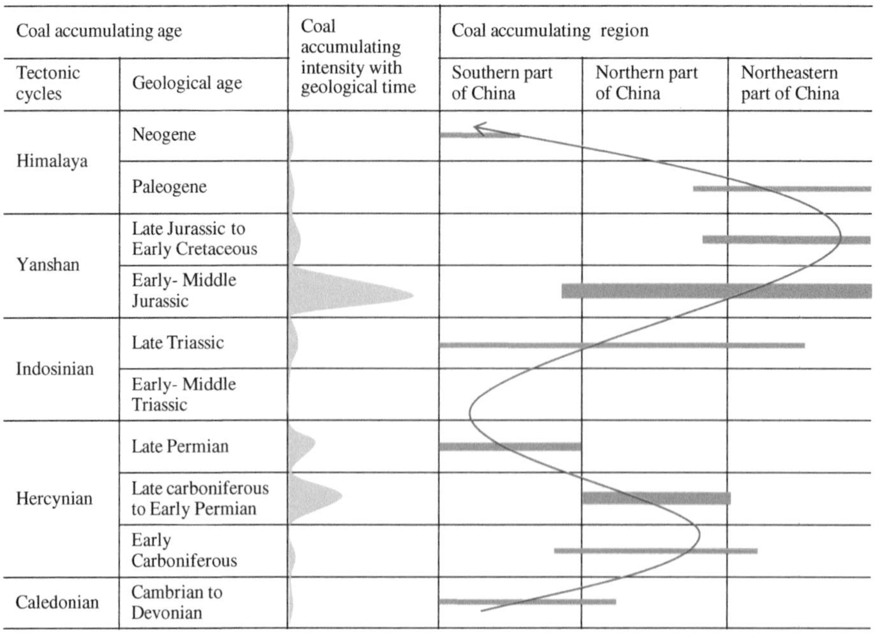

Figure 2. Sketch map of spatiotemporal distribution for coal accumulation in China. Coal accumulating profile is relative scale.

to be 17.93 trillion cubic meters (TCM) for depths less than 2000 m (Table 3). Note that this estimate and other *in-situ* estimates given in this paper are roughly equivalent to what the Petroleum Resource Management System would designate as original gas in-place (Anonymous 2011). From 1985 to 2002, more than 10 evaluations were carried out, with estimates ranging from 10.6 to 36.3 TCM in coal reservoirs buried to a depth of less than 2000 m (Table 3). The range of these estimates occurs for a variety of reasons, but most notably as a result of lack of detailed exploration data and differences in evaluation methodology.

During the period from 2004 to 2006, the Ministry of Land and Resources of China (MLRC) estimated nation-wide CBM resources (from 42 basins) as 36.81 TCM from coal seams more than 0.7 m in thickness and buried to depths of less than 2000 m (Table 3). From 2011 to 2012, the Chinese Academy of Engineering conducted a stra-tegic forecast of what China's CBM development might be for the years 2015 to 2030. The outcome of the study indicated that the total CBM resources available for further exploration and development could be as high as 25 TCM (Table 3). The resources identified all occur in large-to-medium-sized basins such as the Qinshui, Ordos, Junggar, Erenhot, Hailar, and western Guizhou.

Recently, the Langfang Natural Gas Institute of CNPC and CUMT estimated the deep CBM resources in the main basins of China through investigating the geolo-gical controls. It was estimated that these resources might be as high as 19 TCM in the depth range of 2000–3000 m (Li *et al.* 2016b; Qin and Ye 2015). Indeed, the high stress and low permeability of the coal reservoirs with such a great depth make it difficult

to develop the CBM economically (Qin *et al.* 2005). However, some previous successful experiments have indicated that exploitation of deep CBM reservoirs might be possible (Olson *et al.* 2002; Sun *et al.* 2008).

Based on the MLRC's evaluation, conducted in 2006 (Che *et al.* 2008; Liu *et al.* 2009a), the distribution of CBM resources in China is as follow:

(1) Overall, China has *in-situ* CBM resources of 36.81 TCM for reservoirs with depths less than 2000 m (Table 4). Of these *in-situ* resources, the recoverable resources to depths less than 1500 m are estimated to be approximately 10.87 TCM. Thus, on average, a recoverable resource of 0.98 BCM per square kilometre is calculated for CBM basins in China.

(2) CBM resources occur mainly in the areas north of the Qinling–Dabieshan belt, that is North China, including the northeast, north, and northwest provinces, accounting for the 87.34% of total CBM resources (Figure 4). A breakdown of these

Table 4. CBM resources of China in CBM provinces.[a]

CBM province	Geological resources		Recoverable resources		Abundance of geological resources (BCM[c]/km^2)
	TCM[b]	%	TCM	%	
NE	4.62	12.55	2.63	24.20	0.87
North	17.17	46.65	3.68	33.85	0.97
South	4.66	12.66	1.70	15.64	1.06
NW	10.36	28.14	2.86	26.31	1.02
DZ	0.0044	0.00	0.00	0.00	0.07
Total	36.81	100	10.87	100	0.98

[a]Data cited from Che *et al.* (2008) and Liu *et al.* (2009a), simplified; [b]TCM is an acronym for trillion cubic meters; [c]BCM is an acronym for billion cubic meters. NE: Northeast; NW: northwest; DZ: Dian-Zang.

Table 3. Previous evaluation results of *in-situ* CBM resources China.

No.	Source	Resources (trillion m^3)	Brief instruction	Source
1	Institute of Petroleum Geology, Ministry of Geology and Mineral Resources of China	17.93		Feng *et al.* (1995)
2	Chinese National Coal Corporation	31.92	Recoverable coal seams	Yang (1987)
3	Xi'an Branch, Coal Science Research Institute of China	32.15	Recoverable coal seams	Li *et al.* (1990)
4	Xi'an Branch, Coal Science Research Institute of China	30–35	Not including lignite and the early Carboniferous and early Permian coal seams	Zhang *et al.* (1991)
5	Jiaozuo Institute of Mining and Technology	24.75	Recoverable resources in recoverable coal seams	Zhang *et al.* (1992)
6	Petroleum Geology Research Group, Ministry of Geology and mineral resources of China	36.30	Including the recoverable resources of 18.15 trillion m^3	Duan (1992)
7	Institute of petroleum geology, Ministry of Geology and mineral resources of China	10.6–25.23	Resources of 17.06 trillion m^3 at 50% probability value	Li (1994)
8	Xi'an Branch, Coal Science Research Institute of China	32.86	Not including lignite and the early Carboniferous and early Permian coal seams	Zhang *et al.* (2002a
9	China National Administration of Coal Geology	14.34	Recoverable coal seams with gas content of more than 4 m^3 per ton, not including lignite seams	Ye *et al.* (1999)
10	Ministry of land and resources of China	36.81	Recoverable coal seams and including the recoverable resources of 10.87 trillion m^3	Che *et al.* (2008)
11	Chinese Academy of Engineering	25.29	Only including the large basins with recoverable resources of 4.08 trillion m^3	Xie *et al.* (2014)

resources by area is 4.62 TCM in the northeast, 17.17 TCM in the north, 10.36 TCM in the north-west, 4.66 TCM in the south, and 0.0044 TCM in Qinghai–Tibet (Table 4).

(3) By depth, CBM resources less than 1000 m are thought to be approximately 14 TCM, accounting for 39% of all CBM resources nationwide; the CBM resources between 1000 and 1500 m are estimated at 11 TCM, or 29% of all calculated resources; the CBM resources between 1500 and 2000 m are about 12 TCM, or 32% of total estimated resources.

(4) By rank, 53% (or 19 TCM) of total CBM resources are in middle-to-high rank coal reservoirs, and potential recoverable resources from these ranks are thought to be about 6 TCM. The remainder of the CBM resources (18 TCM or 47%) are from low rank coal reservoirs, and the potential recoverable resource is thought to be about 5 TCM (Table 5).

(5) From the basin point of view, CBM resources in China are concentrated in a few large basins. It is estimated that 84% of all CBM resources (i.e. 31 TCM) occur within nine large basins; each of these basins is estimated to have more than 1 TCM of CBM resources. These nine basins are Ordos, Qinshui, Junggar, western Guizhou, Erenhot, Turpan-Hami, Tarim, Tianshan, and Hailar (Figure 1). All other remaining basins have only 16% of the country's total CBM resources, and each of those individual basin's resources is less than 1 TCM.

3. Important CBM regions of China

3.1. Southern region of Qinshui basin

The Qinshui basin is located in the southwest of Shanxi Province, with an area of about 27 thousand square kilometres, and has been the largest CBM production in China since 2003 (Figure 1). CBM in the basin is sourced from Permo-Carboniferous coal bearing strata. In 2014, the basin had more than 6300 CBM wells and

accounts for 74% of CBM production for all of China (Qin and Ye 2015).

3.1.1. Geological setting of CBM

The coal-bearing strata of the Qinshui basin mainly include the lower Permian Shanxi and the upper Carboniferous Taiyuan Formations (Figure 3(a)). In the middle and southern region of the basin, the Taiyuan Formation is 80–93 m thick and is composed of medium-to-fine sandstone, siltstone, mudstone, limestone (four to five intervals), and coal (seven to nine seams). The Taiyuan Formation is interpreted as forming in an epicontinental sea environment with barrier, lagoon, and tidal flat systems in the north and a carbonate platform system in the south (Liu et al. 1998). The Shanxi Formation is 30–50 m thick and is conformable with the underlying Taiyuan Formation. The formation is composed of sandstone, sandy mudstone, mudstone, and coal (two to three seams) and is interpreted to have formed in a deltaic setting, with the delta plain in the north and the delta front in the south (Liu et al. 1998).

Tectonically, the Qinshui basin is a residual part of the giant late Palaeozoic basin in North China (Yang and Han 1979). It is a synclinorium, generally striking north-northeast to south-southwest (Figure 3(b)). The surface geology changes from Triassic age sediments in the synclinorium core to Permian, Carboniferous, Ordovician, and Cambrian on the limbs (Liu et al. 1998; Su et al. 2005). The structural differentiation within the basin is distinct, with a complex gas-controlling structure (Qin et al. 2001a). A series of secondary folds with an axial direction approximately parallel to the synclinorium axis exists in the interior of the basin and is cut by a series of faults with different strikes planes (Figure 3(b)). It should be noted that the NE-extending Shitou normal fault has a strike of N10–60°E and in the southern part of this fault it is down thrown to by as much as 360–560 m to the northwest (Qin et al. 2012a). This fault is thought to have a profound influence on CBM retention and coal reservoir pressure in the east and west sides of the fault (Figure 4). Interestingly, CBM

Table 5. Low-rank CBM resources of China.[a]

Basin (group)	CBM resources (BCM)		Basin (group)	CBM resources (BCM)	
	Geological	Recoverable[b]		Geological	Recoverable
Junggar	3826.816	807.795	Ordos	5277.541	616.445
Tuha	2119.834	410.046	Datong	142.811	47.063
Santanghu	594.214	1752.60	Erenhot	2581.663	2102.638
Qaidam	141.176	57.924	Hailar	1101.203	250.251
Tianshan	1626.154	667.162	Basin group in Northeast China	42.289	21.220

[a]The data ate cited from Qin (2012); [b]the recoverable CBM resources are defined in accordance with the geological and mineral industry standard (DZ/T 0216–2002) of China (Ministry of Land and Resources of China 2001).

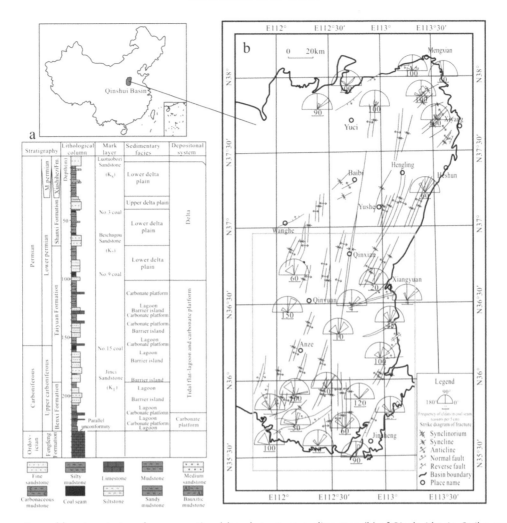

Figure 3. Composite coal-bearing strata columnar section (a) and structure outline map (b) of Qinshui basin. Strike rose diagrams of the fissures in coal seams are included in the structure outline map. Columnar section is simplified (Shao *et al.* 2015). Structure outline map is cited from Qin *et al.* (2012a). Blue box in Figure 3(b) shows the geographical extent of the southern Qinshui basin.

production in the east up-thrown side of the fault is higher than that in the west down-thrown side (Cai *et al.* 2011).

The depth of burial for the coal-bearing strata is ultimately controlled by the structure of the syncline (Qin *et al.* 2001a; Su *et al.* 2005a; Wei *et al.* 2007). The depth of the strata gradually increases from the edge of the basin into the core of the syncline, with a maximum depth from the surface of over 2000 m. The Qinshui basin has two areas where sediments have been buried deeply: one centre is in the south where sediments have been buried to about 1200 m and another centre in the north where sediments have been buried to depths of greater than 2200 m (Figure 5(a)). This tectonic framework defines the basic pattern of CBM accumulation in the basin through the overall control of the coal reservoir pressure and gradient pattern (Figure 5(b)).

Current tectonic stress field influences the degree of opening of the fractures within the coal. Since the

Mesozoic, the Qinshui basin has experienced three stages of tectonic stress. The modern *in-situ* stress field of the basin is under NNW–SSE-trending extension (Xu *et al.* 1998). There are two zones currently with high principal stress differentials (i.e. the difference between horizontal maximum and minimum principal stress directions) in the east limb of the basin. These areas have good coal reservoir permeability (Qin *et al.* 2000). The good permeability is a result of the relative tension or opening of the coalbed fissures parallel to the orientation of the maximum horizontal principal stress under the relative increase of the maximum stress (Qin *et al.* 2012a).

3.1.2. CBM reservoirs

Main coal reservoirs in the southern Qinshui basin are the No. 3 coal seam of the Shanxi Formation and the No. 15 coal seam of the Taiyuan Formation, and these seams are separated by 80–87 m of interburden

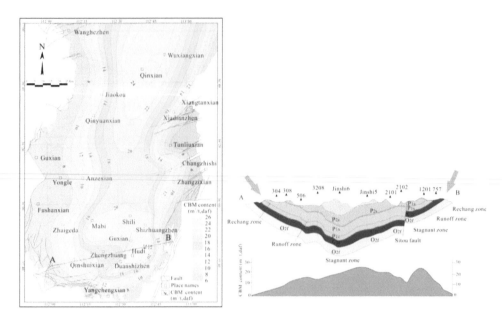

Figure 4. CBM content distribution and its structure control of No. 3 coal reservoir in the southern Qinshui basin. Structural profile is cited (Cai *et al.* 2011). The position of the basin is shown in Figures 1 and 3.

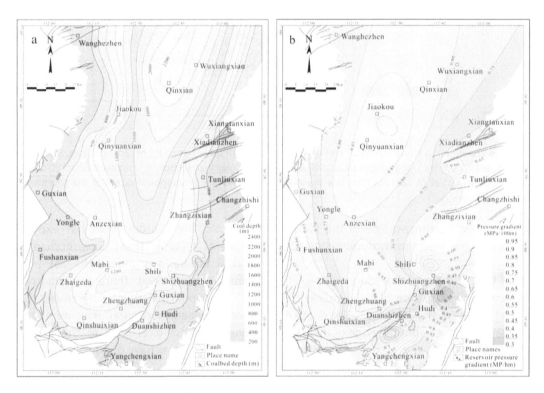

Figure 5. Isoline maps of buried depth (a) and pressure gradient (b) of No. 3 coal reservoir in the southern Qinshui basin. The position of the basin is shown in Figures 1 and 3.

(Figure 3(a)). Locally, the No. 9 coal seam of the Taiyuan Formation is also a CBM target. The No. 3 coal seam has simple structure and ranges in thickness from 0.8 to 6.4 m and averages greater than 5 m in the southeast of the region but less than 2 m in the north (Figure 6(a)). The thickness of the No. 15 coal seam is 1.0–5.0 m, but on average is over 2.5 m, especially in the southern part of the basin (Figure 6(b)). The No. 15 coal seam is overlain by limestone in most places and is locally split. The roof of the No. 3 coal seam is mostly mudstone, with fine to silty sandstone or sandy mudstone locally.

Figure 6. Isoline maps of No. 3 (a) and No. 15 (b) coal reservoir thickness in the southern Qinshui basin. The position of the basin is shown in Figures 1 and 3.

There are three main aspects of coal rank variation in the basin. Maximum vitrinite reflectance ($R_{o,max}$) of the main coal seams is over 2.0% in most areas, especially in the southern areas of the basin. The western and eastern parts of the basin have slightly lower rank, usually ranging from 1.3% to 2% $R_{o,max}$. Except for the southern area (see next paragraph), $R_{o,max}$ increases with depth. $R_{o,max}$ of the No. 15 coal seam is generally 0.2–0.3% higher than that of the No. 3 coal seam in the same area (Qin et al. 2012a).

Gas volume in coals of the region is relatively high and generally ranges from 4 to 22 m³/t (all gas data are presented on a 'daf' basis unless otherwise noted). For example, the CBM content of the No. 3 coal seam is 0.42–27.99 m³/t, but most of the gas contents are between 4 and 16 m³/t, with an average of 10 m³/t (Figure 4). The gas content of the No. 15 coal seam ranges from 0.33 to 37.93 m³/t, though most of the values are between 4 and 22 m³/t, with an average of 11.3 m³/t. The regional distribution pattern of the CBM content has two characteristics: (1) the gas content is higher at the southern, deeper end than in other areas and (2) gas content increases gradually from the edge of the basin into the more deeply buried areas. It means that the CBM content in the south is controlled by the groundwater dynamics to a greater extent (Qin et al. 2001b).

Based on the observations from underground coal mines, there are two principle orientations for fractures in the coal seams throughout the region. These two fracture orientations are nearly perpendicular to each other but are both vertical relative to the bedding surface. Notably, mineral fracture fillings in either of the orientations are rarely seen (Qin et al. 2012a). The frequency of NNE–SSW fractures is significantly higher than that of the WNW–ESE fractures. These fracture orientations and frequencies can be observed throughout the basin with only relatively slight variation (Figure 3(b)). The dominant trends of the fractures in the south of the basin, which occur mostly in the roof material above the coal seams, are predominantly north-northwest to south-southeast. However, towards the north part of the basin, the fractures change orientation to a WNW–ESE direction. Overall, there is a large intersection angle between the main fracture orientation of the coal seams and their roofs. This may signal an important aspect of the tectonic stress control on CBM accumulation in the basin (Qin et al. 2001a), such as a difference in the way strain affects the coal and rock.

Based on 92-well tests from 55 CBM wells, the *in-situ* permeability of the coal reservoirs in the region ranges from 0.0004 to 112.6 mD; the majority of the measurements, however, are less than 2 mD (Qin et al. 2012a). The permeability of the No. 3 coal seam is less than 0.1 mD at a 50% statistical probability, 0.1–1.0 mD at 26% probability, and 1.0–5.0 mD at 20% probability. The permeability of the No. 15 coal seam is lower than 0.1 mD at a 50% statistical probability, while at a 26% probability, the range of permeability is 0.1–1.0 mD. The permeability in the range of 1.0–5.0 mD has a

20% probability. The permeability of the No. 15 coal seam is less than 0.1 mD with a 48% probability; a range of 0.1–1.0 mD has a 33% probability, while a range of 1.0–5.0 mD has a 12% probability. On average, it can be seen that the permeability of the No. 3 coal seam is slightly higher than that of the No. 15 coal seam. Regionally, the coal reservoir permeability tends generally to increase from the north to the south and is commonly over 0.5 mD in the southeast but below 0.3 mD in the north (Figure 7). In addition, the permeability in the region changes in line with the general rule that permeability decreases with increasing depth (McKee *et al.* 1987; Qin *et al.* 2001a; Meng *et al.* 2011).

Based on 76-well tests conducted in 55 CBM wells, the coal reservoir pressure gradient (CRPG) in the region changes from 0.28 to 1.6 MPa per hundred meters (MPa/hm), indicating that the coal reservoirs are either normally pressured or slightly under pressured, which is beneficial to CBM production in general. There are only a few overpressured areas (Table 6). The No. 3 coal reservoir is mainly under-pressured, with a CRPG of 0.153–1.08 MPa/hm and an average of 0.62 MPa/hm. The under-pressure state of the No. 15 coal reservoir is more obvious: the CRPG ranges from 0.28 to 0.97 MPa/hm, averaging 0.64 MPa/hm. Isopachs of the coal reservoir pressure are parallel to the depth contours of the Qinshui synclinorium in a ring-like pattern, that is the reservoir pressure generally increases from the edge to the deeper parts of the basin, reflecting an overall control of burial depth. In the southwestern part

Table 6. Statistics of well-testing pressure gradient of main coal reservoirs in southern Qinshui basin.

| No. of coal seam | Statistical unit | Well-testing pressure gradient (MPa/hm) | | | | | Average (MPa/hm) |
		<0.50	0.50–0.75	0.75–0.90	0.90–1.0	>1.0	
No. 3	Number of wells	11	22	9	6	2	0.62
	Percentage (%)	22	44	18	12	4	
No. 15	Number of wells	8	17	7	7	2	0.64
	Percentage (%)	19.51	41.46	17.07	17.07	4.88	

Note: The 'hm' means100 m.

of the basin, the pressure of the No. 3 coal reservoir is abnormally high relative to depth. This is thought to be the result of local of the strong groundwater dynamics on the reservoir pressure system (Qin *et al.* 2001b; Wei *et al.* 2007; Meng *et al.* 2011), resulting in drainage difficulty and gas productivity decline of the CBM wells.

3.1.3. Geological controls of CBM accumulation and coal reservoirs

CBM reservoirs are commonly characterized based on their gas content, pressure, and permeability as well as their response to production-enhancing technology. Of particular relevance is permeability, which is a function of the stress as well as the aperture, length, height, density, and connectivity of fractures systems (Aydin, 2000; Close 1993; Karacan and Okandan 2000; Titheridge 2014). These

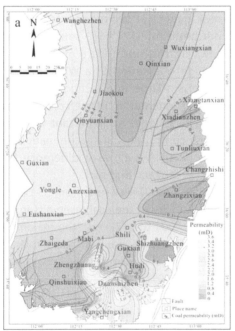 

Figure 7. Isoline maps of No. 3 (a) and No. 15 (b) coal reservoir permeability in the southern Qinshui basin. The permeability was obtained with the injection pressure-drop tests of the CBM wells. The position of the basin is shown in Figures 1 and 3.

fracture systems are formed during burial, compaction, coalification, and structural deformation subsequent to coal formation. Thus, the gas content and overall resources as well as coal reservoir pressure and ultimately gas deliverability are controlled by sedimentological and geological evolution of the individual coal reservoirs. The Qinshui basin has had numerous studies since the 1990s on geological controls on gas properties (e.g. Liu *et al.* 1998; Qin *et al.* 2001b–d; Su *et al.* 2005; Cai *et al.* 2011; Qin *et al.* 2012a; Zhou *et al.* 2012, 2015), on the physical properties of its coal reservoirs (e.g. Qin *et al.* 1999; Fu *et al.* 2005; Wu *et al.* 2007; Yao *et al.* 2009; Shen *et al.* 2011), on the coal reservoir pressure variation (e.g. Ye *et al.* 1999; Wu *et al.* 2014), on CBM reservoir-forming processes (e.g. Qin *et al.* 2012a; Wei *et al.* 2007; Song *et al.* 2012; Zhang *et al.* 2015, 2016), as well as on gas productivity (e.g. Lv *et al.* 2011, 2012; Shu *et al.* 2012; Zhang *et al.* 2013).

A dynamic equilibrium model was proposed to describe the geological process of gas accumulation in coal reservoirs in the Qinshui basin, including gas generation and migration (Wei *et al.* 2007). The results show that this process can be characterized into three stages for the Qinshui basin. The first, or primary, stage occurred during the late Palaeozoic and Triassic. The quantity of gas generation during this initial stage is thought to have been low and only weak diffusion took place (Figure 8(a)). During the second stage, termed the active accumulation–dissipation stage, coalification developed rapidly and was accompanied by strong gas generation under an abnormally high palaeo-geothermal gradient (Qin *et al.* 1997; Ren *et al.* 2005) during the Late Jurassic and Early Cretaceous (Figure 8(b)). This high geothermal gradient resulted in an increase in gas content. The increase in gas content may have resulted in abundant gas migration with significant loss of gas

through diffusion, cap outburst, and seepage (Qin *et al.* 2016c). During the third stage, defined as the absolute dissipation stage, coalification and gas generation no longer occurred, and gas was lost mainly through diffusion and partly by seepage during uplift of the basin (Figure 8(c)). It has been suggested that although high gas generation was accompanied by crustal uplift, which may have led to active gas dissipation from the coal reservoir, gas content remained relatively high in the south and north of the basin as a result of good gas retention conditions (Wei *et al.* 2007).

Within the Qinshui basin, CBM content is positively correlated with depth in a certain depth range (Figure 4), which is consistent with the syncline controlling gas model (Cai *et al.* 2011; Song *et al.* 2012). However, gas contents at the edge of the basin do not completely follow a depth increase gas model (Liu *et al.* 1998; Su *et al.* 2005), which suggests other geological controls. Regionally, groundwater in the coal bearing strata is mainly recharged from the northwest and, less significantly, from other basin margins. However, the NW-tilting Sitou fault in the south results in a hydrodynamically blocked area to the west of the fault, that is this results in a trap retaining gas on the east side of the fault (Figure 9(a)). This hydrodynamic condition leads to slow or stagnant groundwater zones on both sides of the Sitou fault and is characterized by significantly increased total ion concentration of the groundwater, resulting in gas accumulation (Figure 9(b)). The conclusion here is that groundwater hydrodynamics is very important for gas accumulation within the southern Qinshui basin; weak or stagnant groundwater runoff is a key to CBM enrichment. Finally, gas content in the region increases with a concomitant increase in the maximum vitrinite reflectance, and conversely, gas content decreases

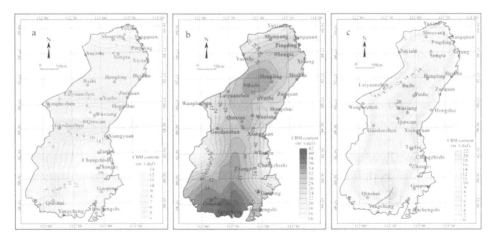

Figure 8. Geological evolution history of CBM content isolines in No. 3 coal reservoir of Qinshui basin with numerical simulation (Wei *et al.* 2007; Qin *et al.* 2012a). CBM content at the end of Middle Jurassic is shown in Figure 8(a), that at the end of Early Cretaceous in Figure 8(b), and that at the end of Neogene in Figure 8(c). The position of the basin is shown in Figures 1 and 3.

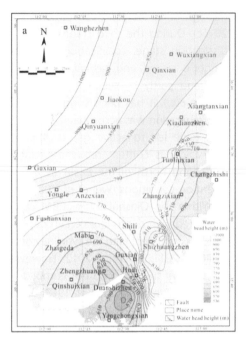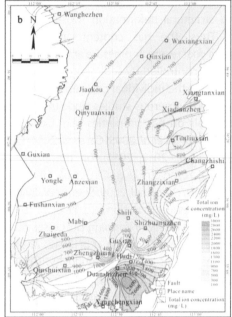

Figure 9. Isoline maps of head height (a) and total ion concentration (b) of ground water in No. 3 coal reservoir of the southern Qinshui basin. The position of the basin is shown in Figure 1.

with an increase in ash yield within the coal reservoirs (Qin et al. 2001b, 2001c).

At the basin scale, in-situ permeability of coal reservoirs in the region is similarly controlled by depth and stress field (Qin et al. 2012a). As a general background, the permeability is closely related to the tectonic stress field, the secondary structure, and the structural deformation of coal. The Qinshui basin has a present-day horizontal compressional stress field in the direction of east-northeast to west-southwest (Xu et al. 1998). It was found that the permeability in the region is directly related with the difference between the current maximum and minimum principal stresses in the basin. The permeability is more than 0.5 mD when the principal stress difference is over 85 MPa and tends to attenuate as the difference reduces (Figure 10). Structural curvature of strata is the result of compression from the palaeo-tectonic stress field and reflects the degree of development of both the secondary folds and derived tensile fractures. The in-situ permeability of the coal reservoirs in the region is relatively high when the curvature of the coal seam is more than 0.1×10^{-4}/m when in proximity to secondary anticline axes (Zhang et al. 2003). Moreover, coal fractures in the region increase in the vicinity of faults resulting in what is termed as 'TDCs'; TDCs have an excessive fracturing or pulverized structure and thus significantly reduced permeability in these areas (Teng et al. 2015).

Favourable blocks for CBM development occur mainly in the south and west side of the region as a result of geological controls. These areas connect into a high CBM production 'corridor' (i.e. 'sweetspot' or 'fairway') having a NNW–SSE-trend. The southern end of the Qinshui synclinorium is in the heartland of the horizontal principal stress difference and has stagnant groundwater runoff, which is conducive to the preservation of thermogenic gas. It is no wonder then that this area includes several of the most important CBM production blocks within the basin and even China, such as the Panzhuang, Panhe, and Fanzhuang blocks (Qin and Ye 2015; Ye and Li 2016). High gas content and coal permeability in several other blocks, such as the Zhengzhuang, Shizhuang, and Qinyuan, also have been confirmed through CBM production in recent years.

3.2. East margin of Ordos basin

The eastern margin of the Ordos basin currently has the second largest CBM production in China. Its area is about 20 thousand square kilometres (Figures 1 and 11). In the first 8 months of 2015, about 0.56 BCM of gas was produced from the 2677 CBM wells in the region and this production was sourced primarily from three blocks: the Baode, Yanchuannan, and Hancheng (Ye and Li 2016).

258

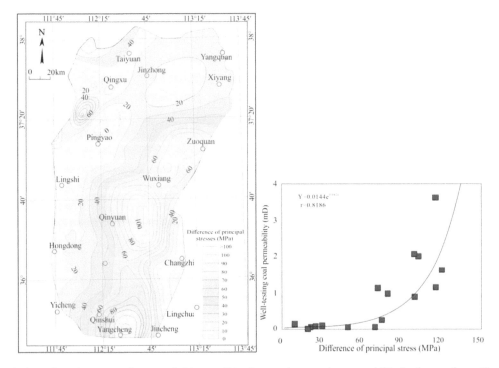

Figure 10. Correlation of modern tectonic stress field to well-testing coal reservoir permeability in the southern Qinshui basin (Qin et al. 2012a). Isoline map of the difference between maximum and minimum horizontal principal stresses can be seen in Figure 10 (a) and the plots of the stress difference to the well-testing permeability in Figure 10(b). The position of the basin is shown in Figures 1 and 3.

3.2.1. Geological setting

A coal-bearing stratum in the region is mainly composed of the lower Permian Shanxi Formation and the upper Carboniferous Taiyuan Formation, which together include more than 20 coal seams (Figure 11). The Shanxi Formation has nine coal seams with 2–3 minable seams (more than 0.7 m in single seam thickness), and the Taiyuan Formation contains 10–13 coal seams, including 3–5 minable seams. Burial depth of the coal seams is controlled by a westward dipping monocline, with a maximum depth greater than 1500 m (Shen 2011; Chen et al. 2015). A series of subordinate N–S and NE–SW-trending folds and faults are overlaid on the monocline (Wang et al. 2013).

CBM reservoirs in the basin have been greatly affected by three tectonic events: (1) the Late Triassic Indosinian movement, (2) the Jurassic-to-Cretaceous Yanshanian movement, and (3) the Cenozoic Himalayan movement (Wang et al. 2010; Wei et al. 2010). The maximum horizontal compressive stress extended in a nearly N–S direction during the Indosinian epoch, but migrated to a NE–NNE direction during the Yanshanian epoch, and finally changed into a NE–SW direction during the Himalayan tectonic phase. In modern times, the maximum horizontal stress trends east-northeast to northeast (Wang et al. 2010),

which is sub-parallel to the strike of the coal fractures and is thus conducive for permeability.

3.2.2. CBM reservoirs

Coal reservoirs in the region are thicker in the north and thinner in the south. Average thickness of the two main coal reservoirs is, respectively, 7.6 and 10.2 m in the Baode block in the north (Tian et al. 2012), 5.51 and 5.11 m in the Liulin block in the middle (Zhang et al. 2010; Liu and Zhang 2016), and 3 and 4.5 m in the Hancheng block in the south (Zhang 2008).

The depth generally controls coal rank in the region. Thus, the maximum vitrinite reflectance generally increases towards the west in coincidence with the deepening of the synclinorium and is highest near Yangchuan, where the coal depth is greatest (Figure 12). Locally, the Mesozoic magmatic events have led to higher coal rank near Linxian in the north where the maximum vitrinite reflectance is up to 2.1% proximal to the Zijinshan igneous rock mass.

Based on the measurements of coal samples from 403 wells, the gas content in the region varies from 3 to 27.2 m^3/t. The gas content in the Taiyuan and Shanxi Formations generally increases towards the west with the plunging of the synclinorium (Figure 13), which is

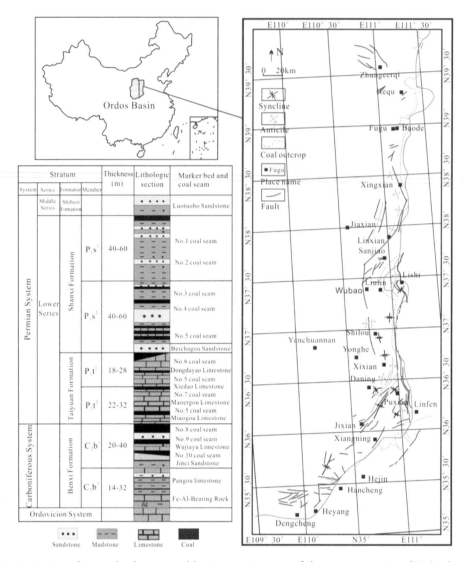

Figure 11. Location, tectonic outline, and columnar coal-bearing strata maps of the eastern margin of Ordos basin. The position of the basin is shown in Figure 1.

consistent with the trends of coal rank distribution. Thus, the deeper the coal reservoir, the higher the rank and larger the volume of CBM (Shen 2011).

Porosity of the coals in the region ranges from 2.55% to 8.43% calculated from true density and apparent density, with an average of 4.83%, and decreases from north to south with increasing coal rank (Zhang 2008; Li *et al.* 2014). Based on the observations of the coal seams in wells and mines, two sets of coal fractures exist in the region and are nearly perpendicular to each other. One fracture system is oriented in a NE–SW direction and the other in an ENE–WSW direction; both are partly filled by calcite (Wei 2015).

Compared with most other regions in China (Ye *et al.* 1999), the permeability of the coal reservoirs in the region is relatively high, ranging from 0.006 to 42.86

mD and averaging 3.39 mD (Qin *et al.* 2012a). The highest permeability is generally in the north and central parts of the region with the south having the lowest permeability coal reservoirs (Zhou *et al.* 2013; Jian *et al.* 2015). The strong heterogeneity in the coal permeability is likely controlled by a combination of depth, local structure, tectonic stresses, and coal rank (Song *et al.* 2013; Sun *et al.* 2013; Li *et al.* 2015b).

Based on 92-well tests, the pressure gradient of coal reservoirs in the region varies systematically (Figure 14). The average pressure gradient for the northern and central parts of the region is 1.08 MPa/hm for the Shanxi and 0.82 MPa/hm for the Taiyuan Formations. However, the pressure decreases in the south-central part to 0.58 and 0.70 MPa/hm and then recovers somewhat further southward to 0.9

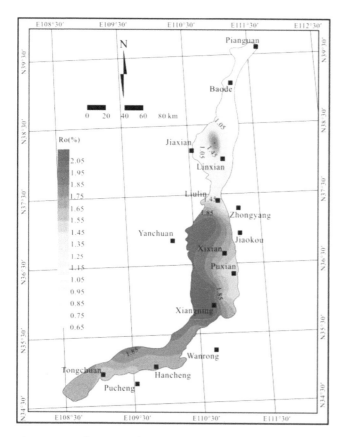

Figure 12. Isoline map of maximum vitrinite reflectance of main coal seams in the eastern and southern margins of Ordos basin. The position of the basin is shown in Figures 1 and 11.

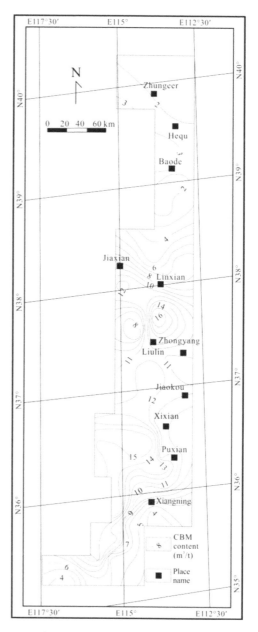

Figure 13. Isoline map of CBM content in No. 4 + 5 coal seam from the eastern margin of Ordos basin (Li *et al.* 2014b). The position of the basin is shown in Figures 1 and 11.

and 0.95 MPa/hm but once again reduces in the far south to 0.82 and 0.91 MPa/hm, for the Shanxi and Taiyuan Formations, respectively (Shen 2011; Xue *et al.* 2012). Note that the different formations have different pressure gradients, suggesting that there are multiple CBM systems within a vertical stratigraphic section (Shen 2011). The differing CBM systems may result from the change of the hydrogeological conditions within the vertical section, possibly linked to the area's sequence stratigraphic architecture (Qin *et al.* 2008).

3.2.3. Geological controls of CBM accumulation and coal reservoirs

CBM volume is controlled by many factors such as coal rank, present, and past burial depth, floor and roof lithology, formation temperature, structure, and hydrodynamics (Ayers *et al.* 1991; Scott *et al.* 1994; Johnson and Flores 1998). All of these influences are probably at work in the eastern margin of the Ordos basin.

CBM content in the region does not just increase linearly with depth. Gas volumes tend to increase from the surface to about 600 or 700 m but after that tend to decrease (Figure 15). This type of trend has been observed

elsewhere, such as the Surat basin in Australia (Hamilton *et al.* 2012). The key causes for this parabolic trend in gas content, in the Ordos basin, are thought to be a combination of formation temperature, reservoir pressure, and coal rank (Qin *et al.* 2005, 2012b; Shen 2011; Song *et al.* 2013). According to the Langmuir equation, shallower coal reservoirs are highly influenced by reservoir pressure; with increasing depth (i.e. hydrostatic pressure), a significant increase in the capacity to absorb methane will result. However, with increased depth comes increased temperature, working against gas-holding capacity, and may result in a decrease in gas content (see Section 4.1 and Moore 2012).

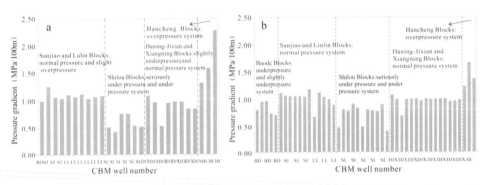

Figure 14. Distribution of coal reservoir pressure gradient in the eastern margin of Ordos basin from Tian *et al.* (2015) and Shen *et al.* (2011). Pressure gradient of the coal reservoirs in Shanxi formation is shown in Figure 14(a) and that in Taiyuan formation in Figure 14(b).

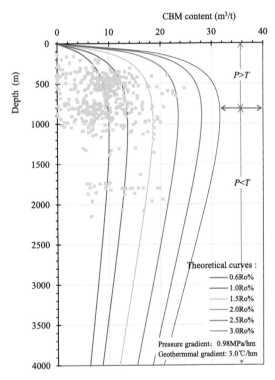

Figure 15. Plots of CBM content to depth of coal reservoirs in eastern Ordos basin. Blue squares are measured CBM content from 427 drilling coal samples and theoretical CBM content curves came from the model suggested by Shen (2011), Qin *et al.* (2012b), and Shen *et al.* (2014). It is emphasized that the change of CBM content is related to coal reservoir pressure and geothermal gradients except for depth.

It is known that low rock permeability will result in stronger sealing ability in relation to CBM reservoirs. Generally, sealing ability reduces successively from mudstone, siltstone, limestone, sandy mudstone to sandstone. As shown in Figure 16, the coal reservoirs in the Shanxi Formation within the Daning–Jixian block of the south part of the Ordos basin have roof and floor composed of thick mudstone and sandy mudstone, with gas contents

of 13–15 m^3/t. In the Sanjiao, Linxian, and Baode blocks of the north, the roof and floor are composed of limestone, mudstone, muddy sandstone, and sandstone, with gas contents ranging between 3 and 10 m^3/t. Interestingly, coal reservoirs within the Taiyuan Formation that are bounded by sediments composed predominantly of muddy sandstone and sandstone have gas contents that are lower than 2.0 m^3/t (Tian *et al.* 2015).

Within the region, CBM contents are generally higher in the syncline and near reverse faults and lower in the anticline and adjacent to normal faults. For example, the CBM content in the Hancheng block varies from 7.54 to 15.47 m^3/t in the syncline, but only from 4.81 to 8.46 m^3/t adjacent to normal faults (Wang 2002; Li *et al.* 2014; Xu and Guo 2014). In the Daning–Jixan block, also in the south, the CBM contents are over 20 m^3/t near reverse faults (Chen et al. 2015), while those near normal faults are often less than 10 m^3/t (Wang *et al.* 2013).

Hydrodynamics will also exert significant influence on gas contents. For example, the gas content in coal reservoirs increases continuously from the eastern recharge zone of the Ordos basin to the middle runoff zone and then on into the western retention zone. This trend is attributed to hydrodynamic sealing of the gas generated within the coal reservoirs. Again, CBM content in the Liulin block is 8.25–12.83 m^3/t, and westward of Liulin is the Wupu block with gas contents that range from 14.37 to 14.95 m^3/t (Li and Zhang 2013). Note that the gas content of the No. 5 coal reservoir in the Daning area of the south Ordos basin is lower than 10 m^3/t in the eastern runoff zone but is up to 20 m^3/t in the western retention zone.

Permeability of the coal reservoirs in the region is dominantly controlled by coal-body structure, fracture frequency, aperture size, *in-situ* stress, and burial depth. In general, permeability is inversely related to depth but, as illustrated by Figure 17, there can be considerable variation. This variation indicates the influence of

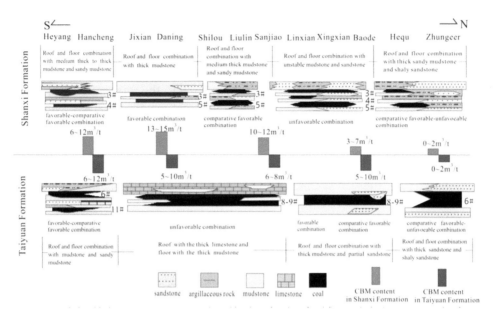

Figure 16. Relationship between CBM content and combination of coal roof and floor rocks in the eastern margin of Ordos basin (Tian *et al.* 2015). The area position of the map is shown in Figure 1, and the places are shown in Figure 11.

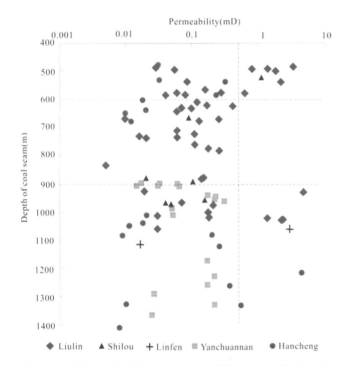

Figure 17. Plots of well-testing permeability to depth of coal reservoirs in the eastern margin of Ordos basin (Li *et al.* 2015b). The location of the four places is shown in Figure 11.

other geological factors on permeability. Generally, there is higher permeability in coal reservoirs that are not intensely deformed, though still fractured, while lower permeability is associated with coal reservoirs that are fragmented or mylonitic. For example, a coal reservoir which is fractured, as reported in well Han-1 of the Hancheng block, has a permeability of 16.17 mD,

but a coal reservoir with mylonitic structures, as reported in the adjacent well Han-2, is only 0.04 mD (Zhang 2008; Xu and Guo 2014).

3.3. Southern margin of Junggar basin

The Junggar basin is located in the northern part of the Xinjiang Uighur Autonomous Region and has an area of approximately 134,000 km^2. The CBM resources have been estimated at about 3.83 TCM (Che *et al.* 2008), thus ranking it as the third largest CBM basin in China. Pilot programs in CBM along the south margin of the basin have resulted in annual CBM production capacity of 30 million cubic meters. This production comes from the western Fukang block and has the highest production (with a peak production of 17,000 m^3/day) from a vertical well (CDS01) in all of China (Qu 2016).

3.3.1. Geological setting

Coal bearing strata in the southern margin of the Junggar basin (Figure 1) include the Lower Jurassic Badaowan and the Middle Jurassic Xishanyao Formations, and these units are interpreted as fan delta and lake environments (Figure 18; see also Fu *et al.* 2016). The Badaowan Formation is composed of mudstone, muddy siltstone, sandstone, conglomerate as well as coals and has a total thickness that ranges between 800 and 1100 m and at least 27 coal seams with an average net coal thickness of 69 m (Tian and Yang 2011). The Xishanyao Formation is similar in rock type to the Badaowan Formation, has total average

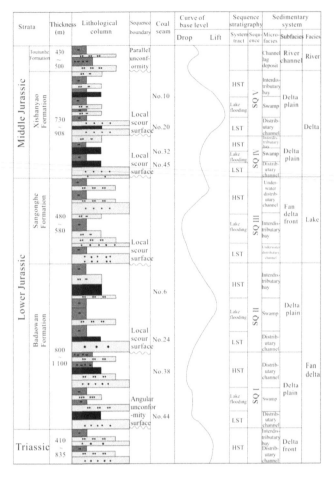

Figure 18. Comprehensive stratigraphic column and sequence scheme of Lower and Middle Jurassic from the southern margin in Junggar basin (Tian and Yang 2011).

thickness between 730 and 908 m with 28–45 coal seams, and an average net coal thickness is about 97 m (Tian and Yang 2011).

Tectonically, the Junggar basin is located at the intersection of the Kazakhstan, Siberian, and Tarim Plates (Jiang *et al.* 2016). The collision of these plates resulted in the Altay and Tianshan orogenic belts (Qi *et al.* 2008), and thus, the tectonic deformation is very strong at the periphery of the basin but weakens towards the centre (Wu *et al.* 2005; Chen *et al.* 2007). The structure in the region is characterized by highly developed thrust belts and compact folds and can be categorized into four E–W-trending folded zones (Wang *et al.* 2016).

3.3.2. CBM reservoirs
Depending on the location, there are between 9 and 53 minable coal seams in the Southern Junggar basin, with a total net coal thickness ranging between 21 and 203 m. Three thick coal seam belts occur in the

southern, southeastern, and western margins, and the coal seams become thin towards the centre of basin (Yu 2002; Cui *et al.* 2007; Yu *et al.* 2008). Net coal thickness of the Badaowan Formation ranges between 15 and 20 m near the margin of the basin (Figure 19(a)), while net coal thickness within the Xishanyao Formation is more than 40 m in the southern margin (Figure 19(b)).

Overall, the coal rank in the Junggar basin increases with increasing depth. Thus, there is a low rank coal zone in the periphery of the basin, with maximum vitrinite reflectance ranging between 0.45% and 0.80% (Figure 20). However, the coal rank in the middle part of the southern margin has a maximum vitrinite reflectance of about 1.3% because of what has been interpreted as a higher palaeo-geothermal gradient resulting from deep-seated faults (Yu 2002; Fu 2006).

Average gas content in coal reservoirs within the Junggar basin is about 6.5 m³/t at the margins of the basin but gradually rises with increasing burial depth towards the centre of the basin (Ye 2007). Gas contents in coal reservoirs at the southern margin of the basin are relatively high, with ranges between 2 and 10 m³/t, with localized contents of up to 10–15 m³/t (Figure 21), again possibly a result of localized higher palaeo-geothermal gradients which would have increased rank and gas generation. As would be expected, at any one particular location, the CBM contents are higher within the underlying Badaowan Formation than the overlying Xishanyao Formation (Sun *et al.* 2012).

Permeability of the coal reservoirs in the region has a wide variance. It is thought that this variation is mainly controlled through the tectonic deformation level of coals, particularly as a result of late Mesozoic nappe style thrusting in the south of the basin (Qi *et al.* 2008). Based on injection-fall off tests, the permeability ranges from 0.01 to 16.4 mD. These data indicate that there may be geographic trends in permeability. The permeability in one well in the southwest ranges from 0.01 to 0.02 mD (Wang 2014), while in a well in the south central part of the basin, the permeability has been recorded as ranging from 0.5 to 13.4 mD. At another well in the southeast, permeability results range from 0.01 to 16.48 mD (Cao *et al.* 2012). The permeability in these areas is based on 10 tests from 10 wells and although the average is only 5.59 mD, it should be noted that this is still higher than most other coal reservoirs in China (Ye *et al.* 1999). The relatively good gas production rates from coal reservoirs in the south central part, especially around the Fukang area, can be attributed to the relatively higher permeability. It should be remembered though that in areas with low

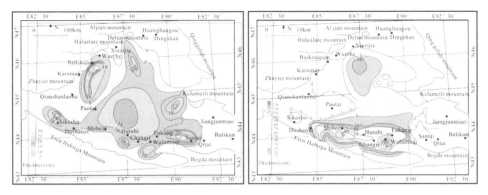

Figure 19. Isoline maps of coal seams thickness in the Badaowan (a) and Xishanyao (b) formations from Junggar basin (Yu 2002). The position of the basin is shown in Figure 1.

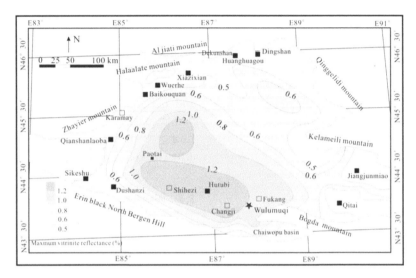

Figure 20. Isoline map of maximum vitrinite reflectance in the Jurassic coals from Junggar basin (Yu 2002). The position of the basin is shown in Figure 1.

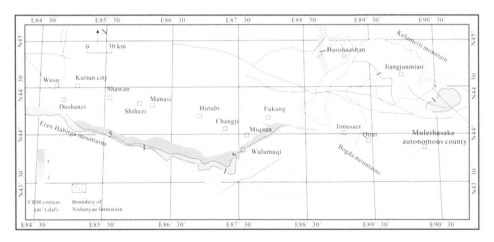

Figure 21. CBM content of coal reservoirs in Xishanyao formation from the southern margin and eastern part of Junggar Basin (Sun et al.2012). The position of the basin is shown in Figure 1.

gas contents (i.e. low gas saturations; see Lamarre 2006; Moore 2012), even with relatively good permeability, good gas production would be unlikely.

Pressure gradients for coal reservoirs in the southern Junggar basin range from 0.49 to 1.08 MPa/hm, with most of the gradients being less than 0.90 MPa/hm (Cao et al. 2012; Wei et al. 2013; Chen 2014). These trends suggest that the coal reservoirs are mostly under-pressured. Pressure gradients tend to be higher in the eastern and western parts of the Southern Junggar basin but lower in the central part. The higher pressure gradient in the eastern and western parts not only is thought to be caused by relatively strong groundwater recharge but also is consistent with the distribution of TDCs (Cao *et al.* 2012). Vertical pressure gradients, as measured in boreholes, of coal reservoirs in the Xishanyao Formation show substantial variability while those in the Badaowan Formation are less variable and display a normal gradient (Cao *et al.* 2012).

With an increase in depth, coal reservoir pressure increases linearly and is greatly influenced both by the hydrodynamics and vertical principal stress (Figure 22). Interestingly, the highest gas flows in the region has been obtained in the western Fukang block in the central part of the southern Junggar basin but this area has the lowest pressure gradient (about 0.50 MPa/hm), though the permeability is the same as other areas in the region. In the western Fukang area, the average CBM production per well for 23 wells has been about 5000 m³/day since 2014. The geological reasons for this abnormal phenomenon need further study.

3.3.3. Geological controls of CBM accumulation and coal reservoirs

Because the coal rank in the region is comparatively low, biogenic gas is likely to be the main gas source (Sun *et al.* 2012). Hydrodynamics has an important influence on CBM in the region as well. For example, the groundwater runoff is strong in the southeastern, northwestern, and northeastern parts of the Junggar basin and may at least be the partial cause of low gas

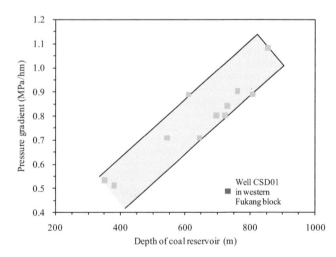

Figure 22. Plots of pressure gradient to depth of the Jurassic coal reservoirs in the southern margin of Junggar Basin drawn from the data of Cao *et al.* (2012).

contents. However, the direction of the groundwater movement in the southern margin of the basin is contrary to that of CBM migration; thus, this condition may help to enhance CBM volumes in this area (Ye 2007). In addition, the Xishanyao Formation in the east of the basin has been heavily weathered to a depth of more than 600 m and the gas content in the coal reservoirs are commonly lower than 1.0 m³/t (Sun *et al.* 2012).

Permeability of the coal reservoirs in the region has no obvious relationship with stratigraphic position but is more related to a combination of the lithological type, tectonic deformation, and the fracture network. It had been shown that the permeability is lowest for the dull coal types and increases successively from semi-dull coal, semi-bright coal to bright coal if the stress remains unchanged (Wei 2003). The south Junggar basin has been subjected to different styles and types of compressive tectonic stresses, starting in the late Mesozoic and continuing into the Cenozoic, and generally, it is thought that these stresses have had an overall effect of worsening permeability due to strong compressional deformation of the coal reservoirs (Chen 2014).

3.4. Western Guizhou Region

Western Guizhou, located in southwest China, has a coal-bearing area of about 28,000 km² (Figure 1). This region has the highest potential for CBM for all of South China. Two coal fields, the Zhina and the Liupanshui, have very active CBM exploration and development programs and have an estimated CBM resource of 2.09 TCM. This volume accounts for about 5.68% of the total estimated CBM resource for all of China (Gao *et al.* 2009). The highest gas production from a single CBM well within a batch of test wells in the region is close to 3000 m³/day (Yi *et al.* 2016).

3.4.1. Geological setting

The main coal-bearing units occur in the Permian-age Longtan and Changxin Formations. These formations are mainly preserved within in a series of small-scale synclines, such as the Bide-Santang, Panguan, Gemudi, Qingshan, and Langdai (Figure 23). These synclines have resulted from late Mesozoic tectonic events (Gao *et al.* 2009).

The coal-bearing strata are thought to have formed within a coastal delta setting along the western margin of the late Palaeozoic South China basin. Lithologically, these units contain thin and multiple coal seams, frequently interbedded with other clastic sediments; these sequences have been interpreted to have formed as a result of frequent sea level fluctuations (Shao *et al.*

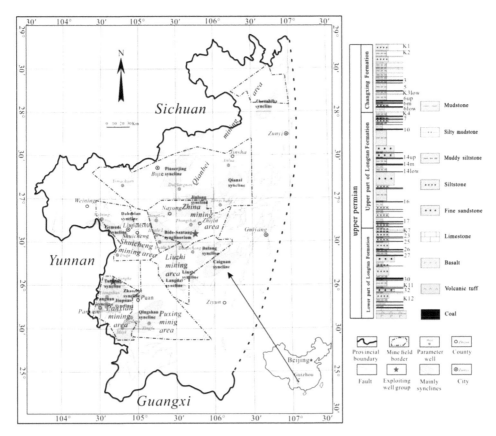

Figure 23. Synclines and comprehensive stratigraphic column of coal-bering strata in the western Guizhou (Yang *et al.* 2015). The position of the basin is shown in Figure 1.

1998; Wang *et al.* 2011). The thickness of the coal-bearing strata varies from 220 to 513 m and may be controlled by syndepositional faulting (Xiong *et al.* 2006). In general, there are more than 10 coal seams but in some areas, individual coal seams can number as many as 83. At any particular location, workable coal seams can number between 1 and 24 (where workable coal thickness is equal to at least 0.7 m). Single coal seam thicknesses are commonly 1.0–2.5 m with a maximum thickness of 6 m (Gao *et al.* 2009).

The maximum vitrinite reflectance in the Western Guizhou region ranges from 1% to 3% (Figure 24). The spatial distribution of coal rank is likely controlled by differential burial and palaeo-hydrothermal activity from deep faults since the Mesozoic (Dai *et al.* 2003; Xu and He 2003; Gao *et al.* 2009; Xu *et al.* 2011).

3.4.2. Coal reservoir and its geological controls
There have been numerous geological investigations since the beginning of this century with the goal of understanding the CBM geology in Western Guizhou region (e.g. Gao *et al.* 2009; Yang 2011; Lei *et al.* 2012; Li et al. 2012; Wu *et al.* 2013; Ge 2015; Guo 2015; Li *et al.*

2015; Shen *et al.* 2016). The results of these studies are summarized below.

CBM occurrence in the Western Guizhou region is primarily controlled by structure, coal rank, stratigraphic sequence structure, and hydrodynamics. CBM contents generally increase with increasing burial depth so that gas content isopachs follow depth isopachs within each syncline (Figure 25). Based on desorption data from 998 coal samples with depths less than 1000 m, the Zhina area with high rank coal reservoirs has quite variable gas contents ranging from 0.24 to 29.21 m^3/t with an average of 13.81 m^3/t, while the Liupanshui area with moderate rank coal reservoirs has gas contents ranging from 3.87 to 29.16 m^3/t with an average of 12.79 m^3/t (Gao *et al.* 2009). Note that the relationship between gas content (on a daf basis) and depth in most of the synclines has a cyclic pattern and has little to do with the composition and content of mineral matter in coals. That is, the 'cycle' generally is related to third-order stratigraphic sequence units (Figure 26). This phenomenon suggests a strong depositional control on gas contents (for details, see Section 4.3.3).

Permeability of the coal reservoirs in the Western Guizhou region is variable but generally low because

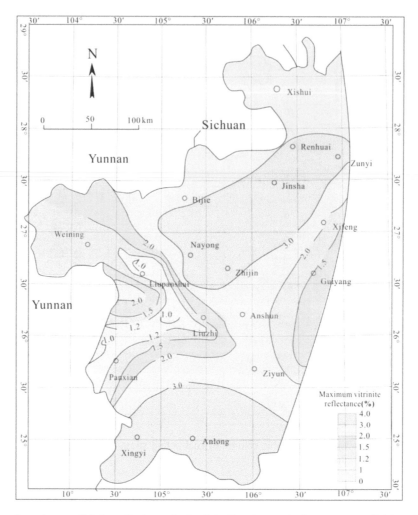

Figure 24. Isoline map of maximum vitrinite reflectance in the late Permian coals from western Guizhou (Xu and He 2003). The position of the basin is shown in Figure 1.

of the combination of the depth, structural deformation, and *in-situ* stress (Xu *et al.* 2014). Data from more than 50 permeability tests taken from 22 wells indicate that the permeability decreases with depth in a stepwise fashion (Figure 27). At depths of less than 300 m, permeability is relatively high. At depths between 300 and 800 m, permeability is lower but relatively stable, whereas in depths of greater than 800 m, permeability is sharply reduced. This sudden reduction is thought to be stress related (Yang and Qin 2015). However, there is considerable variation in the data shown in Figure 27, indicating that permeability is also controlled by other geological factors. With increasing degree of tectonic deformation, coal reservoirs may change from primary banded to fractured banded to fragmented and finally to mylonitic structures and the permeability will concomitantly drop sharply (Jiang *et al.* 2010; Hou *et al.* 2012). Coal reservoir structures are mainly fractured banded and fragmented in the Zhina region and

fragmented and mylonitic in the Liupanshui region (Li *et al.* 2011). Thus, permeability in the Zhina is higher than that in the Liupanshui area because of the different levels of deformation (Figure 28). As also shown in Figure 28, the *in-situ* stress is higher in Liupanshui (2.0–3.3 MPa/hm) and relatively low in Zhina (1.7–2.1 MPa/hm); that is, the coal reservoir permeability is higher in regions with lower minimum stress gradient, indicating that vertical and horizontal stress is a key control of permeability. This characteristic is in agreement with the stress-related models of permeability proposed elsewhere (Palmer and Mansoori 1996; Brooke-Barnett *et al.* 2012, 2015; Flottman *et al.* 2013).

Coal reservoir pressure in the Western Guizhou region is seemingly highly influenced by factors such as stratigraphic framework and *in-situ* stress, as well as groundwater hydrodynamics especially related to folds and faults (Gao *et al.* 2009). Based on well-test data, reservoir pressure increases with depth; the gradient

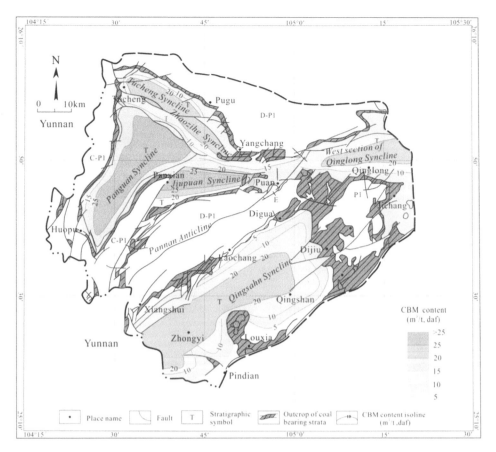

Figure 25. CBM content isoline map of the late Permian coal reservoir in local area of western Guizhou (Dou 2012). The position of the basin is shown in Figure 1. The figure range is shown in Figure 23. Symbol D represents the Devonian strata, C is the Carboniferous strata, P_1 means the early Permian strata, and T is the Triassic strata.

ranges between 0.48 and 1.75 MPa/hm with an average of 1.03 MPa/hm. These variable gradients indicate that some reservoirs are under-pressured and some are over-pressured (Figure 29(a)). It is estimated that about 30% of the coal reservoirs have a pressure coefficient over 1.2 and are thus over-pressured (Yang and Qin 2015). It is noted that although *in-situ* pressure tends to increase with depth, variability in pressure also increases (Figure 29(b)). This implies that there are other controls (i.e. other than depth) on pressure, indicating that the stratigraphic seals and lithology influence gas contents in the Western Guizhou region (Xu *et al.* 2011; Yang and Qin 2015). In addition, pressure coefficients in relation to depth have a cyclic pattern when viewed in the context of a single borehole profile, and each cycle spans two third-order stratigraphic sequence units (Qin *et al.* 2008; Yang 2011; Guo 2015). It was found that mudstone near the maximum transgression surface (MTS) on the top of each third-order sequence unit has very low permeability and blocks the vertical connection of the groundwater and CBM between the overlying and underlying units (Ge 2015; Yang *et al.* 2015). That is, this cyclic pattern in

pressure coefficients on a borehole profile originates from the combined effect of sedimentary sequence and hydrodynamic variations (Shao *et al.* 2015; Ge 2015; Yang *et al.* 2015; Shen *et al.* 2016).

Based on the above geological conditions, it was predicted that the potential for CBM development in the Zhina area might be much superior than those in the Liupanshui (Gao *et al.* 2009). This prediction has been confirmed preliminarily from the outcome of CBM exploration and pilot projects in the region in recent years (Guo 2015). It is also noticed that localized areas with fractured banded coal reservoirs exist within Liupanshui and thus might have reasonable prospects for CBM development. For example, three CBM wells in the Tucheng block within Liupanshui have had commercial CBM production since 2014 (Yi *et al.* 2016).

4. Frontier CBM geology research within China

4.1. Deep CBM pool-forming behaviours

The main basins within China have an estimated CBM resource of about 16.77 TCM between depths of 1200

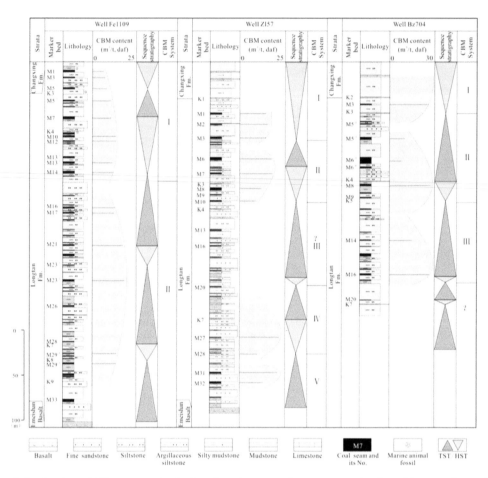

Figure 26. Vertical distribution of CBM content in the late Permian strata from western Guizhou (Shen *et al.* 2016).

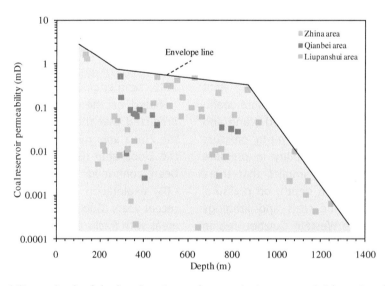

Figure 27. Plots of permeability to depth of the late Permian coal reservoirs in western Guizhou. Area location can been seen in Figure 23.

and 2000 m (Liu *et al.* 2009b), and 18.47 TCM between 2000 and 3000 m (Zhao 2011). At the beginning of this century, a successful pilot test for gas production was conducted to jointly access both deep coal and tight sandstone reservoirs in the Piceance basin of the USA (Olson *et al.* 2002; Nelson 2003). Following this success, new technologies have been trialled in China to also try to produce gas from deep coal reservoirs, and these trials have had some success in the Ordos, Qinshui, and Junggar basins (Ye

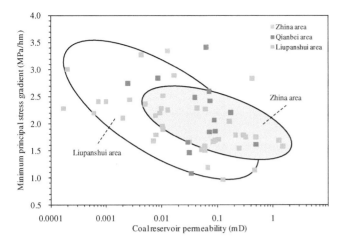

Figure 28. Plots of coal reservoir permeability to minimum principal geo-stress gradient in western Guizhou. Area location can been seen in Figure 23.

2013; Zhang et al. 2014; Chen et al. 2015; Zhang et al. 2015; Li et al. 2016b; Qin and Shen 2016b). These advances have preliminarily demonstrated the commercial potential of deep CBM extraction in China but have also highlighted many geological and technical challenges that still need to be addressed.

Without doubt, deep CBM development will be highly challenged because of low permeability. It was pointed out that the stress-dependent permeability is a major influence on permeability (Brooke-Barnett et al. 2012, 2015; Flottman et al. 2013). The ratio of average horizontal principal stress to vertical stress is termed the lateral pressure coefficient (LPC) (Chen 2006). According to the Heok–Brown's criterion (Hoek and Brown 1980), as the depth increases, the in-situ stress state will vary. Typically, the LPC will transform from more than 1.0 to less than 1.0. The depth at which the LPC is equal to 1.0 is termed the critical conversion depth of an in-situ stress state (Qin et al. 2012b). The transformation will make the properties of deep coals

different from shallow coals and affect gas storage and production behaviours of deep coal reservoirs, such as the porosity, permeability, stress sensitivity, and fracturing properties.

Based on high-temperature overburden tests of 158 coal samples from the Junggar basin, the porosity of coals is the positive power function of permeability (Figure 30(a)) and declines with an increase of effective stress (Figure 30(b)). If the effective stress remains unchanged, the porosity enhances with increasing temperature. If the temperature remains unchanged, the loss rate of the permeability increases with an increase of effective stress (Figure 30(c)). Thus, it is implied that if the geothermal gradient is not high and the fluid pressure is larger, the permeability of the deep coal reservoirs may not be as low as expected, which is supported by the predicted result from a variable pore compressibility model (Figure 30(d)).

Stress sensitivity of coals can be described by the effective stress–permeability law (Li and Xiao 2008):

$$K = f(\sigma_e) = f(\sigma - ap) \qquad (1)$$

where K is the permeability, mD; σ_e is the effective stress, MPa; f is the function of the effective stress to permeability; σ is the confining pressure of strata, MPa; a is the effective stress coefficient of permeability; and p is the formation fluid pressure, MPa.

In formula (1), σ is determined by depth under the influence of a tectonic stress field, and the vertical change of the stress state is bound to lead to great differences in permeability between the deep and shallow coal reservoirs. Further, a is a function among parameters depended on depth, such as mechanical nature of the rock material, pressure, temperature, and fluid properties of the coal reservoir. Changes in these parameters with depth can result in a change of the effective stress coefficient, which can then be the cause of the differences in stress sensitivity between deep and shallow coal reservoirs.

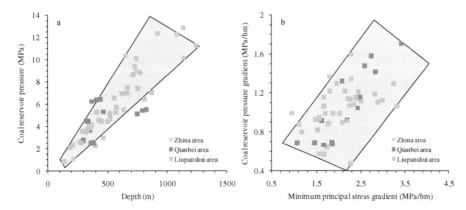

Figure 29. Plots of coal reservoir pressure to depth or geo-stress gradient in western Guizhou. Area location can be seen in Figure 23.

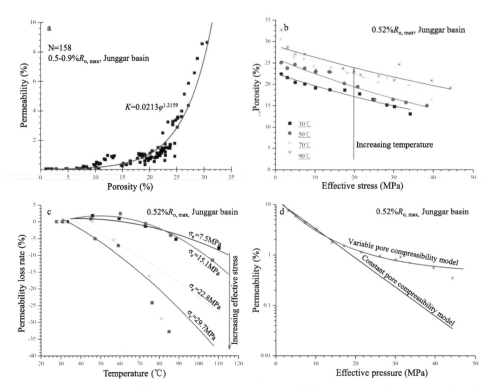

Figure 30. Relationship among the permeability, formation temperature, and effective stress of deep low-rank coal reservoirs by the physical simulation experiment of the samples from eastern Junggar basin (Wang *et al.* 2014).

Traditionally, it is considered that the effective stress coefficient of reservoir rocks can be regarded as 1.0 (Terzaghi 1960; Li and Xiao 2008). However, it is found from experiment that the effective stress coefficient is far less than 1.0 when the confining pressure is very large (Li and Xiao 2008). High stress and confining pressure are the key geological manifestations of deep coal reservoirs. Based on the test data from 26 CBM wells in the southeast margin of the Ordos basin, the mean effective stress coefficient of the coal reservoirs with depths of more than 1000 m is only 0.48, which is substantially less than 1.0 (Meng *et al.* 2013).

According to Langmuir's equation, the adsorbed gas increases monotonically with depth. However, the increase of depth means the increase of coal temperature. In this case, the behaviour of CBM adsorption and desorption can be described by a complete Langmuir model (Yu 1993):

$$q = \left[\frac{abp}{1 + bp} \, e^{n(T_1 - T)} \right] K_1 \qquad (2)$$

where q is the adsorbed CBM content, m³/t; a and b are the isothermal adsorption constants of coals, m³/t; p is the fluid pressure of coal reservoir, MPa; T_1 is the adsorption temperature, K; T is coal reservoir temperature, K; K_1 is the coefficient related to coal quality; and n is the temperature influence coefficient related to pressure. The exponential temperature term in formula (2)

can be neglected in the case of low formation temperatures for shallow coal reservoirs. However, coal reservoir temperature increases with depth because of the geothermal gradient. Temperature increases will reduce remarkably the ability of a coal seam to adsorb methane (Qin *et al.* 2005; Moore 2012).

Coupling of both pressure and temperature will result in a downward trend of adsorbed gas content at a certain depth. This depth is called the critical CBM content depth (Qin *et al.* 2012b, 2016a). With further burial, a proportion of adsorbed gas will be desorbed (because of higher temperatures) and change into free gas and be expulsed from the coal seam. The reason is that the effects of pressure on absorption are stronger than the effects of temperature on gas desorption at relatively shallow depths. For deep coal reservoirs, the effects are opposite. A case from 427 measured data in the Qinshui basin can been seen in Figure 15, here with the CBM data from depths of up to 2000 m.

As discussed above, the depths at which coal seams can be termed 'deep' can feasibly be defined. That is, 'deep' coal reservoirs can be defined as occurring below the critical depths of the LPC and/or when temperature becomes the dominant control (causing gas desorption) over that of pressure (which general increases adsorption) (Qin *et al.* 2016a). Taking the LPC and temperature demarcations into consideration, three types of the critical depth boundaries can be delineated. The

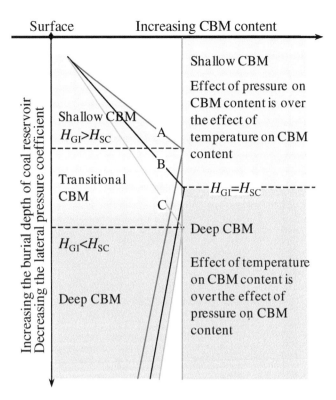

Figure 31. Conceptual diagram of deep coalbed methane definition revised from Qin and Shen (2016b). H_{GI} means the depth corresponding to CBM content inversion point and the H_{SC} represents the depth at which lateral pressure coefficient (γ) is equal to 1.0. Curve A indicates the change of CBM content when H_{GI} is shallower than H_{SC}, Curve C when H_{GI} is deeper than H_{SC}, and Curve B when H_{GI} is equal to H_{SC}. H_{GI} and H_{SC} are, respectively, 200–700 and 700–1200 m in eastern Ordos basin (Shen 2011), respectively, 500–950 and 500–700 m in the southern Qinshui basin (Shen *et al.* 2014; Sun *et al.* 2017), and, respectively, 450–900 and 1500–2300 m in the eastern Junggar basin (Chen 2014; Qin and Shen 2016b).

first is indicated by the curve B in Figure 31; here, the critical LPC depth is equal to the critical temperature depth so that the shallow and deep coal reservoirs can be distinguished clearly. This is an ideal case. If the critical LPC depth is over or below the critical temperature depth, another two boundaries are indicated by the curves A and C in Figure 31. For the latter two cases, there is a transitional belt between the shallow and deep coal reservoirs within a vertical section. This model has been supported by CBM data in other basins, such as the Sydney basin in Australia (Faiz *et al.* 2007), the Qinshui (Figure 15), and Junggar basins in China (Shen *et al.* 2014; Chen *et al.* 2015).

4.2. Superimposed CBM systems

The essence of a gas-bearing system is that there exists a fluid connection between different reservoirs within this system, and this fluid connection can be

recognized by fluid pressure distribution (Qin *et al.* 2008). In this case, CBM content will increase gradually with an increase in depth until the critical depth (see previous section) is reached.

However, it was found that CBM gas contents often have cyclic variation in a borehole profile with multiple coal seams (Figure 26), implying that there exist multiple independent CBM systems in the vertical stratigraphic section. A vertical overlap of multiple and independent CBM or fluid systems is referred to as a 'superimposed CBM system' (SCBMS) (Qin *et al.* 2008). Since 2010, the SCBMS has been identified in some basins such as the Qinshui, Ordos, and Western Guizhou (e.g. Shen 2011; Yang 2011; Yang *et al.* 2011, 2015; Liu and Zhang 2016).

Based on the study of the late Permian coal measures in western Guizhou, Qin *et al.* (2008) had suggested that (1) the impermeable layers within the stratigraphic sequence units of the coal measures provide a sealing foundation for the formation of SCBMS; these layers are called 'key horizons' (Yuan 2014), and (2) these impermeable layers block the formation fluid (water and/or gas) connection vertically between independent overlying and underlying systems. That is, the SCBMS results from a dual control of both lithology and hydrogeology; the stratigraphic sequence framework is a key for the formation of SCBMS (Figure 26). It was further found that the physical properties of rocks within coal measures can vary in a cyclic fashion vertically within a borehole (Figure 32). These physical properties include porosity, permeability, and breakthrough pressure and are no doubt related to the vertical changes represented in the lithology.

Key horizons have low porosity and permeability similar to those of hydrocarbon reservoir cap rocks and are the most important indicator to define the upper and lower boundaries of independent CBM systems within coal measures (Yuan 2014; Qin *et al.* 2016a). The key horizons in the late Permian coal measures of western Guizhou are mudstones with siderite or pyrite. These sediments were deposited associated with a maximum flooding surface (mfs) and are distributed extensively in the sediments near the highest sea surface (Shen *et al.* 2016). Because a key horizon exists inside single third-order sequence units, a single CBM system goes across two three-order sequence units, including highstand system tracts sediments in the lower part of an overlying sequence unit and the transgressive system tract sediments in the upper part of underlying sequence unit (Figures 26 and 33). Thus the vertical overlap of several sequence units results in a SCBMS.

Further study found that the rocks deposited near maximum flooding surfaces are not necessarily key

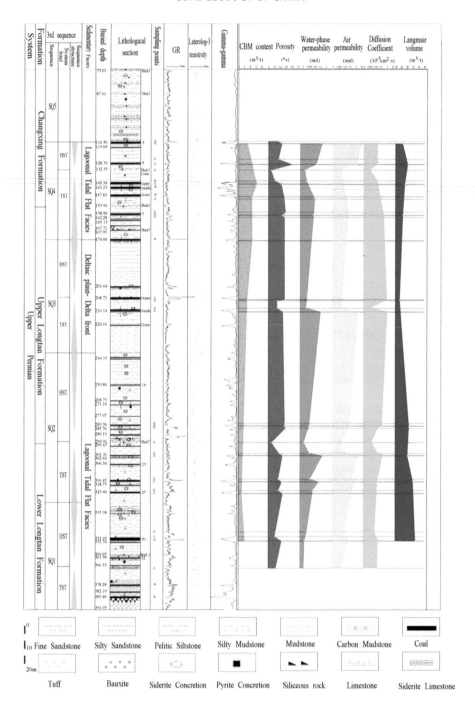

Figure 32. Columnar section of physical properties of rocks in the late Permian coal measure of well 1001 from western Guizhou (Yang 2011).

horizons because the physical properties of these rocks are related to the specific sedimentary facies. For example, the porosity and permeability of the rocks near maximum flooding surfaces increase gradually from the delta front facies to the delta plain facies in the upper Permian coal measures of western Guizhou (Shen *et al.* 2016). In this example, a SCBMS develops mainly in the strata with delta front facies but is not common in the strata with delta plain facies (Figure 33). It was thought that the reason for this is that the mudstones with siderite or pyrite are

mainly deposited in delta front facies rather than delta plain facies (Yuan 2014; Shen *et al.* 2016).

Because of the difference of reservoir fluid pressure and physical properties between different independent systems within a SCBMS, there is often strong fluid interference among coal reservoirs when the gas is extracted simultaneously from multiple coal reservoirs. Such interference has been common in the Qinshui and Ordos basins as well as in Western Guizhou. Production from multiple CBM systems will

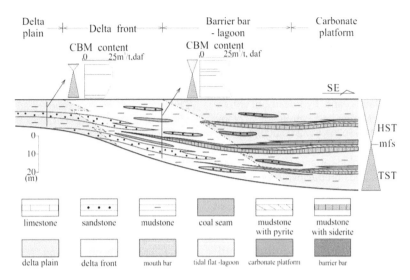

Figure 33. Simplified model of CBM system restricted by paralic sediments (Shen *et al.* 2012, 2016).

seriously constrain gas production (Guo 2015; Yang and Qin 2015; Zhang *et al.* 2015; Cao *et al.* 2016; Yi *et al.* 2016) and is of great concern within the Chinese CBM industry. Furthermore, different CBM systems in a well often require different gas extraction strategies. Innovative anti-fluid interference technology needs to be developed to extract gas from multiple CBM systems simultaneously. This is one of the research targets of China's National CBM Science and Technology Major Projects since 2015 (Qin *et al.* 2016a).

5. Summary remarks

Geological research into CBM in China has been conducted over a long period of time. Since the 1980s, this research has provided a better understanding of CBM geology for application within the CBM industry in China. The outcome of this research has had the most prominent impact for production from the Qinshui and Ordos basins. In order to develop enhanced-CBM production technologies in those two basins, understanding of geological controls on coal reservoirs has been improved to support sweet spot evaluation as well as engineering design of wells and development patterns.

In order to further expand CBM production in China, several additional areas with CBM production potential have been identified. The two main basins with the greatest potential for expansion of CBM production are the southern Junggar and the Western Guizhou basins. However, these basins, and the coal reservoirs within them, are complex and present substantial challenges for production that have not been met yet. The challenges include identification of the resources themselves and an understanding of deep reservoirs as well as superimposed CBM systems. The keys to unlocking the economic potential lie in the future research into these challenges.

Acknowledgements

This research was supported by the National Natural Science Foundation of China: [Grant Numbers U1361207 and 41530314] and the Major National Science and Technology Projects of China: Grant Numbers 2016ZX05044-02 and 2016ZX05066-01]. Especially, the authors thank two anonymous reviewers, who put forward many valuable comments to improve the quality of this manuscript. The authors also thank the Editor-in-Chief Dr. Robert J. Stern and guest editors Dr. Shifeng Dai and Dr. Bob Finkelman of the special volume for their inputs.

Disclosure statement

No potential conflict of interest was reported by the authors.

Funding

This work was supported by the National Natural Science Foundation of China: [Grant Number U1361207], Major National Science and Technology Projects of China: [Grant Numbers 2016ZX05044-02 and 2016ZX05066-01], and National Natural Science Foundation of China: [Grant Number 41530314].

ORCID

Yong Qin http://orcid.org/0000-0002-7478-8828

References

Anonymous, 2011, Guidelines for application of the Petroleum Resources Management System: Society of Petroleum Engineers, p. 221: http://www.spe.org/industry/docs/PRMS_Guidelines_Nov2011.pdf

Aydin, A., 2000, Fractures, faults, and hydrocarbon entrapment, migration and flow: Marine and Petroleum Geology, v. 17, p. 797–814. doi:10.1016/S0264-8172(00)00020-9

Ayers, W.B., Kaiser, W.R., Laubach, S.E., Ambrose, W.A., and Baumgardner, R.W., 1991, Geological and hydrologic controls on the occurrence and producibility of coalbed methane, Fruitland Formation, San Juan Basin: Topical Report, GRI-5087-214-1544, Texas University at Austin and Bureau of Economic Geology: https://www.tib.eu/en/search/id/ntis%3Asid~oai%253Ads2%253Antis%252F531a107e02b803781c02d923/Geologic-and-Hydrologic-Controls-on-the-Occurrence/?tx_tibsearch_search%5Bsearchspace%5D=tn#documentinfo

Brooke-Barnett, S., Flottman, T., Paul, P.K., Busetti, S., Hennings, P., and Reid, R., 2012, Influence of basement structures on stress regimes in the Surat and Bowen Basins, southeast Queensland, in Eastern Australasian Basins Symposium IV: Brisbane, Australia, Petroleum Exploration Society of Australia, p. 4.

Brooke-Barnett, S., Flottman, T., Paul, P.K., Busetti, S., Hennings, P., Reid, R., and Rosenbaum, G., 2015, Influence of basement structures on in situ stresses over the Surat Basin, southeast Queensland: Journal of Geophysical Research: Solid Earth, v. 120, p. 4946–4965.

Cai, Y., Liu, D., Yao, Y., Li, J., and Qiu, Y., 2011, Geological controls on prediction of coalbed methane of No. 3 coal seam in Southern Qinshui Basin, North China: International Journal Coal Geological, v. 88, p. 101–112. doi:10.1016/j.coal.2011.08.009

Cao, D., Liu, K., Liu, J., Xu, H., Li, J., and Qin, G., 2016, Combination characteristics of unconventional gas in coal measure in the west margin of Ordos Basin: Journal of China Coal Society, v. 41, p. 277–365 (in Chinese with English abstract).

Cao, Y., Cui, H., Zhang, Y., and Wei, S., 2012, Evaluation and development technology of low rank CBM for Fukang pilot area, Xinjiang, in Research Report: Jiaozuo, Henan Polytechnic University, p. 182 (in Chinese).

Chakhmakhchev, A., 2007, Worldwide coalbed methane overview, in SPE hydrocarbon economics and evaluation symposium, SPE 106850: Society of Petroleum Engineers, p. 17–23.

Che, C., Yang, H., Li, F., Liu, C., and Zhu, J., 2008, Exploration and development prospects of coalbed methane resources in China: China Mining Magazine, v. 17, p. 1–4 (in Chinese with English abstract).

Chen, G., 2014, Deep low-rank coalbed methane system and reservoiring mechanism-in the case of the Cainan block in Junggar Basin [Doctoral Dissertation]: Xuzhou, China, China University of Mining and Technology, 153 p. (in Chinese with English abstract).

Chen, G., Qin, Y., Hu, Z., Li, W., and Shen, J., 2015, Variations of gas content in deep coalbeds of different coal ranks: Geological Journal of China Universities, v. 21, p. 274–279 (in Chinese with English abstract).

Chen, S., Qi, J., Yu, F., and Yang, Q., 2007, Deformation characteristics in the southern margin of the Junggar basin and their controlling factors: Acta Geologica Sinica, v. 81, p. 151–157.

Chen, Z., 2006, Discussion on change law of different lithology`s side pressure ratio with depths: West China Exploration Engineering, v. 122, p. 99–101 (in Chinese with English abstract).

Cheng, A., Cao, D., and Yuan, T., 2016, Occurrence and evaluation of coal resources in China: Beijing, Science Press, p. 352 (in Chinese with English abstract).

Chi, W., and Yang, L., 2001, Feasibility of coalbed methane exploitation in China: Journal of Petroleum Technology, v. 53, p. 74. doi:10.2118/0901-0074-JPT

Close, J.C., 1993, Natural fractures in coal, in Law, B.E., and Rice, D.D., eds., Hydrocarbons from coal: Tulsa, Oklahoma, American Association of Petroleum Geologists, p. 119–132.

Cui, S., Liu, H., Wang, B., Yang, Y., Ning, N., and Sang, S., 2007, Trapping characteristics of coalbed methane in low-rank coal of Zhungaer basin: Geoscience, v. 21, p. 719–724.

Dai, S., Ren, D., Hou, X., and Shao, L., 2003, Geochemical and mineralogical anomalies of the late Permian coal in the Zhijin coalfield of southwest China and their volcanic origin: International Journal Coal Geological, v. 55, p. 117–138. doi:10.1016/S0166-5162(03)00083-1

Dou, X., 2012, Tectonic evolution and its control on coalbed methane reservoiring in western Guizhou [Doctoral dissertation]: Xuzhou, China University of Mining and Technology, 132 p. (in Chinese with English abstract).

Duan, J., 1992, Analysis on the prospect of coalbed methane resources in China: Natural Gas Industry, v. 12, p. 26–31 (in Chinese with English abstract).

EBDES (Editorial Board of Dictionary of Earth Sciences), 2005, Dictionary of Earth Sciences (Applied Disciplines): Beijing, Geological Publishing House, p. 92 (in Chinese with English terms).

EIA (U.S. Energy Information Administration), 2017, International Energy Outlook 2017: http://www.eia.gov/ieo, #IEO2017.

Faiz, M., Saghafi, A., Sherwood, N., and Wang, I., 2007, The influence of petrological properties and burial history on coal seam methane reservoir characterization, Sydney Basin, Australia: International Journal of Coal Geology, v. 70, p. 193–208. doi:10.1016/j.coal.2006.02.012

Feng, F., Wang, T., and Zhang, S., 1995, Natural gas geology in China: Beijing, Geological Press, p. 337 (in Chinese with English abstract).

Flores, R.M., 1998, Coalbed methane: From hazard to resource: International Journal of Coal Geology, v. 35, p. 3–26. doi:10.1016/S0166-5162(97)00043-8

Flores, R.M., 2014, Coal and coalbed gas: Fueling the future: United States, Elsevier Science Publishing Co Inc, p. 720.

Flottman, T., Brooke-Barnett, S.R.T., Naidu, S.-K., Kirk-Burnnand, E., Paul, P., Busetti, S., and Hennings, P., 2013, Influence of in-situ stresses on fracture stimulations in the Surat Basin, Southeast Queensland, in SPE Unconventional Resources Conference and Exhibition-Asia Pacific, SPE 167064: Society of Petroleum Engineers, p. 14. doi:10.2118/167064-MS

Fu, H., Tang, D., Xu, H., Xu, T., Chen, B., Hu, P., Yin, Z., Wu, P., and He, G., 2016, Geological characteristics and CBM exploration potential evaluation: A case study in the middle of the southern Junggar Basin, NW China: Journal of

Natural Gas Science and Engineering, v. 30, p. 557–570. doi:10.1016/j.jngse.2016.02.024

Fu, X., 2006, Study on the reservoir characteristics and the exploration potential of low rank coal in the western China [Doctoral Dissertation]: Beijing, China University of Geosciences, 167 p. (in Chinese with English abstract).

Fu, X., Qin, Y., Wang, G.G.X., and Rudolph, V., 2009a, Evaluation of gas content of coalbed methane reservoirs with the aid of geophysical logging technology: Fuel, v. 88, p. 2269–2277. doi:10.1016/j.fuel.2009.06.003

Fu, X., Qin, Y., Wang, G.G.X., and Rudolph, V., 2009b, Evaluation of coal structure and permeability with the aid of geophysical logging technology: Fuel, v. 88, p. 2278–2285. doi:10.1016/j.fuel.2009.05.018

Fu, X., Qin, Y., Zhang, W., Wei, C., and Zhou, R., 2005, Fractal classification and natural classification of coal pore structure based on migration of coal bed methane: Chinese Science Bulletin, v. 50, no. S, p. 66–71. doi:10.1007/BF03184085

Gao, D., Qin, Y., and Yi, T., 2009, CBM geology and exploring-developing stratagem in Guizhou Province, China: Procedia Earth and Planetary Science, v. 1, p. 882–887. doi:10.1016/j.proeps.2009.09.137

Ge, Y., 2015, Study on the multi-superimposed aquifer system of coal seams and source discrimination of commingling CBM produced-water: An sxample from Zhucang Syncline in western Guizhou Province [Doctoral dissertation]: Xuzhou, China University of Mining and Technology, 140 p. (in Chinese with English abstract).

Guo, C., 2015, Multi-layer superposed CBM system and srainage model optimization: In the case of the Upper Permian Bide-Santang basin [Doctoral dissertation]: Xuzhou, China University of Mining and Technology, 216 p. (in Chinese with English abstract).

Hamilton, L.H., Esterle, J.S., and Golding, S.D., 2012, Geological interpretation of gas content trends, Walloon Subgroup, eastern Surat Basin, Queensland, Australia: International Journal of Coal Geology, v. 101, p. 21–35. doi:10.1016/j.coal.2012.07.001

Han, D., and Yang, Q., 1979, Coal Geology of China (Volume 2): Beijing, Coal Industry Press, p. 74–92 (in Chinese).

Han, J., Yang, Z., Li, X., Lu, Y., and Zhang, J., 2015, The relevant information of coalbed methane development in China, in Proceedings - 7th International Conference on Measuring Technology and Mechatronics Automation, IEEE: p. 1253–1258 (Open CD-ROM). doi:10.1109/ICMTMA.2015.305

Hoek, E., and Brown, E.T., 1980, Empirical strength criterion for rock masses: Journal of the Geotechnical Engineering Division, v. 106, p. 1013–1035.

Hou, J., 2002, Evaluation of coalbed methane reservoirs from geophysical log data using an improved fuzzy comprehensive decision method and a homologous neural network: Geophysical Prospecting, v. 50, p. 453–462.

Hou, Q., Li, H., Fan, J., Ju, Y., Wang, T., Li, X., and Wu, Y., 2012, Structure and coalbed methane occurrence in tectonically deformed coals: Science China (Earth Sciences), v. 55, p. 1755–1763. doi:10.1007/s11430-012-4493-1

Jian, X., Tang, S., Liu, R., Zhao, G., and Sun, P., 2015, CBM Reservoir characteristics and productivity prediction in Liulin area, Ordos Basin: Xinjiang Petroleum Geology, v. 36, p. 326–329 (in Chinese with English abstract).

Jiang, B., Qu, Z., Wang, G.G.X., and Li, M., 2010, Effects of structural deformation on formation of coalbed methane reservoirs in Huaibei coalfield, China: International Journal of Coal Geology, v. 82, p. 175–183. doi:10.1016/j.coal.2009.12.011

Jiang, D., Robbins, E.I., Wang, Y., and Wang, H., 2016, Petrolipalynology: Springer Geology, p. 263. doi:10.1007/978-3-662-47946-92

Jie, M., Ge, X., Peng, C., Xiong, X., and Xia, F., 2011, Advances in the CBM exploration and development techniques and their developing trend in China: Natural Gas Industry, v. 31, p. 63–65 (in Chinese with English abstract).

Johnson, R.C., and Flores, R.M., 1998, Developmental geology of coalbed methane from shallow to deep in Rocky Mountain basins and in Cook Inlet–Matanuska basin, Alaska, U.S.A. and Canada: International Journal of Coal Geology, v. 35, p. 241–282. doi:10.1016/S0166-5162(97)00016-5

Karacan, C.O., and Okandan, E., 2000, Fracture/cleat analysis of coals from Zonguldak Basin (northwestern Turkey) relative to potential of coalbed methane production: International Journal of Coal Geology, v. 44, p. 109–125. doi:10.1016/S0166-5162(00)00004-5

Keim, S.A., Luxbacher, K.D., and Karmis, M., 2011, A numerical study on optimization of multilateral horizontal wellbore patterns for coalbed methane production in Southern Shanxi Province, China: International Journal of Coal Geology, v. 86, p. 306–317. doi:10.1016/j.coal.2011.03.004

Lamarre, R.A., 2006, Under-saturation in coals: How does it happen and why is it important?: Search and Discovery, p. Article #40195: http://www.searchanddiscovery.com/documents/2006/06034lamarre/index.htm

Lei, B., Qin, Y., Gao, D., Fu, X., Wang, G.G.X., Zou, M., and Shen, J., 2012, Vertical diversity of coalbed methane content and its geological controls in the Qingshan syncline, western Guizhou province, China: Energy Exploration and Exploitation, v. 30, p. 43–58. doi:10.1260/0144-5987.30.1.43

Li, G., and Zhang, H., 2013, The origin mechanism of coalbed methane in the eastern edge of Ordos Basin: Science China: Earth Sciences, v. 56, p. 1701–1706. doi:10.1007/s11430-013-4616-3

Li, M., Jiang, B., Lin, S., Wang, J., Ji, M., and Qu, Z., 2011, Tectonically deformed coal types and pore structures in Puhe and Shanchahe coal mines in western Guizhou: Mining Science and Technology, v. 21, p. 353–357.

Li, M., and Xiao, W., 2008, Experimental study on permeability-effective- stress law in low-permeability sandstone reservoir: Chinese Journal of Rock Mechanics and Engineering, v. 27, p. 3535–3540 (in Chinese with English abstract).

Li, M., and Zhang, W., editors, 1990, Shallow coal-derived gas in main coal field of China: Beijing, Science Press, p. 225 (in Chinese with English abstract).

Li, S., Tang, D., Pan, Z., Xu, H., and Guo, L., 2015, Evaluation of coalbed methane potential of different reservoirs in western Guizhou and eastern Yunnan, China: Fuel, v. 139, p. 257–267. doi:10.1016/j.fuel.2014.08.054

Li, S., Tang, D., Xu, H., Tang, S., Shao, G., and Ren, P., 2015b, Evolution of physical differences in various buried depth of coal reservoir under constraint of stress: Acta Petrolei Sinica, v. 36, p. 68–75 (in Chinese with English abstract).

Li, S., Tang, D., Xu, H., and Yang, Z., 2012a, The differences of physical properties of coal reservoirs and their origin mechanism between Zhijin and Panxian Areas, Western

Guizhou, China: Energy Exploration and Exploitation, v. 30, p. 661–676. doi:10.1260/0144-5987.30.4.661

Li, W., Jiang, B., Moore, T.A., Wang, G., Liu, J.-G., and Song, Y., 2017, Characterisation of the chemical structure of tectonically deformed coals: Energy & Fuels, v. 31, p. 6977–6985. doi:10.1021/acs.energyfuels.7b00901

Li, X., Ju, Y., Hou, Q., and Lin, H., 2012b, Spectra response from macromolecular structure evolution of tectonically deformed coal of different deformation mechanisms: Science China: Earth Sciences, v. 55, p. 1269–1279. doi:10.1007/s11430-012-4399-y

Li, X., Wang, Y., Jiang, Z., Chen, Z., Wang, L., and Wu, Q., 2016b, Progress and study on exploration and production for deep coalbed methane: Journal of China Coal Society, v. 41, p. 24–31 (in Chinese with English abstract).

Li, Y., 1994, Coal bed methane resources and its development of China: China Geology, v. 21, p. 15–18 (in Chinese with English abstract).

Li, Y., Tang, D., Feng, Y., Xu, H., and Meng, Y., 2014, Distribution of stable carbon isotope in coalbed methane from the east margin of Ordos Basin: Science China (Earth Sciences), v. 44, p. 1741–1748. doi:10.1007/s11430-014-4900-x

Liu, C., Zhu, J., and Che, C., 2009a, Methodologies and results of the latest assessment of coalbed methane resources in China: Natural Gas Industry, v. 29, p. 130–132 (in Chinese with English abstract).

Liu, D., Yao, Y., Tang, D., Tang, S., Che, Y., and Huang, W., 2009b, Coal reservoir characteristics and coalbed methane resource assessment in Huainan and Huaibei coalfields, Southern North China: International Journal of Coal Geology, v. 79, p. 97–112. doi:10.1016/j.coal.2009.05.001

Liu, H., Qin, Y., and Sang, S., eds., 1998, Coalbed geology in southern Shanxi, China: Xuzhou, China University of Mining and Technology Press, p. 151 (in Chinese with English abstract).

Liu, Y., and Zhang, L., 2005, Statistical analysis of gas accidents in coal mines from January 2003 to June 2005: Journal of Henan Polytechnic University, v. 24, p. 259–262 (in Chinese with English abstract).

Liu, Y., and Zhang, P., 2016, Analysis on production dynamic and main controlling factors of single coalbed methane well in Liulin area: Coal Geology & Exploration, v. 44, p. 34–38 (in Chinese with English abstract).

Lott, D.A., Coveney, R.M., Jr., Murowchick, J.B., and Grauch, R.I., 1999, Sedimentary exhalative nickel-molybdenum ores in south China: Economic Geology and the Bulletin of the Society of Economic Geologists, v. 94, no. 7, p. 1051–1066. doi:10.2113/gsecongeo.94.7.1051

Luo, J., Yang, Y., and Chen, Y., 2012, Optimizing the drilled well patterns for CBM recovery via numerical simulations and data envelopment analysis: Internation Journal of Mining Science and Technology, v. 22, p. 503–507. doi:10.1016/j.ijmst.2012.01.011

Lv, Y., Tang, D., Xu, H., and Luo, H., 2012, Production characteristics and the key factors in high-rank coalbed methane fields: A case study on the Fanzhuang Block, Southern Qinshui Basin, China: International Journal of Coal Geology, v. 96-97, p. 93–108. doi:10.1016/j.coal.2012.03.009

Lv, Y., Tang, D., Xu, H., and Tao, S., 2011, Productivity matching and quantitative prediction of coalbed methane wells based on BP neural network: Science in China: Technological Sciences, v. 54, p. 1281–1286. doi:10.1007/s11431-011-4348-6

McKee, C.R., Bumb, A.C., and Koenig, R.A., 1987, Stress-dependent permeability and porosity of coal, in Proceedings of coabed methane symposium: Tuscaloosa, Alabama, Alabama State University, p. 183–193.

Meng, Y., Tang, D., Xu, H., Li, C., Li, L., and Meng, S., 2014, Geological controls and coalbed methane production potential evaluation: A case study in Liulin area, eastern Ordos Basin, China: Journal Natural Gas Science and Engineering, v. 21, p. 95–111. doi:10.1016/j.jngse.2014.07.034

Meng, Z., Lan, Q., and Liu, C., 2013, In-situ stress and coal reservoir pressure in Southeast margin of Ordos Basin and their coupling relations: Journal of China Coal Society, v. 38, p. 122–128 (in Chinese with English abstract).

Meng, Z., Zhang, J., and Wang, R., 2011, In-situ stress, pore pressure and stress-dependent permeability in the Southern Qinshui Basin: International Journal of Rock Mechanics and Mining Sciences, v. 48, p. 122–131. doi:10.1016/j.ijrmms.2010.10.003

Ministry of Land and Resources of China, 2001, Specification for coalbed methane resources/reserves (DZ/T 0216-2002): Beijing, China Standard Press, p. 17 (in Chinese).

Moore, T.A., 2012, Coalbed methane: A review: International Journal of Coal Geology, v. 101, p. 36–81. doi:10.1016/j.coal.2012.05.011

NEAC (National Energy Administration of China), 2017, The fourteenth meeting of the inter-ministerial coordination leading group of coal mine gas control was held in Beijing: http://www.nea.gov.cn/2017-01/16/c_135985557.htm (in Chinese).

Nelson, C.R., 2003, Deep coalbed gas plays in the U.S. Rocky Mountain Region, in Proceedings of the AAPG annual meeting: Salt Lake City, UT, AAPG, p. 5–8: http://www.searchanddiscovery.com/pdfz/abstracts/pdf/2003/annual/short/ndx_78975.PDF.html

Olson, T., Hobbs, B., and Brooks, R., 2002, Paying off for Tom Brown in White River Dom Field's tight sandstone, deep coals: The American Oil and Gas Reports, v. 10., p. 67–75.

Palmer, I., and Mansoori, J., 1996, How permeability depends on stress and pore pressure in coalbeds: A new model, in SPE Annual Technical Conference and Exhibition, SPE 36737: Denver, Colorado, Society of Petroleum Engineers.

Qi, J., Chen, S., Yang, Q., and Yu, F., 2008, Characteristics of tectonic deformation within transitional belt between the Junggar Basin and the northern Tianshan Mountain: Oil & Gas Geology, v. 29, p. 252–260.

Qin, Y., 2006, Situation and challenges of China's CBM exploration (I): Current stage of development: Natural Gas Industry, v. 26, p. 4–7 (in Chinese with English abstract).

Qin, Y., 2008, Mechanism of CO_2 enhanced CBM recovery in China: A review: Journal of China University of Mining and Technology, v. 18, p. 406–412. doi:10.1016/S1006-1266(08)60085-1

Qin, Y., 2009, Coalbed methane resources and exploitation & utilization of China, in Proceedings of the 2008 Asia Pacific CBM Symposium: Brisbane, Australia, University of Queensland and China University of Mining and Technology, p. 12–19 (Open CD-ROM).

Qin, Y., 2012, Advances and reviews on coalbed methane reservoir formation in China: Geological Journal of China Universities, v. 18, p. 405–418 (in Chinese with English abstract).

Qin, Y., Fu, X., Jiao, S., Li, G., and Wei, C., 2001b, Key geological controls to formation of coalbed methane reservoirs in Southern Qinshui Basin of China: III, Factor assembly and CBM reservoiring pattern, *in* Proceedings of the '2001 International Coalbed Methane Symposium: Tuscaloosa, Alabama, Alabama State University, p. 367–370.

Qin, Y., Fu, X., Jiao, S., Li, T., Jiao, S., and Wei, C., 2001c, Key geological controls to formation of coalbed methane reservoirs in Southern Qinshui Basin of China: II, Modern tectonic stress field and burial depth of coal reservoirs, *in* Proceedings of the '2001 International Coalbed Methane Symposium: Tuscaloosa, Alabama, Alabama State University, p. 363–366.

Qin, Y., Fu, X., Jiao, S., Li, T., and Wei, C., 2001d, Key geological controls to formation of coalbed methane reservoirs in Southern Qinshui Basin of China: I, Hydrological conditions, *in* Proceedings of the '2001 International Coalbed Methane Symposium: Tuscaloosa, Alabama, Alabama State University, p. 21–28.

Qin, Y., Fu, X., Wei, C., Hou, Q., Jiang, B., and Wu, C., 2012a, Dynamic conditions and pool- controlling effect of CBM reservoir formation: Beiijng, Science Press, 317 (in Chinese with English abstract).

Qin, Y., Fu, X., and Ye, J., 1999, Geological controls and their mechanisms of coal-reservoir petrography and physics of coalbed methane occurrence in China, *in* Proceedings of the 99' International Coalbed Methane Symposium: Tuscaloosa, AL, Alabama State University, p. 187–196.

Qin, Y., and Shen, J., 2016b, On the fundamental issues of deep coalbed methane geology: Acta Petrolei Sinica, v. 37, p. 125–136 (in Chinese with English abstract).

Qin, Y., Shen, J., and Shen, Y., 2016a, Joint mining compatibility of superposed gas-bearing systems: A general geological problem for extraction of three natural gases and deep CBM in coal series: Journal of China Coal Society, v. 41, p. 14–23 (in Chinese with English abstract).

Qin, Y., Shen, J., Wang, B., Yang, S., and Zhao, L., 2012b, Accumulation effects and coupling relationship of deep coalbed methane: Acta Petrolei Sinica, v. 33, p. 48–54 (in Chinese with English abstract).

Qin, Y., Song, D., and Wang, C., 1997, Coalification of the Upper Paleozoic coal and its control to the generation and preservation of coalbed methane in the southern Shanxi: Journal of China Coal Society, v. 22, p. 230–234 (in Chinese with English abstract).

Qin, Y., Song, Q., and Fu, X., 2005, Discussion on reliability for co-mining the coalbed gas and normal petroleum and natural gas: Absorptive effect of deep coal reservoir under condition of balanced water: Natural Gas Geoscience, v. 16, p. 492–498 (in Chinese with English abstract).

Qin, Y., Wei, C., Zhang, Z., Wang, C., and Yang, Z., 2016c, Geological controls of free natural gas reservoirs in coal measures and overlying strata in the central and southern Qinshui basin: Earth Science Frontiers, v. 23, p. 24–35 (in Chinese with English abstract).

Qin, Y., Xiong, M., Yi, T., Yang, Z., and Wu, C., 2008, On unattached multiple superposed coalbed methane system: In a case of the shuigonghe syncline, Zhijin- Nayong coalfield, Guizhou: Geological Review, v. 54, p. 65–70 (in Chinese with English abstract).

Qin, Y., and Ye, J., 2015, A review on development of CBM industry in China, *in* Seminar presentation to AAPG GTW: Brisbane, Australia, AAPG, 2015- 02-12: http://www.searchanddiscovery.com/pdfz/documents/2015/80454yong/ndx_yong.pdf.html

Qin, Y., Ye, J., and Lin, D., 2001a, Geological seeking for potential CBM-accumulation zones and districts in China, *in* Xie Heping, editor-in-chief, Mining Science and Technology '99: Rotterdam, Holland, A.A. Balkema Publisher, p. 243–246.

Qin, Y., Zhang, D., and Fu, X., 2000, Discussion on correlation of modern tectonic stress field to physical properties of coal reservoirs in Central and Southern Qinshui Basin, *in* Proceedings of the 31th International Geological Congress: Rio de Janeiro, Brail, International Federation of Geosciences, p. 457–462.

Qu, P., 2016, Xinjiang ranks among the top third coalbed methane development hot spots. China Energy News, 2016-08-22: http://paper.people.com.cn/zgnyb/html/2016-08/22/content_1705869.htm (in Chinese)

Ren, Z., Xiao, H., Liu, L., Zhang, S., Qin, Y., and Wei, C., 2005, The evidence of fission-track data for the study of yectonic thermal history in Qinshui Basin: Chinese Science Bull, v. 50, no. S, p. 104–110.

Repine, T.E., Jr., 1990, Coalbed methane - A new west virginia industry?, *in* Mountain State Geology: West Virginia Geological and Economic Survey, p. 1, 6–7.

SAWSC (State Administration of Work Safety of China), 2011, Classification of coalification degree by vitrinite reflectance, *in* China Coal Industry Standard (MT/T 1158-2011): Beijing, SAWSC, p. 3 (in Chinese).

Scott, A.R., Kaiser, W.R., Walter, B., and Ayers, J., 1994, Thermogenic and secondary biogenic gases, San Juan Basin, Colorado and Nex Mexico: Implications for coalbed gas producibility: AAPG Bulletin, v. 78, p. 1186–1209.

Shao, L., Yang, Z., Shang, X., Xiao, Z., Wang, S., Zhang, W., Zheng, M., and Lu, J., 2015, Lithofacies palaeogeography of the Carboniferous and Permian in the Qinshui Basin, Shanxi Province, China: Journal of Palaeogeography, v. 4, p. 384–412.

Shao, L., Zhang, P., Ren, D., and Lei, J., 1998, Late Permian coal-bearing carbonate successions in southern China: Coal accumulation on carbonate platforms: International Journal Coal Geological, v. 37, p. 235–256.

Shen, J., 2011, On CBM-reservoiring effect in deep strata [Doctoral Dissertation]: Xuzhou, China, China University of Mining and Technology, 135 p. (in Chinese with English abstract).

Shen, J., Qin, Y., and Fu, X., 2014, Properties of deep coalbed methane reservoir-forming conditions and critical depth discussion: Natural Gas Geoscience, v. 25, p. 1470–1476 (in Chinese with English abstract).

Shen, J., Qin, Y., Wang, G.G.X., Fu, X., and Wei, C., 2011, Relative permeabilities of gas and water for different rank coals: International Journal of Coal Geology, v. 86, p. 266–275.

Shen, Y., Qin, Y., Guo, Y., Ren, H., Wei, Z., and Xie, G., 2012, The Upper Permian coalbed methane bearing system and its sedimentary control in western Guizhou, China: Journal of China Universities, v. 18, p. 427–434 (in Chinese with English abstract).

Shen, Y., Qin, Y., Guo, Y., Yi, T., Yuan, X., and Shao, Y., 2016, Characteristics and sedimentary control of a coalbed methane-bearing system in Lopingian (late permian) coal-bearing strata of western Guizhou Province: Journal Natural Gas Science and Engineering, v. 33, p. 8–17.

Shu, T., Yanbin, W., Dazhen, T., Hao, X., Yumin, L., Wei, H., and Yong, L., 2012, Dynamic variation effects of coal permeability during the coalbed methane development process in the Qinshui Basin, China: International Journal of Coal Geology, v. 93, p. 16–22.

Song, Y., Liu, H., Hong, F., Qin, S., Liu, S., Li, G., and Zhao, M., 2012, Syncline reservoir pooling as a general model for CBM accumulations: Mechanisms and case studies: Journal of Petroleum Science and Engineering, v. 88-89, p. 5–12.

Song, Y., Liu, S., Ju, Y., Hong, F., and Jiang, L., 2013, Coupling between gas content and permeability controlling enrichment zones of high abundance coalbed methane: Acta Petrollei Sinica, v. 34, p. 417–426 (in Chinese with English abstract).

Stevens, S.H., 1999, China coalbed methane reaches turning point: Oil and Gas Journal, v. 97, p. 101–106.

Stevens, S.H., Sheehy, L.D., Zhu, H., and Yuan, B., 1992, Reservoir evaluation of the Tangshan coalbed methane prospect, Hebei province, the People's Republic of China: Proceedings - SPE Annual Technical Conference and Exhibition, v. Sigma, p. 475–490.

Su, X., Lin, X., Zhao, M., Song, Y., and Liu, S., 2005, The upper Paleozoic coalbed methane system in the Qinshui basin, China: AAPG Bulletin, v. 89, p. 81–100.

Sun, L., Kang, Y., Wang, J., Jiang, S., Zhang, B., Gu, J., Ye, J., and Zhang, S., 2017, Vertical transformation of in-situ stress types and its control on coal reservoir permeability: Geological Journal of China Universities, v. 23, no. 1, p. 148–156 (in Chinese with English abstract).

Sun, L., Zhao, Y., Cui, D., and Cai, D., 2013, Prediction of higher-permeability region of coal reservoir and its key control factor in the east of ordos basin: Journal of Shandong Unversity of Science and Technology, v. 32, p. 60–67 (in Chinese with English abstract).

Sun, P., Liu, H., Chao, H., and Wang, Y., 2008, Exploration direction of coalbed methane in lw-rank coals: Natural Gas Industry, v. 28, no. 3, p. 19–22 (in Chinese with English abstract).

Sun, Q., Sun, B., Sun, F., Yang, Q., and Chen, G., 2012, Accumulation and geological controls of low-rank coalbed methane in southeastern Junggar basin: Geological Journal of China Universities, v. 18, p. 460–464 (in Chinese with English abstract).

Teng, J., Yao, Y., Liu, D., and Cai, Y., 2015, Evaluation of coal texture distributions in the southern Qinshui basin, North China: Investigation by a multiple geophysical logging method: International Journal Coal Geological, v. 140, p. 9–22.

Terzaghi, K., 1960, Theoretical soil mechanics (translated by Xu Zhiying): Beijing, Geological Publishing House, p. 509 (in Chinese).

Tian, J., and Yang, S., 2011, Sequence strata and coal accumulation of lower and middle Jurassic formation from southern margin of Junggar Basin, Sinkiang, China: Journal of China Coal Society, v. 36, p. 58–64 (in Chinese with English abstract).

Tian, W., Tang, D., Wang, Z., and Sun, B., 2012, Origin of coalbed methane in Baode, Northeastern Ordos Basin: Geological Journal of China Universities, v. 18, p. 479–484 (in Chinese with English abstract).

Tian, W., Xiao, J., Zhang, J., and Zhao, S., 2015, CBM reservoir-cap formation type and its gas controlling function in the eastern margin of Ordos basin: Coal Geology & Exploration, v. 43, p. 31–35 (in Chinese with English abstract).

Titheridge, D.G., 2014, Drilling induced fractures in coal core from vertical exploration well: A method to determine cleat azimuth, and the angle between cleat and maximum horizontal stress, and its application, in 14th Coal Operators' Conference: Wollongong, Australia, University of Wollongong, Australasian Institute of Mining and Metallurgy & Mine Managers Association of Australia, p. 8–17: http://ro.uow.edu.au/coal/494/

UN-ECE (United Nations - Economic Commission for Europe), 1992, International classification of in seam coals: Geneva, United Nations, p. 11.

Wang, G., Qin, Y., Shen, J., Zhao, L., and Zhao, J., 2014, Experimental studies of deep low-rank coal reservoirs' permeability based on variable pore compressibility: Acta Petrolei Sinica, v. 35, p. 462–468 (in Chinese with English abstract).

Wang, G., Qin, Y., Xie, Y., Shen, J., Han, B., Huang, B., and Zhao, L., 2015, The division and geologic controlling factors of a vertical superimposed coalbed methane system in the northern Gujiao blocks, China: Journal Natural Gas Science and Engineering, v. 24, p. 379–389.

Wang, H., Shao, L., Hao, L., Zhang, P., Ian, J.G., and James, R.W., 2011, Sedimentology and sequence stratigraphy of the Lopingian (Late Permian) coal measures in southwestern China: International Journal Coal Geological, v. 85, p. 168–183.

Wang, L., 2014, Research on CBM development and utilization of Fukang city, Xinjiang [Master Thesis]: Urumqi, Xinjiang Agricultural University, 42 p. (in Chinese with English abstract).

Wang, L., Jiang, B., and Qu, Z., 2013, Structural control on gas content distribution in eastern margin of Ordos basin: Coal Geology & Exploration, v. 41, p. 14–19 (in Chinese with English abstract).

Wang, S., 2002, geologic structure control of CBM in Hancheng mining area: Coal Geology & Exploration, v. 30, p. 21–25 (in Chinese with English abstract).

Wang, T., Feng, F., Jiang, T., Wang, Q., Xia, Y., Wei, B., and Yang, S., 2016, Fundamental structural framework and cognition of Junggar coal basin, Xinjiang: Acta Geologica Sinica, v. 90, p. 628–638 (in Chinese with English abstract).

Wang, X., Zhang, Q., Wang, L., Ge, R., and Chen, J., 2010, Structural features and tectonic stress fields of the Mesozoic and Cenozoic in the eastern margin of the Ordos basin, China: Geological Bulletin of China, v. 29, p. 1168–1176 (in Chinese with English abstract).

Wei, C., Qin, Y., Wang, G.G.X., Fu, X., Jiang, B., and Zhang, Z., 2007, Simulation study on evolution of coalbed methane reservoir in Qinshui basin, China: International Journal of Coal Geology, v. 72, p. 53–69.

Wei, C., Qin, Y., Wang, G.G.X., Fu, X., and Zhang, Z., 2010, Numerical simulation of coalbed methane generation, dissipation and retention in SE edge of Ordos Basin, China: International Journal of Coal Geology, v. 82, p. 147–159.

Wei, G., 2015, Basic characteristics and key factors of gas accumulation in Yanchuannan coalbed gas field: Petroleum Geology & Experiment, v. 37, p. 341–346 (in Chinese with English abstract).

Wei, Q., Wang, Y., Wang, H., Li, Z., Zhang, Q., and Zuo, J., 2013, Coalbed methane exploration prospect and development suggestion in the southern margin of Junggar basjn, in Proceedings of the 2011 China Coalbed Methane

Symposium: Hangzhou, China, China Coal Society, p. 11–17 (in Chinese with English abstract).

Wei, Y., 2003, Preliminary study on low rank coal reservoirs and coalbed methane pool forming in Zhungaer Basin, NW China [Doctoral Dissertation]: Beijing, China University of Geosciences, p. 131 (in Chinese with English abstract).

Wong, S., Macdonald, D., Andrei, S., Gunter, W.D., Deng, X., and Law, D., 2010, Conceptual economics of full scale enhanced coalbed methane production and CO2 storage in anthracitic coals at South Qinshui basin, Shanxi, China: International Journal of Coal Geology, v. 82, p. 280–286.

Wu, C., Qin, Y., and Fu, X., 2007, Stratum energy of coal-bed gas reservoir and their control on the coal-bed gas reservoir formation: Science in China (D), v. 50, p. 1319–1326.

Wu, C., Qin, Y., and Zhou, L., 2014, Effective migration system of coalbed methane reservoirs in the southern Qinshui Basin: Science China Earth Sciences, v. 57, p. 2978–2984.

Wu, C.F., Zhou, L.G., and Lei, B., 2013, Coal reservoir permeability in the gemudi syncline in Western Guizhou, China: Energy Sources, Part A: Recovery, Utilization and Environmental Effects, v. 35, p. 1532–1538.

Wu, K., Zha, M., Wang, X., Qu, J., and Chen, X., 2005, Further researches on the tectonic evolution and dynamic setting of the Junggar basin: Acta Geoscientica Sinica, v. 26, p. 217–222.

Xie, K., Qiu, Z., Jin, Q., Yuan, L., and Zhao, W., 2014, Strategic study on the development and utilization of unconventional natural gas in China: Beijing, Science Press, p. 441 (in Chinese with English abstract).

Xiong, M., Qin, Y., and Yi, T., 2006, Sedimentary patterns and structural controls of Late Permian coal-bearing strata in Guizhou, China: Journal of China University of Mining and Technology, v. 35, p. 778–782 (in Chinese with English abstract).

Xu, B., and He, M., 2003, Geology of coal fields in Guizhou Province: Xuzhou, China University of Mining and Technology, p. 283 (in Chinese with English abstract).

Xu, G., Ge, N., Chen, J., and Shen, S., 1990, Valency study of vanadium in stone coal of Southern China: Rare Metals, v. 9, no. 2, p. 110–116.

Xu, H., Sang, S., Yi, T., Zhao, X., Liu, H., and Li, L., 2014, Control mechanism of buried depth and in-situ stress for coal reservoir permeability in western Guizhou: Earth Science, v. 39, p. 1607–1616 (in Chinese with English abstract).

Xu, H., Tang, D., Qin, Y., Meng, C., Tao, S., and Chen, Z., 2011, Characteristics and origin of coal reservoir pressure in the west Guizhou area: Journal of China University of Mining and Technology, v. 40, p. 556–560 (in Chinese with English abstract).

Xu, S., and Guo, Z., 2014, Influence of structure characteristics to coal seam gas bearing in Hancheng mining area: Shanxi Coking Coal Science & Technology, v. 6, p. 20–22 (in Chinese with English abstract).

Xu, Z., Yun, W., Wang, J., and Qin, Y., 1998, Development and dynamic analysis of tectonic-stress fields during Mesozoic and Cenozoic era in mid-south Shanxi Province: Earth Science Frontiers, v. 5, no. S, p. 152–161 (in Chinese with English abstract).

Xue, G., Liu, H., Yao, H., and Li, W., 2012, Features of coalbed methane reservoir in development zone of Hancheng in Weihe Basin: Journal of Taiyuan University of Technology, v. 43, p. 185–189 (in Chinese with English abstract).

Yang, L., 1987, National coal mine gas geological map of China: Xi'an, Shaanxi Science and Technology Press, p. 248 (in Chinese).

Yang., Q., and Han, D., (editors in chief), 1979, Coal Geology of China (Volume 1): Beijing, China Coal Industry Press, p. 321 (in Chinese).

Yang, Q., Liu, D., Huang, W., Li, Y., Hu, B., and Wei, Y., 2005, Comprehensive evaluation of coalbed methane geology and resources in Northwest China: Beijing, Geological Press, p. 358 (in Chinese with English abstract).

Yang, Z., 2011, Coalbed methane reservoiring process under condition of multi-coalbeds overlay [Doctoral dissertation]: Xuzhou, China University of Mining and Technology, 131 p. (in Chinese with English abstract).

Yang, Z., and Qin, Y., 2015, A study of the unattached multiple superposed coalbed-methane system under stress conditions: Journal of China University of Mining & Technology, v. 44, p. 662–675 (in Chinese with English abstract).

Yang, Z., Qin, Y., Gao, D., and Wang, B., 2011, Coalbed methane reservoir -forming character under conditions of coal seam groups: Coal Geology & Exploration, v. 39, p. 22–26 (in Chinese with English abstract).

Yang, Z., Qin, Y., Wang, G.G.X., and Hui, A., 2015, Investigation on coal seam gas formation of multi-coalbed reservoir in Bide-Santang Basin, Southwest China: Arabian Journal of Geosciences, v. 8, p. 5439–5448.

Yao, Y., Liu, D., Tang, D., Tang, S., Che, Y., and Huang, W., 2009, Preliminary evaluation of the coalbed methane production potential and its geological controls in the Weibei Coalfield, Southeastern Ordos Basin, China: International Journal of Coal Geology, v. 78, p. 1–15.

Ye, J., 2013, Current status of China coalbed methane industry, in 2013 National Symposium on Coalbed Methane: Hangzhou, China, China Coal Society (in Chinese). http://www.ccs-cbm.org.cn/news_detail.asp?news_id=4698.

Ye, J., and Li, X., 2016, Development status and technical progress of China coalbed methane industry: Coal Science and Technology, v. 44, p. 24–28, 46 (in Chinese with English abstract).

Ye, J., Qin, Y., and Lin, D., 1999, Coalbed Methane Resources of China: Xuzhou, China University of Mining and Technology Press, p. 229 (in Chinese with English abstract).

Ye, X., 2007, The research of accumulation of low rank coalbed methane in the north-west of China [Master Thesis]: Chengdu, Chengdu University of Technology, 98 p. (in Chinese with English abstract).

Yi, T., Zhou, X., and Jin, J., 2016, Study on the reservoir forming characteristics and co-exploration and concurrent production technology of Longtan coal measure coalbed methane & tight gas in Songhe field, western Guizhou: Journal of China Coal Society, v. 41, p. 212–220 (in Chinese with English abstract).

Yu, Q., 1993, Mine Gas Control: Xuzhou, China University of Mining & Technology Press, p. 263 (in Chinese).

Yu, Y., 2002, Preliminary study on low-rank coal reservoirs and coalbed methane pool forming in Zhungaer basin, NW China [Doctoral Dissertation]: Beijing, China University of Geosciences, 131 p. (in Chinese with English abstract).

Yu, Y., Wang, H., Yang, Q., Liu, D., Hu, B., Huang, W., and Che, Y., 2008, Adsorption characteristics of low-rank coal reservoirs and coalbed methane development potential,

Junggar Basin: Petroleum Exploration and Development, v. 35, p. 410–416.

Yuan, X., 2014, Recognition of multi-coalbed methane bearing system: A case study of coal-bearing strata of Upper Permian in Western Guizhou [Doctoral Dissertation]: Xuzhou, China University of Mining and Technology, 162 p. (in Chinese with English abstract).

Yun, J., Xu, F., Liu, L., Zhong, N., and Wu, X., 2012, New progress and future prospects of CBM exploration and development in China: Internation Journal of Mining Science and Technology, v. 22, p. 363–369.

Zhang, J., Qin, Y., Wang, H., and Chen, J., 2003, Prediction of the occurrence of high permeability coalbed gas reservoirs with tectonics analysis: Journal of China Universities, v. 9, p. 359–364 (in Chinese with English abstract).

Zhang, P., 2008, Analysis method of coalbed methane reservoiring conditions - case study in Hancheng area: China Coalbed Methane, v. 5, p. 12–15 (in Chinese with English abstract).

Zhang, S., Tang, S., Qian, Z., Pan, Z., and Guo, Q., 2014, Evaluation of geological features for deep coalbed methane reservoirs in the Dacheng Salient, Jizhong Depression, China: International Journal of Coal Geology, v. 133, p. 60–71.

Zhang, S., Tang, S., Tang, D., Pan, Z., and Yang, F., 2010, The characteristics of coal reservoir pores and coal facies in Liulin district, Hedong coal field of China: International Journal of Coal Geology, v. 81, p. 117–127.

Zhang, X., Zhang, S., and Zhong, L., 1991, Coalbed methane of China: Xi'an, Shaanxi Science and Technology Press, p. 145 (in Chinese with English abstract).

Zhang, X., Zhuang, J., and Zhang, S., 2002a, Geology and resource evaluation of coalbed methane in China: Beijing, Science Press, p. 305 (in Chinese with English abstract).

Zhang, Z., Qin, Y., Bai, J., Fu, X., and Liu, D., 2016, Evaluation of favorable regions for multi-seam coalbed methane joint exploitation based on a fuzzy model: A case study in southern Qinshui Basin, China: Energy Exploration and Exploitation, v. 34, p. 400–417.

Zhang, Z., Qin, Y., Fu, X., Yang, Z., and Guo, C., 2015, Multi-layer superposed coalbed methane system in southern Qinshui Basin, Shanxi Province, China: Journal of Earth Science, v. 26, p. 391–398.

Zhang, Z., Qin, Y., Wang, G.G.X., and Fu, X., 2013, Numerical description of coalbed methane desorption stages based on isothermal adsorption experiment: Science China: Earth Sciences, v. 56, p. 1029–1036.

Zhang, Z., Sun, J., Yuan, D., and Zhang, R., 2002b, Regional distribution of Carboniferous-Permian coalbed gas in China, in Xie Heping, editor-in-chief, Mining Science and Technology '99: Rotterdam, Holland, A.A. Balkema Publisher, p. 251–254.

Zhang., Z., and Zhang., Z.X., (editors in chief), 1992, Geological map and its instruction of coal seam gas in China: Xi'an, Map Publishing House, p. 1 (in Chinese with English abstract).

Zhao, Q., 2011, Coalbed methane accumulation and favorable area prediction, in Langfang Branch of China Petroleum Exploration and Development Research Institute, Research Report of the Major National Science and Technology Projects of China: Langfang, Heibei, China, Sun Fenjin, p. 324 (in Chinese with English abstract).

Zhou, F., Allinson, G., Wang, J., Sun, Q., Xiong, D., and Cinar, Y., 2012, Stochastic modelling of coalbed methane resources: A case study in Southeast Qinshui Basin, China: International Journal of Coal Geology, v. 99, p. 16–26.

Zhou, F., and Yao, G., 2014, Sensitivity analysis in permeability estimation using logging and injection-falloff test data for an anthracite coalbed methane reservoir in Southeast Qinshui Basin, China: International Journal of Coal Geology, v. 131, p. 41–51.

Zhou, F., Yao, G., and Tyson, S., 2015, Impact of geological modeling processes on spatial coalbed methane resource estimation: International Journal of Coal Geology, v. 146, p. 14–27.

Zhou, J., Yuan, Y., Wang, X., and Cao, D., 2013, Positive and negative coupling effects of hydrogeological condition gas control in Sanjiao block, Hedong coalfield: Coal Geology of China, v. 25, p. 16–39 (in Chinese with English abstract).

Zhu, S., 1979, Discussion on some problems of stone coal exploration: Coal Geology & Exploration, v. 8, no. 2, p. 99–101 (in Chinese with English abstract).

Index

T - #0205 - 111024 - C300 - 280/210/14 - PB - 9780367571184 - Gloss Lamination